AF581792

Recent Advancements in Radar Imaging and Sensing Technology

Recent Advancements in Radar Imaging and Sensing Technology

Editors

Piotr Samczynski
Elisa Giusti

MDPI • Basel • Beijing • Wuhan • Barcelona • Belgrade • Manchester • Tokyo • Cluj • Tianjin

Editors
Piotr Samczynski
Warsaw University of Technology
Poland

Elisa Giusti
CNIT—National
Inter-University Consortium for Telecommunications
Italy

Editorial Office
MDPI
St. Alban-Anlage 66
4052 Basel, Switzerland

This is a reprint of articles from the Special Issue published online in the open access journal *Sensors* (ISSN 1424-8220) (available at: https://www.mdpi.com/journal/sensors/special_issues/Radar_Imaging_Sensing).

For citation purposes, cite each article independently as indicated on the article page online and as indicated below:

LastName, A.A.; LastName, B.B.; LastName, C.C. Article Title. *Journal Name* **Year**, *Volume Number*, Page Range.

ISBN 978-3-0365-0918-1 (Hbk)
ISBN 978-3-0365-0919-8 (PDF)

Cover image courtesy of Prof. Samczynski.

Contents

About the Editors

Piotr Samczynski (Associate Professor) received his B.Sc. and M.Sc. degrees in electronics and Ph.D. and D.Sc. degrees in telecommunications, all from the Warsaw University of Technology (WUT), Warsaw, Poland in 2004, 2005, 2010 and 2013 respectively. Since 2018, he has been the Associate Professor at the WUT; and since 2014—a member of the WUT's Faculty of Electronics and Information Technology Council. Prior to this, he was Assistant Profesor at WUT (2018–2010), a research assistant at the Przemyslowy Instytut Telekomunikacji S.A. (PIT S.A.) (2010–2005) and the head of PIT's Radar Signal Processing Department (2010–2009). Prof. Samczynski's research interests are in the areas of radar signal processing, passive radar, synthetic aperture radar and digital signal processing. He is the author of over 200 scientific papers. Prof. Samczynski was involved in several projects for the European Research Agency (EDA), Polish National Centre for Research and Development (NCBiR) and Polish Ministry of Science and Higher Education (MiNSW), including the projects on SAR, ISAR and passive radars. Since 2009 he has been a member of several research task groups under the NATO Science and Technology Organization, where he supports the research work in the fields of radar signal processing, modern passive and active radars architectures and noise radars. Since 2018 he is a Chair of NATO SET-258 research task group (RTG) on Deployable Multiband Passive/Active Radar (DMPAR) deployment and assessment in military scenarios. Prof. Samczynski is an IEEE member since 2003, and IEEE Senior member since 2016. He is a member of IEEE AES, SP, and GRS Societies and in 2017–2019 Prof. Samczynski was a Chair of the Polish Chapter of the IEEE Signal Processing Society, and from 2019 he is Vice-Chairman (AES) of the IEEE Poland APS/AESS/MTTS Joint Chapter. He received IEEE Fred Nathanson Memorial Award for outstanding contribution to the field of passive radar imaging, including systems design, experimentation and algorithm development in 2017.

Elisa Giusti received the Telecommunication Engineering Laurea (cum laude) and Ph.D. degrees from the University of Pisa (Italy) in 2006 and 2010, respectively. From June 2009 to November 2014, she was a researcher under contract at the Department of Information Engineering of the University of Pisa. From November 2014 to December 2015 she was a research under contract at the CNIT-RaSS National Laboratory. Since December 2015 she is a permanent researcher at the CNIT-RaSS National Laboratory (Pisa, Italy). Since 2009 she has been involved as a researcher in several international projects funded by Italian ministries (Ministry of Defence, Ministry of Economic Development) and European organisations (EDA, ESA). She is co-founder of a radar systems-related spin-off company, "ECHOES radar technologies" located in Pisa. She is a co-author of more than 60 papers and 3 book chapters. She is editor of a book titled "Radar Imaging for Maritime Observation", CRC press. She has been recipient of the 2016 Outstanding Information Research Foundation Book publication award for the book "Radar Imaging for Maritime Observation". Her research interests are mainly in the field of radar imaging, including active, passive, bistatic, multistatic, and polarimetric radar.

Preface to "Recent Advancements in Radar Imaging and Sensing Technology"

During the last decades, radar imaging and sensing technology has made major scientific and technical progress. The first applications of this technology were devoted mostly to military uses. Nowadays, radar imaging and sensing techniques are widely used in many civilian applications, ranging from medicine, though security, to safety assistance sensors widely used in transportation, including cars, trains, and airplanes, among others. These technologies are starting to be present everywhere around us. With the fast development of new hardware platforms with advanced computational resources that are widely available on the market, novel signal processing techniques—enabling enhanced functionalities of radar systems—have been implemented. This, in turn, makes it possible to apply new technology in radar imaging, such as, for example, passive radar sensing. Just a few years ago, this type of sensing was at a very low technical readiness level, and today it has become a mature technology that will probably be offered on the market within the next few years. Moreover, the ever-wider bandwidth of the currently available receivers allows the creation of very high-resolution radar images utilizing both active and passive radar technology.

The aim of this Printed Edition of Special Issue was to gather the latest research results in the area of modern radar technology using active and/or radar imaging sensing techniques in different applications, including both military use and a broad spectrum of civilian applications. As a result, the 19 papers that have been published highlighted a variety of topics related to modern radar imaging and microwave sensing technology. The papers included in the Printed Edition of Special Issue dealt with wide aspects of different applications of radar imaging and sensing technology. A brief revision of the content of this book is presented below.

The first paper in this Special Issue entitled "A Hybrid SAR/ISAR Approach for Refocusing Maritime Moving Targets with the GF-3 SAR Satellite" proposed to combine Synthetic Aperture radar (SAR) and Inverse SAR (ISAR) to refocus the maritime moving targets. In the paper, the authors describe a novel hybrid SAR/ISAR approach. This approach is based on the improved rank-one phase estimation method (IROPE). The authors proposed to use an iterative two-step convergence approach in the IROPE. As a result, the proposed method achieves accurate phase error, maintains robustness to noise, and performs well in estimating various phase errors. In the presented paper, the proposed method's performance has been compared with other focusing algorithms in terms of processing simulated data and real complex image data acquired by Gaofen-3 (GF-3) in spotlight mode. The results shown in this paper demonstrates the effectiveness of the proposed method and high potentials also for high-resolution long-CPI spaceborne radar.

The authors of the paper "Azimuth Phase Center Adaptive Adjustment upon Reception for High-Resolution Wide-Swath Imaging", propose a method to set the proper PRF in a multichannel SAR system composed of a number of receiving antennas to effectively implement high resolution wide swath (HRWS) SAR imaging. In fact when using sub-apertures at the receiver the used PRF may be not optimum decreasing the quality of the SAR image. Particularly a non uniform sampling of the received along the along track dimension may cause gratings lobes, higher side lobes and even ghost targets. The authors propose a way to automatically adapt the optimum value of the PRF within a certain range by adjusting the phase center spacing of the sub-apertures.

The paper entitled "Focusing Bistatic Forward-Looking Synthetic Aperture Radar Based on an Improved Hyperbolic Range Model and a Modified Omega-K Algorithm" proposes to improve hyperbolic approximation range form with high-order terms to obtain a more accurate compensation result in focusing bistatic forward-looking SAR. Additionally, the authors adopt a modified omega-K algorithm based on the new slant range or parallel bistatic forward-looking SAR imaging. The paper presents several simulation results, which validate the effectiveness of the proposed imaging algorithm.

The paper "Microwave Staring Correlated Imaging Based on Unsteady Aerostat Platform" propose an algorithm to form radar images of the observed area by using an unsteady aerostat platform and the microwave staring correlated imaging algorithm. MSCI has been proven effective when SAR cannot be used (forward looking or staring geometries). However the MSCI algorithm relies on the hypothesis of stationary platform which is difficult to be realized in practice especially when using a tethered aerostat. The authors then propose an algorithm to include the antenna motion and the dynamic beam coverage cause by the instability of the platform in the imaging model. Therefore the real-time position vectors of the antenna are used in the imaging model instead of static position vector. A least square error curve fitting is used to estimate the accurate translational speed and rotational velocity of the array at each pulse.

The paper "Strip-Mode Microwave Staring Correlated Imaging with Self-Calibration of Gain–Phase Errors" proposes to apply microwave staring correlated imaging (MSCI) with strip-mode self-calibration of gain–phase errors. The method solves the problem of MSCI with gain–phase error, which occurs in a large SAR scene. The problem exists in the multi-transmitter array, resulting in an imaging model mismatch and considerably degrading the imaging performance. The authors propose to divide the imaging SAR scene into multiple imaging strips. Then in the next step, the strip target scattering coefficient and the gain–phase errors are combined into a multi-parameter optimization problem that can be solved in the iteration procedure. The error estimation results in each iteration are set as the initial value for the next iteration. As a result, the whole SAR imaging in a large scene is achieved by multi-strip image splicing. The proposed method reduces the time required by the SAR imaging process and improves the imaging quality.

The paper "Geometrical Matching of SAR and Optical Images Utilizing ASIFT Features for SAR-based Navigation Aided Systems" proposes a new approach for the estimation of shift and rotation between optical and SAR images. The estimated shift and rotation can be used to calculate the navigation correction when the drift of the calculated SAR platform trajectory is expected. The method can be used in platforms where there is no satellite navigation signal present. In such a case, the trajectory is calculated only on the basis of an inertial navigation system, which is characterized by a significant error. The proposed method of estimating the navigation error utilizing Affine Scale-Invariant Feature Transform (ASIFT) and Structure from Motion (SfM) is described in this paper. The presented methodology was tested and verified using real-life SAR images. Merged techniques such as ASIFT-based keypoint extraction and SfM-based keypoints matching make this method robust and resistant to noise and interference. Thus, the presented methodology can be successfully integrated with existing systems to enhance their precision and dependability. Additionally, the authors provide a comparison of several filters, including their computational complexity and performances. The detailed results of this analysis are shown in the article.

The paper "Wavelength-Resolution SAR Ground Scene Prediction Based on Image Stack" presents five different statistical methods for ground scene prediction (GSP) in wavelength-resolution SAR images. The predictions are based on image stacks, which are composed of images from the same scene acquired at different instants with the same flight geometry. In the paper, the authors considered the following methods in their study: autoregressive models, trimmed mean, median, intensity mean, and mean calculations. The authors indicate that the median method provided the most accurate representation of the true ground. Additionally, in the paper, a change detection algorithm was considered using the median ground scene as a reference image to show the applicability of the GSP. The obtain results presenting competitive performance when compared with recently published works.

The paper "A Multi-Scale U-Shaped Convolution Auto-Encoder Based on Pyramid Pooling Module for Object Recognition in Synthetic Aperture Radar Images" proposes another way to implement auto-encoder to simultaneously extract global and local target features. More specifically, the proposed approach learns multi-scale features at two levels: the modality level features and the branch level feature. Also a modified objective function is proposed to handle the degradation caused by the speckle. Moreover, a new convolution layer and its counterpart are also developed to reduce the number of trainable parameters in the model in order to alleviate overfitting caused by the limited training samples.

The paper "A MIMO-SAR Tomography Algorithm Based on Fully-Polarimetric Data" presents a fully-polarimetric unitary multiple signal classification (UMUSIC) tomography algorithm to acquire high-resolution 3D radar imagery for a multiple-input multiple-output (MIMO) SAR with a small number of baselines. The authors employ fully-polarimetric data and their conjugation to obtain the sample covariance matrix to mitigate the effect of multi-looking on the range-azimuth resolution. In the presented paper, two algorithms, including the popular distributed compressed sensing (DCS) and UMUSIC, are compared through numeric simulation of different point scatterers. All these comparisons have been made by the authors using the fully-polarimetric data. The MIMO SAR algorithm has been validated using measured data of an aircraft model with six different baselines. The final obtain results show the usefulness of the algorithm for 3D imagery of complex radar targets.

The authors of "Target Localization Using Double-Sided Bistatic Range Measurements in Distributed MIMO Radar Systems" propose a way to enhance the target localization performance using a distributed MIMO radar system. The main novelty with respect to the literature consists in the use of both the target-transmitter and target receiver distances as auxiliary information to enhance the calculation of target time delays and therefore enhance the estimation of the target coordinates.

The authors of the paper "Research of a Radar Imaging Algorithm Based on High Pulse Repetition Random Frequency Hopping Synthetic Wideband Waveform" propose a radar imaging algorithm specifically designed for high PRF and random frequency hopping (RFH) waveforms. The use of high PRF has obvious advantages especially when using very fast moving platforms (e.g. supersonic or hypersonic aircrafts). Moreover the use of RFH make the radar system particularly robust to jammer. On the other end, however the use of RFH makes impossible the use of Fourier based approaches to form the radar image of the observed scene because the received echoes are not uniformly spaced in the data domain. Algorithms have been proposed in the literature but a branch of these algorithm requires many constraints on the structural characteristics of the non-uniform sampling signal therefore making them poor versatile, the other branch of algorithm that is based on the use of compressed sensing while showing good performance requires however high computational burden. The authors therefore propose an algorithm based on Doppler preprocessing and 2D generalized matched filter (GMF) to try overcoming the limitations of the algorithms proposed in the literature. Moreover several RFH modes are designed along with the corresponding imaging algorithm.

"Compressed Sensing Radar Imaging: Fundamentals, Challenges, and Advances" is a review of the modern concept of compressed sensing applied to radar imaging. The authors present the main technical achievements and the technical background of the most used techniques, such as the minimum variance unbiased estimation, least squares (LS) estimation, Bayesian maximum a posteriori (MAP) estimation. Moreover the main challenges and still open problems are analyzed in this paper These include the sampling scheme, the computational complexity, the sparse representation, the influence of clutter and the model error compensation.

"Compressive Sensing-Based Bandwidth Stitching for Multichannel Microwave Radars" address the topic of forming high range resolution range profiles (HRRP) using different frequency bands and compressive sensing. Phase errors due to incorrect timing synchronization and antenna phase's center relative locations make it complicated to form HRRPs. In this paper this challenges is addressed proposing two methods based on CS theory: the pruned orthogonal matching pursuit (POMP) and using a L1-norm regularization algorithm to jointly estimate the range profile and the phase errors.

"Compressive Sensing for Tomographic Imaging of a Target with a Narrowband Bistatic Radar" proposes a method to form high resolution 2D radar images using narrow band radar but large synthetic aperture and compressive sensing (CS) based algorithm instead of standard tomographic approach. Particularly the authors of this paper propose the use of the parameter refined orthogonal matching pursuit (PROMP) algorithm. A key feature of this algorithm is that it can address the dictionary mismatch problem that may arise because of the presence of off-grid scatterers. The algorithm performance are the compared to that of standard tomographic approaches and of the orthogonal matching pursuit (OMP) by using simulated and data acquired in an anechoic chamber in a fully controlled experiment.

"Two-Dimensional Augmented State–Space Approach with Applications to Sparse Representation of Radar Signatures" propose a sub-space based method to reconstruct a sparse 2D radar image of man-made targets. Among all the approaches that have been proposed in literature a group of those are the subspace-based approaches, such as MUSIC, MEMP, etc. These demonstrate to have good performance but still some challenges to address, such as the model order that is a requirement for the MUSIC and that is difficult to be a priori known. The authors of this paper propose a 2D augmented state-space approach (ASSA) to try answering these challenges adequately.

The paper "On the Slow-Time k-Space and its Augmentation in Doppler Radar Tomography" presents enabling signal processing technique as a combination of Doppler Radar Tomography (DRT) and a sparse reconstruction technique such as Orthogonal Matching Pursuit (OMP), with a unifying mathematical framework based on the slow-time k-space. DRT relies on spatial diversity from the rotational motion of a target rather than spectral diversity from wide bandwidth signals. The slow-time k-space is a novel form of the spatial frequency space generated by the relative rotational motion of a target at a single radar frequency, which can be exploited for high-resolution target imaging by a narrowband radar with Doppler tomographic signal processing. In the paper, the authors demonstrated the ability to improve image resolution using a rotating target with an ultra-narrowband radar. The proposed technique has been validated using real measurements. As a result, it has been shown that closely spaced scatterers can be resolved by illustrating the creeping wave effect when the scatterer size is similar to the radar wavelength. The proposed method offers a unique and interesting characteristic of the slow-time k-space, which can be augmented and significantly enhance imaging resolution by signal processing and provide more information to identify unknown targets detected by the radar.

The paper "Target Doppler Rate Estimation Based on the Complex Phase of STFT in Passive Forward Scattering Radar" presents a novel approach to estimating target motion parameters in passive forward scattering radars (FSR). In the proposed method, the modulation factor, also called the Doppler rate, is estimated in the time-frequency (TF) domain. The approach proposed by the authorsutilizes the idea of the complex phase of the short-time Fourier transform (STFT) and its modification known from the literature. Additionally, in this paper, the accuracy of the considered estimators were verified using the Cramer-Rao lower bound (CRLB). The authors validate the proposed method using simulations and signals collected during real measurement from a radar operated in passive FSR geometry. The accuracy of the considered tools has been verified by statistical analysis and the comparison of results to the CRLB. The obtained results showed the differences between the estimators as well as the expected accuracy.

The paper "The Use of the Reassignment Technique in the Time-Frequency Analysis Applied in VHF-Based Passive Forward Scattering Radar" presents the application of the time-frequency (TF) reassignment technique in passive forward scattering radar (FSR) using Digital Video Broadcasting – Terrestrial (DVB-T) transmitters of opportunity operating in the Very High Frequency (VHF) band. The authors propose to use this method to enhance the readability of the energy distribution in the TF domain, which improved the result of the Hough transform and finally the precision of the Doppler rate estimation in the passive FSR system. The algorithm has been validated using real-life signals collected by the passive radar demonstrator during a measurement campaign. Additionally, in the described experiment, the authors tested the possibility of utilizing FSR geometry in foliage penetration conditions taking advantage of the VHF band of a DVB-T illuminator of opportunity. The final obtained results show that the concentrated (reassigned) energy distribution of the signal in the TF domain allows a more precise target Doppler rate to be estimated using the Hough transform.

The paper "Noise Suppression for GPR Data Based on SVD of Window-Length-Optimized Hankel Matrix" presents a novel method based on singular value decomposition (SVD) of a window-length-optimized Hankel matrix in application to improve the noise suppression performance in ground-penetrating radar (GPR). The effectiveness of the proposed method has been verified by authors using simulated and real measurement GPR data. The experimental results show that the proposed method can effectively improve noise removal performance under different detection scenarios in GPR applications.

The "New Concept of Combined Microwave Delay Lines for Noise Radar-Based Remote Sensors" paper focuses on the implementation of an optimized analogue microwave tunable delay line to be used in a noise radar to detect micro-movement with range determination. Despite the fact that current development of noise radars mainly concerns the use of advanced techniques of digital signal processing in order to obtain fully-digital correlation receivers, the use of analog correlation based receiver and tunable reference delay line may still be an interesting solutions, especially when dealing with detection of millimeters movement, like vital activity of the human body. To address this issue the paper comprise the concept of a digital controlled delay line with a set of fine distance gates. This concept assumes the use of a combined set of three lines, including a new version of a tapped delay line.

In this section, the short introduction of the published articles in the Special Issue entitled "Recent Advancements in Radar Imaging and Sensing Technology" has been presented. All published papers show new directions of developing algorithms in the area of topics including high-resolution radar imaging, novel Synthetic Apertura Radar (SAR) and Inverse SAR (ISAR) imaging techniques, passive radar imaging technology, modern civilian applications of using radar technology for sensing, multiply-input multiply-output (MIMO) SAR imaging, tomography imaging, among others.

Piotr Samczynski, Elisa Giusti
Editors

Article

A Hybrid SAR/ISAR Approach for Refocusing Maritime Moving Targets with the GF-3 SAR Satellite

Zhishuo Yan [1,2], Yi Zhang [1] and Heng Zhang [1,*]

[1] Department of Space Microwave Remote Sens. Systems, Aerospace Information Research Institute, Chinese Academy of Sciences, Beijing 100190, China; yanzhishuo17@mails.ucas.ac.cn (Z.Y.); zhangyi@mail.ie.ac.cn (Y.Z.)

[2] School of Electronics, Electrical and Communication Engineering, University of Chinese Academy of Sciences, Beijing 100049, China

* Correspondence: zhangheng@aircas.ac.cn; Tel.: +86-1850-190-1425

Received: 18 February 2020; Accepted: 2 April 2020; Published: 4 April 2020

Abstract: Due to self-motion and sea waves, moving ships are typically defocused in synthetic aperture radar (SAR) images. To focus non-cooperative targets, the inverse SAR (ISAR) technique is commonly used with motion compensation. The hybrid SAR/ISAR approach allows a long coherent processing interval (CPI), in which SAR targets are processed with ISAR processing, and exploits the advantages of both SAR and ISAR to generate well-focused images of moving targets. In this paper, based on hybrid SAR/ISAR processing, we propose an improved rank-one phase estimation method (IROPE). By using an iterative two-step convergence approach in the IROPE, the proposed method achieves accurate phase error, maintains robustness to noise and performs well in estimating various phase errors. The performance of the proposed method is analyzed by comparing it with other focusing algorithms in terms of processing simulated data and real complex image data acquired by Gaofen-3 (GF-3) in spotlight mode. The results demonstrate the effectiveness of the proposed method.

Keywords: synthetic aperture radar (SAR); moving targets; inverse SAR (ISAR); motion compensation; hybrid SAR/ISAR; improved rank-one phase estimation (IROPE); Gaofen-3 (GF-3)

---

1. Introduction

Synthetic aperture radar (SAR) provides all-weather, day–night, wide-range high-resolution imaging capabilities for a wide range of applications in Earth science and climate change research, marine detection and imaging, and disaster monitoring [1]. Nevertheless, SAR uses the motion of the radar, ignoring target motion, to coherently synthesize a large aperture that provides a narrow synthesized beam, and it thus has a high resolution across the range. Therefore, in the ship detection and classification scenario, ships are always defocused with conventional SAR processing because of individual motion and sea waves. Inverse SAR (ISAR) uses the rotational motion of the targets, ignoring radar motion, to distinguish different relative velocities through coherent processing and Doppler effects and to form the synthetic aperture. Thus, ISAR processing is more adaptable to the moving scenario since it is superior in imaging moving targets undergoing complex unknown motions [2]. However, due to the unpredictability of non-cooperative targets, motion compensation in ISAR imaging is a challenging task and usually includes two steps: range tracking and Doppler tracking, i.e., coarse phase compensation and fine phase compensation [3]. This paper focuses on the study of fine phase compensation, which is more sensitive than range tracking. Moreover, during the operation of SAR for a moving target, both the radar and target are in motion, which means that the processing of the moving target must combine the processes of SAR (radar motion) and ISAR (target motion) [4,5]. Hybrid SAR/ISAR [6,7] processing is such an approach to optimally process SAR data by treating target and radar platform motions on an equal footing, which takes advantage of ISAR processing to generate the focused image of the moving target in SAR.

Ideally, the phase of the echo signal in the range and azimuth profiles varies linearly during the processing time. However, due to various factors, there exist undesired phase changes in the echo signal, which are collectively referred to as phase error [8]. Phase error, which is divided into low-frequency, high-frequency, and random phase errors, causes geometric distortion, resolution degradation, false targets, and reduced signal-to-noise ratio (SNR), thus resulting in poor image quality. Low-frequency errors encompass linear phase errors, quadratic phase errors (QPEs), and so on. The low-frequency phase errors primarily affect the main lobe of the system impulse response, while high-frequency errors affect the sidelobe regions. Random phase errors cause multiple pairs of echoes around the target, and the main lobe energy is reduced [9].

To estimate the phase error and refocus the images, many fine phase compensation algorithms have been proposed, which are roughly divided into parametric and nonparametric algorithms. The parametric algorithms include the Mapdrift (MD) method [10,11], the phase difference (PD) method [12] and methods of parameter estimation [13]. MD and PD methods are easy to implement, but they only compensate for QPEs, which limits their applications. Furthermore, Chen et al. [14] proposed a parametric sparse representation method. The acceleration and third-order phase were considered in [15,16]. Tang et al. [17] achieved 2D velocity estimation of moving targets and refocusing based on back projection and velocity SAR (another multichannel SAR-GMTI technique). Nevertheless, these methods introduce nonlinear operations, which degrade the performance in the case of low SNR. [18–21] presented a method for imaging moving targets via the compressive sensing (CS) method, which is capable of generating images with better target focusing, especially with low SNR and high undersampling ratios.

The nonparametric algorithms mainly include the maximum contrast (MC) [22], minimum entropy (ME) [23,24], weighted least-squares (WLS) [25], sharpness optimization [26,27], Doppler centroid tracking (DCT) [28], phase gradient autofocus (PGA) [29] and rank-one phase estimation (ROPE) [30,31] methods. Because the MC, ME, WLS and sharpness optimization methods do not make any assumptions about the characteristics of the target itself, they are highly adaptable. However, since the synthetic aperture process is non-stationary and random, such algorithms generally have local extremum problems. The DCT, PGA, and ROPE methods align the range envelopes and successively adjust the phase to compensate for the translational motion. The DCT and PGA [32] methods are not model-based and exhibit robust performance. Nevertheless, their compensation accuracy is unsatisfactory if there are high-frequency and random phase errors. Furthermore, the model-based ROPE method assumes that each range bin contains no more than one scattering center. The main idea of the method is to use the phase finite difference to estimate the phase error, to find the phase average by alternating along the range direction and the azimuth direction, and to estimate both the phase error and Doppler frequency. In addition, the high azimuth resolution in ISAR processing is generated using a Doppler frequency gradient generated by the rotation of the target relative to the radar line of sight (RLOS) [33]. By averaging all range units of the estimated Doppler frequency, the phase error will be obtained more accurately with influences of the rotational phase weakened, ultimately owning a more precise compensation for the translational phase error. The most remarkable feature of the method is that it estimates not only the phase error of arbitrary order but also the wideband phase error. However, the ROPE method still includes some flaws: The model-based ROPE algorithm strictly requires that there be at most one strong scattering point for each range bin, which limits its application to many images that do not approximate the model; the performance of the method with respect to the phase error estimation will be greatly reduced under low SNR; and the ROPE method uses zero as the initial estimation for the Doppler frequency, which is blind and may lead to unsatisfactory estimates.

Motivated by these aforementioned observations, in this paper, a refocusing method named IRPOE is proposed to solve the above problems existing in ROPE. Our contributions are summarized as follows:

- We use the DCT algorithm in the IROPE method to improve the SNR of the data and render the subsequent estimation more accurate, which also makes the data more consistent with the model of the ROPE method.
- We avoid the averaging of phase vectors with different linear components and maximize the accuracy of the phase compensation by using the circular shift of the prominent point in each range bin to zero frequency to rationalize the initialization of the Doppler center.
- Moreover, better estimates are obtained through iteration. Multiple iterative algorithms improve the SNR, which further improves the accuracy of the Doppler circular shift and the estimation of phase error.
- The proposed method focuses the blurred images well and exhibits the superiority of robustness, reduced sidelobes, and suitability for various phase errors, which does not require time-consuming parameter adjustment procedures to achieve improved performance and allows a long coherent processing interval.

The Gaofen-3 (GF-3) satellite is the first Chinese C-band multi-polarization high-resolution SAR imaging satellite [34]. As one of the most important satellites in China's Earth observation systems, GF-3's features include high resolution, large imaging swath, multiple imaging modes, and long operating life [35,36]. GF-3 plays an essential role in the fields of marine environment monitoring, land resource investigation, and disaster prevention, providing high-quality data for scientific experiments. This paper uses the ocean data of the GF-3 satellite to demonstrate the proposed method and the work has the merit to show its potentialities against satellite data.

This paper is organized as follows. The moving signal model in SAR and ISAR systems is presented in Section 2. Section 3 proposes a phase estimation algorithm named IROPE and elaborates the performance. Extensive experimental results on both simulated and real data are presented in Section 4 to demonstrate the effectiveness and robustness of the proposed method. Finally, Section 5 concludes the paper.

2. Moving Signal Model

In this section, the signal models of SAR and ISAR are presented. When introducing the SAR signal model, the moving echo signal characteristics and the influence of motion parameters are analyzed. Furthermore, the translation and rotation Doppler shifts involved in the ISAR signal model are analyzed.

2.1. SAR Signal Model

2.1.1. Analysis of Moving Echo Characteristics

The geometry of SAR imaging of the moving target is shown in Figure 1. The target is described in Cartesian coordinates with the initial position at $P(x_0, y_0, 0)$. The SAR platform moves along the predetermined track, where v_a and h represent the velocity and height, respectively. v_x, a_x, v_y, a_y, v_r, and a_r are the observed target's velocities and accelerations in azimuth, range and radar RLOS directions. t is the slow time, and the distance from point P to the radar platform is R_c, $R_c^2 = y_0^2 + h^2$. R_0 is the distance between SAR and the target at the initial time, and $R_0^2 = x_0^2 + R_c^2$. The target moves to $P(x_t, y_t, 0)$ at time t, and the distance between SAR and the target is $R(t)$.

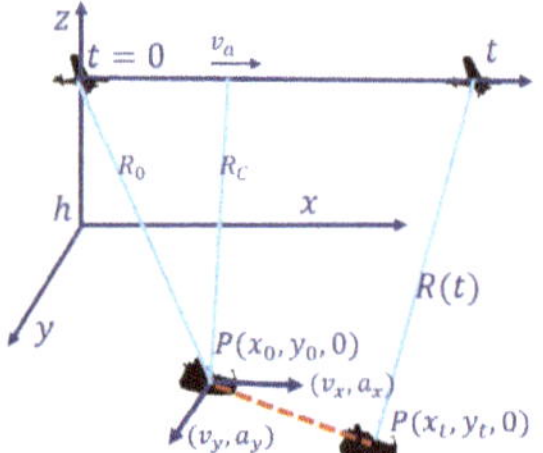

Figure 1. Moving target SAR geometry.

Letting $\widehat{v} = v_a - v_x$, the square of the slant range $R(t)$ is described as

$$\begin{aligned} R(t)^2 &= h^2 + (v_a t - x_0 - v_x t - \frac{1}{2}a_x t^2)^2 + (y_0 + v_y t + \frac{1}{2}a_y t^2)^2 \\ &= R_0^2 + (\widehat{v}t - \frac{1}{2}a_x t^2)^2 - 2x_0(\widehat{v}t - \frac{1}{2}a_x t^2) + (v_y t + \frac{1}{2}a_y t^2)^2 + 2y_0(v_y t + \frac{1}{2}a_y t^2) \end{aligned} \tag{1}$$

Equation (2) gives Taylor series expansions of $R(t)$ and ignores high-order items (cubic or higher), where $\frac{v_r}{v_y} = \frac{y_0}{R_0}$ and $\frac{a_r}{a_y} = \frac{y_0}{R_0}$. Equation (1) is simplified as

$$R(t) = R_0 + \frac{1}{2R_0}((\widehat{v}^2 + v_y^2 + a_r R_0 + x_0 a_x)t^2 - 2x_0\widehat{v}t) + v_r t \tag{2}$$

Accordingly, the Doppler phase $\phi(t)$ is

$$\begin{aligned} \phi(t) &= \frac{4\pi}{\lambda}R(t) \\ &= \frac{4\pi}{\lambda}(R_0 + \frac{1}{2R_0}((\widehat{v}^2 + v_y^2 + a_r R_0 + x_0 a_x)t^2 - 2x_0\widehat{v}t) + v_r t) \end{aligned} \tag{3}$$

where λ represents the wavelength of the transmitted signal.

$$\begin{cases} f_c = \left.\frac{-1}{2\pi}\frac{d\phi}{dt}\right|_{t=0} = \frac{-2v_r}{\lambda} + \frac{2x_0\widehat{v}}{\lambda R_0} \\ f_r = \left.\frac{df_c}{dt}\right|_{t=0} = \frac{-2}{\lambda R_0}(\widehat{v}^2 + v_y^2 + x_0 a_x + a_r R_0) \end{cases} \tag{4}$$

where f_c represents the Doppler centroid frequency, and f_r is the azimuth FM rate.

For stationary targets, $v_x = v_y = v_r = 0, a_x = a_y = a_r = 0$. Equation (4) is expressed as

$$\begin{cases} f_{rc} = \frac{2x_0 v_a}{\lambda R_0} \\ f_{rr} = \frac{-2v_a^2}{\lambda R_0} \end{cases} \tag{5}$$

Then, the Doppler centroid frequency and the azimuth FM rate generated by the target's motion are

$$\begin{cases} f_{lc} = \frac{-2v_r R_0 - 2x_0 v_x}{\lambda R_0} \\ f_{lr} = \frac{-2}{\lambda R_0}(v_x^2 - 2v_a v_x + v_y^2 + a_r R_0 + x_0 a_x) \end{cases} \tag{6}$$

2.1.2. Analysis of Moving Target Response

Equation (7) gives Taylor series expansions of the phase error:

$$\begin{aligned}\Delta\phi(t) &= \frac{4\pi}{\lambda}\Delta R(t) \\ &\approx \frac{4\pi}{\lambda}(R + \left.\frac{d\Delta R(t)}{dt}\right|_{t=0} t + \left.\frac{d^2\Delta R(t)}{dt^2}\right|_{t=0} \frac{t^2}{2} + ...)\end{aligned} \tag{7}$$

As [37] mentioned, the first-order phase errors cause azimuth positional offset of the target scattering point, i.e., position deviation. Quadratic phase errors cause target defocusing. Third-order phase errors mainly cause the asymmetry of the sidelobe levels on both sides of the main lobe; in the strong target condition, the image appears ghost-like. The fourth phase errors mainly cause the sidelobe level to increase. The higher-order phase errors will increase the integrated sidelobe level and have little effect on the main lobe width. Generally, due to the Doppler effect, the azimuthal motion of the ship results in blurred defocusing, and the range motion results in an additional shift of the image.

2.2. ISAR Signal Model

Assuming that the number of scattering centers is K, the range-compressed data are represented as [37]

$$s(\tau, t) = \sum_{k=1}^{K} A_k \rho_r(\tau - 2r(t)/c)\omega_a(t - t_c)e^{-j\frac{4\pi f_0 r(t)}{c}} \tag{8}$$

Here, A_k represents the backscattered coefficients of the scatterer k. f_0 is the carrier frequency of the system, and τ, t and t_c denote the fast time, slow time, and beam center offset time, respectively. The distance of point P from the radar is $r(t)$. ρ_r represents range envelope (a sinc function) and ω_a represents azimuth envelope (a sinc-squared function) [38].

The Doppler effect of the target's motion is described in the geometry in Figure 2. From Equation (8), motion compensation removes the phase term $exp(-j4\pi f_0 r(t)/c)$. Assuming that the middle of the target is the origin O, $r(t)$ is expressed as

$$r(t) \approx R(t) + x cos\theta(t) + y sin\theta(t) \tag{9}$$

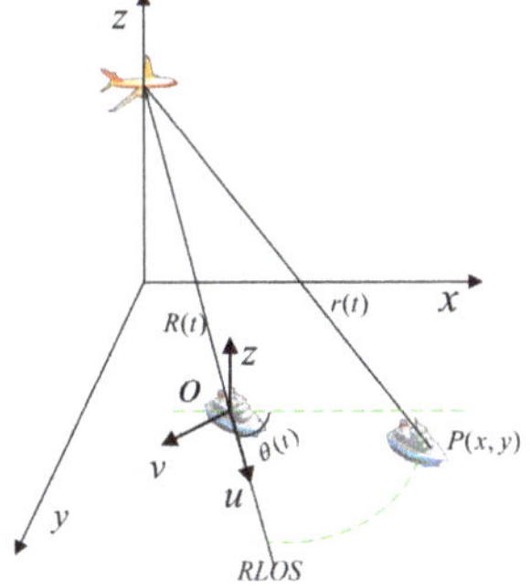

Figure 2. Geometry of the ISAR system.

Here, $R(t)$ is the target's translational range distance from the radar and $\theta(t)$ represents the rotational angle of the target with respect to the RLOS axis, u. Equation (10) gives the Taylor series expansions of $R(t)$ and $\theta(t)$ and ignores high-order items (cubic or higher):

$$\begin{aligned} R(t) &\approx R_0 + v_t t + 1/2 a_t t^2 + \cdots \\ \theta(t) &\approx \theta_0 + \omega_r t + 1/2 \alpha_r t^2 + \cdots \end{aligned} \tag{10}$$

R_0 is the initial range of the target, and v_t and a_t are the target's translational velocity and acceleration, respectively. Similarly, θ_0 is the initial angle of the target with respect to the RLOS axis. ω_r and α_r are the angular velocity and acceleration of the target, respectively.

The echoes of the kth range bin are expressed as:

$$s_k(\tau, t) = A_k \rho_r(\tau - 2r(t)/c)\omega_a(t - t_c)\phi_t\phi_r \tag{11}$$

where ϕ_t and ϕ_r are the phase terms caused by the translational and rotational movement of the target, respectively.

$$\begin{cases} \phi_t = e^{\frac{-j4\pi f_0}{c}(R_0 + v_t t + 1/2 a_t t^2 + \cdots)} \\ \phi_r = e^{\frac{-j4\pi f_0 \sqrt{x^2+y^2}}{c} sin(\beta + \theta_0 + \omega_r t + 1/2 \alpha_r t^2 + \cdots)} \end{cases} \tag{12}$$

Here, $sin\beta = \frac{x}{\sqrt{x^2+y^2}}$. The imaging process of the following section removes the influence of the translational phase, including v_t and a_t, which offers no contribution to ISAR imaging [39].

3. Improved Rank-One Phase Estimation Algorithm

3.1. Problems of the Rank-One Phase Estimation Algorithm

The ROPE method, first developed in [30], estimates and removes the phase error, which guarantees that range-Doppler (RD) imaging can proceed in the usual manner. The algorithm has been modified by [31] to extend its scope of application (including ISAR processing), but some limitations remain. First, the model-based ROPE algorithm strictly requires that there be at most one strong scattering point for each range bin, which limits its application to many images that do not fit the model. Second, the performance of the algorithm decreases sharply under low SNR. Next, blind initialization of the Doppler frequency to zero results in inaccurate estimation. These problems limit the performance and application of the ROPE algorithm. To solve the above problems, the IROPE algorithm is proposed and explained in detail as follows.

3.2. Principle of IROPE

Combine formulas to explain the improvements of IROPE and the reasons for the improvement in detail.

I. Preliminary Phase Compensation

First, the range-aligned echo signal $e(r)$ is subjected to preliminary phase estimation and compensation using the DCT algorithm.

Multiply the conjugate of the ith echo with the next echo to find the average phase difference between adjacent range units:

$$e^{j\varphi} = \frac{\int e_i^*(r)e_{i+1}(r)dr}{\int |e_i(r)e_{i+1}(r)|\, dr} \tag{13}$$

Use $e^{j\varphi}$ to adjust the phase shift of $e_{i+1}(r)$ so that the average phase shift with respect to the adjacent one-dimensional range direction is zero, which is equivalent to aligning the target to a phase center, and the average Doppler shift of the target rotating around the center is zero, thus eliminating the effect of the remaining phase difference.

Through the preliminary phase correction, the SNR is improved, the subsequent estimation becomes more accurate, and the processed image is more consistent with the signal model. After the initial phase correction in this step, the peak value of the special point obtained by Fourier transform will be sharper than the original, which improves the effect of the subsequent circular shifting step.

II. Two-step Convergence

Next, there are J range units and that each range cell contains no more than one scattering center. The signal of the jth range cell is given by

$$s_{k,j} = a_j \exp\left[i\left(\omega_j k\Delta T + \varepsilon_k + \alpha_j\right)\right] + n_{k,j} \tag{14}$$

where a_j is the amplitude of the signal, k is the azimuth pulse number, ΔT is the pulse period, ε_k is the phase error, $n_{k,j}$ is the additional complex noise, and $\omega_j/2\pi$ is the Doppler position after imaging.

The phase in each pixel of the range-time array is

$$\varphi_{k,j} = \omega_j k\Delta T + \varepsilon_k + \alpha_j \tag{15}$$

The difference of Equation (15) is

$$\hat{\varphi_{k,j}} = \varphi_{k+1,j} - \varphi_{k,j} = \omega_j \Delta T + \varepsilon_{k+1} - \varepsilon_k \tag{16}$$

Assume $\hat{\omega_j} = \omega_j \Delta T$ so that

$$\hat{\varphi_{k,j}} = \varphi_{k+1,j} - \varphi_{k,j} = \hat{\omega_j} + \varepsilon_{k+1} - \varepsilon_k \tag{17}$$

when the SNR is infinite, i.e., $n_{k,j} = 0$.

$$D_{k,j} = \frac{s_{k+1,j}s^*_{k,j}}{\left|s_{k+1,j}\right|\left|s^*_{k,j}\right|} = \exp\left[i\left(\hat{\omega_j} + \varepsilon_{k+1} - \varepsilon_k\right)\right] = \exp\hat{\varphi_{k,j}} = \exp\left[i\hat{\omega_j}\right]\exp\left[i\left(\varepsilon_{k+1} - \varepsilon_k\right)\right] \tag{18}$$

$D_{k,j}$ is the product of a column vector and a row vector, and $D = \left[D_{k,j}\right]$ is a rank-one matrix.

Let $\hat{\varepsilon}_k^{(p)} = \varepsilon_{k+1} - \varepsilon_k$. After initialization, the ROPE method consists of the following two operations

$$\begin{cases} \hat{\varepsilon}_k^{(p)} = \angle \sum\limits_{j=1}^{J} D_{k,j} \exp\left(-i2\pi\hat{\omega_j}^{(p-1)}\right) \\ \hat{\omega_j}^{(p)} = \angle \sum\limits_{k=1}^{K} D_{k,j} \exp\left(-i2\pi\hat{\varepsilon}_k^{(p)}\right) \end{cases} \tag{19}$$

The superscript p represents the pth operation. When the maximum value of the two estimated changes is less than the small threshold value T, the process leads to convergence. When the SNR is high, although the influence of noise causes the rank of matrix D to not be equal to one but approaching one, the two-dimensional alternative estimation is still reasonable and feasible. Nevertheless, when the SNR is low, $n_{k,j}$ cannot be ignored, D is not a rank-one matrix, and the ROPE algorithm fails. The final phase error estimate is

$$\hat{\varepsilon} = \sum_{k=1}^{K} \hat{\varepsilon}_k^{(p)} \tag{20}$$

Equation (19) estimates the phase error and Doppler frequency simultaneously, which estimates the phase error more accurately and avoids the influence of the phase rotation component.

However, the initial phase error $\hat{\mathcal{E}}_1$ in Equation (20) of the original ROPE algorithm is simply set to zero, which is relatively blind and results in unsatisfactory estimation.

III. Circular Shifting

The maximum value of the range bin still represents the Doppler frequency corresponding to the strong scattering point, but the energy of the scattering point diffuses in the azimuth direction. Moving the strongest scattering point to zero Doppler frequency eliminates the deficiency of setting the initial Doppler frequency to zero to some extent. The circular shift operation not only aligns the strong scatterers but also improves the SNR of the phase compensation; subsequently, the processed data are more consistent with the model.

IV. Iteration

Nevertheless, it is difficult to accurately align the Doppler circular shift when the SNR is low, which affects the estimation accuracy. Therefore, multiple iterative algorithms are then used to improve the SNR, thereby further improving the accuracy of the Doppler circular shift and the estimation of phase error.

The algorithm is summarized in Algorithm 1.

Algorithm 1: The IROPE algorithm for phase compensation

Input: The range-aligned echo $e(r)$, $[Nr, Na] = size(e(r))$, $p = 0$, **number of iterations** l, **threshold value** T
I. Preliminary Phase Compensation
for $i = 1 : Na - 1$
  $e_{i+1}(r) = e_{i+1}(r). * e^{j\varphi}$
end
IV. Iteration
for $l = 1 : l$ **(Image entropy is further applied to control the iteration process)**
  II. Two-step Convergence
  III. Circular Shifting
  while $\hat{\varepsilon}_k^{(p)}$ - $\hat{\varepsilon}_k^{(p-1)}$ **> T**
    for $k = 1 : Na - 1$
      Update $\hat{\varepsilon}_k^{(p)}$ calculated by Equation (19)
    end
    for $j = 1 : N_r$
      Update $\omega_j^{(p)}$ calculated by Equation (19)
    end
    $p = p + 1$
  end while
  $\hat{\varepsilon} = \sum_{k=1}^{K} \hat{\varepsilon}_k^{(p)}$
  $e(r) = e(r). * exp(-1i * \hat{\varepsilon})$
end
Output: Compensated range-Doppler echo $e(r)$, **phase error** $\hat{\varepsilon}$

The flow chart of the IROPE procedure is indicated in Figure 3 and described in detail as follows.

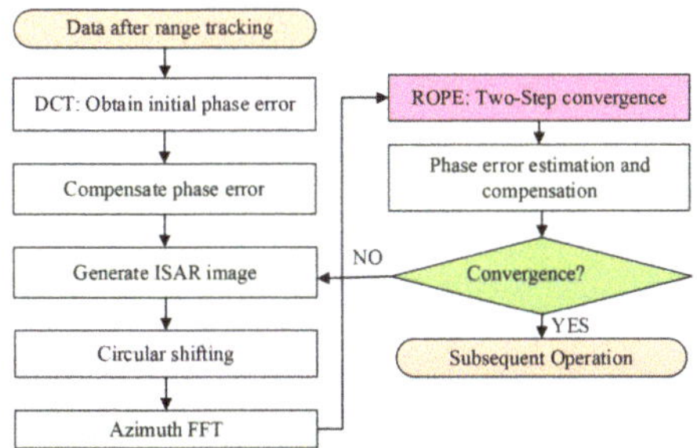

Figure 3. Block diagram of the IROPE procedure.

Step 1: Use the DCT method to perform initial phase correction on the echo data after range tracking for preprocessing.

Step 2: Perform IFFT transform in the azimuth direction to generate an ISAR image.

Step 3: Find the maximum amplitude and set the initial zero Doppler to the circular shift of the prominent point in each range bin.

Step 4: By performing azimuth FFT, the data are transformed to the range-Doppler domain.

Step 5: Use two-step convergence approach to obtain and compensate for the phase error.

Step6: If the effect of refocusing is not sufficient, repeat the process from Step2 to Step5.

3.3. Performance of IROPE

This subsection presents the experimental results based on range-aligned echo to illustrate the performance of IROPE. The radar operates in the X band. The transmitted signal bandwidth and the synthetic aperture time are 100 MHz and 3.32 s, respectively. The translational motion of the target with range velocity of v_y = 6 m/s, azimuth velocity of v_x = 15 m/s, and azimuth acceleration of a_x = 2 m/s^2 is determined. Assume that there is a sinusoidal error term caused by the target's rotation, which is chosen as $\frac{0.5\pi}{180}sin(0.6t)$ rad. The rotational velocity is 0.6 rad/s and t represents the azimuth observation time. The pulse repetition frequency (PRF) and the radar velocity are 600 Hz and 250 m/s, respectively.

Example 1 Figure 4 shows the experimental results corresponding to a single moving point target. It can be seen that the point is well focused by the ROPE and IROPE methods. Comparing the interpolated contour and the value of PSLR, IROPE exhibits slight superiority over ROPE. (The technical indicators shown in the figure are explained as follows: impulse response width (IRW), namely the 3 dB main lobe width of impulse response; peak sidelobe ratio (PSLR), the height ratio of the maximum sidelobe to the main lobe; and integrated sidelobe ratio (ISLR)).

Example 2 Figure 5 shows the experimental results for a simulated ship, i.e., multiple-point targets, in which each range bin of the ship's hull has three strong scattering points with the same strength. It can be seen from Figure 5b that the ROPE algorithm fails because of not satisfying the model in which each range cell contains no more than one scattering center. According to the above subsection analysis, IROPE compensates for the deficiency of ROPE and obtains a good focusing effect.

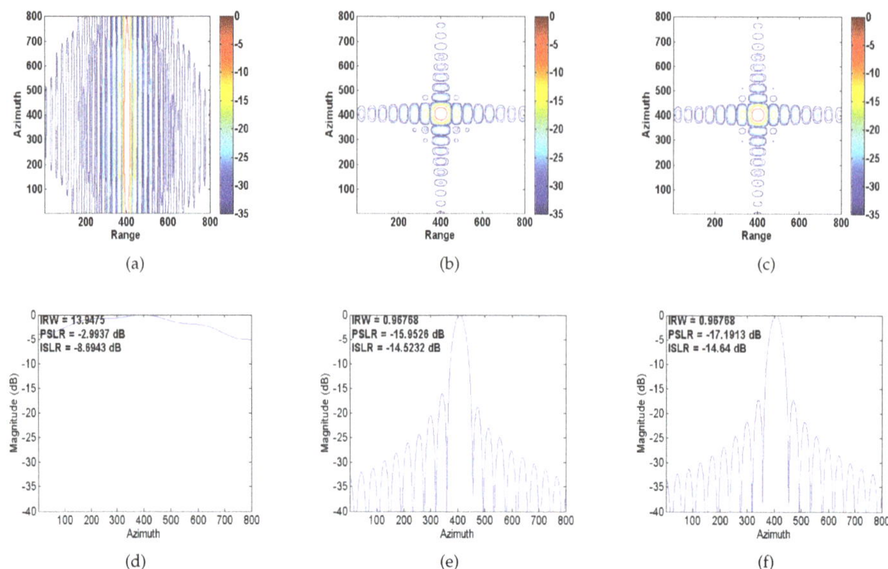

Figure 4. The interpolated contour and azimuth profile. (**a**,**d**) Conventional SAR processing; (**b**,**e**) ROPE; (**c**,**f**) IROPE.

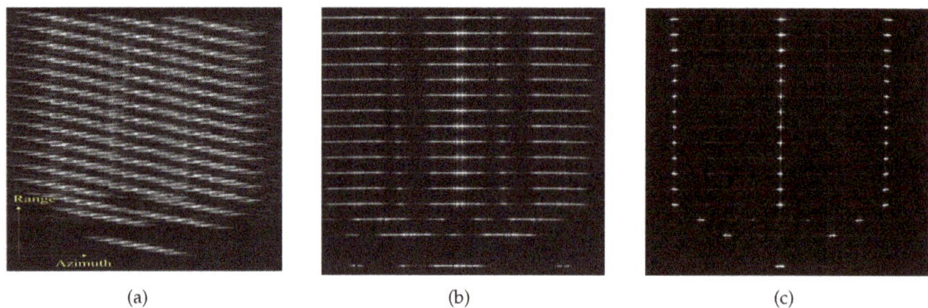

Figure 5. Refocused performance. (**a**) Conventional SAR processing; (**b**) ROPE; (**c**) IROPE.

Figure 6a shows that the image entropy (defined in Equation (21)) decreases with the increase of IROPE's iteration times, which proves that iteration improves the image focusing effect. Figure 6b exhibits the image entropy processed by different algorithms, which changes with SNR. The results prove the robust performance with respect to the noise of the IROPE algorithm.

The phase error estimation performance is provided in the next section.

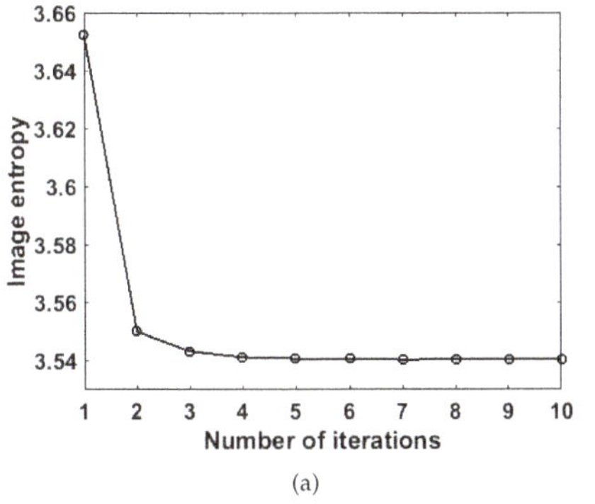

(a)

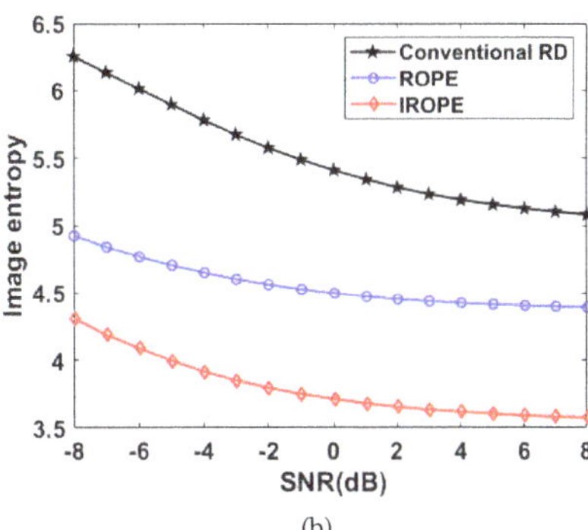

(b)

Figure 6. Variation in image entropy in terms of iterations (**a**) and SNRs (**b**).

3.4. The Whole Process of the Refocusing Method

The whole imaging flow chart is shown in Figure 7.

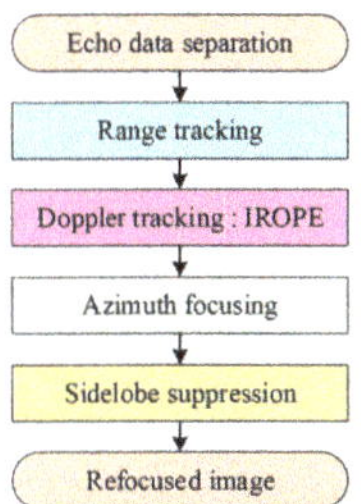

Figure 7. Block scheme of the whole process of the refocusing method.

First, the separated echo signal is obtained. The echo data are obtained from the original raw echo data or complex image data. The former approach implements range compression on the original raw echo data, while the latter performs azimuth inverse compression on the selected complex image data of the target of interest to acquire the required echo data for subsequent operations.

Second, range tracking is performed on echo data. Using the cross-correlation [40] of the average range profile, a real-time and efficient method, the ship echo is correlated with the first echo in the imaging time. In addition, through range alignment, the range units of the echoes are aligned, and the amplitude and phase changes of the echo range sequence of each range unit are normal. Eventually, the phase change process generated by the target translation is retained.

Furthermore, we apply IROPE to estimate phase error $\hat{\varepsilon}_k$ and obtain the compensated range-Doppler echo $e(r)$.

Next, we combine the conventional RD algorithm and use the Hamming window to achieve well-focused images and suppress sidelobes. The conventional RD algorithm is applied to obtain focused ISAR images if the maritime target moves smoothly. However, if the target maneuvers or undergoes significant angular motions (roll, pitch, and yaw), the RD technique does not function properly, and the time-frequency analysis [41] method is a better choice.

4. Experiments and Performance Comparisons

In this section, the robustness and effectiveness of the proposed method are verified by simulation experiments. Then, the results based on the spaceborne SAR data acquired by the GF-3 SAR system are demonstrated.

4.1. Results of Spotlight Simulation

The proposed method is applied to spotlight simulation data and analyzed for different motions of the ship target by comparing it with other refocusing algorithms. The basic parameters of the spotlight simulation are shown in Table 1.

Table 1. Parameters of spotlight simulation.

Parameter	Value
Mode	Spotlight
Radar Center Frequency (GHz)	5.4
Wavelength (m)	0.0555
Bandwidth (MHz)	50
Azimuth Resolution (m)	1.5
Range Resolution (m)	2.6562
PRF (Hz)	3125
Upsampled PRF (Hz)	9950.2398
Upsampled Doppler Bandwidth (Hz)	6188.7356
Slant-Range (m)	1,067,731.2395
Synthetic Aperture Time (s)	2.3342
SAR Velocity (m/s)	7500
squint angle (°)	0

4.1.1. Ship Target with Velocity and Acceleration

The target is moving away from the radar with range velocity of $v_y = 3$ m/s, azimuth velocity of $v_x = 15$ m/s and azimuth acceleration of $a_x = 2$ m/s^2. As mentioned in Sections 1 and 2, the quadratic phase errors caused by velocity lead to image defocusing; cubic phase error introduced by acceleration mainly causes the asymmetry of the sidelobe levels on both sides of the main lobe [37]. The conventional SAR image and the recovered images of MD, ROPE and the proposed algorithm are shown in Figure 8. It can be seen from Figure 8c,d that the MD algorithm only compensates for the quadratic phase errors caused by velocity but cannot compensate for the cubic phase error caused by acceleration. The image processed by the ROPE algorithm has a high energy of the sidelobes, as shown in Figure 8e,f. Based on the above analysis, the image quality of the proposed method is superior to the other methods.

A discussion of what amount of non-uniform motion can be effectively analyzed during the long CPI, i.e., the limitation in azimuth linear acceleration, follows. Azimuth linear acceleration varies from -2 to 6 m/s^2 according to a step size of 2 m/s^2 [42]. The variations in the magnitudes of PSLR and ISLR are shown in Figure 9. It can be seen that the magnitudes of PSLR and ISLR fluctuate with increasing acceleration, and the overall trend is upward. The maximum magnitude of PSLR is below -14 dB, while the maximum magnitude of ISLR is below -9 dB, which means that the algorithm is suitable for practical situations.

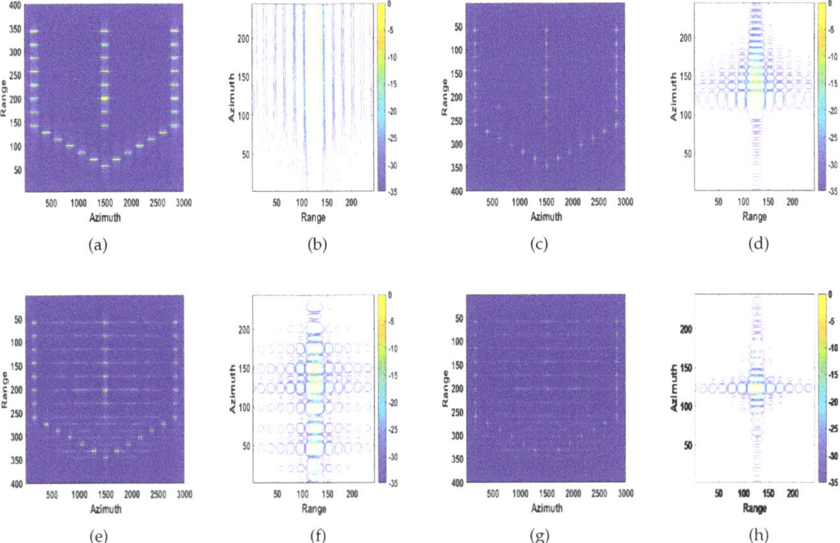

Figure 8. Recovered images of defocused ship target with velocity and acceleration. (**a,b**) Conventional SAR processing system; (**c,d**) MD method; (**e,f**) ROPE method; (**g,h**) The proposed method.

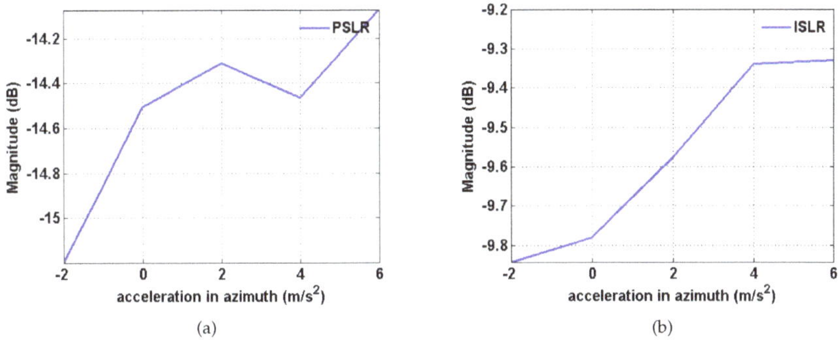

Figure 9. Variations in the magnitudes of PSLR and ISLR in terms of the linear acceleration based on the proposed method. (**a**) PSLR; (**b**) ISLR.

4.1.2. Ship target with Translation and Rotation

The target is moving away from the radar with range velocity of v_y = 3 m/s and azimuth velocity of v_x = 15 m/s. We assume that there is a sinusoidal error term caused by the target's rotation, which is chosen as $\frac{0.5\pi}{180} sin(0.6t)$ rad. High-frequency and wideband phase errors introduced by rotational velocity mainly affect the sidelobe region and increase the sidelobe level [37]. The conventional SAR image and the recovered images of MD, ROPE and the proposed algorithm are shown in Figure 10. It can be seen from Figure 10c-d that the MD algorithm only compensates for the quadratic phase errors caused by velocity and cannot compensate for the high-frequency and wideband phase errors introduced by rotational velocity. From Figure10e-f, processed by the ROPE method, the main lobe is almost submerged by the sidelobes. Comparing the values of PSLR and ISLR for different algorithms, the image quality of the proposed algorithm is also superior to the other approaches.

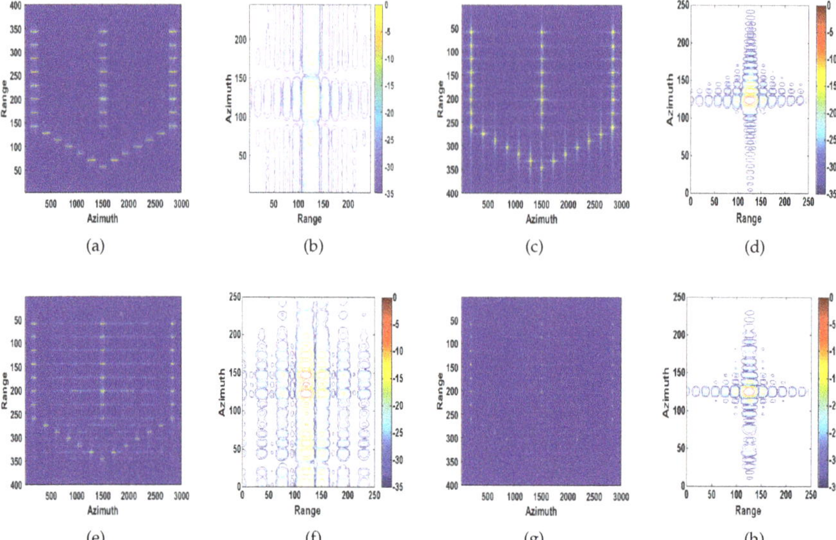

Figure 10. Recovered images of defocused ship target with translation and rotation speed. (**a**,**b**) Conventional SAR processing system; (**c**,**d**) MD method; (**e**,**f**) ROPE method; (**g**,**h**) The proposed method.

Moreover, assume that $\omega_r = \frac{0.5\pi}{180} sin(wt)$ rad. w varies according to a step size of 0.1 rad/s from 0 to 1 rad/s [43]. The experimental results are shown in Figure 11. It can be seen that the magnitudes of PSLR and ISLR fluctuate with increasing rotational angular velocity, and the overall trend is upward. It is observed that the magnitudes of PSLR and ISLR fluctuate with increasing acceleration, and the overall trend is upward. The maximum magnitude of PSLR is below −14 dB, while the maximum magnitude of ISLR is below −9.5 dB, which means that the algorithm is suitable for practical situations.

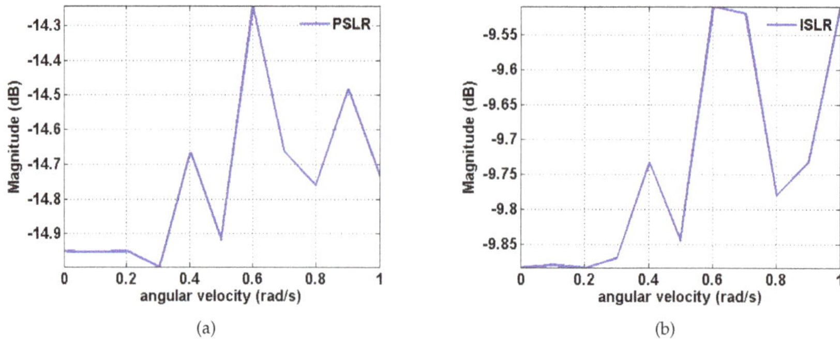

Figure 11. Variation in the magnitudes of PSLR and ISLR in terms of the rotational angular velocity based on the proposed method. (**a**) PSLR; (**b**) ISLR.

4.2. Spaceborne SAR Data Experiments

In this subsection, the results based on the spaceborne SAR data acquired by the GF-3 SAR system are demonstrated. The parameters are shown in Table 2. The synthetic aperture time has a relatively large value of 8.58 s. In addition to visual inspection, image quality is also evaluated by entropy, contrast and the peak value of the intensity image [44], which are referred to as image quality evaluation metrics (IQEMs).

Table 2. Parameters of the GF-3 SAR System.

Parameter	Value
Mode	Spotlight
Radar Center Frequency (GHz)	5.400012
Bandwidth (MHz)	240.000000
Azimuth Resolution (m)	1
Range Resolution (m)	0.6
PRF (Hz)	3125.164062
Synthetic Aperture Time (s)	8.58
Satellite Velocity (m/s)	7570.962970

Let $I(m,n)$ be the absolute value of a two-dimensional complex image, where m is the range sample number and n is the azimuth number. The image entropy (IE) [23] is written as follows:

$$IE(I) = -\sum_{m=1}^{M}\sum_{n=1}^{N}\frac{|I(m,n)|^2}{\alpha_I}\ln\frac{|I(m,n)|^2}{\alpha_I} \tag{21}$$

where α_I is the total energy of the image, explained as follows:

$$\alpha_I = \sum_{m=1}^{M}\sum_{n=1}^{N}|I(m,n)|^2 \tag{22}$$

When the image is well focused, the entropy value is small because of the uniform distribution.

The image contrast (IC) is defined as follows [45]:

$$IC(I) = \frac{\sqrt{E\{[I(m,n) - E\{I(m,n)\}]^2\}}}{E\{I(m,n)\}} \tag{23}$$

when the image is focused correctly, it comprises several significant peaks, which enhances the contrast.

The peak value of the intensity image (IP) is an indicator of the image focusing of a local area of the image, and the calculation expression is

$$IP(I) = 10 \cdot log10(\max(I(m,n))) \tag{24}$$

The larger the IP is, the better the image focus is.

Hence, to present a more intuitive and quantitative comparison, Tables 3 and 4 provide the difference of IQEMs between the processed and the original images—contrast increase, entropy reduction, and IP increase. The higher the value is, the better the image quality is.

4.2.1. Real Data Corrupted by Phase Error

Figures 12 and 13 show the experimental image results for phase error and demonstrate the capability to estimate the phase error of the proposed method. The phase errors of the quadratic, third and fourth superpositions of the same weight are added to Figure 12a. Figure 12b shows the corrupted image. There is no improvement in the image after the MD method refocusing, as can be seen from Figure 12c. The ROPE method exhibits a better focusing effect than the MD method does, but it is visually worse than the proposed method. A comparison of the image recovered using the proposed method with the original image shows very good agreement, although there are some slight differences. Figure 13 demonstrates that the phase error estimated by IROPE is closest to the introduced phase error. MD only compensates for the quadratic phase errors, so the estimated and true phase errors display considerable deviation. The difference phase error also exhibits a large deviation of ROPE because the actual image does not satisfy the model. The success of the IROPE algorithm is obvious; a bias exists, but the bias does not affect the quality of the recovered image from Table 3.

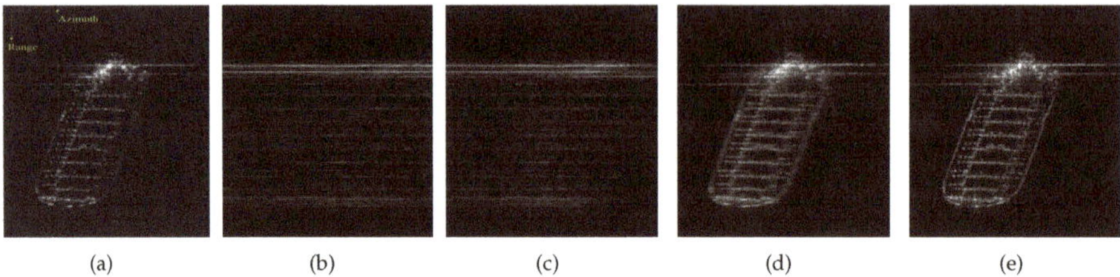

Figure 12. Nominal, corrupted, and recovered images. (**a**) Nominal; (**b**) Corrupted by phase error; (**c**) Image recovered with MD method; (**d**) Image recovered with ROPE method; (**e**) Image recovered with the proposed method.

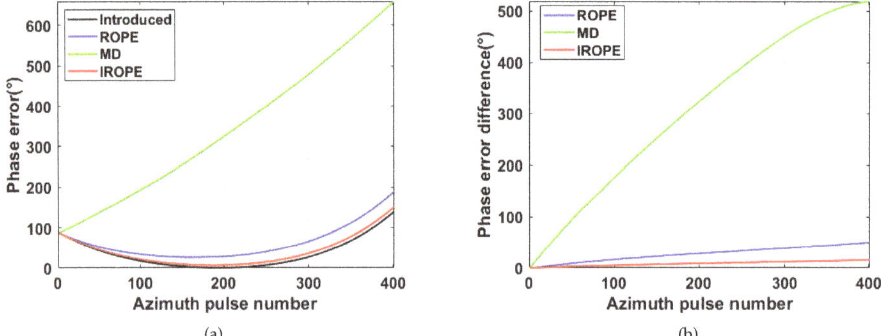

Figure 13. Phase error curve. (**a**) Introduced and estimated phase error; (**b**) Phase error difference.

Table 3. IQEMs of the corrupted and recovered images.

IQEMs	Corrupted image	MD	ROPE	IROPE
Contrast increase	−4.08	−3.86	−0.47	−0.09
Entropy reduction	−1.69	−1.58	−0.21	0.08
IP increase	−20.01	−16.82	−2.58	0.77

4.2.2. Intrinsically Corrupted Real Data

Figure 14 illustrates C-band, spotlight GF3 SAR images with azimuth resolution of 1 m.

The center latitude and longitude of the photographed area are (E104.0, N1.3) and (E104.1, N1.3), respectively, located at the Port of Singapore near Changi Airport. This location is at the southern end of the Malay Peninsula, the entrance to the Straits of Malacca.

Three representative defocused ships, marked as Ship1, Ship2 and Ship3, are selected from Figure 14 for the experiment.

Figure 15 gives the original and refocused images, in which the first column shows the original images and the second, third, and fourth columns present the images obtained by MD, ROPE, and IROPE, respectively. From the defocused images of Ship1 and Ship2, the outlines of these large petroleum tankers are vague, and the details are unrecognizable. The wake of Ship3 is obvious due to the smooth sea conditions and the rapid speed. It is determined that the visual quality of every original image after refocusing is improved; the details and contours of the targets are apparent, which is more conducive to subsequent use. Nevertheless, the images processed by the ROPE method are not well focused and remain slightly blurry. A detailed look reveals that the edge of every ship is processed worse by MD than by the proposed method. The values of the quality metrics in Table 4 indicate that the ship reconstructed by the proposed method, is better focused than that reconstructed by the other methods. The entropy, contrast, and IP are superior in our case.

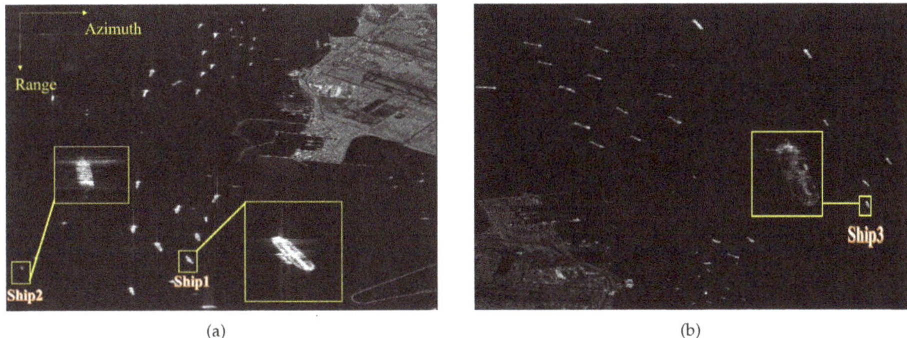

Figure 14. GF-3 SAR image of the Port of Singapore. The yellow rectangles are the enlarged defocused sub-images. (**a**) (E104.0, N1.3); (**b**) (E104.1, N1.3).

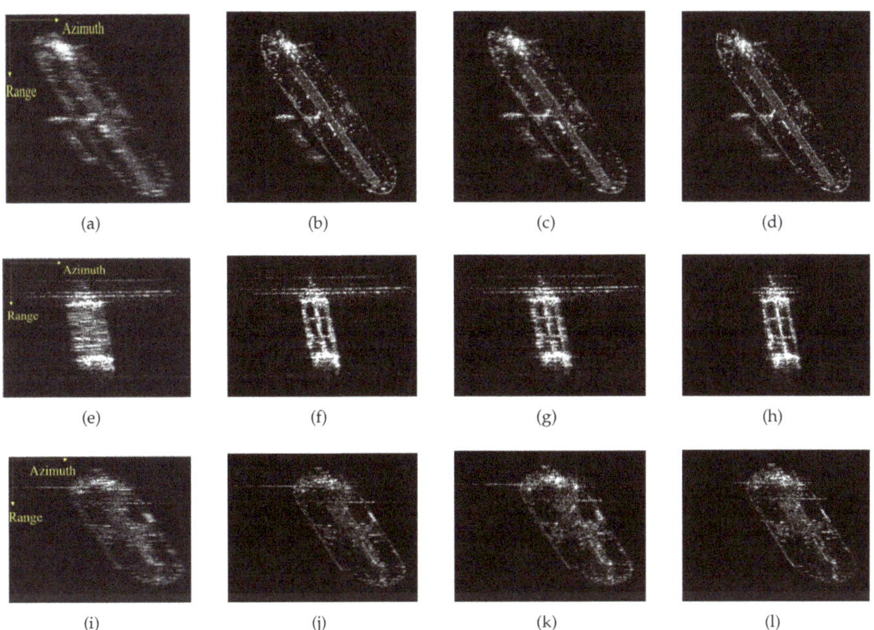

Figure 15. Original and refocused images. (**a–d**) Ship1; (**e–h**) Ship2; (**i–l**) Ship3.

The experimental results for real data demonstrate the effectiveness and superiority of the IROPE algorithm. The three ships used in the experiment are characterized by various phase errors, with multiple strong scattering points in each range bin under low SNR condition. In these cases, the MD and ROPE algorithms fail, and the proposed method exhibits a better refocusing effect.

Table 4. IQEMs of the GF-3 ship

Ship	Ship1			Ship2			Ship3		
IQEMs	**MD**	**ROPE**	**IROPE**	**MD**	**ROPE**	**IROPE**	**MD**	**ROPE**	**IROPE**
Contrast increase	0.42	0.23	0.49	0.01	0.01	0.02	0.23	−0.12	0.93
Entropy reduction	0.40	0.25	0.47	0.16	0.15	0.24	0.20	−0.02	0.47
IP increase	6.08	5.21	6.78	3.05	2.60	4.82	4.18	1.53	4.69

5. Discussion

In this paper, we studied the topic of how to refocus and accurately image a moving ship which cannot be focused by using conventional SAR. The proposed method is compared with the MD [10,11] method and the ROPE [30,31] method and the comparison result is shown in Table 5: Experimental results show that the performance of MD and ROPE in phase error estimation and accuracy are unsatisfactory. The sub-aperture correlation operation of the MD method only compensates for quadratic phase errors, and real SAR images do not fit well with the narrowly defined ROPE model, which limit their applications.

In view of above-mentioned problems, we have made improvements in preprocessing, circular shifting and iteration based on two-step convergence, which is embodied in the following aspects: improving the SNR and the accuracy of estimation through the DCT method for preliminary phase compensation; eliminating the shortcoming of setting the initial zero Doppler through the center shift of the strongest scattering point of each range bin and enhancing the performance of the method through several iterations. With these improvements, our proposed IROPE can achieve more complete image recovery for SAR images. The application scope and further development of the ROPE method are promoted through our improvement. Meanwhile, the experiment on adding phase error to the real data shows that IROPE estimates the phase error and compensates for arbitrary phase errors more accurately.

Moreover, many studies have been taken on this topic [15,16,19,22,24]. Nevertheless, these methods are based on low-resolution, short-CPI simulation or airborne SAR data, and their application to high-resolution, long-CPI spaceborne SAR images requires further verification. However, the proposed method processes high-resolution spaceborne SAR images, which is verified with GF-3 satellite data. Next, our results suggest a possibility of applying to the spaceborne SAR system with a long CPI (8.58 s). Furthermore, our work demonstrates great potency of the application of the high-resolution (1 m) SAR images.

However, the proposed method also has shortcomings. On the one hand, the dimensional classes of the ships in the experiments need further confirmation. On the other hand, although the proposed method allows a long coherent processing interval and performs well for maritime targets in stable sea conditions, it still needs further research and improvement for maneuvering targets that are experiencing high sea conditions. Therefore, in the future, we will focus our efforts on solving these problems like engaging the verification of the stated ship type in the AIS signals [27] and the optical photographs [46], and exploring the algorithm of maneuvering targets.

Table 5. Comparison with other refocusing algorithms.

Works	Schemes	Application	Phase Error Estimation	*Accuracy*
MD	Sub-aperture correlation	Limited	Unsatisfactory	Unsatisfactory
ROPE	Two-step convergence	Limited	Unsatisfactory	Unsatisfactory
IROPE	Preprocessing+ Circular Shifting +Two-step convergence+Iteration	Wide	Good	Good

6. Conclusions

During the detection of marine moving targets, both SAR and targets are in motion, and thus, conventional SAR processing that only images stationary targets achieves unsatisfactory performance. In this paper, we combined SAR and ISAR techniques and proposed a hybrid SAR/ISAR approach, as the core of the IROPE, to refocus the maritime moving targets. The proposed IROPE method overcomes the weakness of the original ROPE method. Experiments based on simulation and GF-3 measured data demonstrate the effectiveness of the proposed method. In summary, our work makes contributions to the improvement of the unsatisfactory method and the proposed method is also suitable for high-resolution long-CPI spaceborne radar.

Author Contributions: Conceptualization, Z.Y., H.Z. and Y.Z.; methodology, Z.Y., H.Z. and Y.Z.; software, Z.Y.; validation, Z.Y., H.Z. and Y.Z.; formal analysis, Z.Y.; investigation, Z.Y.; resources, Z.Y. and H.Z.; data curation, Z.Y. and H.Z.; writing—original draft preparation, Z.Y; writing—review and editing, Z.Y., H.Z. and Y.Z.; visualization, Z.Y.; supervision, H.Z. and Y.Z.; project administration, H.Z. and Y.Z.; funding acquisition, Y.Z. All authors have read and agreed to the published version of the manuscript.

Funding: This work was supported in part by the National Key Research and Development Program of China under Grant 2017YFB0502700 and in part by the Talent Fund of the Aerospace Information Research Institute for Distinguished Young Scholars under Grant Y9G01903AF.

Acknowledgments: The authors would like to thank the anonymous reviewers for their valuable suggestions and comments. The authors also thank the National Satellite Ocean Application Center in China for providing the GF-3 images. Many thanks to Da Liang, who provided great help and many valuable comments on the revision of this paper.

Conflicts of Interest: The authors declare no conflict of interest.

References

1. Moreira, A.; Prats-Iraola, P.; Younis, M.; Krieger, G.; Papathanassiou, K.P. A Tutorial on Synthetic Aperture Radar. *IEEE Geosc. Rem. Sen. M.* **2013**, *1*, 6–43. [CrossRef]
2. Rong, J.; Wang, Y.; Han, T. Iterative Optimization-Based ISAR Imaging With Sparse Aperture and Its Application in Interferometric ISAR Imaging. *IEEE Sens. J.* **2019**, *19*, 8681–8693. [CrossRef]
3. Noviello, C.; Fornaro, G.; Martorella, M.; Reale, D. IS. In Proceedings of Geoscience and Remote Sens. Symposium (IGARSS), Quebec City, QC, Canada, 13–18 July 2014; pp. 926–929. [CrossRef]
4. Jiang, Y.; Wang, H. Hybrid SAR/ISAR imaging of ship targets based on parameter estimation. *Remote Sens. Lett.* **2017**, *8*, 657–666. [CrossRef]
5. Martorella, M.; Giusti, E.; Berizzi, F.; Bacci, A.; Mese, E.D. ISAR based techniques for refocusing non-cooperative targets in SAR images. *IET Radar Sonar Navig.* **2012**, *6*, 332–340. [CrossRef]
6. Chen, V.C.; Liu, B. Hybrid SAR/ISAR for distributed ISAR imaging of moving targets. In Proceedings of the 2015 IEEE Radar Conference (RadarCon), Arlington, VA, USA, 10–15 May 2015; pp. 658–663.
7. Sun, H.; Zhou, L.; Ma, J.; Liu, X. The hybrid SAR-ISAR imaging algorithm applied to SAR moving target imaging. In Proceedings of the 2012 9th International Conference on Fuzzy Systems and Knowledge Discovery, Sichuan, China, 29–31 May 2012; pp. 1985–1988. [CrossRef]
8. Kim, J.; Hwang, I.; Jo, H.; Kim, G.; Yoo, J.S.; Yu, J. Phase error compensation in fourier domain for fast autofocus of spotlight SAR. In Proceedings of the 2017 USNC-URSI Radio Science Meeting (Joint with AP-S Symposium), San Diego, CA, USA, 9–14 July 2017; pp. 31–32. [CrossRef]
9. Brown, W.M.; Palermo, C.J. Effects of Phase Errors on Resolution. *IEEE Trans. Mil. Electron.* **2007**, *9*, 4–9. [CrossRef]
10. Calloway, T.M.; Donohoe, G.W. Subaperture autofocus for synthetic aperture radar. *IEEE Trans. Aerosp. Electron. Syst.* **1994**, *30*, 617–621. [CrossRef]
11. Samczynski, P.; Kulpa, K. The use of the Coherent MapDrift technique for SAR image focusing on the sea surface. In Proceedings of the 2010 IEEE Radar Conference,Washington, DC, USA, 10–14 May 2010; pp. 816–820.
12. Koo, V.C.; Chan, Y.K.; Chuah, H.T. Multiple Phase Difference Method for Real-Time SAR Autofocus. *J. Electromagn. Waves Appl.* **2006**, *20*, 375–388. [CrossRef]

13. Pan, Z.; Fan, H.; Zhang, Z. Nonuniformly-Rotating Ship Refocusing in SAR Imagery Based on the Bilinear Extended Fractional Fourier Transform. *Sensors* **2020**, *20*, 550. [CrossRef]
14. Chen, Y.; Li, G.; Zhang, Q.; Sun, J. Refocusing of Moving Targets in SAR Images via Parametric Sparse Representation. *Remote Sens.* **2017**, *9*, 795. [CrossRef]
15. Wang, Y.; Jiang, Y. Inverse Synthetic Aperture Radar Imaging of Maneuvering Target Based on the Product Generalized Cubic Phase Function. *IEEE Geosci. Remote Sens. Lett.* **2011**, *8*, 958–962. [CrossRef]
16. Wan, J.; Zhou, Y.; Zhang, L.; Chen, Z.; Yu, H. Efficient Algorithm for SAR Refocusing of Ground Fast-Maneuvering Targets. *Remote Sens.* **2019**, *11*, 2214. [CrossRef]
17. Tang, X.; Zhang, X.; Shi, J.; Wei, S.; Tian, B. Ground Moving Target 2-D Velocity Estimation and Refocusing for Multichannel Maneuvering SAR with Fixed Acceleration. *Sensors* **2019**, *19*, 3695. [CrossRef] [PubMed]
18. Giusti, E.; Wei, Q.; Bacci, A.; Tomei, S.; Martorella, M. Super resolution ISAR imaging via Compressing Sensing. In Proceedings of the EUSAR 2014; 10th European Conference on Synthetic Aperture Radar, Berlin, Germany, 3–5 June 2014; pp. 1–4.
19. Zhao, L.; Wang, L.; Bi, G.; Yang, L. An Autofocus Technique for High-Resolution Inverse Synthetic Aperture Radar Imagery. *IEEE Trans. Geosci. Remote Sens.* **2014**, *52*, 6392–6403. [CrossRef]
20. Liu, Y.; Quan, Y.; Li, J.; Zhang, L.; Xing, M. SAR imaging of multiple ships based on compressed sensing. In Proceedings of the 2009 2nd Asian-Pacific Conference on Synthetic Aperture Radar, Xian, China, 26–30 October 2009; pp. 112–115. [CrossRef]
21. Zhang, Y.; Xing, M.; Ye, P. Joint Rotation Parameters Estimation and High Resolution Imaging for Maneuvering Targets. In Proceedings of the EUSAR 2018, 12th European Conference on Synthetic Aperture Radar, Aachen, Germany, 4–7 June 2018; pp. 1–3.
22. Jing, Y.; Li, Y. Moving target parameter estimation algorithm using contrast optimization. In Proceedings of the International Bhurban Conference on Applied Sciences and Technology, Islamabad, Pakistan, 12–16 January 2016.
23. Xi, L.; Guosui, L.; Ni, J. Autofocusing of ISAR images based on entropy minimization. *IEEE Trans. Aerosp. Electron. Syst.* **1999**, *35*, 1240–1252. [CrossRef]
24. Huang, X.; Ji, K.; Leng, X.; Dong, G.; Xing, X. Refocusing Moving Ship Targets in SAR Images Based on Fast Minimum Entropy Phase Compensation. *Sensors* **2019**, *19*, 1154. [CrossRef] [PubMed]
25. Hui, Y.; Wenying, W.; Long, Z.; Wanming, L.; Xin, N.; Pin, L. Improved weighted least-squares phase unwrapping method for interferometric SAR processing. *J. Eng.* **2019**, *2019*, 6471–6474. [CrossRef]
26. Morrison, R.L.; Do, M.N.; Munson, D.C. SAR image autofocus by sharpness optimization: A theoretical study. *IEEE Trans. Image Process.* **2007**, *16*, 2309–2321. [CrossRef]
27. Sommer, A.; Ostermann, J. Backprojection Subimage Autofocus of Moving Ships for Synthetic Aperture Radar. *IEEE Trans. Geosci. Remote Sens.* **2019**, *57*, 8383–8393. [CrossRef]
28. Ye, C.M.; Jia, X.; Peng, Y.N.; Wang, X.T. Improved Doppler centroid tracking for ISAR based on target extraction. In Proceedings of the 2008 IEEE Radar Conference, Rome, Italy, 26–30 May 2008.
29. Wahl, D.E.; Eichel, P.; Ghiglia, D.; Jakowatz, C. Phase gradient autofocus-a robust tool for high resolution SAR phase correction. *IEEE Trans. Aerosp. Electron. Syst.* **1994**, *30*, 827–835. [CrossRef]
30. Dyson, F.; Garwin, R.; Horowitz, P.; Katz, J.; Muller, R.; Press, W.; Weinberger, P. *Recovery of SAR Images with One Dimension of Unknown Phases*; Report No. JSR-91-175, DTIC AD-B164; The MITRE Corporation Jason Program Office: McLean, VA, USA 1992.
31. Snarski, C.A. Rank one phase error estimation for range-Doppler imaging. *IEEE Trans. Aerosp. Electron. Syst.* **1996**, *32*, 676–688. [CrossRef]
32. Ji-Ming, Q.; Xu, H.Y.; Liang, X.D.; Li, Y.L. An Improved Phase Gradient Autofocus Algorithm Used in Real-time Processing. *J. Radars* **2015**, *4*, 600–607.
33. Gao, Y.; Xing, M.; Zhang, Z.; Guo, L. ISAR Imaging and Cross-Range Scaling for Maneuvering Targets by Using the NCS-NLS Algorithm. *IEEE Sens. J.* **2019**, *19*, 4889–4897. [CrossRef]
34. Han, B.; Ding, C.; Zhong, L.; Liu, J.; Qiu, X.; Hu, Y.; Lei, B. The GF-3 SAR Data Processor. *Sensors* **2018**, *18*, 835. [CrossRef]
35. Sun, J.; Yu, W.; Deng, Y. The SAR Payload Design and Performance for the GF-3 Mission. *Sensors* **2017**, *17*, 2419. [CrossRef] [PubMed]
36. Zheng, M.; Yan, H.; Zhang, L.; Yu, W.; Deng, Y.; Wang, R. Research on Strong Clutter Suppression for Gaofen-3 Dual-Channel SAR/GMTI. *Sensors* **2018**, *18*, 978. [CrossRef] [PubMed]

37. Quegan, S.; Carrara, W.G.; Goodman, R.S.; Majewski, R.M., *Spotlight Synthetic Aperture Radar: Signal Processing Algorithms*; Artech House: Boston, MA, USA; London, UK, 1995; 554p, ISBN 0-89006-634-5.
38. Cumming, I.G.; Wong, F.H. Digital Processing of Synthetic Aperture Radar Data: Algorithms and Implementation. Available online: https://pdfs.semanticscholar.org/23eb/33af4f0495edff01402ec8eb019e80717897.pdf (accessed on 20 March 2020).
39. Zheng, J.; Liu, H.; Liu, Z.; Liu, Q.H. ISAR Imaging of Ship Targets Based on an Integrated Cubic Phase Bilinear Autocorrelation Function. *Sensors* **2017**, *17*, 498. [CrossRef] [PubMed]
40. Xing, M.D.; Zheng, B.; Ming, Z.Y. Range Alignment Using Global Optimization Criterion in ISAR Imaging. *Acta Electron. Sin.* **2001**, *29*, 1807–1811.
41. Vehmas, R.; Jylha, J.; Vaila, M.; Vihonen, J.; Visa, A. Data-Driven Motion Compensation Techniques for Noncooperative ISAR Imaging. *IEEE Trans. Aerosp. Electron. Syst.* **2018**, *54*, 295–314. [CrossRef]
42. Rong, J.; Wang, Y.; Han, T. Interferometric ISAR Imaging of Maneuvering Targets With Arbitrary Three-Antenna Configuration. *IEEE Trans. Geosci. Remote Sens.* **2020**, *58*, 1102–1119. [CrossRef]
43. Zheng, J.; Su, T.; Zhang, L.; Zhu, W.; Liu, Q.H. ISAR Imaging of Targets With Complex Motion Based on the Chirp Rate Quadratic Chirp Rate Distribution. *IEEE Trans. Geosci. Remote Sens.* **2014**, *52*, 7276–7289. [CrossRef]
44. Martorella, M.; Berizzi, F.; Haywood, B. Contrast maximization based technique for 2-D ISAR autofocusing. *IEE P-Radar Son. Nav.* **2005**, *152*, 253–262. [CrossRef]
45. Wu, W.; Hu, P.; Xu, S.; Chen, Z.; Chen, J. Image registration for InISAR based on joint translational motion compensation. *IET Radar Sonar Navig.* **2017**, *11*, 1597–1603. [CrossRef]
46. Hajduch, G.; Le Caillec, J.; Garello, R. Airborne high-resolution ISAR imaging of ship targets at sea. *IEEE Trans. Aerosp. Electron. Syst.* **2004**, *40*, 378–384. [CrossRef]

Article

Azimuth Phase Center Adaptive Adjustment upon Reception for High-Resolution Wide-Swath Imaging

Wei Xu [1,2,*], Jialuo Hu [1,2], Pingping Huang [1,2,*], Weixian Tan [1,2] and Yifan Dong [1,2]

[1] College of Information Engineering, Inner Mongolia University of Technology, Hohhot 010051, China; hujialuo166@163.com (J.H.); wxtan@imut.edu.cn (W.T.); yfdong@imut.edu.cn (Y.D.)

[2] Inner Mongolia Key Laboratory of Radar Technology and Application, Hohhot 010051, China

* Correspondence: xuwei1983@ imut.edu.cn (W.X.); hwangpp@imut.edu.cn (P.H.); Tel.: +86-1581-1597-439 (W.X.); +86-0471-360-1821 (P.H.)

Received: 3 September 2019; Accepted: 30 September 2019; Published: 2 October 2019

Abstract: A spaceborne azimuth multichannel synthetic aperture radar (SAR) system can effectively realize high resolution wide swath (HRWS) imaging. However, the performance of this system is restricted by its two inherent defects. Firstly, non-uniform sampling is generated if the pulse repetition frequency (PRF) deviates from the optimum value. Secondly, multichannel systems are very sensitive to channel errors, which are difficult to completely eliminate. In this paper, we propose a novel receive antenna architecture with an azimuth phase center adaptive adjustment which adjusts the phase center position of each sub-aperture to improve multichannel SAR system performance. On one hand, the optimum value of the PRF can be adaptively adjusted within a certain range by adjusting receiving phase centers to obtain uniform azimuth sampling. On the other hand, false targets introduced by residual channel errors after azimuth multichannel error compensation can be further suppressed. The effectiveness of the proposed method to compensate for non-uniform sampling and suppress false targets is verified by simulation experiments.

Keywords: synthetic aperture radar (SAR); high resolution wide swath (HRWS); azimuth multichannel reconstruction; phase center adaptation; false targets suppression

---

1. Introduction

Synthetic aperture radar (SAR) is an extremely important device for the application of earth observation, especially where there is cloud or poor atmosphere conditions [1–3]. It is also widely used in military surveillance and civilian remote sensing [4–7]. The azimuth multichannel spaceborne SAR systems, usually working with a single transmit antenna and multiple received sub-apertures, can effectively overcome the contradiction between azimuth resolution and range mapping swath width [8–10]. However, in practical multichannel high resolution wide swath (HRWS) SAR systems, there are two problems which affect SAR system performance. Firstly, azimuth non-uniform sampling is generated if the pulse repetition frequency (PRF) is not selected as the optimum value [11–13]. Secondly, the multichannel system is very sensitive to channel errors, which are difficult to completely eliminate.

Many researchers have done a lot of research on the two mentioned issues, but some problems still exist. Firstly, azimuth non-uniform sampling in HRWS SAR can be overcome by the recent proposed azimuth multichannel reconstruction algorithms, but these are only prepared for azimuth band-limited signals [11,14–16]. However, the azimuth echo signal received by a practical SAR system is non-band-limited, and false targets still exist after reconstruction. To solve this problem, a novel transmit antenna architecture which allows for the adjustment of the transmit phase center position through the activation of a specific number of elements on the corresponding location of the transmit antenna was proposed in [17]. However, a decreased number of activated elements reduces the transmit antenna gain and the transmit signal power. Secondly, azimuth channel errors can be compensated for by many recently proposed compensation methods [18–20]. However, since factors such as the manufacturing process, temperature and radiation that affect the characteristics of antenna are not fixed, these methods are unable to perfectly estimate or compensate for multichannel errors, especially when the signal-to-noise ratio (SNR) of the obtained raw data is not high enough. The residual channel errors still cause high false targets which cannot be ignored and need to be further suppressed.

In this paper, a novel receiving antenna architecture that allows for the adjustment of the receiving phase center position of each sub-aperture through the closing of the corresponding elements on the side of each sub-antenna is proposed. In this approach, the transmitted signal power is not reduced, but the reduced receive antenna gain is compensated for by a narrow transmit antenna beam with a high antenna gain. Compared with traditional multichannel SAR systems, the novel proposed approach brings two benefits. Firstly, the optimum value of the PRF can be adaptively adjusted within a certain range by adjusting the phase center spacing of the sub-apertures. Uniform samples can be obtained if the PRF is taken in that range. Secondly, false targets caused by the residual channel errors could be further suppressed by adaptively adjusting phase center upon reception. The novel antenna structure is an improvement on the widely used conventionally phased array antenna. This improvement will hardly increase costs.

Based on the ideas above, this paper is structured as follows. Section 2 analyzes the influence of azimuthal non-uniform sampling and channel imbalance on azimuth multichannel SAR imaging. In Section 3, the basic principle of the proposed azimuth phase center adaptive adjustment upon reception is presented, and its effects on SAR system performance improvement and false targets suppression are analyzed. Simulation experiments are carried out to validate the proposed method in Section 4. Finally, this paper is concluded in Section 5.

2. Influence of Azimuthal Non-Uniform Sampling and Channel Imbalance

2.1. Influence of Azimuthal Non-Uniform Sampling

For simplicity, only the azimuth signal of multichannel SAR is analyzed in this paper. Assuming that the number of azimuth receiving channels is an odd N and the slant range from the radar to the target is R_0 and the intermediate channel is used as the reference channel, the received azimuth echo signal of receiving channel i can be written as

$$s_i(t) = exp\left[-j\frac{2\pi}{\lambda}\left(\sqrt{R_0^2 + (v_s t)^2} + \sqrt{R_0^2 + (v_s t - \Delta x_i)^2}\right)\right] \tag{1}$$

where Δx_i is the phase center position of channel i, $\Delta x_i = \left(\frac{(N+1)}{2} - i\right) \cdot d_{az}$, $i = 1, 2, \cdots, N$, d_{az} is the phase center spacing of receiving channel, λ is the wavelength of the carrier, v_s is the velocity of SAR platform, and t is the azimuth time.

Figure 1 illustrates the principle of multichannel system sampling. The equivalent sampling position can be approximately regarded as the midpoint of the transmitting position and the receiving position. In order to uniformly distribute the equivalent single-channel sampling centers, the PRF must meet Equation (2).

$$\text{PRF}_{\text{opt}} = \frac{2 \cdot v_s}{N \cdot d_{az}} \tag{2}$$

If (2) is violated, azimuth non-uniform sampling will be induced, and false targets in azimuth will be generated [21].

The multichannel reconstruction algorithm introduced in [12] can effectively compensate for the non-uniform sampling in azimuth. However, for strong deviations from the optimum PRF in Equation (2), the inverse character of such an algorithm might result in a degraded system performance.

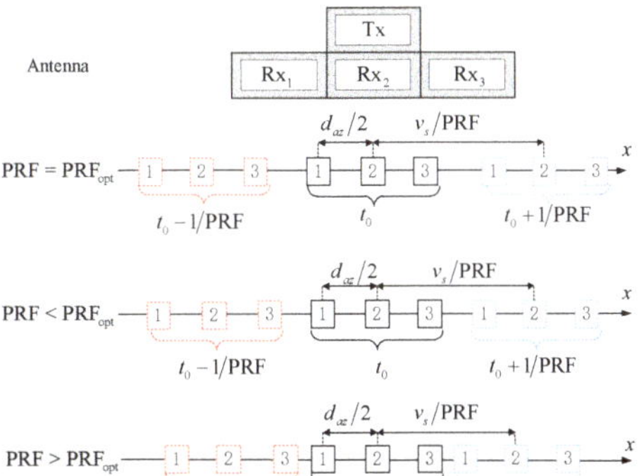

Figure 1. Reasons of uniform and non-uniform sampling.

The multichannel SAR system performance of such a system with azimuth non-uniform sampling is based on multichannel reconstruction algorithms [22], and the ratio of the input to output SNR, normalized to the ratio obtained for uniform sampling is expressed as Equation (3) [12]:

$$\Phi_{\mathrm{bf}}(\mathrm{PRF}) = \frac{\left(\frac{\mathrm{SNR_{in}}}{\mathrm{SNR_{out}}}\right)}{\left(\frac{\mathrm{SNR_{in}}}{\mathrm{SNR_{out}}}\right)\Big|_{\mathrm{PRF_{opt}}}} = N \cdot \sum_{j=1}^{N} E\left[\left|P_j(f, \mathrm{PRF})\right|^2\right], \tag{3}$$

where $E[\cdot]$ represents the calculation of the mean value, and $P_j(f, \mathrm{PRF})$ is the filter function of channel j.

Assume that the reconstruction filter matrix is $\boldsymbol{P}(f)$, which is obtained by inverting matrix $\boldsymbol{H}(f)$ according to [14]. The azimuth ambiguity-to-signal ratio (AASR) multi-channel system can be written as Equation (4) [12]:

$$AASR_m = \frac{\int_{-\frac{B_d}{2}}^{\frac{B_d}{2}} \left|2 \cdot \sum_{k=1}^{\infty} \sum_{m=m_0}^{N} \sum_{j=1}^{N} U_{jk}(f) \cdot P_{jm}(f)\right|^2 df}{\int_{-\frac{B_d}{2}}^{\frac{B_d}{2}} \left|U(f)\right|^2 df} \tag{4}$$

with

$$U_{jk}(f) = U(f) \cdot H_{jk}(f) \tag{5}$$

$$m_0 = max\{N - k + 1, 1\}, \tag{6}$$

where B_d is the Doppler bandwidth, $U(f)$ is the spectrum of the equivalent monostatic SAR signal, $P_j(f)$ is the reconstruction filter of channel j, $P_{jm}(f)$ is m-th filter of $P_j(f)$, $H_j(f)$ is the pre-filter of channel j, and $H_{jk}(f)$ is the k-th filter in $H_j(f)$.

According to (3) and (4), both the SNR scaling factor and the AASR are related to the reconstruction matrix $\boldsymbol{P}(f)$. A PRF deviating strongly from the optimum value will lead to an irreversible $\boldsymbol{H}(f)$ or very different eigenvalues of $\boldsymbol{H}(f)$. As a result, the multichannel SAR system performance would be significantly declined.

Using the simulation parameters listed in Table 1, Figure 2 shows that the reconstruction filter causes a degraded imaging performance. Figure 2a,b show the spectrum and pulse compression, respectively, result of the equivalent single-channel signal when the PRF is the optimum value. Since the azimuth SAR echo signal is non-band-limited, several false targets caused by azimuth ambiguity appear in the compression result. When the actual PRF of the system takes a non-optimum value, the reconstruction result of the non-band-limited signal is shown in Figure 2c,d. It can be seen that the tiny false targets in Figure 2b are significantly enlarged in Figure 2d by the non-optimum PRF.

Table 1. System simulation parameters.

Parameter (Azimuth)	Symbol	Value
Carrier frequency	f_c	9.6 GHz
Carrier wavelength	λ	0.031
Orbit height	R_0	895 km
Sensor velocity	v_s	7560 m/s
Overall Rx antenna length	L_a	12.25 m
Rx sub-aperture length	l_a	1.75 m
Elements number on each sub-aperture	K	60
Element size	d	0.03 m
Channels number	N	7
Desired PRF range	PRF	1100 Hz–1500 Hz

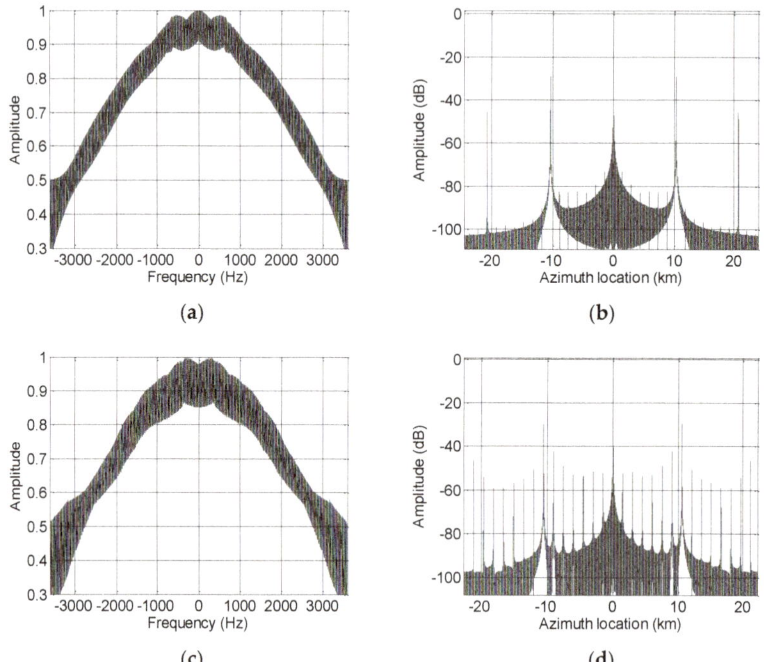

Figure 2. Influence of the pulse repetition frequency (PRF) deviating from the optimum value on imaging performance. (**a**) Reconstructed spectrum with the optimum PRF; (**b**) reconstructed pulse compression result with the optimum PRF; (**c**) reconstructed spectrum with the non-optimum PRF; and (**d**) reconstructed pulse compression result with the non-optimum PRF.

Assuming that the desired PRF range of the system is 1100–1500 Hz and the optimum PRF is 1234.3 Hz according to Table 1, the resulting azimuth ambiguity-to-signal ratio ($AASR_N$) and SNR scaling factor (Φ_{bf}) are shown in Figure 3. With the growth of the PRF, the AASR continues to decline when the PRF is lower than the optimum value of 1234.3 Hz, but the AASR rises when the PRF exceeds the ideal value, as is shown in Figure 3a. From Figure 3b, it can be seen that sufficiently low values of Φ_{bf} are shown from 1100 up to 1370 Hz, but unacceptably high values are generated for the PRF range above 1400 Hz. The phase center adjustment method proposed in this article can obviously decrease the AASR and significantly suppress the unacceptably high values of Φ_{bf}.

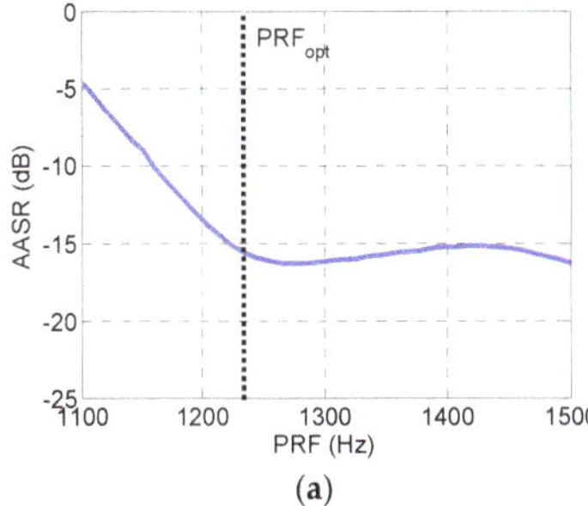

(**a**)

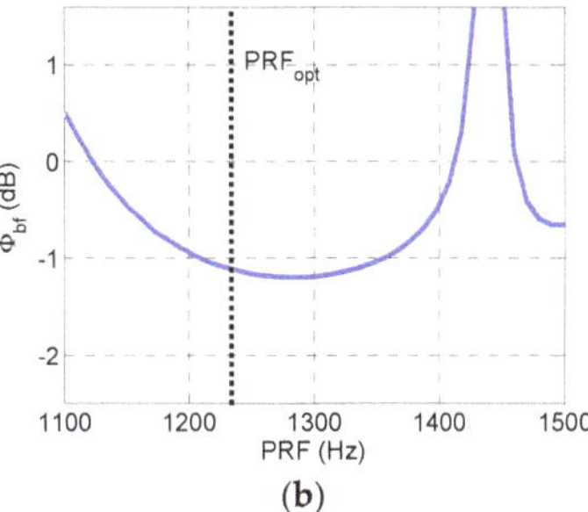

(**b**)

Figure 3. Simulated ambiguity-to-signal ratio (AASR) and signal-to-noise ratio (SNR) scaling factor of conventional reconstruction approach. (**a**) Simulated AASR of conventional reconstruction approach and (**b**) simulated SNR scaling factor Φ_{bf} of conventional reconstruction approach.

2.2. *Influence of Channel Imbalance*

In actual multi-channel SAR systems, there are always channel errors among channels. These errors cause false targets that seriously reduce the quality of imaging. Assuming that the phase error of the n-th channel is ϕ_n and the amplitude error is a_n, the echo signal model of channel n can be expressed as follows:

$$s_n(t) = a_n exp(j\phi_n) exp\left[-j\frac{2\pi}{\lambda}\left(\sqrt{R_0^2 + (v_s t)^2} + \sqrt{R_0^2 + (v_s t - \Delta x_n)^2}\right)\right]. \tag{7}$$

In the process of multi-channel signal combination, $N-1$ zeros need to be added between each sampling point of each channel, and the equivalent single channel signal can be written as:

$$s(m) = \sum_{n=0}^{N-1} s_n^0(m)\,, \tag{8}$$

where m represents integers associated with sampling time and $s_n^0(m)$ represents signals after adding zeros of channel n.

According to (8), the discrete time Fourier transform (DTFT) [23] of $s_n^0(m)$ is:

$$S_n^0\left(e^{j\omega}\right) = \frac{1}{N}\sum_{k=0}^{N-1} e^{-j2\pi\frac{nk}{N}} S\left(e^{j(\omega - \frac{2\pi k}{N})}\right), \tag{9}$$

where k represents integers associated with sampling frequency. Denote the digital frequency of (9) using analog frequency, and the spectrum of signals after adding zeros of channel n can be derived as:

$$\begin{aligned} S_n^0(f) &= a_n exp(j\phi_n) rect\left(\frac{f}{B_d}\right) exp\left(-j\pi\frac{f^2}{K_a}\right) \\ &+ a_n exp(j\phi_n) \sum_{l=0}^{1}\sum_{k=1}^{N-1} W(f) exp\left(\frac{-j2\pi nk}{N}\right) exp\left(-j\pi\frac{\left(f + l\cdot \mathrm{PRF} - \frac{k\cdot \mathrm{PRE}}{N}\right)^2}{K_a}\right), \end{aligned} \tag{10}$$

where K_a is the azimuth frequency modulation (FM) rate, and f is the frequency variable in Hz. $W(f)$ is given by:

$$W(f) = \mathrm{rect}\left[\frac{f - \frac{\left(\frac{k\cdot \mathrm{PRE}}{N} - l\cdot \mathrm{PRF}\right)}{2} - (-1)^l \frac{(\mathrm{PRF} - B_d)}{4}}{(-1)^{l+1}\left(\frac{k\cdot \mathrm{PRF}}{N} - l\cdot \mathrm{PRF}\right) + \frac{(B_d + \mathrm{PRF})}{2}}\right]. \tag{11}$$

Due to the spectrum components corresponding to other l values move to the right and fall out, the value of l is only 0 and 1.

According to (8) and (10), the pulse compression result of the combined signal can be derived as:

$$\begin{aligned} s_{out}(t) = B_d \sin c(B_d t) \sum_{n=0}^{N-1} a_n exp(j\phi_n) + \sum_{l=0}^{1}\sum_{k=1}^{N-1} f_{l,k}(t)\cdot \\ \sum_{n=0}^{N-1} exp\left(\frac{-j2\pi nk}{N}\right) a_n exp(j\phi_n), \end{aligned} \tag{12}$$

where $f_{l,k}(t)$ is a function introduced to simplify the written, and it has nothing to do with amplitude and phase errors. According to (12), there are $2(N-1)$ false targets, and the position of each false target is shown as follows:

$$\text{POS}_{l,k} = \frac{\left(l \cdot \mu - \frac{k \cdot \mu}{N}\right)}{T_a}, \tag{13}$$

where $l = 0, 1$, $k = 1, 2, \ldots, N-1$, μ is the oversampling rate, and T_a is the synthetic aperture interval. Assuming that the channel characteristics are independent of frequency, the false target-to-peak ratio of the false target at $\text{POS}_{l,k}$ is as follows:

$$\text{PGR}_{l,k} = 20log_{10}\left(\frac{\left|\sum_{n=0}^{N-1} a_n exp(j\phi_n) exp\left(\frac{-j2\pi nk}{N}\right)\right|}{\left|\sum_{n=0}^{N-1} a_n exp(j\phi_n)\right|}\right) + 20log_{10}\left(\frac{g_{l,k}}{B_d}\right), \tag{14}$$

in which $g_{l,k}$ is given by $g_{l,k} = (-1)^{l+1}\left(\frac{k \cdot \text{PRF}}{N} - l \cdot \text{PRF}\right) + \frac{(B_d + \text{PRF})}{2}$ and is independent with amplitude and phase errors.

With parameters listed in Table 2, a simulation experiment to reconstruct the echo signal with channel errors is shown in Figure 4. A reconstructed spectrum and compression result of the system with no channel errors is shown in Figure 4a,b. Figure 4c,d show the reconstruction results in the presence of amplitude errors within 0.5 dB and phase errors of 1–5 degrees. It can be seen that these tiny channel errors that may be residual after channel error compensation can cause false targets of around −35 dB. False targets of such intensity still cause a reduction in image quality and should be further suppressed.

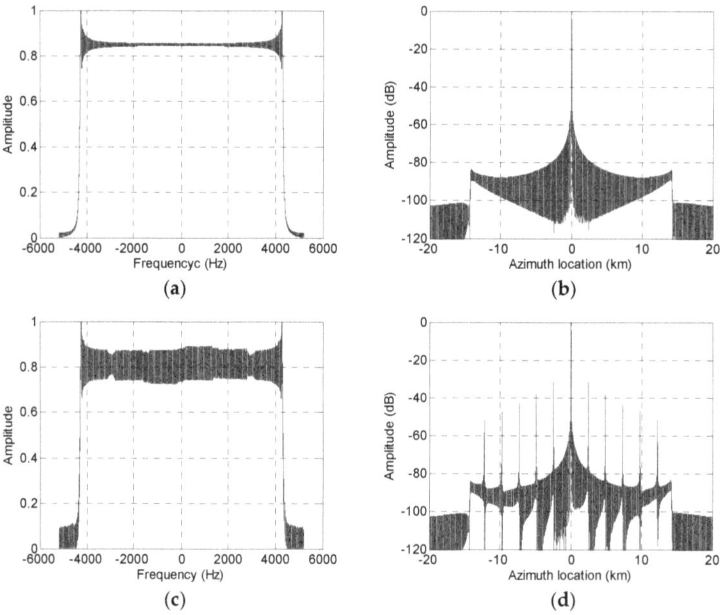

Figure 4. Influence of channel imbalance. (**a**) Reconstructed spectrum with no channel errors; (**b**) reconstructed pulse compression result with no channel errors; (**c**) reconstructed spectrum with channel errors; and (**d**) reconstructed pulse compression result with no channel errors.

3. Phase Center Adjustment upon Reception

This section introduces an innovative receive antenna architecture in azimuth which allows for the compensation of the non-optimum PRF values and the suppression of false targets by phase center adjustment upon reception. The working process of the innovative system is shown in Figure 5.

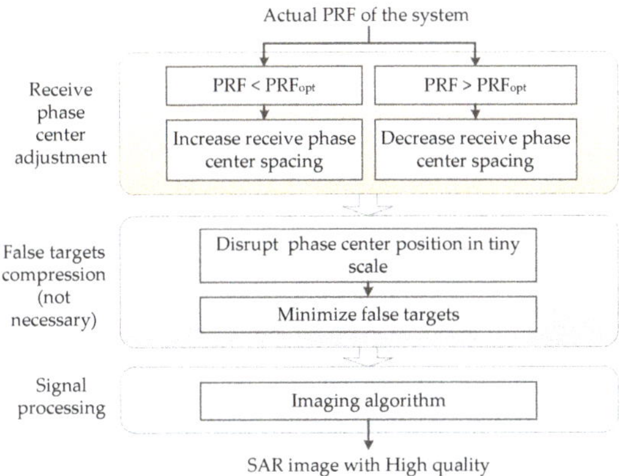

Figure 5. The working process of the innovative system.

3.1. System Architecture and Basic Principle

Compared with traditional multi-channel SAR systems, such as systems used by satellites RADARSAT-2 (Canada, 2007), TerraSAR-X (Germany, 2007) and Sentinel-1A (launched by European Space Agency, in 2014), each receiving sub-antenna of the innovative system consists of a large number of individually controllable elements. Such an aperture permits the changing of the position and the length of the effective receiving sub-apertures on the antenna by activating respective elements. The phase center position of the receiving sub-apertures can be adaptively adjusted by turning off a part of elements on the antenna. As an example, a system with three receive apertures is shown in Figure 6.

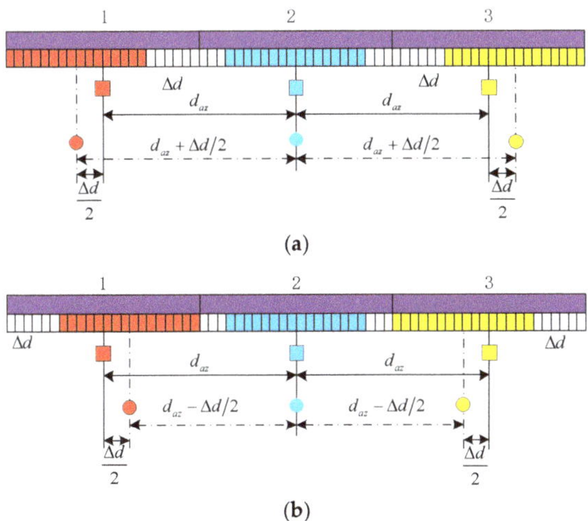

Figure 6. System architecture and basic principle. (**a**) Adjustment to increase phase center spacing and (**b**) adjustment to reduce phase center spacing.

Closing a part of elements inside a sub-aperture can move its phase center outward and increase the phase center spacing, as shown in Figure 6a, while the inward movement of the phase center can be achieved by closing the elements on the outside of the sub-aperture, as shown in Figure 6b. To ensure the effective length of each antenna is equal, the same number of elements on each aperture should be turned off. The phase center position can be adjusted by controlling the amount of closed elements on each side of the aperture. After adjustment, the phase center spacing can be uniform or non-uniform, which should be decided according to practical needs.

3.2. Effect of Receiving Sub-Aperture Phase Center Adjustment on Azimuth Non-Uniform Sampling

Take the case where the phase center spacing is reduced as an example. If the length of the Δd antenna on left side of channel 1 and the right side of channel N are invalid, the phase center of the two sub-apertures moves inward by $\frac{\Delta d}{2}$ and the distance between the phase centers of channel 1 and channel N is reduced by Δd. In the case where the phase center is evenly distributed, the new phase center spacing is:

$$d_{\text{az\_new}} = \frac{(N-1)\cdot d_{\text{az}} - \Delta d}{N-1} = d_{\text{az}} - \frac{\Delta d}{N-1}. \tag{15}$$

To ensure that the phase centers are uniformly distributed, the phase center of channel n should move inward by:

$$x_n = \left|\frac{N+1}{2} - n\right| \cdot \frac{\Delta d}{N-1}, n = 1, 2, \ldots, N. \tag{16}$$

The length of the invalid antenna on the outer side of channel n should be $2x_n$ longer than the invalid antenna on the inner side:

$$\Delta d_{n,\text{outer}} - \Delta d_{n,\text{inner}} = 2x_n = \frac{2\Delta d}{N-1} \cdot \left|\frac{N+1}{2} - n\right|. \tag{17}$$

Since the number of closed elements in each channel must be equal (assumed to be Δd), the length of the invalid part on both sides of channel n should satisfy:

$$\Delta d_{n,\text{outer}} + \Delta d_{n,\text{inner}} = \Delta d. \tag{18}$$

Combine Formulas (17) and (18) together, and then $\Delta d_{n,\text{outer}}$ and $\Delta d_{n,\text{inner}}$ can be written as:

$$\Delta d_{n,\text{outer}} = \frac{\Delta d}{2} + \frac{\Delta d}{N-1} \cdot \left|\frac{N+1}{2} - n\right| \tag{19}$$

$$\Delta d_{n,\text{inner}} = \frac{\Delta d}{2} - \frac{\Delta d}{N-1} \cdot \left|\frac{N+1}{2} - n\right|. \tag{20}$$

Because of the size of individually controllable antenna elements, the receive center of antenna in practical system cannot be adjusted arbitrarily but can only be selected in a series of discrete positions. Assuming that each receive antenna consists of a number of K elements, the number of closed elements on each antenna can be written as:

$$p = \text{round}\left\{\Delta d \cdot \frac{K}{l_a}\right\}, \tag{21}$$

in which the operator round $\{\cdot\}$ indicates the calculation of rounding integers. The number of invalid elements on the outer side and the inner side of the antenna can be written as:

$$p_{n,\text{outter}} = \text{round}\left\{\left[\frac{\Delta d}{2} + \frac{\Delta d}{N-1} \cdot \left|\frac{N+1}{2} - n\right|\right] \cdot \frac{K}{l_a}\right\}. \tag{22}$$

$$p_{n,\text{inner}} = \text{round}\left\{\left[\frac{\Delta d}{2} - \frac{\Delta d}{N-1} \cdot \left|\frac{N+1}{2} - n\right|\right] \cdot \frac{K}{l_a}\right\} \tag{23}$$

If the actual value of the PRF is $\mathrm{PRF_{act}}$, which is higher than the optimum value, the phase center spacing should be reduced by Δd_{az}, which is given by:

$$\Delta d_{az} = d_{az} - \frac{2v_s}{N \cdot \mathrm{PRF_{act}}}. \tag{24}$$

According to (15), the length of closed elements on each receive antenna should be:

$$\Delta d = \left(d_{az} - \frac{2v_s}{N \cdot \mathrm{PRF_{act}}}\right) \cdot (N-1). \tag{25}$$

The number of closed elements on each antenna can be derived as:

$$p = \mathrm{round}\left\{\frac{K}{l_a} \cdot (N-1) \cdot \left(d_{az} - \frac{2v_s}{N \cdot \mathrm{PRF_{act}}}\right)\right\}. \tag{26}$$

The respective number of closed elements on the outer side and inner side on channel n is given by:

$$p_{n,\mathrm{outter}} = \mathrm{round}\left\{\left[\left(d_{az} - \frac{2v_s}{N \cdot \mathrm{PRF_{act}}}\right) \cdot \frac{(N-1)}{2} + \left(d_{az} - \frac{2v_s}{N \cdot \mathrm{PRF_{act}}}\right) \cdot \left|\frac{N+1}{2} - n\right|\right] \cdot \frac{K}{l_a}\right\} \tag{27}$$

$$p_{n,\mathrm{inner}} = \mathrm{round}\left\{\left[\left(d_{az} - \frac{2v_s}{N \cdot \mathrm{PRF_{act}}}\right) \cdot \frac{(N-1)}{2} - \left(d_{az} - \frac{2v_s}{N \cdot \mathrm{PRF_{act}}}\right) \cdot \left|\frac{N+1}{2} - n\right|\right] \cdot \frac{K}{l_a}\right\}. \tag{28}$$

Using the method discussed above, if the number of closed elements on each channel is given by p, the optimum PRF of the system can be written as:

$$\mathrm{PRF_{opt}} = \frac{2v_s}{N \cdot \left(d_{az} \pm \frac{p \cdot l_a}{K \cdot (N-1)}\right)}. \tag{29}$$

The "±" indicates the increase or decrease of the phase center spacing. In practice, in order to ensure the receiving performance of the antenna, a limited number of elements can be turned off. If the maximum ratio of the closed elements number to the total number on each antenna is η, the number of closed elements can be written as:

$$p_m = \mathrm{round}\{\eta \cdot K\} \tag{30}$$

The range of the adjusted optimum PRF is:

$$\frac{2v_s}{N \cdot \left(d_{az} + \frac{p_m \cdot l_a}{K \cdot (N-1)}\right)} \le \mathrm{PRF_{opt}} \le \frac{2v_s}{N \cdot \left(d_{az} - \frac{p_m \cdot l_a}{K \cdot (N-1)}\right)}. \tag{31}$$

The minimum value of $\mathrm{PRF_{opt}}$ is:

$$\mathrm{PRF_{opt\_min}} = \frac{2v_s}{N \cdot \left(d_{az} + \frac{p_m \cdot l_a}{K \cdot (N-1)}\right)}; \tag{32}$$

and the maximum value is:

$$\mathrm{PRF_{opt\_max}} = \frac{2v_s}{N \cdot \left(d_{az} - \frac{p_m \cdot l_a}{K \cdot (N-1)}\right)}. \tag{33}$$

This is to say, if the practical PRF of the system varies between $\mathrm{PRF_{opt\_min}}$ and $\mathrm{PRF_{opt\_max}}$, a uniformly (at least approximately uniformly) sampled signal can be obtained by adjusting the phase center spacing. Consequently, the multichannel reconstruction algorithm can be omitted, and the computational complexity can be reduced compared with conventional systems. If the PRF is outside the optimum PRF range, the phase center spacing should be adjusted to the minimum or maximum value that makes the optimum PRF closest to the actual PRF of the system. In this case, although the non-uniformity of the samples is reduced, the sampling is still non-uniform, and the reconstruction algorithm is still needed. Therefore, the complexity of processing is approximately equal to that of traditional systems.

The effect of the largest proportion of closed elements on the range of the optimum PRF is shown in Figure 7, where Γ is as follows:

$$\Gamma = \frac{\text{PRF}_{\text{opt\_max}} - \text{PRF}_{\text{opt\_min}}}{\text{PRF}_{\text{opt}}}. \tag{34}$$

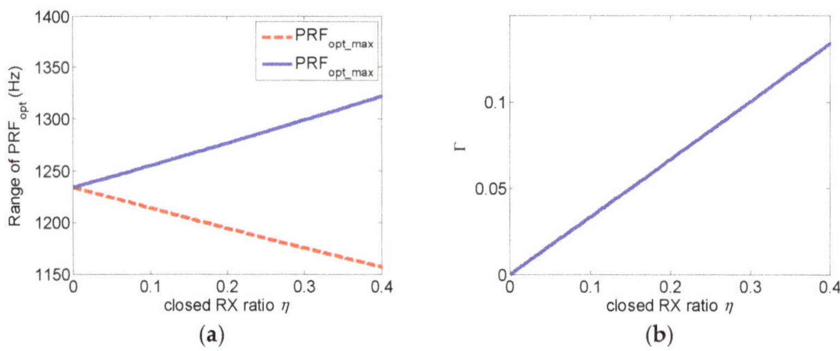

Figure 7. Relationship between the optimum PRF and the invalid receiving antenna ratio η. (**a**) Maximum (solid blue line) and minimum (red dotted line) value of the optimum PRF and (**b**) range expansion ratio of the optimum PRF.

3.3. Suppression of False Target by Adjusting Receiving Sub-Aperture Phase Center

In Section 2.2, the expressions of the position and intensity of false targets caused by channel imbalance are analyzed. For the convenience of analysis, the false target-to-peak ratio of the false target at $\text{POS}_{l,k}$ given by (14) can be rewritten as:

$$\text{PGR}_{l,k} = 20log_{10}\left(\frac{\left|\sum_{n=0}^{N-1} a_n exp(j\phi_n) exp\left(\frac{-j4\pi k|X_n|}{L}\right)\right|}{\left|\sum_{n=0}^{N-1} a_n exp(j\phi_n)\right|}\right) + 20log_{10}\left(\frac{g_{l,k}}{B_d}\right), \tag{35}$$

where L represents the length of the antenna and X_n represents the phase center position of channel n. If the phase centers of sub-apertures are adjusted, the antenna length L can be written as:

$$L = L_a - \frac{l_a \cdot q_{1,l}}{K} - \frac{l_a \cdot q_{N,r}}{K}, \tag{36}$$

where L_a is the length of receive antenna before adjusting, $q_{n,l}$ is the number of inactive elements on the left side of the antenna n, and $q_{n,r}$ is the number of inactive elements on the right side of the antenna n. The value of X_n can be written as:

$$X_n = \left(\frac{N+1}{2} - n\right) \cdot d_{az} + \frac{l_a \cdot \left(q_{n,r} + q_{\frac{(N+1)}{2},l} - q_{\frac{(N+1)}{2},r} - q_{n,l}\right)}{2K}. \tag{37}$$

The value of X_n can be adaptively adjusted to generate an additional phase which can cancel a part of the phase error. Based on this, the molecule of the first item in Formula (35) can be minimized by adjusting the phase center position of each channel, and the false-target-to-peak-ratio can be suppressed. The false-target-to-peak ratio can be reduced to:

$$\mathrm{PGR}_{l,k} = 20log_{10}\left(\frac{\min\left\{\left|\sum_{n=0}^{N-1} a_n exp(j\phi_n) exp\left(\frac{-j4\pi k|X_n|}{L}\right)\right|\right\}}{\left|\sum_{n=0}^{N-1} a_n exp(j\phi_n)\right|}\right) + 20log_{10}\left(\frac{g_{l,k}}{B}\right) \tag{38}$$

Though false targets can be suppressed by simply scrambling the uniformly distributed phase center position on a tiny scale, it is often necessary to compare several different results for optimal suppression. As a result, computational complexity and time-consumption is increased, and the increased complexity and time-consumption are related to the number of comparisons. Assuming that n different suppression results are compared, the computational complexity is n times higher than that of a traditional multichannel system.

The reconstruction result of error-free signals after phase center adaptive adjustment is shown in Figure 8. It can be seen that adjusting phase center non-uniformly does not affect the reconstruction results for multichannel signals.

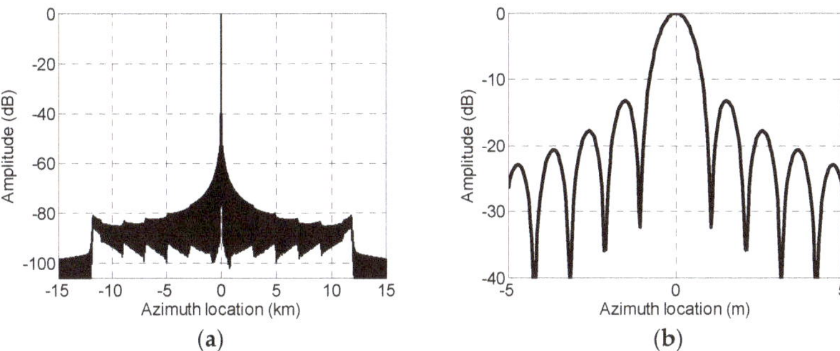

Figure 8. Reconstruction of non-uniform phase center echo signal without channel error. (**a**) Reconstructed pulse compression result with no channel errors and (**b**) magnified reconstructed pulse compression result with no channel errors.

4. Simulation and Performance Analysis

To validate the proposed receiving phase center adaptive adjustment approach, simulations were carried out. Simulation parameters are listed in Table 1.

4.1. *Effect on Azimuth Non-Uniform Sampling*

The seven-channel system was used in the simulation experiment. The optimum PRF of the system without adjusting receiving phase centers was 1234.3 Hz. Assuming that up to 30% of elements on each antenna was allowed to be turned off, the optimum value of the PRF could be adaptively adjusted within 1175.5–1299.2 Hz by invalidating elements at both ends of the antenna. Figure 9 gives the reconstruction result of a non-band-limited signal with 1300 Hz of the PRF in the case of conventional phase center spacing and adaptively adjusted phase center spacing. With respect to Figure 9a, false targets in Figure 9b were significantly reduced by phase center adaptive adjustment upon reception leaving only false targets caused by the non-band-limitation of the signal.

The corresponding AASR and SNR scaling factor Φ_{bf} are shown in Figure 10. When the PRF value fell into the range 1175.5–1299.2 Hz, the AASR was consistent with the value of equivalent single channel SAR. For PRF values below 1175.5 Hz or higher than 1299.2 Hz, because of the non-uniform sampling, the AASR was higher than that of the equivalent monostatic signal but was obviously decreased with respect to the conventional multichannel reconstruction approach.

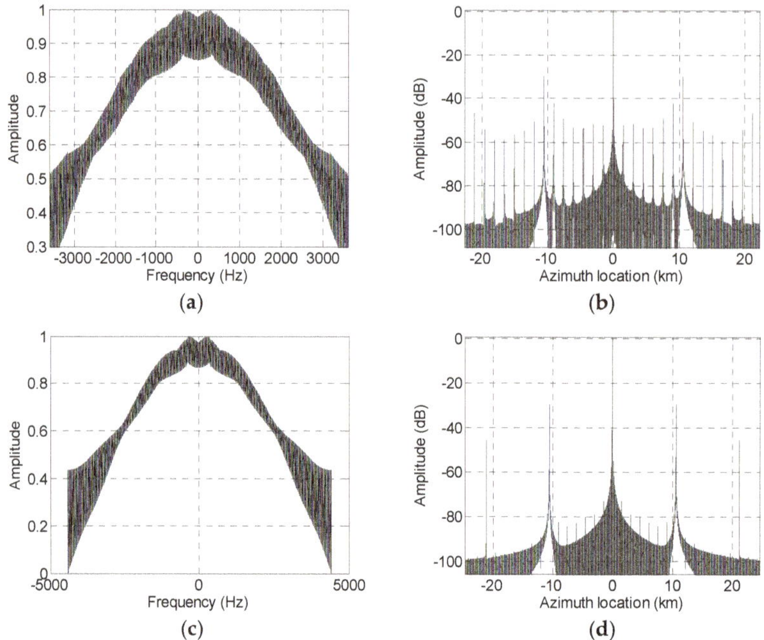

Figure 9. Reconstruction results of non-uniform sampling and adaptive phase center adjustment. (**a**) Reconstructed spectrum in the non-optimum PRF; (**b**) reconstructed pulse compression result in the non-optimum PRF; (**c**) reconstructed spectrum after adaptive phase center adjustment; and (**d**) reconstructed pulse compression result after adaptive phase center adjustment.

Regarding the SNR, when the PRF fell into 1175.5–1299.2Hz, the value of Φ_{bf} remained constant independently of the PRF since the uniform sampling was ensured. For other values of the PRF, because of the non-uniform sampling, the AASR was higher than that in uniform sampling but was obviously lower than the conventional reference, as is shown in Figure 10b.

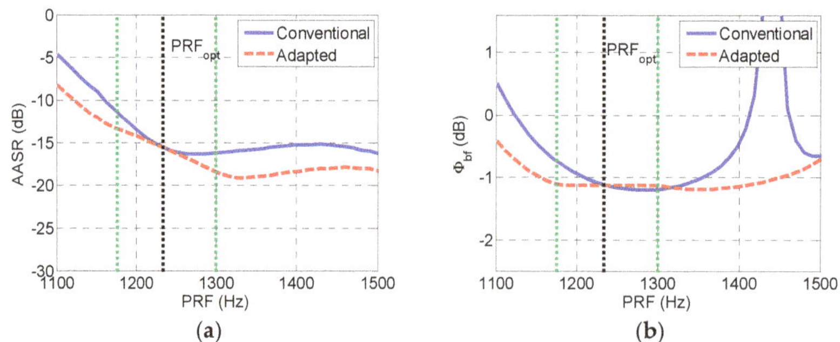

Figure 10. Simulated AASR and SNR scaling factor. (**a**) Simulated AASR and (**b**) simulated SNR scaling factor Φ_{bf}.

4.2. Effect of False Target Suppression

An example of phase center adjusting to suppress false targets is given in this section. The phase center positions before and after adjustment are listed in Table 2. Figure 11 verifies the suppression effect of phase center adaptively adjusting on the false targets. Figure 11a shows the reconstructed pulse compression results with the presence of channel errors; Figure 11b is an enlargement of Figure 11a at the top of a false target. It can be seen that false targets could be suppressed for about 2 dB after adaptive phase center adjustment. Though the suppression effect was not significant, the proposed method could further suppress the false targets after channel error compensation. This can also be regarded as an advantage of presented receiving phase center adjustment.

Table 2. Phase center position before and after adjusting.

Channels	1	2	3	4	5	6	7
X_n before adjusting (m)	5.250	3.500	1.750	0	−1.750	−3.500	−5.250
X_n after adjusting (m)	5.565	3.752	1.813	0	−1.855	−3.710	−5.502

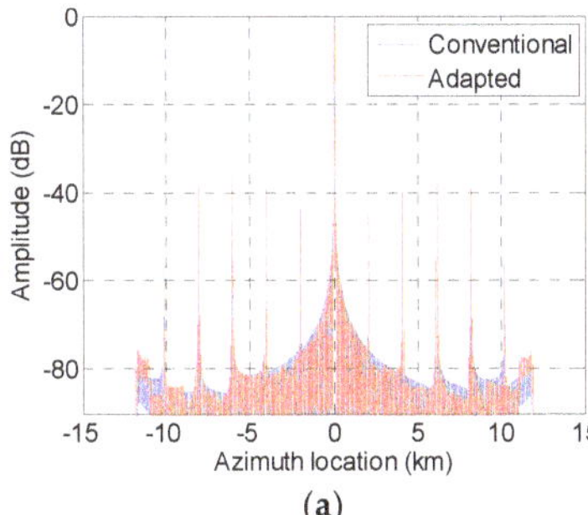

(a)

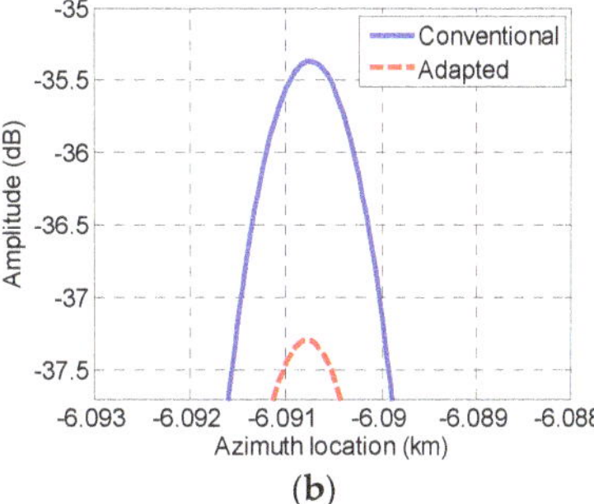

(b)

Figure 11. The effect of phase center adjusting on the suppression of false targets. (**a**) Reconstructed pulse compression results for uniform and non-uniform phase center and (**b**) the top of a fake target is zoomed in.

5. Conclusions

An advanced azimuth antenna architecture which allows for the adjustment of the phase center position of sub-apertures was proposed in this paper. Benefiting from receiving phase center adjustment, the performance of the HRWS SAR system can be improved in following three aspects. Firstly, the PRF of the system in a certain range can be regarded as the optimum value, so complex signal reconstruction algorithms will be omitted in novel systems. Secondly, non-uniform sampling that results in the severe degradation of imaging performance is avoided by adjusting the phase center if it falls into the optimum-PRF-range. Thirdly, by adaptively adjusting the phase center position of each channel, false targets caused by residual channel error after compensation can be further suppressed to some degree, and the quality of the resulting SAR image can be further improved.

In conclusion, receiving phase center adjustment is an effective method for compensating for non-uniform sampling and can suppress the false targets caused by channel error to a certain extent. However, since the number of elements that can be turned off cannot exceed a certain ratio, the optimum PRF can only be adjusted within a range, which restricts the usable PRF range of the system. In further research, the joint adjustment of transmit and receive phase centers will be considered to compensate for a wider range of the non-uniform PRF. Furthermore, the presented method can be extended to multiple-input multiple-output (MIMO) SAR systems to further improve the performance.

Author Contributions: All the authors made contributions to this work. W.X. and J.H. proposed the idea and wrote the paper; P.H. conceived and designed the experiments; W.T. performed the experiments; and Y.D. revised the manuscript.

Funding: This research was funded by National Equipment Pre-Research Foundation of China, grant number JZX7Y20190253041401 and JZX7Y20190253040501, Inner Mongolia Science and Technology Innovation Guidance Project, grant number KCBJ2018014, and National Natural Science Foundation of China, grant number 61631011, 61701264 and 61661043.

Conflicts of Interest: The authors declare no conflict of interest.

References

1. Ciuonzo, D. On Time-Reversal Imaging by Statistical Testing. *IEEE Signal Process. Lett.* **2017**, *24*, 1024–1028. [CrossRef]
2. Newey, M.; Benitz, G.R.; Barrett, D.J.; Mishra, S. Detection and Imaging of Moving Targets With LiMIT SAR Data. *IEEE Trans. Geosci. Remote Sens.* **2018**, *56*, 3499–3510. [CrossRef]
3. Ciuonzo, D.; Carotenuto, V.; Maio, A.D. On Multiple Covariance Equality Testing with Application to SAR Change Detection. *IEEE Trans. Signal Process.* **2017**, *65*, 5078–5091. [CrossRef]
4. Pei, J.; Huang, Y.; Huo, W.; Miao, Y.; Zhang, Y.; Yang, J.J.S. Synthetic Aperture Radar Processing Approach for Simultaneous Target Detection and Image Formation. *Sensors* **2018**, *18*, 3377. [CrossRef] [PubMed]
5. Bi, H.; Zhang, B.; Zhu, X.X.; Hong, W.; Sun, J.; Wu, Y. L_1-Regularization-Based SAR Imaging and CFAR Detection via Complex Approximated Message Passing. *IEEE Trans. Geosci. Remote Sens.* **2017**, *55*, 3426–3440. [CrossRef]
6. Wu, Q.; Zhang, Y.D.; Amin, M.G.; Himed, B. High-Resolution Passive SAR Imaging Exploiting Structured Bayesian Compressive Sensing. *IEEE J. Sel. Top. Signal Process.* **2015**, *9*, 1484–1497. [CrossRef]
7. Zhang, Y.; Xiong, W.; Dong, X.; Hu, C.; Sun, Y. GRFT-Based Moving Ship Target Detection and Imaging in Geosynchronous SAR. *Remote Sens.* **2018**, *10*, 2002. [CrossRef]
8. Currie, A.; Brown, M.A. Wide-swath SAR. *IEE Proc. F Radar Signal Process.* **1992**, *139*, 122–135. [CrossRef]
9. Currie, A. Wide-swath SAR imaging with multiple azimuth beams. In Proceedings of the IEEE Colloquium on Synthetic Aperture Radar, London, UK, 29 November 1989.
10. Li, Z.; Wang, H.; Tao, S.; Zheng, B.J.I.G.; Letters, R.S. Generation of wide-swath and high-resolution SAR images from multichannel small spaceborne SAR systems. *IEEE Geosci. Remote Sens. Lett.* **2005**, *2*, 82–86. [CrossRef]
11. Gebert, N.; Krieger, G.; Moreira, A. Digital Beamforming for HRWS-SAR Imaging: System Design, Performance and Optimization Strategies. In Proceedings of the IEEE International Conference on Geoscience & Remote Sensing Symposium, Denver, CO, USA, 31 July–4 August 2006.
12. Gebert, N.; Krieger, G.; Moreira, A. Digital Beamforming on Receive: Techniques and Optimization Strategies for High-Resolution Wide-Swath SAR Imaging. *IEEE Trans. Aerosp. Electron. Syst.* **2009**, *45*, 564–592. [CrossRef]
13. Wu, X.; Xu, J.Z.; Zhang, C.; Yi, J. A novel spectrum reconstruction algorithm for high resolution and wide swath space-borne SAR. In Proceedings of the Asian and Pacific Conference on Synthetic Aperture Radar, Huangshan, China, 5–9 November 2007.
14. Krieger, G.; Gebert, N.; Moreira, A. SAR Signal Reconstruction from Non-Uniform Displaced Phase Centre Sampling. In Proceedings of the IEEE International Geoscience & Remote Sensing Symposium, Anchorage, AK, USA, 20–24 September 2004.
15. Tan, W.; Xu, W.; Huang, P.; Huang, Z.; Qi, Y.; Han, K.J.S. Investigation of Azimuth Multichannel Reconstruction for Moving Targets in High Resolution Wide Swath SAR. *Sensors* **2017**, *17*, 1270. [CrossRef] [PubMed]
16. Zhou, R.; Sun, J.; Hu, Y.; Qi, Y. Multichannel High Resolution Wide Swath SAR Imaging for Hypersonic Air Vehicle with Curved Trajectory. *Sensors* **2018**, *18*, 411. [CrossRef] [PubMed]
17. Gebert, N.; Krieger, G.J.I.G.; Letters, R.S. Azimuth Phase Center Adaptation on Transmit for High-Resolution Wide-Swath SAR Imaging. *IEEE Geosci. Remote Sens. Lett.* **2009**, *6*, 782–786. [CrossRef]
18. Sun, G.; Xiang, J.; Xing, M.; Yang, J.; Guo, L. A Channel Phase Error Correction Method Based on Joint Quality Function of GF-3 SAR Dual-Channel Images. *Sensors* **2018**, *18*, 3131. [CrossRef] [PubMed]
19. Yang, T.; Li, Z.; Liu, Y.; Suo, Z.; Zheng, B. Channel error estimation methods for multi-channel HRWS SAR systems. In Proceedings of the Geoscience & Remote Sensing Symposium, Melbourne, VIC, Australia, 21–26 July 2013.
20. Guo, X.; Gao, Y.; Wang, K.; Liu, X. Improved channel error calibration method for the azimuth multichannel SAR. *IEEE Geosci. Remote Sens. Lett.* **2016**, *13*, 1022–1026.
21. Berens, P. SAR with ultra-high range resolution using synthetic bandwidth. In Proceedings of the IEEE International Geoscience & Remote Sensing Symposium, Hamburg, Germany, 28 June–2 July 1999.

22. Ciuonzo, D.; Romano, G.; Solimene, R. Performance Analysis of Time-Reversal MUSIC. *IEEE Trans. Signal Process.* **2015**, *63*, 2650–2662. [CrossRef]
23. Gangshu, H. *Digital Signal Processing: Theoretical Algorithms and Implementation*, 2nd ed.; Tsinghua University Press: Beijing, China, 2003; pp. 410–412.

Article

Focusing Bistatic Forward-Looking Synthetic Aperture Radar Based on an Improved Hyperbolic Range Model and a Modified Omega-K Algorithm

Chenchen Wang, Weimin Su *, Hong Gu and Jianchao Yang

School of Electronic and Optical Engineering, Nanjing University of Science and Technology, Nanjing 210094, China

* Correspondence: suweimin@njust.edu.cn

Received: 15 July 2019; Accepted: 30 August 2019; Published: 1 September 2019

Abstract: For parallel bistatic forward-looking synthetic aperture radar (SAR) imaging, the instantaneous slant range is a double-square-root expression due to the separate transmitter-receiver system form. The hyperbolic approximation provides a feasible solution to convert the dual square-root expression into a single-square-root expression. However, some high-order terms of the range Taylor expansion have not been considered during the slant range approximation procedure in existing methods, and therefore, inaccurate phase compensation occurs. To obtain a more accurate compensation result, an improved hyperbolic approximation range form with high-order terms is proposed. Then, a modified omega-K algorithm based on the new slant range form is adopted for parallel bistatic forward-looking SAR imaging. Several simulation results validate the effectiveness of the proposed imaging algorithm.

Keywords: bistatic synthetic aperture radar (SAR); hyperbolic approximation; phase compensation; modified omega-K

---

1. Introduction

Synthetic aperture radar (SAR) attracts massive research enthusiasm among researchers due to its excellent ability to detect targets without the limitation of the external environment [1]. The penetration ability of SAR makes it irreplaceable compared with optical imaging, while it is challenging in traditional monostatic SAR to obtain excellent imaging performance in forward-looking imaging mode, which limits the application of SAR technology. To solve the problem, bistatic SAR has been widely used for forward-looking imaging due to its particular system configuration. The separate transmitter and receiver configuration provides extra advantages like reliable hiding power and system flexibility [2].

One-stationary bistatic SAR, as a special form of general bistatic SAR, was first studied for forward-looking imaging. Several methods have been proposed, such as the squint minimization [3,4], the keystone transform [5], and the ellipse model [2,4]. The Doppler frequency is decided by the moving transmitter or the moving receiver, which is similar to monostatic SAR. Then, the bistatic SAR was proposed where both the transmitter and the receiver are moving. The azimuth resolution is determined by both platforms. For bistatic forward-looking SAR, the difficulty of imaging algorithms lies in the solution of the two-dimensional spectrum because of its unique double-square-root form of echo signal expression [6,7]. Some basic studies of bistatic SAR were proposed to illustrate the advantages [6,8]. Compared with the monostatic situation, the principle of stationary phase (POSP) cannot be applied to solve the derivative zero point when performing azimuth Fourier transform. Several methods have been proposed to solve the problem. Loffeld's bistatic formula (LBF) was proposed to solve the double-square-root expression [6]. Respective stationary points of the transmitter and receiver are obtained first to transform the double-square-root expression into Taylor expansion form. Then, the ultimate spectrum is solved based on the joint stationary point of the Taylor expression. The contributions of the transmitter and receiver are assumed to be the same, which leads to approximation errors. The extended Loffeld's bistatic formula (ELBF) [9] and the modified Loffeld's bistatic formula (MLBF) [10] were proposed later to improve the solution process of stationary points. These two methods assign different weights on the transmitter and receiver. However, all three LBF methods need to solve the stationary points three times, which leads to deduction complexity. The method of series reversion (MSR) [11] is a widely-used method for precisely solving those equations with series terms. In SAR imaging algorithms, Taylor expansion is regarded as a common operation, and MSR can be applied to solve the Fourier transform composed of Taylor expansion. However, it is still challenging to conduct imaging algorithm deduction due to the series form.

To simplify the solution of the spectrum, the hyperbolic approximation was utilized to transform the echo expression with the double-square-root form into the expression with the single-square-root form. In the first version, a parameter named the equivalent speed was defined [12]. In the improved version, two more parameters (the equivalent slant range and the equivalent squint angle) [13] were added in the hyperbolic function to approximate the range more accurately. Moreover, an improved hyperbolic approximation model with additional parameters was proposed for residual compensation [14]. However, considering the solution process in the methods mentioned above, the defined parameters are solved by setting the constant term, the linear term, and the quadratic term of the Taylor expansion of echo equal, which means the influence of the cubic term, the quartic term, and the remaining terms is ignored. In this article, we propose a new model to finish the hyperbolic approximation.

As for imaging algorithms, range Doppler (RD) imaging algorithms, chirp scaling (CS) imaging algorithms, back-projection (BP) imaging algorithms, and omega-K imaging algorithms based on the LBF spectrum, the MSR spectrum, and the hyperbolic approximation spectrum have been proposed in the past few years [9,14–16]. For RD imaging algorithms, it is too fundamental to handle the complex situation of bistatic SAR system. The calculation time consumption is a severe problem for real-time processing when applying BP imaging algorithms. For CS imaging algorithms, it is difficult for researchers to conduct formula derivation. Thus, the omega-K imaging algorithm is selected in this article to finish imaging.

To approximate the slant range more accurately, the cubic term and the quartic term are taken into account in this article. An equivalent hyperbolic range model is introduced first to lay the foundation of the imaging algorithm. The range error analysis is provided to demonstrate the approximation ability of the proposed range model immediately. Then, the modified omega-K imaging algorithm including the signal model and detailed processing steps are presented. Finally, some experimental simulations are given to prove the efficiency of the proposed algorithm.

This article is organized as follows. Section 2 gives the geometry of the bistatic forward-looking SAR and the equivalent hyperbolic range model corresponding to the bistatic system. Section 3 gives the detailed modified omega-K imaging algorithm. Simulation results are given in Section 4 to validate the proposed algorithm. Section 5 provides the conclusion.

2. Geometry and Equivalent Slant Range Model

The parallel bistatic forward-looking SAR system diagram in the Cartesian coordinate system and the derived equivalent slant range model are established first. Then, the analysis of range error based on the equivalent slant range model is provided.

2.1. Equivalent Slant Range Model

Figure 1 shows the geometry of parallel bistatic forward-looking SAR. The transmitter T and the receiver Rmove along the parallel red lines parallel to the x-axis. η_{pc} is the synthetic aperture center time of the imaging scene. $(x_c, y_c, 0)$ is the location coordinate of the imaging center, and $(x_p, y_p, 0)$ is the location of an arbitrary target P in the imaging scene. R_{Tc} is the slant range between the transmitter and the target P at the phase center crossing time η_{pc}, and R_{Rc} is the range between the receiver and the target P at η_{pc}. The approximate forward-looking angle of the receiver is θ_R, and the approximate squint angle of the transmitter is θ_T. V_T and V_R represent the speed of the transmitter and the receiver, respectively. It is assumed that both the transmitter and the receiver can cover the imaging scene during the aperture synthesis.

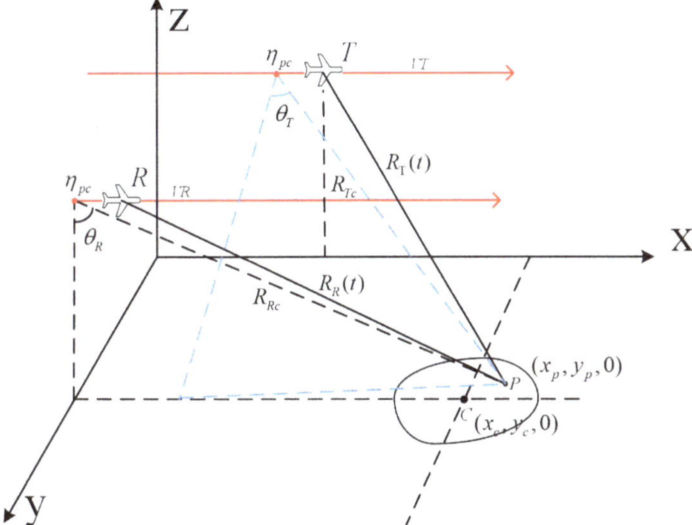

Figure 1. Geometry of forward-looking bistatic SAR.

The instantaneous slant ranges from the transmitter and the receiver to the target P are:

$$\begin{aligned} R_T(\eta) &= \sqrt{R_{Tc}^2 + V_T^2(\eta-\eta_{pc})^2 - 2R_{Tc}V_T(\eta-\eta_{pc})\sin\theta_T}, \\ R_R(\eta) &= \sqrt{R_{Rc}^2 + V_R^2(\eta-\eta_{pc})^2 - 2R_{Rc}V_R(\eta-\eta_{pc})\sin\theta_R}, \end{aligned} \tag{1}$$

where η is the slow time.

Thus, the total range is:

$$R(\eta) = R_T(\eta) + R_R(\eta). \tag{2}$$

It is challenging to solve the two-dimensional spectrum due to the double-square-root expression form of $R(\eta)$. The hyperbolic approximation [12] can be used to convert the double-square-root form to a single-square-root form by defining the equivalent speed and equivalent angle. Traditional hyperbolic approximation [12–14] ignored the high-order terms of the Taylor expansion of $R(\eta)$. To realize a more accurate compensation, an improved equivalent slant range with high-order terms is proposed. The range model is expressed as:

$$R_e(\eta) = \sqrt{R_e^2 + V_e^2(\eta-\eta_{pc})^2 - 2R_eV_e(\eta-\eta_{pc})\sin\theta_e} + E(\eta-\eta_{pc})^3 + F(\eta-\eta_{pc})^4, \tag{3}$$

$$R(\eta) = 2R_e(\eta), \tag{4}$$

where R_e, V_e, and θ_e are the new equivalent slant range at phase crossing time, the new equivalent speed, and the new equivalent squint angle. Compared with existing hyperbolic approximation algorithms, the proposed range model adds two additional high-order terms for range error compensation. To solve the unknown variables, we first expand Equations (1) and (3) into a fourth-order Taylor series at $\eta = \eta_{pc}$. Then, we get:

$$\begin{aligned} R_T(\eta) = & R_{Tc} - V_T \sin\theta_T (\eta - \eta_{pc}) + \frac{V_T^2 \cos\theta_T^2}{2R_{Tc}} (\eta - \eta_{pc})^2 + \\ & \frac{V_T^3 \sin\theta_T \cos^2\theta_T}{2R_{Tc}^2} (\eta - \eta_{pc})^3 + \frac{V_T^4 \cos^2\theta_T (5\sin^2\theta_T - 1)}{8R_{Tc}^3} (\eta - \eta_{pc})^4, \end{aligned} \tag{5}$$

$$\begin{aligned} R_R(\eta) = & R_{Rc} - V_R \sin\theta_R (\eta - \eta_{pc}) + \frac{V_R^2 \cos\theta_R^2}{2R_{Rc}} (\eta - \eta_{pc})^2 + \\ & \frac{V_R^3 \sin\theta_R \cos^2\theta_R}{2R_{Rc}^2} (\eta - \eta_{pc})^3 + \frac{V_R^4 \cos^2\theta_R (5\sin^2\theta_R - 1)}{8R_{Rc}^3} (\eta - \eta_{pc})^4, \end{aligned} \tag{6}$$

$$\begin{aligned} R_e(\eta) = & R_e - V_e \sin\theta_e (\eta - \eta_{pc}) + \frac{V_e^2 \cos\theta_e^2}{2R_e} (\eta - \eta_{pc})^2 + \\ & \frac{V_e^3 \sin\theta_e \cos^2\theta_e}{2R_e^2} (\eta - \eta_{pc})^3 + \frac{V_e^4 \cos^2\theta_e (5\sin^2\theta_e - 1)}{8R_e^3} (\eta - \eta_{pc})^4 + \\ & E(\eta - \eta_{pc})^3 + F(\eta - \eta_{pc})^4. \end{aligned} \tag{7}$$

Substituting Equations (5)–(7) into Equations (2) and (4) and letting the first five terms of Taylor expansion be equal, then we get:

$$\begin{cases} R_{Tc} + R_{Rc} = 2R_e \\ V_T \sin\theta_T + V_R \sin\theta_R = 2V_e \sin\theta_e \\ \frac{V_T^2 \cos\theta_T^2}{2R_{Tc}} + \frac{V_R^2 \cos\theta_R^2}{2R_{Rc}} = 2\frac{V_e^2 \cos\theta_e^2}{2R_e} \\ \frac{V_T^3 \sin\theta_T \cos^2\theta_T}{2R_{Tc}^2} + \frac{V_R^3 \sin\theta_R \cos^2\theta_R}{2R_{Rc}^2} = 2\left(\frac{V_e^3 \sin\theta_e \cos^2\theta_e}{2R_e^2} + E\right) \\ \frac{V_T^4 \cos^2\theta_T (5\sin^2\theta_T - 1)}{8R_{Tc}^3} + \frac{V_R^4 \cos^2\theta_R (5\sin^2\theta_R - 1)}{8R_{Rc}^3} = 2\left[\frac{V_e^4 \cos^2\theta_e (5\sin^2\theta_e - 1)}{8R_e^3} + F\right]. \end{cases} \tag{8}$$

Solving the five equations in Equation (8), then we get:

$$\begin{cases} R_e = \frac{1}{2}(R_{Tc} + R_{Rc}) \\ V_e = \sqrt{A^2 + B} \\ \theta_e = \arcsin(A/V_e) \\ E = C - \frac{V_e^3 \sin\theta_e \cos^2\theta_e}{2R_e^2} \\ F = D - \frac{V_e^4 \cos^2\theta_e (5\sin^2\theta_e - 1)}{8R_e^3}, \end{cases} \tag{9}$$

where:

$$\begin{cases} A = (V_T \sin\theta_T + V_R \sin\theta_R)/2 \\ B = \left(\frac{V_T^2 \cos^2\theta_T}{R_{Tc}} + \frac{V_R^2 \cos^2\theta_R}{R_{Rc}}\right) R_e/2 \\ C = \frac{V_T^3 \sin\theta_T \cos^2\theta_T}{4R_{Tc}^2} + \frac{V_R^3 \sin\theta_R \cos^2\theta_R}{4R_{Rc}^2} \\ D = \frac{V_T^4 \cos^2\theta_T (5\sin^2\theta_T - 1)}{16R_{Tc}^2} + \frac{V_R^4 \cos^2\theta_R (5\sin^2\theta_R - 1)}{16R_{Rc}^2}. \end{cases} \tag{10}$$

At this point, all defined variables are solved. The range error analysis based on the new equivalent range model is presented next.

2.2. Range Error Analysis

To evaluate the proposed equivalent range model, an analysis of the range error based on an X-band bistatic SAR system is given. The simulated parameters are listed in Table 1. The results of the equivalent hyperbolic slant range error are shown in Figure 2.

Table 1. Simulation parameters.

Parameters	Values	Parameters	Values
Carrier frequency	9 GHz	Transmitter center slant range	4300 m
Pulse duration	2 μs	Transmitter squint angle	7°
Bandwidth	200 MHz	Receiver center slant range	3600 m
Sampling frequency	300 MHz	Receiver forward-looking angle	33°
Pulse repetition frequency	1 kHz	Sensor speed	200 m/s

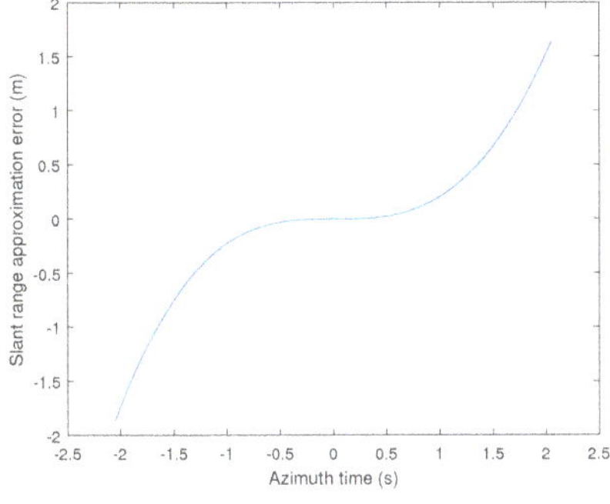

(a) Approximation error of the traditional range model.

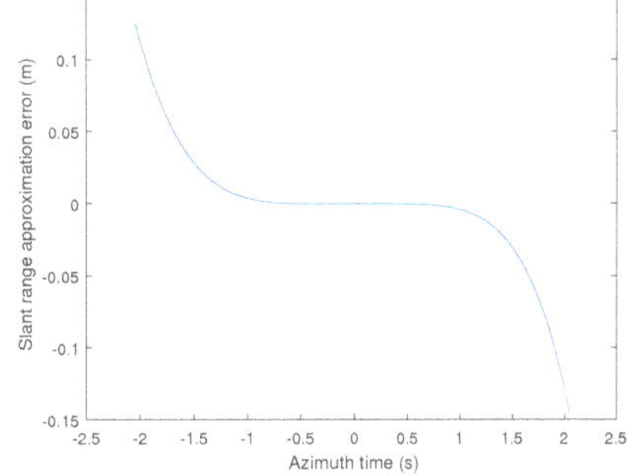

(b) Approximation error of the proposed range model.

Figure 2. Approximation error of the bistatic slant range. (**a**) Approximation error of the traditional range model. (**b**) Approximation error of the proposed range model.

Figure 2a is the approximation error of the traditional hyperbolic approximation range model [13], where the high-order terms are ignored. Figure 2b is the approximation error of the proposed hyperbolic range model. The constant term, the linear term, and the quadratic term in Equations (5)–(7) are used to solve the defined variables. Thus, the residual terms lead to the approximation slant range error. To prove that the proposed model can reduce the range error compared with the traditional model, we first give the expression of the traditional model and its corresponding Taylor expansion, which are:

$$R_t(\eta) = \sqrt{R_t^2 + V_t^2(\eta - \eta_{pc})^2 - 2R_tV_e(\eta - \eta_{pc})\sin\theta_t}, \tag{11}$$

$$\begin{aligned} R_t(\eta) =& R_t - V_t \sin\theta_t(\eta - \eta_{pc}) + \frac{V_t^2 \cos\theta_t^2}{2R_t}(\eta - \eta_{pc})^2 + \\ & \frac{V_t^3 \sin\theta_t \cos^2\theta_t}{2R_t^2}(\eta - \eta_{pc})^3 + \frac{V_t^4 \cos^2\theta_t(5\sin^2\theta_t - 1)}{8R_t^3}(\eta - \eta_{pc})^4 \end{aligned} \tag{12}$$

where $R_t(\eta)$, R_t, V_t, and θ_t are the variables in traditional range model. The error in Figure 2a is the difference between the sum of the cubic terms, the quartic terms, and the residual terms in Equations (5) and (6) and the sum of the cubic term, the quartic term, and the residual term in Equation (12), while the error in Figure 2b is the difference between the sum of the residual terms in Equations (5) and (6) and the residual term in Equation (7). The error caused by the cubic and quartic terms is eliminated. From Figure 2, it can be found that the error in Figure 2a is up to 1.9 m, while the error in Figure 2b is less than 0.15 m. According to the parameters listed in Table 1, the approximation slant range error of the proposed model is much less than a range solution cell. Therefore, the proposed equivalent slant range model is more accurate than the traditional range model. The following imaging algorithm is derived based on the proposed range model.

3. Imaging Algorithm

According to the previous analysis, the improved hyperbolic approximation model can equal the true slant range better than traditional hyperbolic approximate models. In this section, a modified omega-K algorithm based on the improved equivalent range model is proposed for the parallel bistatic forward-looking SAR imaging.

3.1. Signal Model

Assume that a linear frequency-modulated signal is transmitted from the transmitter to the receiver. Then, the base-band echo signal of an arbitrary target P is given as:

$$S_1(t_r,\eta) = \exp\left\{j\pi\gamma\left[t_r - \frac{2R_e(\eta)}{c}\right]\right\}\exp\left[-j\frac{4\pi R_e(\eta)}{\lambda}\right] \tag{13}$$

where γ is the range chirp rate, c is the light speed, λ is the wavelength, t_r is the fast time, and η is the slow time. To simplify the expression and further derivation, the envelopes of the range and azimuth are ignored.

Transforming Equation (13) into the range-frequency azimuth-time domain yields:

$$S_2(f_r,\eta) = \exp\left(-j\frac{\pi f_r^2}{\gamma}\right)\exp\left[-j\frac{4\pi(f_r+f_c)}{c}R_e(\eta)\right] \tag{14}$$

where f_r is the frequency domain variable corresponding to t_r and f_c is the carrier frequency. From Equation (14), it can be easily found that the first exponential term is the range frequency modulation term. This term can be compensated by multiplying its complex conjugate in the range frequency domain. Thus, the first frequency modulation compensation function is:

$$H_{1FM}(f_r,\eta) = \exp\left(j\frac{\pi f_r^2}{\gamma}\right). \tag{15}$$

Multiplying Equation (14) by Equation (15) yields:

$$S_3(f_r,\eta) = \exp\left[-j\frac{4\pi(f_r+f_c)}{c}R_e(\eta)\right]. \tag{16}$$

The exponential term in Equation (16) indicates the severe coupling between range and azimuth. To finish the phase focusing, a modified omega-K algorithm based on the signal model is presented.

3.2. Modified Omega-K Imaging Algorithm

To analyze the exponential term in Equation (16), Equation (3) is substituted into Equation (16) firstly. Then, we get:

$$\begin{aligned} S_4(f_r,\eta) = \exp\Big\{-j\frac{4\pi(f_r+f_c)}{c}\Big[&\sqrt{R_e^2+V_e^2(\eta-\eta_{pc})^2-2R_eV_e(\eta-\eta_{pc})\sin\theta_e} \\ &+E(\eta-\eta_{pc})^3+F(\eta-\eta_{pc})^4\Big]\Big\}. \end{aligned} \tag{17}$$

Equation (17) shows that the signal consists of the traditional hyperbolic term and high-order terms. The traditional omega-K can handle the hyperbolic term well, but cannot handle the high-order terms. The first step of the omega-K algorithm is the compensation of the cubic term and the quartic term. Variable substitution is performed on Equation (17), and then, we get:

$$S_5(k_r, X) = \exp\left\{-jk_r\left[\sqrt{R_e^2 + (X - X_{pc})^2 - 2R_e(X - X_{pc})\sin\theta_e} + \frac{E}{V_e^3}(X - X_{pc})^3 + \frac{F}{V_e^4}(X - X_{pc})^4\right]\right\} \tag{18}$$

where $k_r = \frac{4\pi(f_r+f_c)}{c}$ is the wavenumber, $X = V_e\eta$, and $X_{pc} = V_e\eta_{pc}$. Then, we get $R_e(\eta) = R_e(X)$.

Transforming Equation (18) into two-dimensional wavenumber domain yields:

$$\begin{aligned} S_6(k_r, k_x) &= \int S_5(k_r, X)\exp(-jk_xX)\,dX \\ &= \int \exp\{-jk_rR_e(X)\}\exp(-jk_xX)\,dX \\ &= \int \exp\{-j\phi(k_r, k_x, X)\}\,dX \end{aligned} \tag{19}$$

where $k_x = \frac{2\pi f_a}{V_e}$, f_a is the azimuth frequency, and:

$$\phi(k_r, k_x, X) = k_r\left[\sqrt{R_e^2 + (X - X_{pc})^2 - 2R_e(X - X_{pc})\sin\theta_e} + \frac{E}{V_e^3}(X - X_{pc})^3 + \frac{F}{V_e^4}(X - X_{pc})^4\right] + k_xX. \tag{20}$$

To solve Equation (19), the stationary phase point of $\phi(k_r, k_x, X)$ should be obtained firstly. However, the existence of high-order terms complicates the solution process. For further analysis, the phase is first rewritten as:

$$\phi(k_r, k_x, X) = \phi_t(k_r, k_x, X) + k_r\left[\frac{E}{V_e^3}(X - X_{pc})^3 + \frac{F}{V_e^4}(X - X_{pc})^4\right] \tag{21}$$

where $\phi_t(k_r, k_x, X)$ is the traditional phase term. It is widely accepted that if the phase error is smaller than $\pi/4$ [1], the imaging performance will not be affected much by the approximation. The phase error simulation is given in Figure 3.

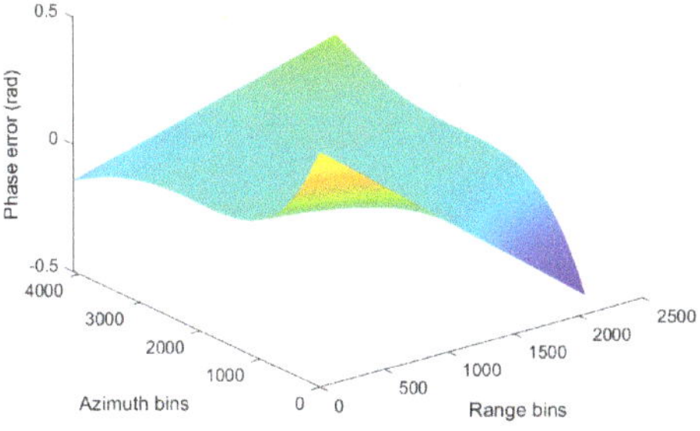

Figure 3. Phase error simulation.

From Figure 3, it can been seen that all absolute phase errors are less than $\pi/4$. Thus, the stationary phase point of $\phi_t(k_r, k_x, X)$ is regarded as the approximate stationary phase point of $\phi(k_r, k_x, X)$. The approximate stationary phase point of $\phi(k_r, k_x, X)$ is:

$$X^* = -\frac{k_xR_e\sin\theta_e}{\sqrt{k_r^2 - k_x^2}} + R_e\sin\theta_e + X_{pc}, \tag{22}$$

where X^* is only a designation of the solution and $(*)$ is not an operator.

Substituting Equation (22) in Equation (19) and applying POSP yield the two-dimensional wavenumber domain signal as:

$$S_7(k_r,k_x) = \exp\left\{-j\sqrt{k_r^2-k_x^2}R_e\cos\theta_e - jk_x\left(R_e\sin\theta_e + X_{pc}\right)\right.$$
$$\left. -jk_r\left[\frac{E}{V_e^3}\left(X^*-X_{pc}\right)^3 + \frac{F}{V_e^4}\left(X^*-X_{pc}\right)^4\right]\right\}. \quad (23)$$

The cubic term and quartic term in Equation (23) can be easily compensated by multiplying its conjugate form. Therefore, the high-order filter is:

$$H_2(k_r,k_x) = \exp\left\{jk_r\left[\frac{E}{V_e^3}\left(X^{**}-X_{pc}\right)^3 + \frac{F}{V_e^4}\left(X^{**}-X_{pc}\right)^4\right]\right\} \quad (24)$$

where X^{**} is the value of X^* at the reference range and $(**)$ is not an operator.

Multiplying Equations (23) and (24), we get the compensated signal for the further omega-K imaging algorithm. The signal is:

$$S_8(k_r,k_x) = \exp\left\{-j\sqrt{k_r^2-k_x^2}R_e\cos\theta_e - jk_x\left(R_e\sin\theta_e + X_{pc}\right)\right\}. \quad (25)$$

A two-step omega-K is performed on Equation (25) to finish the imaging focusing.

The first step is the bulk focusing. A reference function is designed based on the reference range to finish coarse focusing. This filter can compensate the phase of signals of those points at the reference range. The reference function is:

$$H_{rf}(k_r,k_x) = \exp\left\{j\sqrt{k_r^2-k_x^2}R_{ref}\cos\theta_e + jk_x\left(R_{ref}\sin\theta_e + X_{pc}\right)\right\}. \quad (26)$$

Multiplying Equations (25) and (26) gets:

$$S_9(k_r,k_x) = \exp\left\{-j\sqrt{k_r^2-k_x^2}\cos\theta_e\left(R_e-R_{ref}\right) - jk_x\sin\theta_e\left(R_e-R_{ref}\right)\right\}. \quad (27)$$

After bulk focusing, the residual phase at the reference range is removed. However, the residual phase of points not at the reference range remains. Moreover, the phase contains coupling terms between range and azimuth. For precise focusing of all points, the Stolt interpolation function is given as:

$$k_y = \sqrt{k_r^2-k_x^2}\cos\theta_e + k_x\sin\theta_e. \quad (28)$$

After Stolt interpolation, the resampled signal becomes:

$$S_{10}(k_r,k_x) = \exp\left[-jk_y\left(R_e-R_{ref}\right)\right]. \quad (29)$$

From Equation (29), it is evident that the coupling between range and azimuth has been removed. The phase is a linear function of k_y. Then, the inverse fast Fourier transform is implemented on Equation (29) to complete imaging.

According to the analysis mentioned above, the whole imaging process is shown in Figure 4.

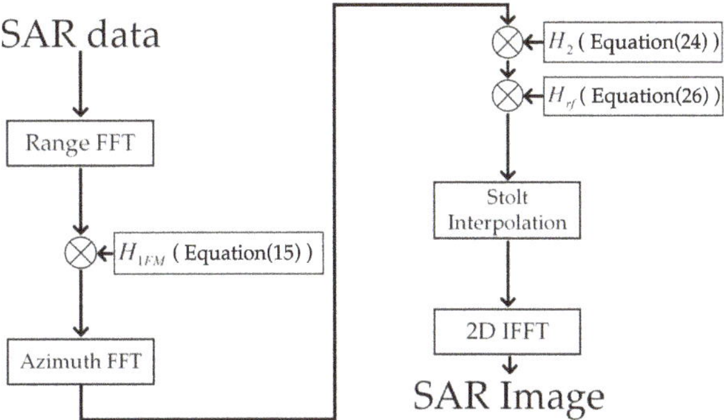

Figure 4. Flowchart of modified omega-K.

The specific steps are as follows:

(1) Performing range fast Fourier transform (FFT) on SAR data gets $S_2(f_r, \eta)$.
(2) Multiplying Equation (15) and $S_2(f_r, \eta)$ gets $S_3(f_r, \eta)$.
(3) Performing azimuth fast Fourier transform (FFT) on $S_3(f_r, \eta)$ gets $S_7(k_r, k_x)$.
(4) Multiplying Equation (24) and $S_7(k_r, k_x)$ gets $S_8(k_r, k_x)$.
(5) Multiplying Equation (26) and $S_8(k_r, k_x)$ gets $S_9(k_r, k_x)$.
(6) Performing Stolt interpolation on $S_9(k_r, k_x)$ gets $S_{10}(k_r, k_x)$.
(7) Performing 2D-IFFT on $S_{10}(k_r, k_x)$ gets output SAR focusing results.

4. Simulation Results

In this section, to demonstrate the effectiveness of the proposed imaging algorithm, experimental simulations of parallel bistatic forward-looking SAR are carried out. The system parameters are listed in Table 1. Four points at different locations were chosen to compare the imaging performance. They were $P_0(0,0)$, $P_1(0,500)$, $P_2(200,0)$, and $P_3(200,500)$. The unit of the coordinates is meters. The omega-K imaging algorithm based on the traditional three-parameters hyperbolic range model [13] was selected as the reference.

Figure 5 is the comparison of the overall imaging performance before geometric correction. Figure 5a is the result of the traditional imaging algorithm, and Figure 5b is the result of the proposed imaging algorithm. In Figure 5a, although the four points can be successfully focused, the quality of the right two points has distortion. In contrast, Figure 5b shows that the proposed algorithm achieves a better focus quality on the right two points than the traditional algorithm.

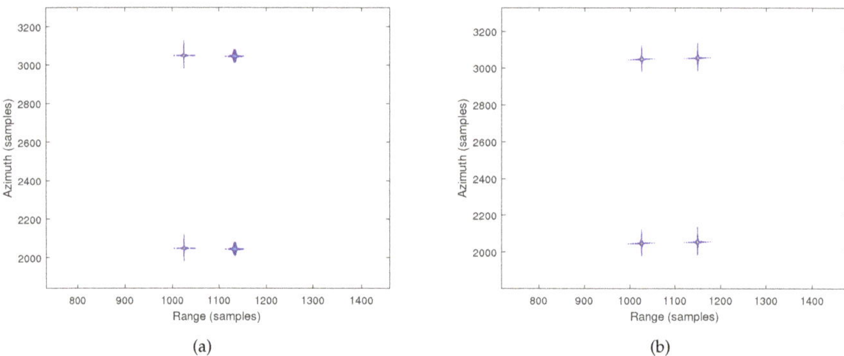

Figure 5. Imaging results. (**a**) Imaging results of the traditional hyperbolic omega-K algorithm. (**b**) Imaging results of the proposed hyperbolic omega-K algorithm.

To observe the imaging performance more intuitively, the sub-images of the four points extracted from Figure 5 are given by Figure 6. Figure 6a–c presents the imaging results of P_0, P_2, andP_3 achieved by the traditional hyperbolic range model given in [13], respectively. Figure 6e,f shows the imaging quality of the three targets obtained by the proposed modified omega-K imaging algorithm. From Figure 6a,d, both algorithms can obtain an excellent focusing quality of the scene center P_0. For the omega-K algorithm, the scene center is always chosen as the reference point to perform bulk focusing. For the points away from the center (P_2 and P_3), it is evident that the proposed algorithm performs much better than the traditional algorithm. For further analysis, the azimuth impulse response of the farthest point P_3 is given in Figure 7. Table 2 gives out the peak sidelobe ratio (PSLR) and the integrated sidelobe ratio (ISLR) of targets P_3.

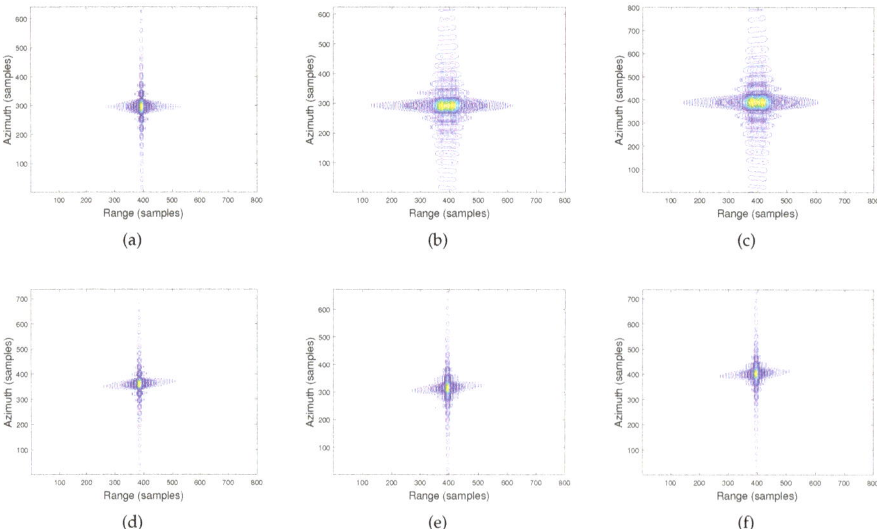

Figure 6. Imaging results. (**a**) Imaging result of P_0 by the traditional algorithm. (**b**) Imaging result of P_2 by the traditional algorithm. (**c**) Imaging result of P_3 by the traditional algorithm. (**d**) Imaging result of P_0 by the proposed algorithm. (**e**) Imaging result of P_2 by the proposed algorithm. (**f**) Imaging result of P_3 by the proposed algorithm.

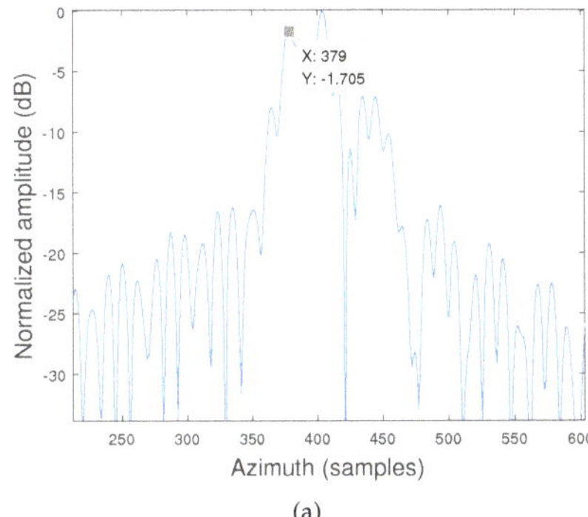

(a)

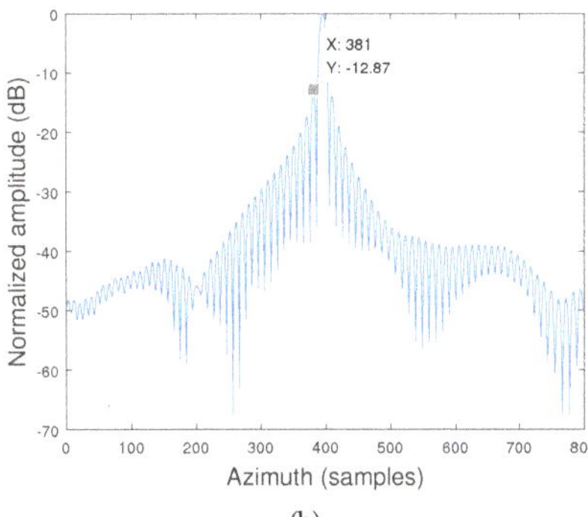

(b)

Figure 7. Azimuth impulse response of P_3. (**a**) Traditional hyperbolic omega-K algorithm. (**b**) Proposed hyperbolic omega-K algorithm.

Table 2. Image quality parameters of P_3. PSLR, peak sidelobe ratio; ISLR, integrated sidelobe ratio.

Targets	PSLR (dB)		ISLR (dB)	
	Azimuth	Range	Azimuth	Range
Traditional omega-K algorithm	−1.705	-	-	-
Proposed omega-K algorithm	−12.87	−13.33	−8.86	−9.9558

Figure 7a is achieved by the traditional hyperbolic omega-K algorithm. Figure 7b is achieved by the proposed hyperbolic omega-K algorithm. Compared with the traditional omega-K algorithm, the proposed omega-K algorithm can improve the performance of the azimuth impulse response. The objective image quality values demonstrated the effectiveness of the proposed omega-K algorithm.

5. Conclusions

In this article, an improved hyperbolic range model was proposed to deal with the particular form of the echo of bistatic forward-looking SAR. The modified omega-K imaging algorithm based on the hyperbolic range model was used to finish focusing. The high-order terms were taken into account to reduce the range approximation error. Extra phase compensation benefited the focusing of the omega-K algorithm. Compared with the range model without high-order compensation terms, the proposed method showed the effectiveness of imaging quality by simulation results.

Author Contributions: Conceptualization, W.S. and H.G.; methodology, C.W.; software, C.W.; validation, all authors; formal analysis, C.W.; investigation, J.Y.; writing–original draft preparation, C.W.; writing—review and editing, W.S. and J.Y.; supervision, H.G.; project administration, W.S.; funding acquisition, W.S., H.G. and J.Y.

Funding: This research was funded by the National Natural Science Foundation of China under Grant Numbers 61471198 and 61671246 and the Natural Science Foundation of Jiangsu Province under Grant Numbers BK20160847 and BK20170855.

Conflicts of Interest: The authors declare no conflict of interest.

References

1. Cumming, I.G.; Wong, F.H. *Digital Processing of Synthetic Aperture Radar Data*; Artech House: Norwood, MA, USA, 2005.
2. Zhong, H.; Yan, A.; Sun, M.; Zhang, X.; Zhang, J. An improved ENLCS algorithm based on a quadratic ellipse model for high-resolution one-stationary bistatic SAR. *Remote. Sens. Lett.* **2018**, *9*, 867–876. [CrossRef]
3. Ma, C.; Gu, H.; Su, W.; Zhang, X.; Li, C. Focusing one-stationary bistatic forward-looking synthetic aperture radar with squint minimisation method. *IET Radar Sonar Navig.* **2015**, *9*, 927–932. [CrossRef]

4. Liang, M.; Su, W.; Ma, C.; Gu, H.; Yang, J. Focusing one-stationary bistatic forward-looking synthetic aperture radar based on squint minimization and modified nonlinear chirp scaling algorithm. *Int. J. Remote Sens.* **2018**, *39*, 7830–7845. [CrossRef]
5. Wu, J.; Li, Z.; Huang, Y.; Yang, J.; Yang, H.; Liu, Q.H. Focusing Bistatic Forward-Looking SAR with Stationary Transmitter Based on Keystone Transform and Nonlinear Chirp Scaling. *IEEE Geosci. Remote Sens. Lett.* **2014**, *11*, 148–152. [CrossRef]
6. Loffeld, O.; Nies, H.; Peters, V.; Knedlik, S. Models and useful relations for bistatic SAR processing. *IEEE Trans. Geosci. Remote Sens.* **2004**, *42*, 2031–2038. [CrossRef]
7. Walterscheid, I.; Ender, J.H.G.; Brenner, A.R.; Loffeld, O. Bistatic SAR Processing and Experiments. *IEEE Trans. Geosci. Remote Sens.* **2006**, *44*, 2710–2717. [CrossRef]
8. Wang, R.; Loffeld, O.; Ul-Ann, Q.; Nies, H.; Ortiz, A.M.; Samarah, A. A Bistatic Point Target Reference Spectrum for General Bistatic SAR Processing. *IEEE Geosci. Remote Sens. Lett.* **2008**, *5*, 517–521. [CrossRef]
9. Wang, R.; Loffeld, O.; Neo, Y.L.; Nies, H.; Dai, Z. Extending Loffeld's bistatic formula for the general bistatic SAR configuration. *IET Radar Sonar Navig.* **2010**, *4*, 74–84. [CrossRef]
10. Ma, C.; Gu, H.; Su, W.; Li, C.; Chen, J. Focusing bistatic forward-looking synthetic aperture radar based on modified Loffeld's bistatic formula and chirp scaling algorithm. *J. Appl. Remote Sens.* **2014**, *8*, 083586. [CrossRef]
11. Neo, Y.L.; Wong, F.; Cumming, I.G. A Two-Dimensional Spectrum for Bistatic SAR Processing Using Series Reversion. *IEEE Geosci. Remote Sens. Lett.* **2007**, *4*, 93–96. [CrossRef]
12. Bamler, R.; Boerner, E. On the use of numerically computed transfer functions for processing of data from bistatic SARs and high squint orbital SARs. In Proceedings of the IEEE International Geoscience and Remote Sensing Symposium, Seoul, Korea, 29 July 2005.
13. Qiu, X.; Hu, D.; Ding, C. Focusing Bistaitc Images use RDA based on Hyperbolic Approximating. In Proceedings of the CIE International Conference on Radar, Shanghai, China, 16–19 October 2006.
14. Ma, C.; Gu, H.; Su, W.; Li, C. Focusing bistatic forward-looking synthetic aperture radar based on modified hyperbolic approximating. *Acta Phys. Sin.* **2014**, *63*, 028403.
15. Chen, S.; Yuan, Y.; Zhang, S.; Zhao, H.; Chen, Y. A New Imaging Algorithm for Forward-Looking Missile-Borne Bistatic SAR. *IEEE J. Sel. Top. Appl. Earth Obs. Remote Sens.* **2016**, *9*, 543–1552. [CrossRef]
16. Feng, D.; An, D.; Huang, X. An Extended Fast Factorized Back Projection Algorithm for Missile-Borne Bistatic Forward-Looking SAR Imaging. *IEEE Trans. Aerosp. Electron. Syst.* **2018**, *54*, 2724–2734. [CrossRef]

Article

Microwave Staring Correlated Imaging Based on Unsteady Aerostat Platform

Zheng Jiang , Yuanyue Guo *, Jie Deng, Weidong Chen and Dongjin Wang

Key Laboratory of Electromagnetic Space Information, Chinese Academy of Sciences, University of Science and Technology of China, Hefei 230026, China; jiangz10@mail.ustc.edu.cn (Z.J.); dengjie@mail.ustc.edu.cn (J.D.); wdchen@ustc.edu.cn (W.C.); wangdj@ustc.edu.cn (D.W.)

* Correspondence: yuanyueg@ustc.edu.cn; Tel.: +86-138-6613-5598

Received: 3 May 2019; Accepted: 21 June 2019; Published: 24 June 2019

Abstract: Microwave staring correlated imaging (MSCI), with the technical capability of high-resolution imaging on relatively stationary targets, is a promising approach for remote sensing. For the purpose of continuous observation of a fixed key area, a tethered floating aerostat is often used as the carrying platform for MSCI radar system; however, its non-cooperative random motion of the platform caused by winds and its unbalance will result in blurred imaging, and even in imaging failure. This paper presents a method that takes into account the instabilities of the platform, combined with an adaptive variable suspension (AVS) and a position and orientation system (POS), which can automatically control the antenna beam orientation to the target area and measure dynamically the position and attitude of the stochastic radiation radar array, respectively. By analyzing the motion feature of aerostat platform, the motion model of the radar array is established, then its real-time position vector and attitude angles of each antenna can be represented; meanwhile the selection matrix of beam coverage is introduced to indicate the dynamic illumination of the radar antenna beam in the overall imaging area. Due to the low-speed discrete POS data, a curve-fitting algorithm can be used to estimate its accurate position vector and attitude of each antenna at each high-speed sampling time during the imaging period. Finally, the MSCI model based on the unsteady aerostat platform is set up. In the simulations, the proposed scheme is validated such that under the influence of different unstable platform movements, a better imaging performance can be achieved compared with the conventional MSCI method.

Keywords: microwave staring correlated imaging; unsteady aerostat platform; motion parameter fitting; position error

---

1. Introduction

Microwave remote sensing has the ability to work in all day and all weather conditions [1], thus it has been used in many civilian and military fields, such as disaster monitoring and military reconnaissance [2]. The conventional high-resolution microwave remote sensing commonly applies Synthetic Aperture Radar (SAR) which is based on Range-Doppler (RD) principle [3]. However relative motion between radar and target is necessary for SAR and the revisit period is long. In forward-looking or staring imaging geometry, SAR cannot work effectively and encounters great challenges to obtain high-resolution imaging.

Microwave staring correlated imaging is a novel high-resolution staring imaging technique without the relative motion limit of target [4–6]. The essence of MSCI is to construct temporal-spatial stochastic radiation field (TSSRF) in the imaging region, which is typically realized by a multi-transmitters configuration emitting independent stochastic waveforms [7,8]. By correlation process (CP) between the target scattering echo and the TSSRF, targets within the antenna beam can be resolved. Due to its superior imaging performance without target relative motion, MSCI has attracted increasing attention and made progress in many aspects such as random radiation source optimization [9–11], imaging algorithm [12–14] and outfield imaging experiment [15].

At present, research on MSCI depends on the premise of an ideal stable imaging platform, i.e., the system platform of the MSCI radar is assumed to be stationary. However, it is not guaranteed in practical applications. To observe a fixed area, the MSCI radar needs to be raised to a certain height. A tethered aerostat is suitable to serve as the platform of MSCI radar with advantages of long-stay time in the air, wide coverage area and low cost [16,17], but it cannot keep absolutely stationary in the air because of the non-cooperative motion caused by wind and unbalance. The platform instability will result in imaging system errors and the imaging performance will be seriously degraded when the random motion of platform becomes intense.

The imaging system errors in MSCI have been investigated by many studies, since it generally exists in practice. For example, to compensate the gain–phase error in MSCI, Zhou et al. propose a sparse auto-calibration method, which is a cyclic iteration processing combined target reconstruction with gain–phase error estimation [18]. In reference [19], the MSCI with phase error is formulated as a Bayesian hierarchical prior modeling, and self-calibration variational message passing (SC-VMP) algorithm is proposed, which estimates the scattering coefficient and phase error iteratively by VMP and Newton's method to improve the performance of MSCI with phase error. To estimate the gain–phase error and the synchronization error under high SNR, Tian et al. add a reference receiver to the MSCI system to receive the direct wave signal and the gain–phase error and the synchronization error are estimated by the direct wave signals [20]. In reference [21], a method of strip-mode MSCI with self-calibration of gain–phase errors is proposed to solve the problem of MSCI with gain–phase errors in a large scene. Reference [22] considers the off-grid problem in MSCI and an algorithm based on variational sparse Bayesian learning (VSBL) is developed to solve the MSCI with off-grid problem. Reference [23] focuses on sparsity-driven MSCI with array position error (APE) and propose two sparse auto-calibration imaging algorithms in sparse Bayesian learning framework to compensate the APE. Li et al. analyzes the target-motion-induced error and provides an applicable approach for MSCI in the presence of target-motion-induced error [24]. Hitherto, research on MSCI system error generally concentrated on gain–phase error, off-grid error, APE, and target-motion-induced error. There is no study on the imaging system error caused by instability of the platform which is an important issue in practice applications.

Aiming at the above problems, this paper proposes a MSCI method based on unsteady aerostat platform. In the proposed method, the antenna array with multiple transmitters and one receiver is mounted on the aerostat platform combined with an adaptive variable suspension (AVS), and the position and orientation system (POS) located at the center of the array, controlling its antenna beam orientation to the target area and measuring dynamically its position and attitude during imaging process. The effects of antenna motion and dynamic beam coverage caused by instability of the platform are considered in imaging model to reduce the imaging model error. For antenna motion, the real-time position vectors of antenna are used in imaging model in place of static position vector. The calculation of real-time position vector of antenna depends on the translational speed and the rotational angular velocity of the array in each signal pulse, then based on the low-speed discrete POS data, a least square curve-fitting method is employed to estimate the accurate translational speed and rotational angular velocity of the array at every sampling time. For dynamic beam coverage, the selection matrix of beam coverage calculated by the position and the attitude of the array is introduced to indicate the illuminated area at each pulse.

The rest of this paper is organized as follows. Section 2 presents the MSCI method based on unsteady aerostat platform. In Section 3, estimation of translational speed and rotational angular velocity of antenna array is given. In Section 4, serval simulations are demonstrated to show the effectiveness of the proposed method. Section 5 concludes this paper.

2. MSCI Method Based on Unsteady Aerostat Platform

2.1. Imaging Scene

MSCI can be realized by using a multi-transmitter configuration to transmit time-independent and group-orthogonal waveforms. To realize observation of targets on ground, MSCI radar can be raised to the air by a tethered aerostat. As shown in Figure 1, the antenna array with N transmitters and one receiver at its array center is carried by AVS which is able to control the antenna beam orientation, and POS is placed at the center of the array to dynamically measure its position and the attitude during the imaging process.

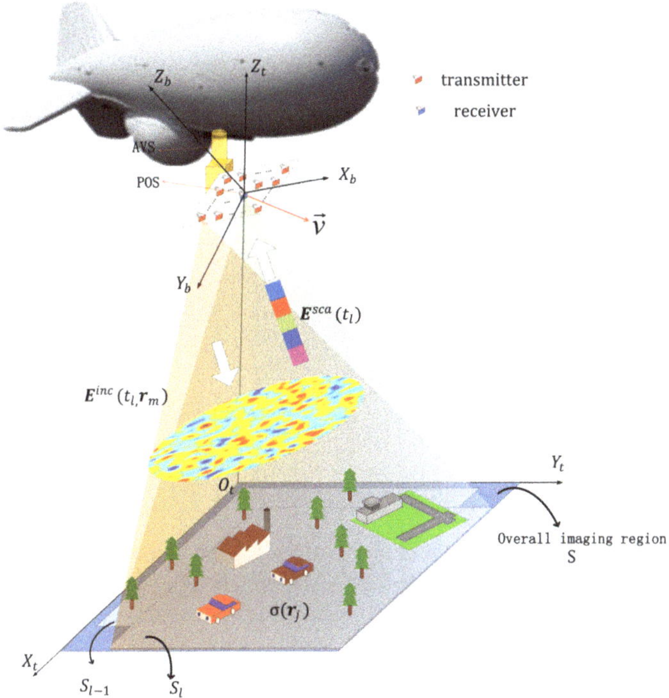

Figure 1. Imaging geometry of MSCI based on unsteady aerostat platform.

To illustrate the geometry of the imaging scene, as the earth-surface inertial reference frame, the coordinate system $O_tX_tY_tZ_t$ is established, with its origin O_t located in the projection point of the array center on the ground on $O_tX_tY_t$ plane at the beginning imaging time, its X_t axis pointing to the east along local latitude line, its Y_t axis pointing to the north along local meridian and its Z_t axis pointing upward along the local geographic vertical line.

The independent signal of random frequency hopping transmitted synchronously by all transmitters and the signal transmitted by the n-th transmitter is denoted as

$$s_n(t) = \sum_{l=1}^{L} rect[\frac{t-(l-1)T_p}{T}] \exp\{j2\pi f_{nl}[t-(l-1)T_p]\}, \tag{1}$$

where f_{nl} is the frequency of the l-th pulse emitted by the n-th transmitter and randomly selected within the system bandwidth. $rect(t)$ is rectangular function. L is the total number of pulses. T_P denotes pulse repetition interval and T is pulse width.

During the imaging process, POS will dynamically record the position and the attitude of the antenna array. The attitude of the array Euler angles includes yaw angle, pitch angle, and roll angle. To give definition of these angles, the aerostat coordinate system $O_bX_bY_bZ_b$ is established on the array with its origin O_b located at its array center, its X_b axis pointing to the right along the horizontal axis of the array, its Y_b axis pointing forward along the longitudinal axis of the array and its Z_b axis perpendicular to the radar array plane. The yaw angle θ is defined as the angle between the projection of Y_b on the $O_tX_tY_t$ plane and the Y_t axis, with the Y_b axis right side being positive. The pitch angle φ is defined as the angle between the Y_b axis and the $O_tX_tY_t$ plane, with Y_b axis up side being positive. The roll angle ϕ is defined as the angle between the Z_b axis and the vertical plane containing the Y_b axis, with Z_b axis right side being positive. The graphical diagram for the altitude angles is shown in Figure 2. Y'_b is the projection of Y_b on the $O_tX_tY_t$ plane and Z'_b is the projection of Z_b on the $O_tZ_tY_b$ plane.

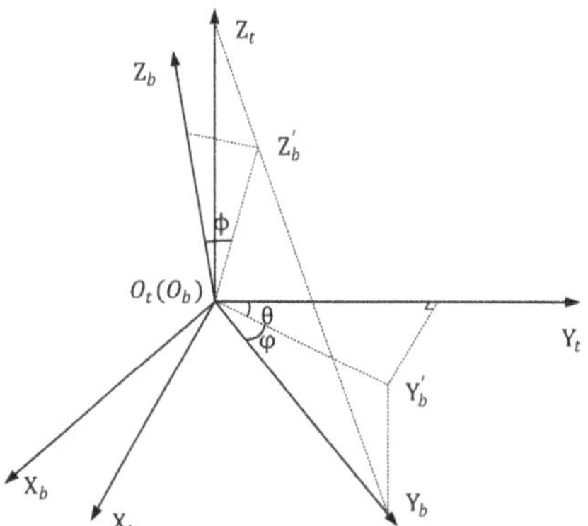

Figure 2. Graphical diagram for the altitude angles.

2.2. Real-Time Position Vector of Antenna

To eliminate the influence of antenna motion, the real-time position vector $\mathbf{r}^n(t_l)$ and $\mathbf{r}^s(t_l)$ are introduced to the MSCI model based on unsteady aerostat platform, where $\mathbf{r}^n(t_l)$ and $\mathbf{r}^s(t_l)$ denote the real-time position vector of the n-th transmitter and the receiver at t_l in the l-th pulse in $O_tX_tY_tZ_t$ respectively.

The complicated motion of the antenna array is decomposed into three-dimensional translations and three rotational components. The three-dimensional translations are along X_t, Y_t and Z_t respectively. The rotational components are rotation of yaw angle, pitch angle, and roll angle, respectively. As the pulse repetition interval T_P is short, the translational speed and the rotational angular velocity will not change drastically during such a short period, so the assumption on the array motion is made that the antenna array motion is uniform translation and uniform rotation during each pulse repetition interval T_P. Hence the translational speed of the antenna array during the l-th pulse is denoted as $\mathbf{v}_l = \left[v_{l,x}, v_{l,y}, v_{l,z}\right]$, where $v_{l,x}, v_{l,y}, v_{l,z}$ are the speeds of the three-dimensional translations along X_t axis, Y_t axis and Z_t axis respectively. The rotational angular velocity of the antenna array during the l-th pulse is denoted as $\boldsymbol{\omega}_l = \left[\omega_{l,\theta}, \omega_{l,\varphi}, \omega_{l,\phi}\right]$, where $\omega_{l,\theta}, \omega_{l,\varphi}, \omega_{l,\phi}$ are rotational angular velocity of the yaw angle, the pitch angle, and the roll angle respectively.

If the motion of antenna array during each pulse is known, the real-time position vector of the antenna can be determined. T_{pos} denotes the repetition period of POS recording data. As Figure 3 shows, the pulse repetition interval T_P is far shorter than T_{pos} of POS, so there are many transmitting pulses between adjacent POS data. Assuming that the recorded time $t_{i,pos}$ of the i-th POS data is in the l'-th pulse and the recorded time $t_{i+1,pos}$ of the next POS data is in the l''-th pulse, the real-time position vector of n-th transmitter $\mathbf{r}^n(t_l)$ at t_l in the l-th $(l' \leq l \leq l'')$ pulse can be expressed as

$$\mathbf{r}^{n}\left(t_{l}\right)=\mathbf{r}^{n}\left(t_{i,pos}\right)+\Delta\mathbf{r}_{v}\left(t_{l}-t_{i,pos}\right)+\Delta\mathbf{r}_{\omega}^{n}\left(t_{l}-t_{i,pos}\right), \tag{2}$$

where $\Delta\mathbf{r}_{v}\left(t_{l}-t_{i,pos}\right)$ and $\Delta\mathbf{r}_{\omega}^{n}\left(t_{l}-t_{i,pos}\right)$ are the displacement vectors of the n-th transmitter caused by translation and rotation during $t_l - t_{i,pos}$, respectively. $\mathbf{r}^{n}\left(t_{i,pos}\right)$ is the position vector of the n-th transmitter at $t_{i,pos}$ and can be calculated by the following formula

$$\mathbf{r}^{n}\left(t_{i,pos}\right)=\mathbf{r}^{s}\left(t_{i,pos}\right)+\mathbf{C}\left(\theta_{t_{i,pos}},\varphi_{t_{i,pos}},\phi_{t_{i,pos}}\right)\mathbf{r}_{b}^{n}, \tag{3}$$

where $\mathbf{r}_{b}^{n}$ is the position vector of the n-th transmitter in $O_bX_bY_bZ_b$. $\mathbf{r}^{s}\left(t_{i,pos}\right)$ is the position vector of the receiver measured by the POS at $t_{i,pos}$ in $O_tX_tY_tZ_t$. $\mathbf{C}\left(\theta_{t_{i,pos}},\varphi_{t_{i,pos}},\phi_{t_{i,pos}}\right)$ is the Direction Cosine Matrix (DCM) that transforms the coordinate from $O_bX_bY_bZ_b$ to $O_tX_tY_tZ_t$. The DCM can be expressed as

$$\begin{aligned}\mathbf{C}\left(\theta_{t_{i,pos}},\varphi_{t_{i,pos}},\phi_{t_{i,pos}}\right)=&\begin{bmatrix}\cos\left(\theta_{t_{i,pos}}\right) & \sin\left(\theta_{t_{i,pos}}\right) & 0\\ -\sin\left(\theta_{t_{i,pos}}\right) & \cos\left(\theta_{t_{i,pos}}\right) & 0\\ 0 & 0 & 1\end{bmatrix}\times\\ &\begin{bmatrix}1 & 0 & 0\\ 0 & \cos\left(\varphi_{t_{i,pos}}\right) & -\sin\left(\varphi_{t_{i,pos}}\right)\\ 0 & \sin\left(\varphi_{t_{i,pos}}\right) & \cos\left(\varphi_{t_{i,pos}}\right)\end{bmatrix}\times\begin{bmatrix}\cos\left(\phi_{t_{i,pos}}\right) & 0 & \sin\left(\phi_{t_{i,pos}}\right)\\ 0 & 1 & 0\\ -\sin\left(\phi_{t_{i,pos}}\right) & 0 & \cos\left(\phi_{t_{i,pos}}\right)\end{bmatrix}.\end{aligned} \tag{4}$$

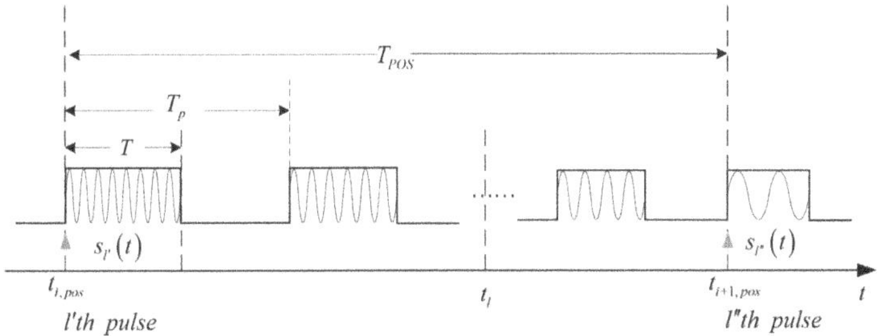

Figure 3. Pulse and POS data timing diagram.

As the receiver is at the center of the antenna array, its position vector at t_l is only affected by the translation of the antenna array during the period of $t_l - t_{i,pos}$ and can expressed as

$$\mathbf{r}^{s}\left(t_{l}\right)=\mathbf{r}^{s}\left(t_{i,pos}\right)+\Delta\mathbf{r}_{v}\left(t_{l}-t_{i,pos}\right). \tag{5}$$

$\Delta\mathbf{r}_{v}\left(t_{l}-t_{i,pos}\right)$ and $\Delta\mathbf{r}_{\omega}^{n}\left(t_{l}-t_{i,pos}\right)$ can be calculated by the translational speed and the rotational angular velocity of the antenna array:

$$\Delta\mathbf{r}_{v}\left(t_{l}-t_{i,pos}\right)=\left[\min\left\{t_{l},l'T_{p}\right\}-t_{i,pos}\right]\mathbf{v}_{l'}+\sum_{k=l'+1}^{l}\left[\min\left\{t_{l},kT_{p}\right\}-\left(k-1\right)T_{P}\right]\mathbf{v}_{k}, \tag{6}$$

$$\Delta\mathbf{r}_{\omega}^{n}\left(t_{l}-t_{i,pos}\right)=\mathbf{C}\left(\Delta\theta_{t_{l}},\Delta\varphi_{t_{l}},\Delta\phi_{t_{l}}\right)\mathbf{r}_{b}^{n}-\mathbf{r}_{b}^{n}. \tag{7}$$

The function $\min\{x,y\}$ returns the minimum of x and y. $\Delta\theta_{t_l}$, $\Delta\varphi_{t_l}$ and $\Delta\phi_{t_l}$ are the changes of the altitude angles during $t_l - t_{i,pos}$ and can be calculated by the following formula

$$\alpha_{t_l} = \left[\min\left\{t_l, l'T_p\right\} - t_{i,pos}\right]\omega_{l',\alpha} + \sum_{k=l'+1}^{l}\left[\min\left\{t_l, kT_p\right\} - (k-1)\,T_P\right]\omega_{k,\alpha}, \tag{8}$$

where $\alpha_{t_l} \in \left(\Delta\theta_{t_l}, \Delta\varphi_{t_l}, \Delta\phi_{t_l}\right)$.

2.3. Influence of Platform Motion on Beam Coverage

The aerostat platform instability not only causes the antenna motion, but also changes the beam coverage in the overall imaging region S. All echo data contains the information of all beam covered areas, therefore as the union of all beam coverages, the overall imaging region S is considered in imaging. The selection matrix of beam coverage is introduced to indicate the dynamically illuminated area of each pulse within the overall imaging region.

The beam coverage of the l-th pulse is denoted as S_l, and the coordinate $\left(x_l^c, y_l^c\right)$ is the beam coverage center on the $O_tX_tY_t$ plane of the l-th pulse:

$$x_l^c = x_l^s - (tan\phi_l cos\varphi_l / cos\theta_l - sin\varphi_l tan\theta_l)z_l^s, \tag{9}$$

$$y_l^c = y_l^s + (tan\phi_l sin\varphi_l / cos\theta_l + cos\varphi_l tan\theta_l)z_l^s, \tag{10}$$

where $\left(x_l^s, y_l^s, z_l^s\right)$, $(\theta_l, \varphi_l, \phi_l)$ are its center position coordinate and its attitude angles of the antenna array at the start time of the l-th pulse.

The overall imaging region S is the union of all beam covered areas during imaging, i.e., $S = S_1 \bigcup S_2 \bigcup ... \bigcup S_L$. The size of the beam covered area of a single pulse is denoted as $w_x \times w_y$, where w_x, w_y are the side length. The size of the overall imaging region is

$$W_x \times W_y = \left(x_{max}^c - x_{min}^c + w_x\right) \times \left(y_{max}^c - y_{min}^c + w_y\right), \tag{11}$$

where W_x,W_y are the side length of S. $x_{\max}^c$ and x_{min}^c are the maximum value and the minimum value of $x_l^c, l = 1, 2, \cdots, L$. $y_{\max}^c$ and y_{min}^c are the maximum value and the minimum value of $y_l^c, l = 1, 2, \cdots, L$.

The overall imaging region will be discretized into $M = P \times Q$ discrete grids, where P is the row number of azimuth resolution cells, and Q is the column number of range resolution cells. In $O_tX_tY_tZ_t$, the position vectors of the m-th grid is denoted as $\mathbf{r}_m$, $m = 1, 2, \cdots, M$.

Selection matrix of beam coverage is as below

$$\mathbf{D} = \begin{bmatrix} D_1(1) & D_1(2) & \cdots & D_1(M) \\ D_2(1) & D_2(2) & \cdots & D_2(M) \\ \vdots & \vdots & \vdots & \vdots \\ D_L(1) & D_L(2) & \cdots & D_L(M) \end{bmatrix}. \tag{12}$$

The element $D_l(m)$ indicates whether the m-th grid is illuminated by the l-th pulse beam:

$$D_l(m) = \begin{cases} 1 & if\,(\mathbf{r}_m \in S_l) \\ 0 & if\,(\mathbf{r}_m \notin S_l) \end{cases} \tag{13}$$

2.4. Imaging Equation

Since the whole imaging region S has been divided into $M = P \times Q$ discrete grids. The scattering coefficient of the m-th grid is $\sigma(\mathbf{r}_m)$. At the beginning of the l-th pulse, each transmitter simultaneously transmits independent and stochastic signal. All signals are superimposed in S to generate TSSRF. The radiation field at $\mathbf{r}_m$ can be expressed as

$$E^{inc}(t_l, \mathbf{r}_m) = \sum_{n=1}^{N} D_l(m) \frac{F_n(\hat{\mathbf{R}}_n) s_n(t_l - |\mathbf{r}_m - \mathbf{r}^n(t_{l,0})|/c)}{4\pi |\mathbf{r}_m - \mathbf{r}^n(t_{l,0})|}, \tag{14}$$

where $\hat{\mathbf{R}}_n = [\mathbf{r}_m - \mathbf{r}^n(t_{l,0})] / |\mathbf{r}_m - \mathbf{r}^n(t_{l,0})|$. $F_n(\hat{\mathbf{R}}_n)$ denotes the radiation pattern of the n-th transmitter antenna. $t_{l,0} = (l-1)T$ denotes the initial time of the l-th pulse.

The radiation field interacts with the targets and the received echo can be expressed as

$$E^{sca}(t_l) = \sum_{m=1}^{M} \sigma(\mathbf{r}_m) \frac{E^{inc}(t_l - \frac{|\mathbf{r}^s(t_l) - \mathbf{r}_m|}{c}, \mathbf{r}_m)}{4\pi |\mathbf{r}^s(t_l) - \mathbf{r}_m|} F_s(\hat{\mathbf{R}}_s) + n(t_l), \tag{15}$$

where $\hat{\mathbf{R}}_s = [\mathbf{r}_m - \mathbf{r}^s(t_l)] / |\mathbf{r}_m - \mathbf{r}^s(t_l)|$. $F_s(\hat{\mathbf{R}}_s)$ denotes the radiation pattern of the receiver antenna. $n(t_l)$ denotes the additive noise.

Considering the round-trip propagation of the electromagnetic field in the free space, the modified radiation field is defined as

$$E^{rad}(t_l, \mathbf{r}_m) = \sum_{n=1}^{N} \left\{ \frac{F_s(\hat{\mathbf{R}}_s) F_n(\hat{\mathbf{R}}_n) s_n[t_l - (|\mathbf{r}_m - \mathbf{r}^n(t_{l,0})| + |\mathbf{r}^s(t_l) - \mathbf{r}_m|)/c]}{(4\pi)^2 |\mathbf{r}_m - \mathbf{r}^n(t_{l,0})| |\mathbf{r}^s(t_l) - \mathbf{r}_m|} D_l(m) \right\}. \tag{16}$$

Let $t_{l,k}$, $l = 1, 2, \cdots, L$ be the sampling time in the l-th pulse, thus the imaging equation in the matrix vector form can be written as

$$\mathbf{E}^{sca} = \mathbf{E}^{rad} \cdot \boldsymbol{\sigma} + \mathbf{n}, \tag{17}$$

where $\mathbf{E}^{sca} = [E^{sca}(t_{1,k}), E^{sca}(t_{2,k}), \cdots, E^{sca}(t_{L,k})]^T$ is the echo vector, $\boldsymbol{\sigma} = [\sigma(\mathbf{r}_1), \sigma(\mathbf{r}_2), \cdots, \sigma(\mathbf{r}_M)]^T$ is the scattering coefficient vector, $\mathbf{n} = [n(t_{1,k}), n(t_{2,k}), \cdots n(t_{L,k})]^T$ is the noise vector, $\mathbf{E}^{rad}$ is the modified radiation field matrix with $[\mathbf{E}^{rad}]_{lm} = E^{rad}(t_{l,k}, \mathbf{r}_m)$.

The scattering coefficient vector $\boldsymbol{\sigma}$ can be reconstructed by the correlated processing between $\mathbf{E}^{sca}$ and $\mathbf{E}^{rad}$, which can be described as

$$\hat{\boldsymbol{\sigma}} = \zeta \left[\mathbf{E}^{rad}, \mathbf{E}^{sca} \right], \tag{18}$$

where ζ denote the correlated operator.

Common correlated imaging algorithms include Pseudo-Inverse algorithm, Tikhonov regularization, TV regularization, and sparse reconstruction algorithms, such as Orthogonal Matching Pursuit, sparse Bayesian learning, etc. This paper adopts Tikhonov regularization algorithm because it is robust to noise and does not require a priori of the target. Tikhonov regularization can be formulated as the following optimization problem

$$\hat{\boldsymbol{\sigma}} = \arg\min_{\boldsymbol{\sigma}} \left\{ \left\| \mathbf{E}^{sca} - \mathbf{E}^{rad} \cdot \boldsymbol{\sigma} \right\| + \lambda \|\boldsymbol{\sigma}\|_2^2 \right\}, \tag{19}$$

where λ is the regularization parameter.

3. Estimation of Translational Speed and Rotational Angular Velocity

POS system sets inertial navigation technology and satellite navigation technology in one body, and adopts the real-time and post-process information fusion respectively to get high precision positioning and orientation information. For MSCI, the imaging time is very short, so the error accumulation of the INS is negligible, and the INS is more accurate in a short time. Hence the measured position and angular data for estimation of translational speed and rotational angular velocity are very accurate.

The calculation of the real-time position vector of each antenna at high-speed sampling time requires the translational speed and the rotational angular velocity of the antenna array. The low-speed discrete POS data will be used to estimate the translational speed and the rotational angular velocity. Since the data rate of POS is usually less than the pulse repetition frequency as Figure 3 shows, there are many pulses between two adjacent POS data. To obtain the translational speed and the rotational angular velocity of the antenna array during each pulse, third-order polynomial curve fitting to the position and attitude is employed and the least squares method is used to obtain the coefficients of the fitting polynomial. The fitting polynomial of the position or the attitude can be expressed as

$$\mu(t) = \sum_{k=0}^{3} a_{\mu,k} t^k, \tag{20}$$

where $a_{\mu,k}$ is the coefficient of the polynomial, and $\mu(t)$ is the fitting curve.

The initial time of the l-th pulse is denoted as $t_{l,0}$, and the end time of the l-th pulse is denoted as $t_{l+1,0} = t_{l,0} + T_p$. By substituting $t_{l,0}$ and $t_{l+1,0}$ into the fitting curve $\mu(t)$, we can get the position or the attitude parameters at the beginning and the end of each pulse. Based on the assumption that the antenna array is uniformly translated and rotated during each pulse, the translational speed and rotational angular velocity in each pulse can be solved by

$$\boldsymbol{\omega}_l = \begin{bmatrix} \dfrac{\theta(t_{l+1,0}) - \theta(t_{l,0})}{T} \\ \dfrac{\varphi(t_{l+1,0}) - \varphi(t_{l,0})}{T} \\ \dfrac{\phi(t_{l+1,0}) - \phi(t_{l,0})}{T} \end{bmatrix}, \tag{21}$$

$$\mathbf{v}_l = \frac{\mathbf{r}^s(t_{l+1,0}) - \mathbf{r}^s(t_{l,0})}{T}. \tag{22}$$

4. Simulation

In this section, simulations are demonstrated to verify the proposed method based on unsteady aerostat platform. The scenario for the simulation is shown in Figure 1. An X-band MSCI radar system with carrier frequency of 10 GHz is considered. The randomly radiating radar array with 25 transmitters and 1 receiver is raised to 350 m height by a tethered aerostat. The main system simulation parameters are given in Table 1, and the target model is shown in Figure 4. In simulations, the measurement errors of position and altitude angles are assumed to be independent and subject to Gauss distribution with zero mean and 1 mm standard deviation for position and 0.05° standard deviation for altitude angles.

Table 1. Simulation parameter.

Simulation parameter	Value
Aperture of antenna array	1.5 m × 1.5 m
Number of transmitting antenna	25
Height of array center	350 m
Slanting angle	$\theta_0 = 45°$
The overall observation area	120 m × 120 m
Beam coverage area by single pulse	105 m × 105 m
Number of grid	40 × 40
Grid spacing	3 m
Signal form	Random frequency hopping
Pulse repetition interval	5 µm
Pulse width	600 ns
Bandwidth	500 MHz
Carrier frequency	10 GHz
Total imaging time	10 ms

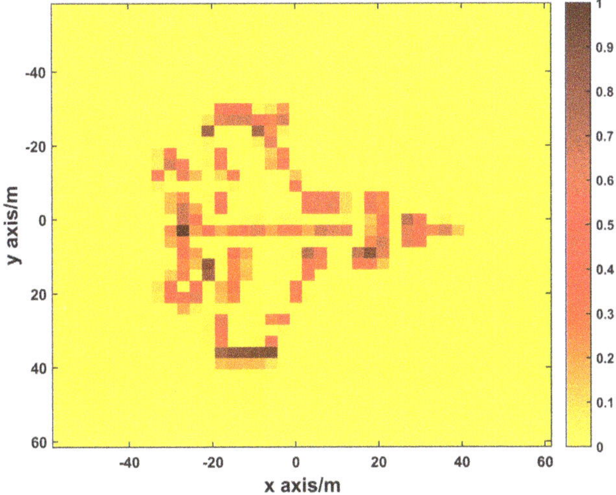

Figure 4. Target image.

To illustrate the effectiveness of the proposed method, the trajectories in Figure 5 are used as the three-dimensional translations and the rotational components of the antenna array caused by unsteady platform.

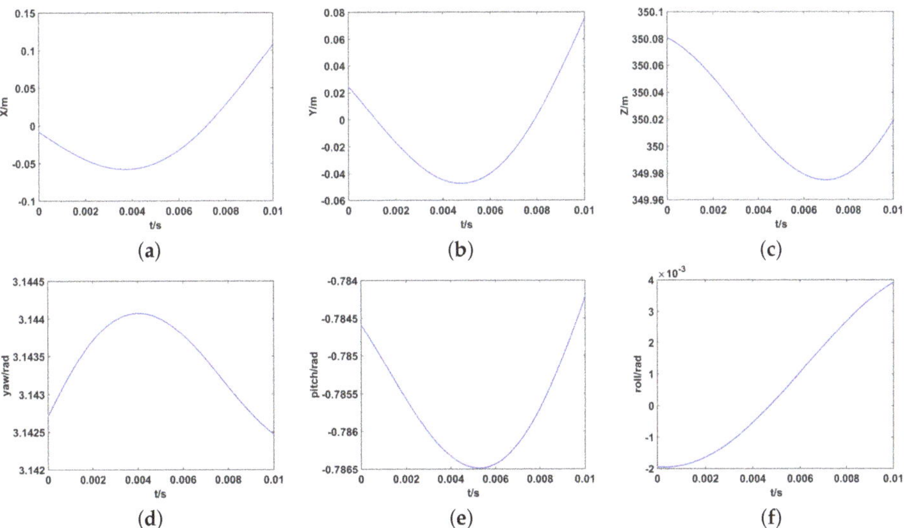

Figure 5. The motion trajectory of each component. (**a**) the translational component along the X_t; (**b**) the translational component along the Y_t; (**c**) the translational component along the Z_t; (**d**) the rotation of yaw; (**e**) the rotation of pitch; (**f**) the rotation of roll.

4.1. Verification of the Proposed Model

In this subsection, simulations are taken to compare the imaging performance of different imaging models. The proposed imaging model based on unsteady aerostat platform (UPIM) will be compared with the imaging model for stationary platform (SPIM) and imaging model which only uses the discrete POS data (DPDIM). SPIM ignores the motion of antenna array and assumes that the position vector of antenna and the beam coverage do not change during the imaging process. When calculating the radiation field, SPIM uses the first recorded POS data as the position and the attitude angles of the array, i.e., $\mathbf{r}^n(t_l) = \mathbf{r}^n(t_{1,pos})$ $(l = 1, 2, \cdots, L)$. DPDIM does not fit the discrete POS data and uses the closest POS data for each pulse. The normalized mean square error (NMSE) is used to quantity the reconstruction performance, with the definition as: $NMSE = \|\hat{\mathbf{x}} - \mathbf{x}\|_2 / \|\mathbf{x}\|_2$, where $\hat{\mathbf{x}}$ and $\mathbf{x}$ denote the reconstructed and true value of target.

The imaging results are depicted in Figure 6. As shown in Figure 6a, Comparably, the image reconstructed by UPIM is focused with quite a few spurious scatters, whose better imaging performance benefits from the fact that the UPIM has the minimal motion estimation error. In Figure 6b, apart from strong scatters, the image reconstructed by DPDIM has many spurious scatters. In Figure 6c, the reconstructed image by SPIM is defocused and blurry, and the target is hard to recognize.

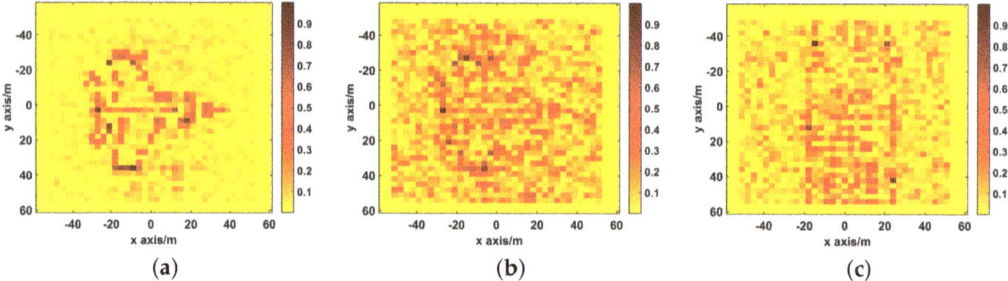

Figure 6. The imaging results of UPIM, DPDIM, and SPIM. (**a**) the reconstructed image by UPIM, the NMSE is 0.28; (**b**) the reconstructed image by DPDIM, the NMSE is 0.94; (**c**) the reconstructed image by SPIM, the NMSE is 1.21.

The point spread functions (PSF) of UPIM, DPIM, and SPIM are illustrated in Figure 7 and the X-axis and Y-axis profiles of the PSF are shown in Figure 8. It can be seen from Figure 7, the number and the level of the side lobes is minimum for UPIM while the other two methods both have more side lobes and higher level of side lobes. Figure 8 shows that these methods have almost the same width of the main lobe in X-axis profile and the Y-axis profile. The above simulation results demonstrate that UPIM indeed reduces the number and the level of side lobes caused by platform instability, but it does not improve the imaging resolution of MSCI.

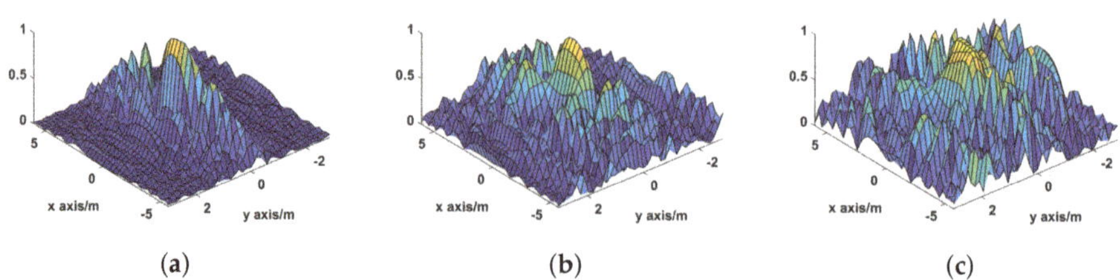

Figure 7. The point spread function of UPIM, DPIM, and SPIM. (**a**) UPIM; (**b**) DPIM; (**c**) SPIM.

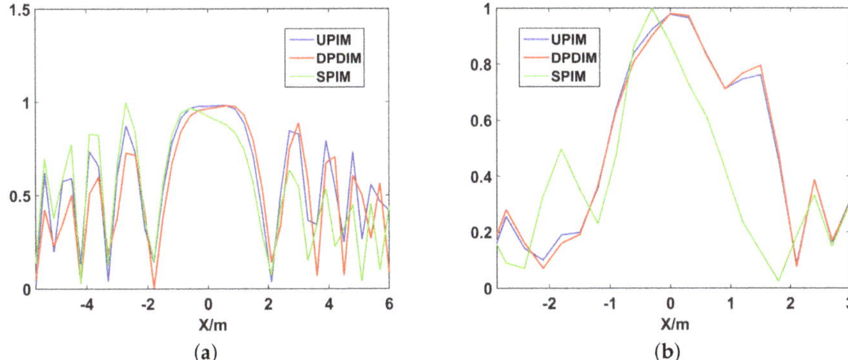

Figure 8. The profile of the point spread function of UPIM, DPIM, and SPIM. (**a**) the X-axis profile of the point spread function; (**b**) the Y-axis profile of the point spread function.

For the three imaging models mentioned above, Figure 9 shows the fitting effect on the translational motion trajectory of the aerostat platform along the X_t. It can be seen that the proposed UPIM model has the best fitting effect that the estimated translational trajectory is almost the same with the real one with the most minimum motion estimation error, which obviously benefits a better imaging performance in UPIM.

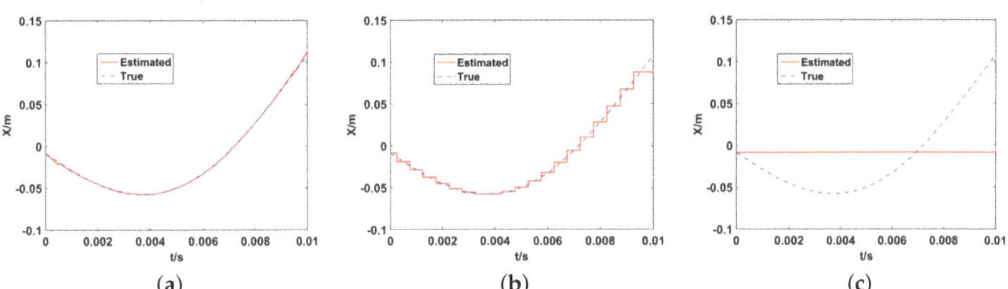

Figure 9. The estimated and the real trajectory of the translation along the X_t. (**a**) UPIM; (**b**) DPDIM; (**c**) SPIM.

To verify the effectiveness of the proposed method with the translation amplitude increasing, the relationship between the imaging quality and the translation amplitude for three imaging models is presented in Figure 10. The amplitude of three-dimensional translations gradually increased by the step of 0.5 times the original amplitude shown in Figure 5, while the rotation amplitudes keep constant. The coordinate of the horizontal axis in Figure 10 represents the multiple of the original translation amplitude. As seen from Figure 10, the imaging performance of UPIM is still better than the other imaging models when the amplitude of translation increases.

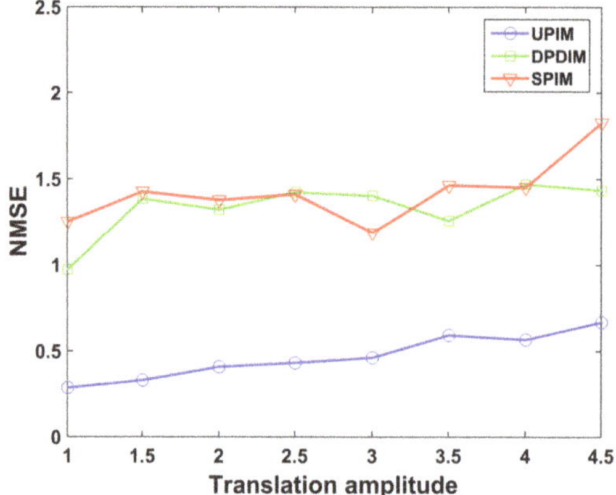

Figure 10. NMSE of the imaging results by UPIM, DPDIM, and SPIM at different amplitudes of translation.

The imaging quality under different rotation amplitudes is depicted in Figure 11. The amplitude of all rotational components is gradually increased by step of 0.5 times the original rotational amplitude, while the translation amplitudes keep constant. Figure 11 shows that the proposed method has better performance under all rotation amplitudes.

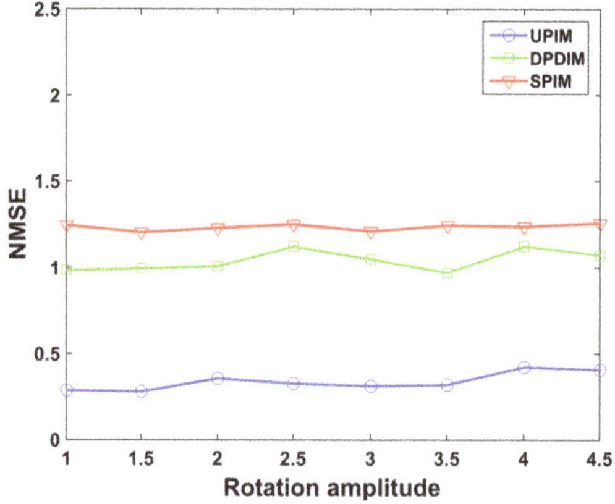

Figure 11. NMSE of the imaging results by UPIM, DPDIM, and SPIM at different amplitudes of rotation.

4.2. Effect of Different Translational Components on Imaging Performance

This section is to study the effect of independent translational component on imaging performance. In simulations, all independent translational components use the same motion trajectory as shown in Figure 5a. Figure 12 shows the imaging results reconstructed by UPIM when only one translational component exists.

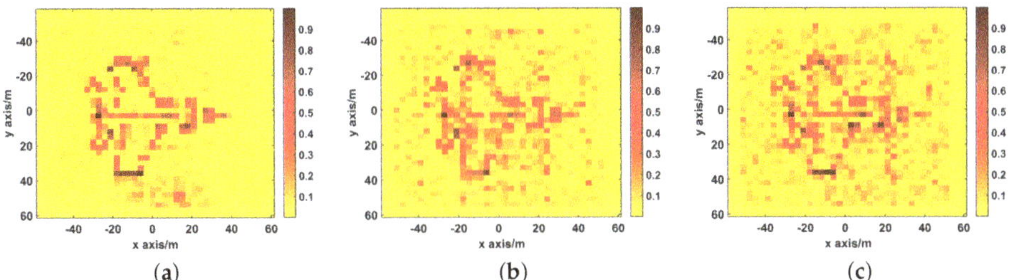

Figure 12. The imaging results of UPIM when only one translation component exists. (**a**) Only translational component along the X_t exists, the NMSE is 0.21; (**b**) Only translational component along the Y_t exists, the NMSE is 0.58; (**c**) Only translational component along the Z_t exists, the NMSE is 0.59.

The imaging quality for each translational component under different translation amplitudes is presented in Figure 13. As shown in Figures 12 and 13, the translation along the X_t has the minimal influence on imaging performance, while both translation along the Y_t and Z_t has almost the same influence on imaging performance. Therefore, for improving the image performance, the position of the antenna array along the Y_t and the Z_t should be estimated more accurate.

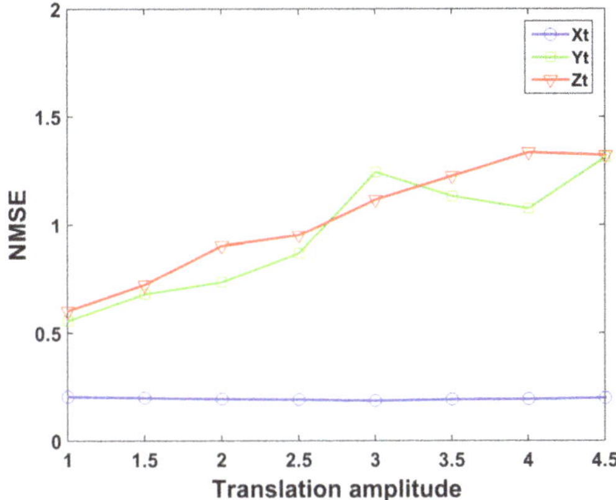

Figure 13. NMSE of the imaging results when only one translational component exists at different amplitudes.

Next, we investigate the reason for the different imaging results in three-dimensional translations. Although the proposed method compensates in part of the non-cooperative motion, due to the limited number of POS data, the estimated translation errors cannot be totally eliminated. Figure 14 shows the estimation error in three-dimensional translations. The accurate calculation of the radiation field is directly related to the round-trip propagation delay of the electromagnetic wave between antenna and target. The propagation time delay error caused by estimation error of array position will lead to the calculated radiation field error. Because the translational estimation errors in three dimensions have different effect on the propagation time delay, the influence of different translational components on imaging is not the same.

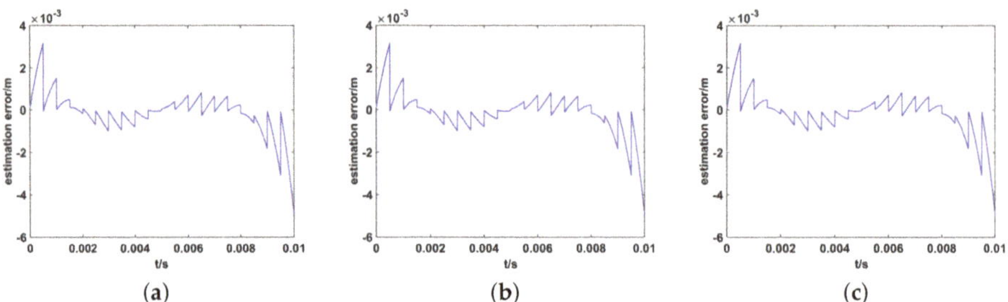

Figure 14. The estimation error of antenna position in three-dimensional translations. (**a**) Only along the X_t; (**b**) Only along the Y_t; (**c**) Only along the Z_t.

Assuming at t time, the coordinate of the i-th antenna is (x_a, y_a, z_a) and the coordinate of any point m in the imaging region is (x, y, z). After Δt, the coordinate of the antenna become $(x_a + \Delta x_a, y_a + \Delta y_a, z_a + \Delta z_a)$.

The distance from the i-th antenna to the point m in imaging region is

$$S_{im} = \sqrt{(x - x_a)^2 + (y - y_a)^2 + (z - z_a)^2}. \tag{23}$$

The partial differential of the propagation path along three coordinate dimensions is

$$\frac{\partial S_{im}}{\partial x_a} = \frac{x_a - x}{\sqrt{(x - x_a)^2 + (y - y_a)^2 + (z - z_a)^2}}, \tag{24}$$

$$\frac{\partial S_{im}}{\partial y_a} = \frac{y_a - y}{\sqrt{(x - x_a)^2 + (y - y_a)^2 + (z - z_a)^2}}, \tag{25}$$

$$\frac{\partial S_{im}}{\partial z_a} = \frac{z_a - z}{\sqrt{(x - x_a)^2 + (y - y_a)^2 + (z - z_a)^2}}. \tag{26}$$

In the simulation scenario, the height of the antenna array is 350 m and the antenna is squint observation with the slanting angle 45°. For any point in the imaging area, its coordinate satisfies $291.5 \le y \le 408.5$, $-58.5 \le x \le 58.5$ and $z = 0$. Because the size of antenna array is much smaller than the size of imaging region, therefore for most points in the imaging region, it is satisfied that $|x| \gg |x_a|$ and $|y| \gg |y_a|$, and the value of partial differential function satisfy that $|\partial S_{im}/\partial y_a| > |\partial S_{im}/\partial x_a|$. As $z_a - z \approx 350$, the partial differential of S_{im} satisfies $|\partial S_{im}/\partial z_a| > |\partial S_{im}/\partial x_a|$. Therefore the same estimation error along the X_t axis will cause less propagation delay error than the other two components, which explains the reason that under the same translational trajectory, the reconstructed image with the translation only along the X_t has the best imaging result.

4.3. Effect of Different Rotation Components on Imaging Performance

This section is to study the effect of independent rotational component on imaging performance. In the simulation, three rotational components have the same rotational trajectory as shown in Figure 5d. Figure 15 shows the imaging results reconstructed by UPIM when only one rotational component exists. As three rotational components gradually increased by the step of 0.5 times the original rotational amplitude shown in Figure 5, the imaging quality for each rotational component under different rotation amplitudes is presented in Figure 16.

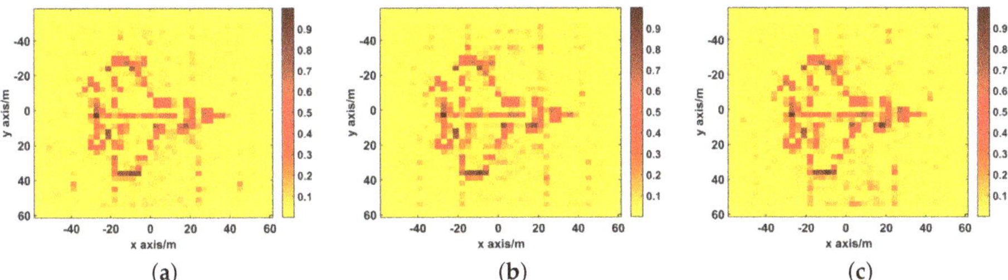

Figure 15. The imaging results of UPIM when only one rotation component exists. (**a**) Only yaw angle, the NMSE is 0.27; (**b**) Only pitch angle, the NMSE is 0.28; (**c**) Only roll angle, the NMSE is 0.28.

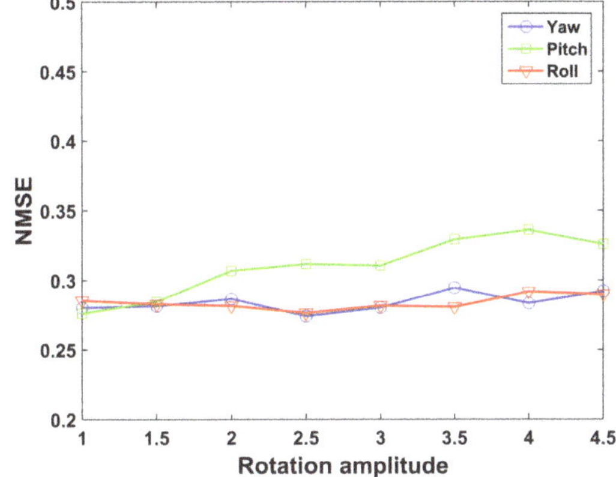

Figure 16. NMSE of the imaging results when only one rotation component exists at different amplitudes.

From Figures 15 and 16, it can be seen that three rotational components have almost the same effect on imaging performance under different rotation amplitudes.

4.4. Effect of the Position and Angular-Measuring Accuracy on Imaging Performance

This section is to study the effect of the measuring accuracy of position and attitude parameters on imaging. The imaging performance under different position accuracy and angular accuracy is simulated. In simulations, the measurement error is assumed to be independent and subject to Gauss distribution with zero mean and different variances. The smaller the variance, the higher the measuring accuracy. Figure 17 shows the imaging quality under different position-measuring accuracy and different angular-measuring accuracy, respectively. The results show that the imaging performance is very sensitive to the measuring accuracy of position and attitude parameters which means that the proposed method has a high demand of accurate measurement of position and attitude parameters.

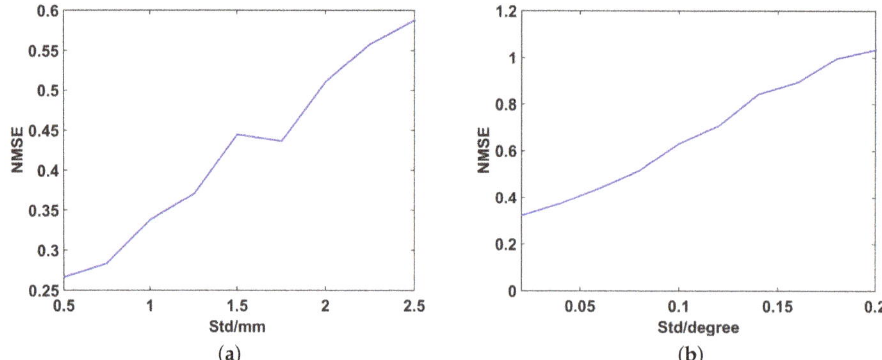

Figure 17. The NMSE of imaging result under different measuring accuracy. (**a**) position-measuring accuracy; (**b**) angular-measuring accuracy.

5. Conclusions

In this paper, a novel MSCI method based on unsteady aerostat platform is proposed, where the MSCI radar array is carried by AVS to keep its antenna beam orientation to the target in the non-cooperative motion of the platform caused by the wind etc., and the POS is used to dynamically measure the position and the attitude of the antenna array. By decomposing of the platform motion to its translation and rotation, the motion model of unsteady aerostat platform in air has been built, and for each antenna, its real-time position vector can be calculated by its translational speed and its rotational angular velocity in each pulse, replacing the static position vector in the traditional MSCI model. For the dynamic beam coverage in the whole observation region, a selection matrix of beam coverage is introduced to indicate the illuminated area at each pulse. By analyzing the modified stochastic radiation field and its scattered echo, the MSCI model based on unsteady aerostat platform is established. Furtherly, based on low-speed POS data, a polynomial curve-fitting algorithm is used to eliminate the position error of the radar array. Simulation experiments demonstrate that under its different random translations and rotations of unsteady aerostat platform, the position and attitude of the antenna array at different time can be estimated well, and better imaging performance can be achieved by the proposed scheme, which provides a feasible technical approach for the floating-observation-platform to realize the microwave staring remote sensing observation in the near space.

Author Contributions: All authors contributed extensively to the work presented in this paper. Y.G. proposed the original idea. Z.J. and J.D. designed the study, performed the simulations and wrote the paper; Y.G., W.C. supervised the analysis, edited the manuscript and D.W. provided their valuable suggestions to improve this study.

Funding: This work has been supported by the National Natural Science Foundation of China under contact No. 61771446 and No. 61431016.

Conflicts of Interest: The authors declare no conflict of interest.

References

1. Ausherman, D.A.; Kozma, A.; Walker, J.L.; Jones, H.M.; Poggio, E.C. Developments in radar imaging. *IEEE Trans. Aerosp. Electron. Syst.* **1984**, *AES-20*, 363–400. [CrossRef]
2. Dunkel, R.; Saddler, R.; Doerry, A. Synthetic aperture radar for disaster monitoring. *Radar Sens. Technol. XV* **2011**, *8021*, 80211R.
3. Brown, W.M.; Porcello, L.J. An introduction to synthetic-aperture radar. *IEEE Spectr.* **1969**, *6*, 52–62. [CrossRef]
4. He, X.; Liu, B.; Chai, S.; Wang, D. A novel approach of high spatial-resolution microwave staring imaging. In Proceedings of the Asia-Pacific Conference on Synthetic Aperture Radar (APSAR), Tsukuba, Japan, 23–27 September 2013; pp. 75–78.

5. Guo, Y.; Wang, D.; He, X.; Liu, B. Super-resolution staring imaging radar based on stochastic radiation fields. In Proceedings of the IEEE MTT-S International Microwave Workshop Series on Millimeter Wave Wireless Technology and Applications, Nanjing, China, 18–20 September 2012; pp. 1–4.
6. Li, D.; Li, X.; Qin, Y.; Cheng, Y.; Wang, H. Radar coincidence imaging: An instantaneous imaging technique with stochastic signals. *IEEE Trans. Geosci. Remote Sens.* **2014**, *52*, 2261–2277.
7. Guo, Y.; He, X.; Wang, D. A novel super-resolution imaging method based on stochastic radiation radar array. *Meas. Sci. Technol.* **2013**, *24*, 074013. [CrossRef]
8. Zhu, S.; Zhang, A.; Xu, Z.; Dong, X. Radar coincidence imaging with random microwave source. *IEEE Antennas Wirel. Propag. Lett.* **2015**, *14*, 1239–1242. [CrossRef]
9. Liu, B.; Wang, D. Orthogonal radiation field construction for microwave staring correlated imaging. *Prog. Electromagnet. Res.* **2017**, *57*, 139–149. [CrossRef]
10. Meng, Q.; Qian, T.; Yuan, B.; Wang, D. Random Radiation Source Optimization Method for Microwave Staring Correlated Imaging Based on Temporal-Spatial Relative Distribution Entropy. *Prog. Electromagn. Res.* **2018**, *63*, 195–206. [CrossRef]
11. Yuan, B.; Guo, Y.; Chen, W.; Wang, D. A Novel Microwave Staring Correlated Radar Imaging Method Based on Bi-Static Radar System. *Sensors* **2019**, *19*, 879. [CrossRef] [PubMed]
12. Zhou, X.; Wang, H.; Cheng, Y.; Qin, Y. Radar coincidence imaging by exploiting the continuity of extended target. *IET Radar Sonar Navig.* **2016**, *11*, 60–69. [CrossRef]
13. Meng, Q.; Liu, B.; Tian, C.; Guo, Y.; Wang, D. Correlation algorithm of Microwave Staring correlated imaging based on multigrid and CGLS. In Proceedings of the IEEE International Conference on Communication Problem-Solving (ICCP), Guilin, China, 16–18 October 2015; pp. 359–362.
14. Wang, G.; Lu, G.; Qian, T. A novel correlation method of microwave staring correlated imaging based on multigrid and LSQR. In Proceedings of the 10th International Conference on Digital Image Processing (ICDIP), Shanghai, China, 11–14 May 2018; p. 108064W.
15. Cheng, Y.; Zhou, X.; Xu, X.; Qin, Y.; Wang, H. Radar coincidence imaging with stochastic frequency modulated array. *IEEE J. Sel. Top. Sign. Proces.* **2017**, *11*, 414–427. [CrossRef]
16. Doyle, G.R., Jr.; Vorachek, J.J. *Investigation of Stability Characteristics of Tethered Balloon Systems*; Technical Report for Goodyear Aerospace Corp.: Akron, OH, USA, 1971.
17. Vierling, L.A.; Fersdahl, M.; Chen, X.; Li, Z.; Zimmerman, P. The Short Wave Aerostat-Mounted Imager (SWAMI): A novel platform for acquiring remotely sensed data from a tethered balloon. *Remote Sens. Environ.* **2006**, *103*, 255–264. [CrossRef]
18. Zhou, X.; Wang, H.; Cheng, Y.; Qin, Y. Sparse auto-calibration for radar coincidence imaging with gain-phase errors. *Sensors* **2015**, *15*, 27611–27624. [CrossRef] [PubMed]
19. Zhou, X.; Wang, H.; Cheng, Y.; Qin, Y. Radar coincidence imaging with phase error using Bayesian hierarchical prior modeling. *J. Electron. Imaging* **2016**, *25*, 013018. [CrossRef]
20. Tian, C.; Yuan, B.; Wang, D. Calibration of gain-phase and synchronization errors for microwave staring correlated imaging with frequency-hopping waveforms. In Proceedings of the IEEE Radar Conference. Oklahoma City, OK, USA, 23–27 April 2018; pp. 1328–1333.
21. Xia, R.; Guo, Y.; Chen, W.; Wang, D. Strip-Mode Microwave Staring Correlated Imaging with Self-Calibration of Gain—Phase Errors. *Sensors* **2019**, *19*, 1079. [CrossRef] [PubMed]
22. Zhou, X.; Wang, H.; Cheng, Y.; Qin, Y. Off-grid radar coincidence imaging based on variational sparse Bayesian learning. *Math. Prob. Eng.* **2016**, *2016*, 1782178. [CrossRef]
23. Zhou, X.; Fan, B.; Wang, H.; Cheng, Y.; Qin, Y. Sparse Bayesian Perspective for Radar Coincidence Imaging With Array Position Error. *IEEE Sens. J.* **2017**, *17*, 5209–5219. [CrossRef]
24. Li, D.; Li, X.; Cheng, Y.; Qin, Y.; Wang, H. Radar coincidence imaging in the presence of target-motion-induced error. *J. Electron. Imaging* **2014**, *23*, 023014. [CrossRef]

Article

Strip-Mode Microwave Staring Correlated Imaging with Self-Calibration of Gain–Phase Errors

Rui Xia , Yuanyue Guo *, Weidong Chen and Dongjin Wang

Key Laboratory of Electromagnetic Space Information, Chinese Academy of Sciences, University of Science and Technology of China, Hefei 230026, China; xrke928@mail.ustc.edu.cn (R.X.); wdchen@ustc.edu.cn (W.C.); wangdj@ustc.edu.cn (D.W.)

* Correspondence: yuanyueg@ustc.edu.cn

Received: 8 January 2019; Accepted: 25 February 2019; Published: 3 March 2019

Abstract: Microwave staring correlated imaging (MSCI) can realize super resolution imaging without the limit of relative motion with the target. However, gain–phase errors generally exist in the multi-transmitter array, which results in imaging model mismatch and degrades the imaging performance considerably. In order to solve the problem of MSCI with gain–phase error in a large scene, a method of MSCI with strip-mode self-calibration of gain–phase errors is proposed. The method divides the whole imaging scene into multiple imaging strips, then the strip target scattering coefficient and the gain–phase errors are combined into a multi-parameter optimization problem that can be solved by alternate iteration, and the error estimation results of the previous strip can be carried into the next strip as the initial value. All strips are processed in multiple rounds, and the gain–phase error estimation results of the last strip can be taken as the initial value and substituted into the first strip for the correlated processing of the next round. Finally, the whole imaging in a large scene can be achieved by multi-strip image splicing. Numerical simulations validate its potential advantages to shorten the imaging time dramatically and improve the imaging and gain–phase error estimation performance.

Keywords: microwave staring correlated imaging (MSCI); gain–phase errors; strip; self-calibration

---

1. Introduction

Radar imaging technology [1,2] has enabled radars to have the ability to obtain a panoramic image of an observation scene, which has been widely used in military warning, disaster detection and other fields. In these application scenarios, long-term continuous monitoring of large areas is an important application requirement.

Synthetic aperture radar (SAR) has high azimuth resolution imaging ability by forming large virtual synthetic aperture through relative motion between the target and radar, but its long revisiting period means that it cannot be applied to the staring imaging [3,4].

Traditional real aperture microwave staring imaging has the characteristics of high real-time, but limited by the actual aperture of the antenna; its azimuth resolution is low, so it is difficult to achieve high resolution imaging. As a new staring imaging method, microwave staring correlated imaging (MSCI) [5–7] can realize super resolution imaging without the limit of the target relative motion. The essence of MSCI is to construct a temporal–spatial stochastic radiation field in the imaging region, which is typically realized by a multi-transmitter array transmitting independent stochastic waveforms [5,6] such as the signals with random amplitude and frequency between different pulses. The radiation field interacts with the target so that the target scattering points at different locations scatter the independent time-varying echoes. Finally, the target information can be obtained by the correlated imaging process between the echoes and the preset radiation field. In [7], two point targets in a small scene are imaged by outfield experiments based on MSCI. Accurate imaging is based on the premise of accurate preset radiation field. However, gain–phase errors generally exist in the multi-transmitter array, so there is a deviation between the actual radiation field and the preset radiation field that is calculated based on transmitted waveform, which results in the imaging model mismatch and degrades the imaging performance considerably. In [8,9], the methods are propose for model mismatch in radar coincidence imaging (RCI), but the gain–phase error model was not analysed.

The studies on calibration of gain–phase error mainly focus on a radar system with multiple transmissions or multiple receptions, including the field of angle estimation of array signals [10,11] and radar imaging [12–17]. In [10], a method based on eigenstructure is proposed for simultaneously estimating the direction of arrival(DOA) and the unknown (or imprecisely known) gain and phase parameters, which applies to arrays with arbitrary sensor geometries. The method is based on the eigendecomposition of the sample covariance matrix of the vector of received signals. In [11], an estimation of signal parameters via a rotational invariance techniques (ESPRIT)-based method is proposed to estimate the gain–phase errors of both transmission and reception arrays and signal angles in bistatic MIMO radars, in which both transmitter and receiver are equipped with uniform linear array, and the first two sensors of transmit array and receive array are well calibrated to obtain a reference channel. In the field of angle estimation of array signal, the method of gain–phase error calibration is generally to ensure the consistency of the gain–phase characteristics of each channel; in contrast, there is no requirement of uniform gain–phase characteristics between the multi-transmitter array channels in an MSCI system.

In the SAR imaging field, a subspace algorithm of calibrating channel gain–phase errors for high-resolution and wide-swath (HRWS) SAR imaging is presented [12]. The proposed method is based on the fact that the signal subspace obtained from the eigendecomposition of covariance matrix equals the space spanned by the practical steering vectors. Channel gain–phase errors can be obtained through eigendecomposition of a special matrix which is the calculation result of the nominal steering vectors and the signal eigenvectors of the covariance matrix.

All the above methods on gain–phase errors calibration make use of the characteristics of eigen-subspace and estimate the gain and phase errors by matrix eigendecomposition. The basic feature of these methods is that the signals of multiple transmitting–receiving channels are separated during processing, but the received echoes are not separated by multiple channels in MSCI, so the channel gain–phase error calibration method based on subspace decomposition cannot be directly adopted in MSCI.

Without subspace decomposition, a method is proposed for joint SAR imaging and phase error correction in [13]. The problem is set up as an optimization problem in a non-quadratic regularization-based framework, and phase error correction is performed during the image formation process. The method involves an iterative algorithm, where each iteration includes consecutive steps of image formation and model error correction. A method for RCI with phase errors is proposed in [14], which adopts the sparse Bayes learning (SBL) framework and jointly estimates target scattering coefficients and phase error during the iterative steps. Soon after, in [15], a method is proposed for sparse auto-calibration for RCI with gain–phase errors(SACRCI), which transforms the imaging into the parameter estimation problem, and then estimates target scattering coefficient and gain–phase errors jointly. In [16], an auto-calibration expansion–compression variance-component (AC-ExCoV)-based auto-focusing method in a sparse Bayesian learning framework is proposed. These methods all take the gain–phase errors as unknown parameters and adopt an iterative procedure to jointly estimates target scattering coefficients and gain–phase errors. The targets in [15,16] are all sparse in small scenes. In other respects, the calibration of the gain–phase and synchronization errors is focused on for MSCI in [17], but a reference receiver is required to receive the direct signals of the transmitters to estimate the errors.

Considering large imaging scenes in MSCI, which means a large number of grid cells, results in very large computational complexity, limit the application for the above methods in large scenes. In [18,19], the problem of MSCI in a large scene is solved by dividing the large scene into strips. In [18], the echoes of the discrete clustered targets are detected to locate the strips with targets and only the regions of interest are discretized to a fine grid.

In order to solve the problem of MSCI with gain–phase error in a large scene, a method of MSCI based on strip-mode self-calibration of gain–phase errors is proposed in this paper. By dividing the target scene into strips, for each strip, the scattering coefficient and the gain–phase errors are combined into a multi-parameter optimization problem, which can be estimated by alternate iteration. Simultaneously, the gain–phase error estimation results of the previous strip can be carried into the next strip as the initial value. All strip imaging results, which can be obtained by correlated processing in turn, are spliced to obtain the image inversion results of the whole scene. To further improve gain–phase error estimation and imaging performance, after all the strips are processed in one round, the gain–phase error estimation results of the last strip can be taken as the initial value and substituted into the first strip for the correlated processing of the next round. In this way, all strips are processed in multiple rounds to obtain the final results.

The rest of the report is organized as follows. In Section 2, the strip-mode MSCI model with gain–phase errors is presented. Section 3 presents strip-mode MSCI algorithm with self-calibration of gain–phase errors. The analysis of the computation of the algorithm is discussed in Section 4. In Section 5, the performance of the proposed method is verified by numerical examples. Finally, Section 6 concludes this paper.

2. Strip-Mode MSCI with Gain–Phase Errors

As shown in Figure 1, a rectangular coordinate system is established with the center of the transmitting array as the origin; the MSCI system located in stationary platforms is composed of N transmitters and one receiver, whose position vectors are denoted as $\vec{r}_n$ and $\vec{r}_s$. The height of the transmitting array is H and θ is the squint angle. The independent narrow-pulse signals of random frequency hopping (RHF) which are transmitted synchronously by each antenna in multi-transmitter array can be expressed as:

$$f_{\mathrm{n}}(t) = \sum_{\mathrm{l}=1}^{L} rect\left[\frac{t-(l-1)\,T_p}{\tau}\right] a_n \mathrm{A}_n e^{j\varphi_n} \exp\{j2\pi f_{nl}\left[t-(l-1)\,T_p\right]\} \tag{1}$$

where f_{nl} is its frequency of the l-th, $l = 1,2,\cdots,L$ pulse emitted by the n-th, $n = 1,2,\cdots,N$ transmitter, and randomly selected within the bandwidth B, and τ and T_p is its narrow pulse width and period. $a_n \mathrm{A}_n$ is the gain of the n-th transmitter, and a_n denotes the gain error coefficient of the n-th transmitter, which equals 1 when there is no gain error, φ_n denotes the phase error of the n-th transmitter, which equals 0 when there is no phase error. For simplicity, in the case that the bandwidth is narrow compared with the central frequency, we consider that the gain–phase errors are fixed in the imaging process.

According to the feature of radar range-gate, the random narrow pulse signals transmitted simultaneously by the multi-transmitter array can divide two-dimensional imaging area S into multiple different strips $S_k, k = 1\cdots K$ in the range direction [19]. The imaging strip S_k has been divided into discrete $J = P \times Q$ grids, where P is the row number of azimuth resolution cells, and Q is the column number of range resolution cells, and position vectors of the center of j-th grid is denoted as $\vec{r}_{k,j}$, whose scattering coefficient is $\sigma\left(\vec{r_{k,j}}\right), j = 1,2,\cdots,J$. According to electromagnetic field propagation in free space, stochastic radiated fields at $\vec{r}_{k,j}$ in the $k-$th strip can be expressed as:

$$E_k{}^{in}(t,\vec{r}_{k,j}) = \sum_{n=1}^{N} \frac{f_n\left(t - \left(|\vec{r}_{k,j} - \vec{r}_n|\right)/c\right)}{4\pi|\vec{r}_{k,j} - \vec{r}_n|} \tag{2}$$

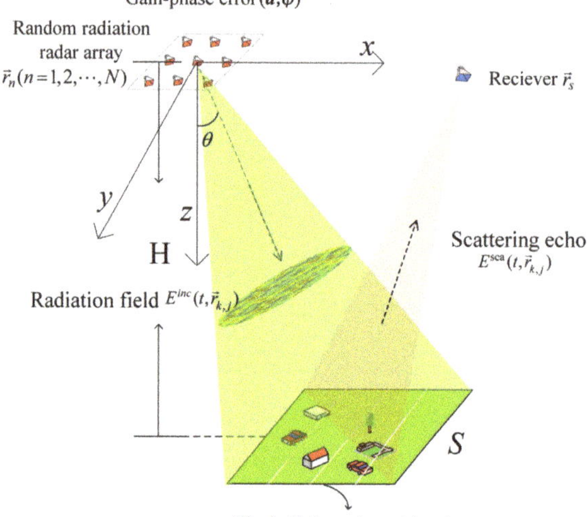

Figure 1. Geometry of MSCI.

The radiation field interacts with the k-th strip targets and the received signal of the k-th strip is:

$$\begin{aligned} E_k^{sca}(t,\vec{r}_{k,j}) &= \sum_{j=1}^{J}\sum_{n=1}^{N} \frac{f_n\left(t - \left(|\vec{r}_{k,j} - \vec{r}_n| + |\vec{r}_s - \vec{r}_n|\right)/c\right)}{(4\pi)^2|\vec{r}_{k,j} - \vec{r}_n||\vec{r}_s - \vec{r}_n|}\sigma(\vec{r}_{k,j}) \\ &= \sum_{j=1}^{J} E_k^{rad}(t,\vec{r}_{k,j})\sigma(\vec{r}_{k,j}) \end{aligned} \tag{3}$$

Define the modified radiation filed of S_k by considering the round-trip time of transmission after target reflection, which can be denoted as:

$$E_k^{rad}\left(t,\vec{r}_{k,j}\right) = \sum_{i=1}^{N} \frac{f_n\left(t - \left(\left|\vec{r}_{k,j} - \vec{r}_n\right| + \left|\vec{r}_s - \vec{r}_{k,j}\right|\right)/c\right)}{(4\pi)^2\left|\vec{r}_{k,j} - \vec{r}_n\right|\left|\vec{r}_s - \vec{r}_{k,j}\right|} \tag{4}$$

The scattered echoes in strip-mode can be written as matrix vector:

$$\mathbf{E}_k^{sca} = \mathbf{E}_k^{rad} \cdot \sigma_k \tag{5}$$

Because of the unknown gain–phase errors, the equation can be rewritten as:

$$\mathbf{E}_k^{sca} = \mathbf{E}_k^{rad}(\boldsymbol{a}, \boldsymbol{\varphi}) \cdot \sigma_k \tag{6}$$

where $\boldsymbol{a} = [a_1, a_2, \cdots, a_N]^T$ is the vector of the gain errors coefficient of the multi-transmitter array, and $\boldsymbol{\varphi} = [\varphi_1, \varphi_2, \cdots, \varphi_N]^T$ is the vector of the phase errors.

Strip-mode MSCI can obtain the target information $\hat{\sigma}_k$ in S_k by the correlated processing between $\mathbf{E}_k^{sca}$ and $\mathbf{E}_k^{rad}$, which can be described as:

$$\hat{\sigma}_k = \wp\left[\mathbf{E}_k^{rad}, \mathbf{E}_k^{sca}\right] \tag{7}$$

where $\wp$ indicates the first-order correlated operator. Common correlated imaging algorithms include LS algorithm, TSVD regularization, Tikhonov regularization, TV regularization, sparse Bayesian learning, and etc.

Each strip is processed in turn to obtain all the MSCI results, and then all the strip images are spliced to obtain the whole scene imaging results. Since the reconstruction result is a one-dimensional vector deformed by the two-dimensional mesh of the strip, all the reconstruction results $\hat{\sigma}_k$ need to be converted into the corresponding two-dimensional form $\hat{\sigma}_k'$. The imaging result of the whole scene can be expressed as:

$$\hat{\sigma}' = \left[\hat{\sigma}_1', \hat{\sigma}_2', \cdots, \hat{\sigma}_K'\right] \tag{8}$$

The whole imaging process mainly includes: transmitting signal, interaction between radiation field and target to form scattering echo, receiving echo, dividing strip, MSCI with self-calibration of gain–phase errors of each strip, and obtaining image results of whole scene by splicing all strips' imaging results. The flow chart of the whole imaging process is as Figure 2.

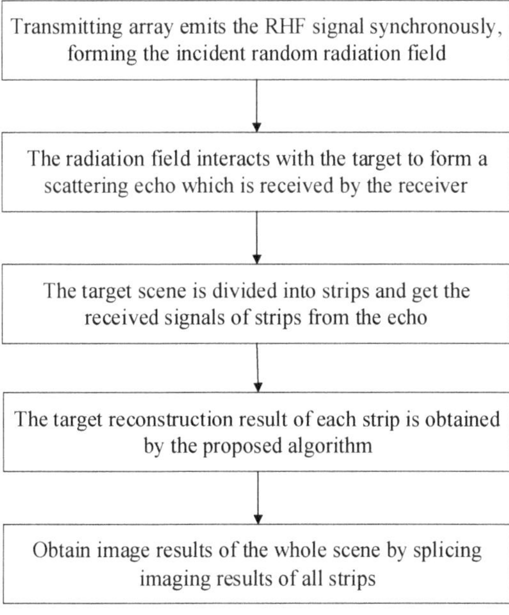

Figure 2. Imaging process flow chart.

3. Strip-Mode MSCI Algorithm with Self-Calibration of Gain–Phase Errors

According to strip-mode MSCI method, the echo corresponding to each strip can be obtained from the received echo according to the distance gate. Therefore, the correlated imaging with gain–phase errors can be carried out separately for each strip. The modified radiation filed is unknown due to the gain–phase errors. The gain–phase error estimation and target reconstruction can be combined as an optimization problem, the cost function can be expressed as:

$$F(\sigma_k, \boldsymbol{a}, \boldsymbol{\varphi}) = ||\boldsymbol{E}_k^{sca} - \boldsymbol{E}_k^{rad}(\boldsymbol{a}, \boldsymbol{\varphi})||_2^2 + \lambda||\sigma_k||_1 \tag{9}$$

where λ is the regularization parameter.

Then the $k-th$ strip gain–phase errors calibration and target reconstruction can be converted into the following optimization problem:

$$[\sigma_k, \boldsymbol{a}, \boldsymbol{\varphi}] = \underset{\sigma_k, \boldsymbol{a}, \boldsymbol{\varphi}}{\arg\min} F(\sigma_k, \boldsymbol{a}, \boldsymbol{\varphi}) \tag{10}$$

In order to solve the above problems, a strip-mode MSCI algorithm based on self-calibration of gain–phase errors is proposed for the whole target scene. The algorithm is used to divide the whole scene into strips, and then the joint iterative solution of target reconstruction and gain–phase error estimation is carried out for each strip. In the process of one iteration, the target reconstruction results are obtained by minimizing of cost function through the given gain–phase errors. Then the gain–phase errors are estimated according to the target reconstruction results, and the modified radiation filed matrix is updated with the gain–phase error estimation for the next iteration. We terminate the iteration if $\left\|\sigma_k^{i+1} - \sigma_k^i\right\|_2^2 / \left\|\sigma_k^i\right\| < \eta$ or the maximum number of iterations $I_{\max}$ is reached, where η is a predetermined threshold and the superscript i refers to the iteration. Key steps of the algorithm include target reconstruction and gain–phase error estimation.The concrete realization course of key steps is as follows.

3.1. Target Reconstruction

For a single strip, the target is reconstructed when the gain–phase errors is given. The initial gain–phase errors of the first strip $\boldsymbol{a} = \mathbf{1}$, $\boldsymbol{\varphi} = \mathbf{0}$. Target reconstruction can be expressed as:

$$\sigma_k^{i+1} = \underset{\sigma_k}{\arg\min} ||\mathbf{E}_k^{sac} - \mathbf{E}_k^{rad}(\boldsymbol{a}^i, \boldsymbol{\varphi}^i) \cdot \sigma_k||_2^2 + \lambda||\sigma_k||_1 \tag{11}$$

The above formula is a standard compressed sensing reconstruction model. There are many existing methods for this problem, such as Basis pursuit (BP) algorithm [20], orthogonal matching pursuit (OMP) algorithm [21], Sparse Bayesian Learning (SBL) [22,23] , etc. In this paper, OMP algorithm is adopted because it is simple in structure and easy to implement and analyze.

3.2. Gain–Phase Error Estimation

The gain and phase errors are estimated in an alternate iteration manner. The gain error is estimated as:

$$\boldsymbol{a}^{i+1} = \underset{\boldsymbol{a}}{\arg\min} ||\mathbf{E}_k^{sca} - \mathbf{E}_k^{rad}(\boldsymbol{a}, \boldsymbol{\varphi}^i) \cdot \sigma_k^{i+1}||_2^2 + \lambda||\sigma_k^{i+1}||_1 \tag{12}$$

Since is $\left\|\sigma_k^{i+1}\right\|_1$ a constant in the iteration, Equation can be rewritten as:

$$\boldsymbol{a}^{i+1} = \underset{\boldsymbol{a}}{\arg\min} ||\mathbf{E}_k^{sca} - \mathbf{E}_k^{rad}(\boldsymbol{a}, \boldsymbol{\varphi}^i) \cdot \sigma_k^{i+1}||_2^2 \tag{13}$$

The above formula is a nonlinear least-squares problem, thus we use Newton's method [24] to solve the problem.

Define $g_k(\boldsymbol{a},\boldsymbol{\varphi}) = \left\|\mathbf{E}_k^{sca} - \mathbf{E}_k^{rad}\left(\boldsymbol{a},\boldsymbol{\varphi}^i\right)\cdot\sigma_k^{i+1}\right\|_2^2$, the updated $\boldsymbol{a}^{i+1}$ estimation denoting by $\boldsymbol{a}^i$ is computed as:

$$\boldsymbol{a}^{i+1} = \boldsymbol{a}^i - \left[\nabla_a^2 g_k\left(\boldsymbol{a}^i,\boldsymbol{\varphi}^i\right)\right]^{-1} / \left[\nabla_a g_k\left(\boldsymbol{a}^i,\boldsymbol{\varphi}^i\right)\right] \tag{14}$$

where $\nabla_a g_k\left(\boldsymbol{a}^i,\boldsymbol{\varphi}^i\right)$ and $\nabla_a^2 g_k\left(\boldsymbol{a}^i,\boldsymbol{\varphi}^i\right)$ represent the gradient and Hessian with respect to the gain error respectively. After derivation and simplification, we have:

$$\nabla_a g_k(\boldsymbol{a}^i,\boldsymbol{\varphi}^i) = -2\mathrm{Re}((\boldsymbol{B}^k(\boldsymbol{a}^i,\boldsymbol{\varphi}^i))^H\widehat{\boldsymbol{w}}) \tag{15}$$

$$\nabla_a^2 g_k(\boldsymbol{a}^i,\boldsymbol{\varphi}^i) = 2\mathrm{Re}((\boldsymbol{B}^k(\boldsymbol{a}^i,\boldsymbol{\varphi}^i))^H\boldsymbol{B}^k(\boldsymbol{a}^i,\boldsymbol{\varphi}^i)) \tag{16}$$

$$\widehat{\boldsymbol{w}} = \mathbf{E}_k^{sca} - \mathbf{E}_k^{rad}(\boldsymbol{a}^i,\boldsymbol{\varphi}^i)\cdot\sigma_k^{i+1} \tag{17}$$

$$\boldsymbol{B}^k(\boldsymbol{a}^i,\boldsymbol{\varphi}^i) = [\boldsymbol{b}_1^k(\boldsymbol{a}^i,\boldsymbol{\varphi}^i),\cdots\boldsymbol{b}_N^k(\boldsymbol{a}^i,\boldsymbol{\varphi}^i)] \tag{18}$$

where Re() denotes the real part,

$$\boldsymbol{b}_{\mathrm{n}}^k(\boldsymbol{a}^i,\boldsymbol{\varphi}^i)=\mathrm{e}^{j\varphi_n^i}\begin{bmatrix} S_n(t_1,\vec{r}_{k,1}) & \cdots & S_n(t_1,\vec{r}_{k,J}) \\ \vdots & \ddots & \vdots \\ S_n(t_L,\vec{r}_{k,1}) & \cdots & S_n(t_L,\vec{r}_{k,J}) \end{bmatrix}\cdot\sigma_k^{i+1} \tag{19}$$

$$S_n(t,\vec{r}_{k,j}) = \frac{\hat{f}_n\left(t-\left(|\vec{r}_{k,j}-\vec{r}_n|+|\vec{r}_s-\vec{r}_{k,j}|\right)/c\right)}{(4\pi)^2|\vec{r}_{k,j}-\vec{r}_n||\vec{r}_s-\vec{r}_{k,j}|} \tag{20}$$

$$\hat{f}_{\mathrm{n}}(t) = \sum_{\mathrm{l}=1}^{L} rect\left[\frac{t-(l-1)\,T_p}{\tau}\right] A_{\mathrm{n}}\exp\{j2\pi f_{nl}\,[t-(l-1)\,T_p]\} \tag{21}$$

In the same way, the phase error is estimated as:

$$\boldsymbol{\varphi}^{i+1} = \arg\min_{\boldsymbol{\varphi}} ||\mathbf{E}_k^{sca} - \mathbf{E}_k^{rad}(\boldsymbol{a}^{i+1},\boldsymbol{\varphi})\cdot\sigma_k^{i+1}||_2^2 + \lambda||\sigma_k^{i+1}||_1 \tag{22}$$

The updated $\boldsymbol{\varphi}^{i+1}$ estimation denoting by $\boldsymbol{\varphi}^i$ is computed as:

$$\boldsymbol{\varphi}^{i+1} = \boldsymbol{\varphi}^i - [\nabla_\varphi^2 g_k(\boldsymbol{a}^{i+1},\boldsymbol{\varphi}^i)]^{-1} / \left[\nabla_\varphi g_k(\boldsymbol{a}^{i+1},\boldsymbol{\varphi}^i)\right] \tag{23}$$

The gradient and Hessian with respect to the phase error can be computed as:

$$\nabla_\varphi g_k(\boldsymbol{a}^{i+1},\boldsymbol{\varphi}^i) = -2\mathrm{Im}((\boldsymbol{D}^k(\boldsymbol{a}^{i+1},\boldsymbol{\varphi}^i))^H\widehat{\boldsymbol{w}}) \tag{24}$$

$$\nabla_\varphi^2 g_k(\boldsymbol{a}^{i+1},\boldsymbol{\varphi}^i) = 2\mathrm{diag}(\mathrm{Re}((\boldsymbol{D}^k(\boldsymbol{a}^{i+1},\boldsymbol{\varphi}^i))^H\widehat{\boldsymbol{w}})) + 2\mathrm{Re}((\boldsymbol{D}^k(\boldsymbol{a}^{i+1},\boldsymbol{\varphi}^i))^H\boldsymbol{D}^k(\boldsymbol{a}^{i+1},\boldsymbol{\varphi}^i)) \tag{25}$$

$$\boldsymbol{D}^k(\boldsymbol{a}^i,\boldsymbol{\varphi}^i) = [\boldsymbol{d}_1^k(\boldsymbol{a}^i,\boldsymbol{\varphi}^i),\cdots\boldsymbol{d}_N^k(\boldsymbol{a}^i,\boldsymbol{\varphi}^i)] \tag{26}$$

where Im() denotes the imaginary part, diag()is the diagonalization operation.

$$d_{\mathrm{n}}^{k}\left(\boldsymbol{a}^{i},\boldsymbol{\varphi}^{i}\right)=\mathrm{e}^{j\varphi_{n}^{i}}a_{n}^{i+1}\begin{bmatrix} S_{n}(t_{1},\vec{r}_{k,1}) & \cdots & S_{n}(t_{1},\vec{r}_{k,J}) \\ \vdots & \ddots & \vdots \\ S_{n}(t_{L},\vec{r}_{k,1}) & \cdots & S_{n}(t_{L},\vec{r}_{k,J}) \end{bmatrix}\cdot\sigma_{k}^{i+1} \tag{27}$$

The above is about the single iteration process of gain–phase error estimation by Newton's method. In the $i-th$ iteration, $\left(\boldsymbol{a}^{i},\boldsymbol{\varphi}^{i}\right)$ will be updated to $\left(\boldsymbol{a}^{i+1},\boldsymbol{\varphi}^{i+1}\right)$. The initial gain–phase errors of the first strip $\boldsymbol{a}=\mathbf{1},\boldsymbol{\varphi}=\mathbf{0}$. The gain–phase error estimation results of the former strip are taken as the initial value and substituted into the latter strip, which makes the estimation results of the latter strip more accurate. After the first round of correlated processing with calibration of gain–phase error of all strips is completed, the gain–phase error estimation results of the last strip are brought into the first strip for the next round, and the whole imaging area divided into strips is processed in multiple rounds to obtain the final results.

The whole process of the algorithm is as follows:

Algorithm 1: Strip-Mode MSCI Algorithm with Self-Calibration of Gain–phase Errors

Input: $\mathbf{E}^{sca}, I_{\max}, U_{\max}, \eta$

Initialization: $\mathrm{k}=1,\quad \mathrm{i}=0,\quad \mathrm{u}=0,\quad \boldsymbol{a}=1,\quad \boldsymbol{\varphi}=0$;

while $u<U_{\max}$ **do**

  for $k<K$ **do**

    Get $\mathbf{E}_{k}^{sca}$ from $\mathbf{E}^{\mathrm{sca}}$;

    if $i<I_{\max}$ *and* $\left\|\sigma_{k}^{i+1}-\sigma_{k}^{i}\right\|_{2}^{2}/\left\|\sigma_{k}^{i}\right\|>\eta$ **then**

      The $k-th$ target reconstruction : $\sigma_{k}^{i+1}=\arg\min\limits_{\sigma_{k}}F\left(\sigma_{k},\boldsymbol{a}^{i},\boldsymbol{\varphi}^{i}\right)$;

      Gain error estimation: $\boldsymbol{a}^{i+1}=\arg\min\limits_{\boldsymbol{a}}F\left(\sigma_{k}^{i+1},\boldsymbol{a},\boldsymbol{\varphi}^{i}\right)$;

      Phase error estimation: $\boldsymbol{\varphi}^{i+1}=\arg\min\limits_{\boldsymbol{\varphi}}F\left(\sigma_{k}^{i+1},\boldsymbol{a}^{i+1},\boldsymbol{\varphi}\right)$;

      $i=i+1$;

    else

      $i=0$;

      $k=k+1$ and update $\boldsymbol{a},\boldsymbol{\varphi}$;

    end

  end

  $k=1$;

  $u=u+1$;

end

Output: Multi-strip reconstructed scattering coefficient vectors.

4. Analysis

The proposed strip-mode MSCI method based on self-calibration of gain–phase errors can greatly reduce the computational cost of the imaging process. The total grid number of the target scene is M, divided into K strips, and the number of grid in each strip is J. The main operations of an iteration during imaging process include updating the modified radiation filed matrix, target reconstruction, gain–phase error estimation by Newton's method. According to the characteristics of MSCI, generally the narrow pulse number L should satisfy $L>M$. Compared to no strip division, after the target scene is divided into K strips, the number of grids with in each strip is decreased to M/K, the number of narrow pulse is decreased to L/K, so the scale of the modified radiation filed matrix is reduced to ML/K^2 . Therefore, the computation required for updating the modified radiation filed matrix and Newton's method is reduced significantly. For the OMP algorithm in the target reconstruction process, in the case that the sparsity is d, the computation is $O(d \cdot L \cdot M)$ [21] when there are no strips, in contrast, when dividing into K bands, the computation is $K \cdot O(d/K \cdot L/K \cdot M/K)$. The above discussion is about the change of the computation in an iteration. In the actual process, due to the strip division, the target scene and the operation process are simplified, the average number of iterations required in the imaging process is also decreased, and the operation time is further reduced.

5. Simulations

The effectiveness of proposed method is verified by several simulations in this section. An X-band MSCI radar system with center frequency 10 GHz is considered. The scenario for simulation is shown in Figure 1. The height of the transmitter array is 300 m , which consists of 25 elements to form a uniform array of 3 × 3 m in size. The distance of target scene is 450 m, and the size of target scene is discretized into 40 × 40 grids with grid size of 2 × 2 m. We initialize $\boldsymbol{a} = 1, \boldsymbol{\varphi} = 0, I_{\max} = 100, \eta = 10^{-4}$. Some system parameters are given in Table 1, and the parameters of gain–phase errors are given in Table 2.

Table 1. System parameters.

Parameter	Value
Center Frequency	10 GHz
Bandwidth	500 MHz
Transmitting signal mode	Frequency hopping
Number of transmitters	25
Aperture of transmitter array	3 × 3 m
Imaging grids	40 × 40
Size of imaging grids	2 × 2 m
Height of transmitter array	300 m
The squint angle	45°

Table 2. Gain–phase error parameters.

Index	1	2	3	4	5	6	7	8	9	10	11	12
a	1	0.85	0.95	1.20	0.80	1.25	1.10	0.90	1.20	0.75	0.90	1.15
$\varphi/\circ$	0	−30	−25	10	−20	−10	20	30	15	20	−25	−10
13	**14**	**15**	**16**	**17**	**18**	**19**	**20**	**21**	**22**	**23**	**24**	**25**
0.80	1.20	1.10	0.80	1.05	0.85	1.05	1.25	0.80	1.20	1.15	0.95	0.80
−25	−10	20	40	15	−30	−45	10	−20	10	25	45	30

5.1. Performances Under Different Number of Strips

In this subsection, simulations are taken to compare the performances with different strips. The normalized mean square error (NMSE) is used to quantify the reconstruction effect and gain–phase error estimation, with the definition as: $NMSE_{dB} = 20\lg(\|\hat{x} - x\|_2/\|x\|_2)$,where x denotes the target imaging or gain–phase errors, accordingly, $\hat{x}$ denotes the target reconstruction or gain–phase error estimation results.

It can be seen in Figure 3b that the image is defocused and many spurious scatterers exist with for the OMP algorithm. In Figure 3c–f, it can be seen that the image become clearer and clearer with increase in the number of strips. The NMSEs of the reconstruction images under different strips are given in Figure 4, and it shows that NMSEs are decreased as the number of strips increases, which means the quality of imaging is getting better. Compared to no strip, proposed method with eight strips improves the imaging performance by about 20 dB from the NMSE perspective.

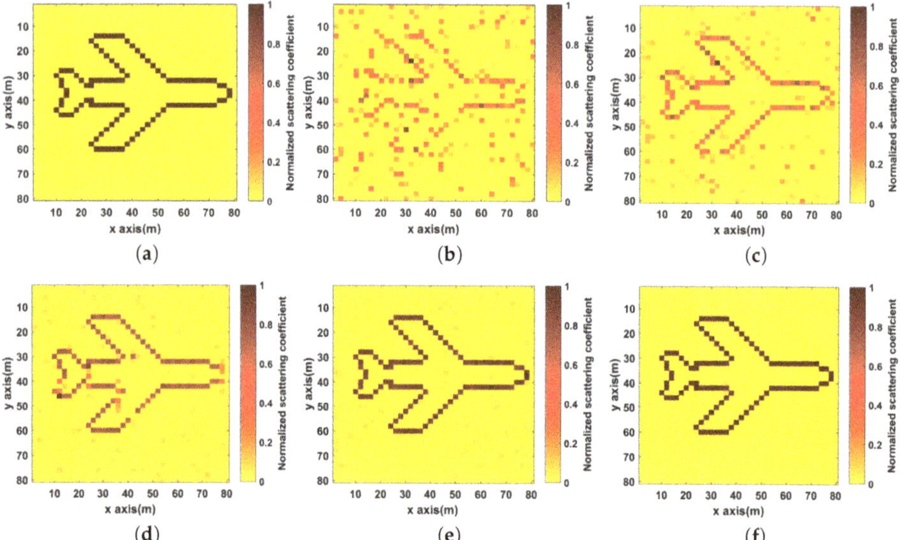

Figure 3. Imaging results (**a**) objective model; (**b**) Imaging results of OMP ; (**c–f**) Imaging results under different number of strips (**c**) no strip; (**d**) two strips; (**e**) four strips; (**f**) eight strips.

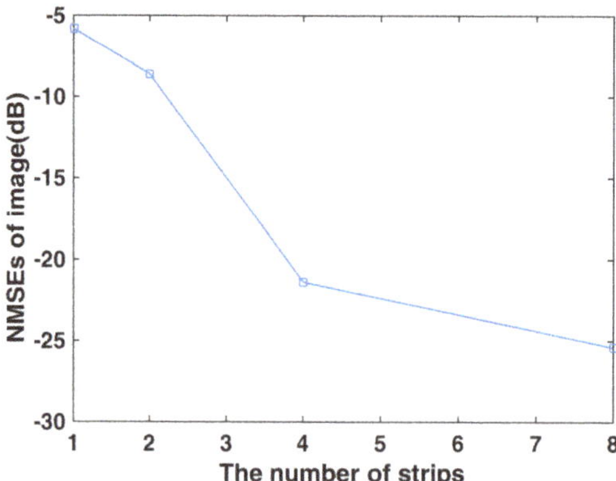

Figure 4. NMSE of target reconstructions under different number of strips.

In Figures 5 and 6, it can be seen that the estimates of gain and phase error are closer to the actual value as the number of strips increases. As shown in Figure 7, the NMSEs of gain–phase error estimation are getting lower as the number of strips increases, which means estimation errors are getting lower, and it is proved that the proposed method in this paper can improve the accuracy of gain–phase error estimation effectively.

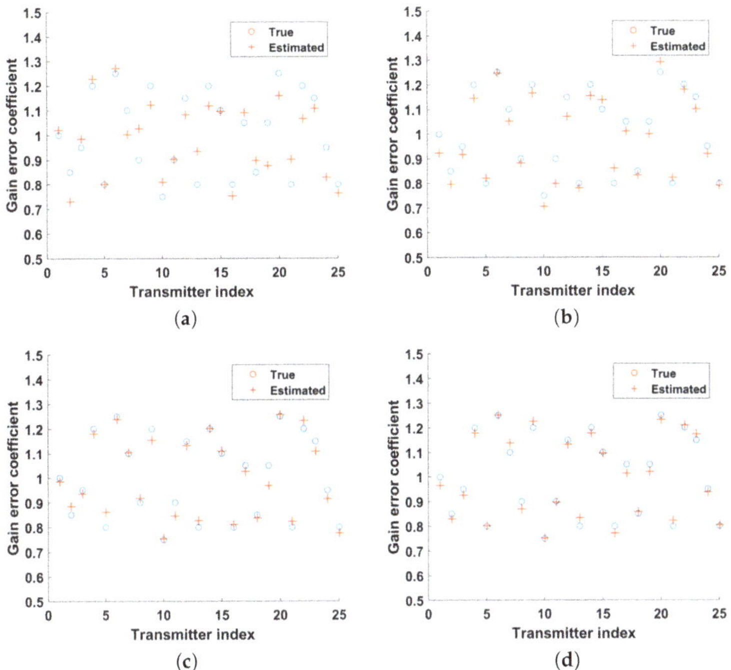

Figure 5. Gain error estimation under different number of strips (**a**) no strip; (**b**) two strips; (**c**) four strips; (**d**) eight strips.

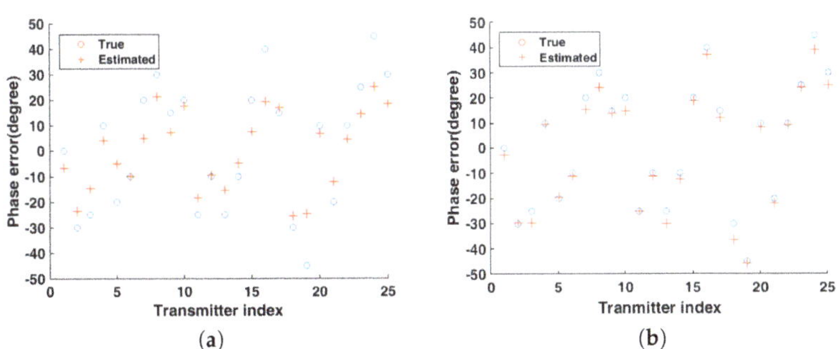

Figure 6. *Cont.*

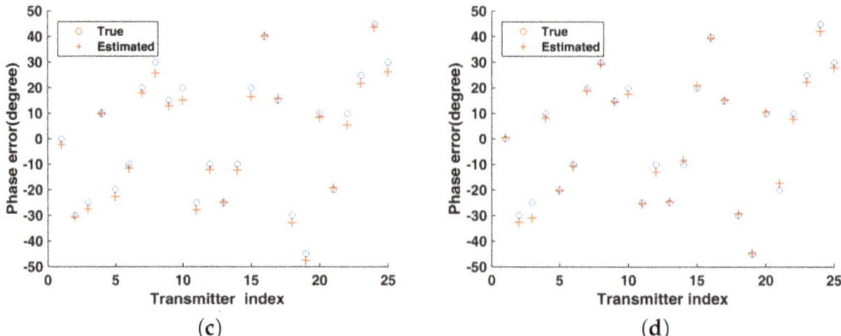

Figure 6. Phase error estimation under different number of strips (**a**) no strip; (**b**) two strips; (**c**) four strips; (**d**) eight strips.

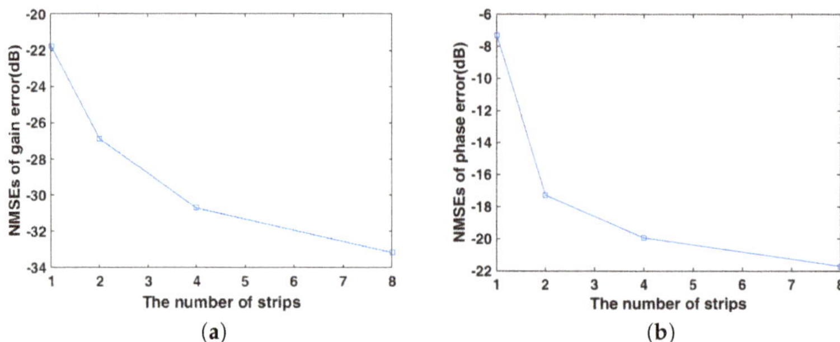

Figure 7. Gain–phase error estimation performance under different strips (**a**) NMSE of gain error estimation; (**b**) NMSE of phase error estimation.

In Figure 8, as the strip increases, the imaging time decreases significantly, which is consistent with the analysis in this paper. It takes less than 1/15 of time by divided into eight strips compared with no strip. It is proved that the strip division can greatly reduce the time required for the correlated imaging process.

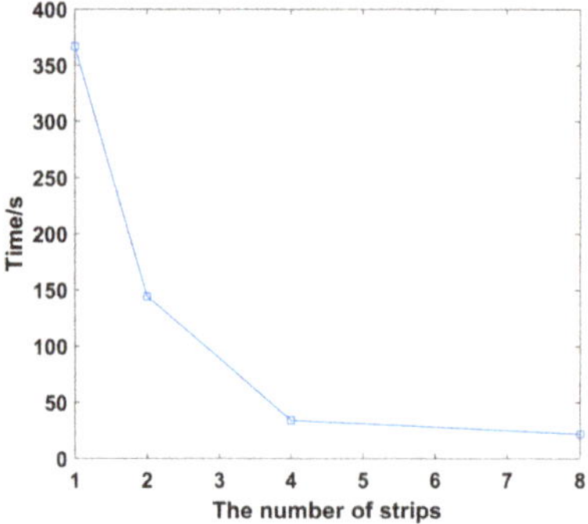

Figure 8. Imaging time under different number of strips.

5.2. Performance under Different SNRs

In this subsection, we compare the performance of algorithms under different SNRs, for the proposed method and SACRCI [15]. As shown in Figure 9, the imaging quality is improved significantly as the SNR increases, which means the two method are sensitive to noise. The proposed method improves the imaging performance by more than 10 dB compared with SACRCI from the NMSE perspective. In Figure 10, it can be seen that the gain–phase error estimation is also sensitive to noise.

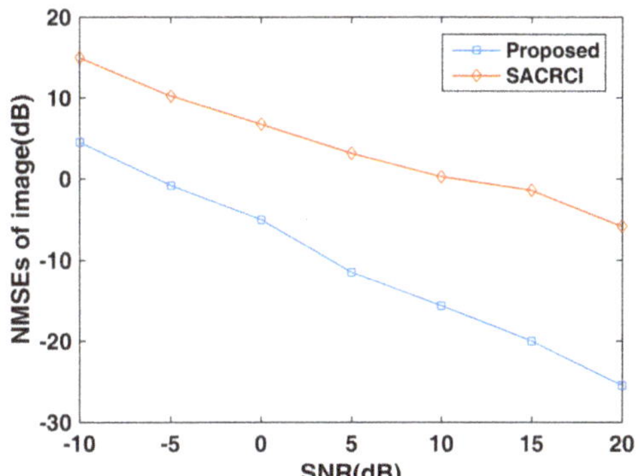

Figure 9. NMSE of target reconstructions under different SNRs.

5.3. Performance under Different Transmitting Array Configurations

In MSCI, transmitting array configurations can influence imaging effect, and considering this, we perform simulations in this subsection to compare the performance under different transmitting array configurations. In Figure 11a, the transmitting array is a array with its aperture of 3 m. In Figure 11b, the array elements are randomly distributed on the plane. In Figure 11c, the aperture of the uniform planar array is reduced to 1.5 m. From the imaging results, it can be seen that the size of the array aperture influences the target reconstruction significantly, which is consistent with the relationship between the array aperture size and the imaging resolution.

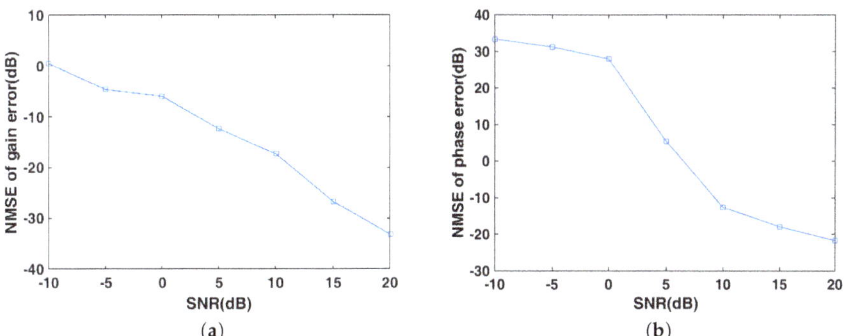

Figure 10. gain–phase error estimation performance under different SNRs (**a**) NMSE of gain error estimation; (**b**) NMSE of phase error estimation.

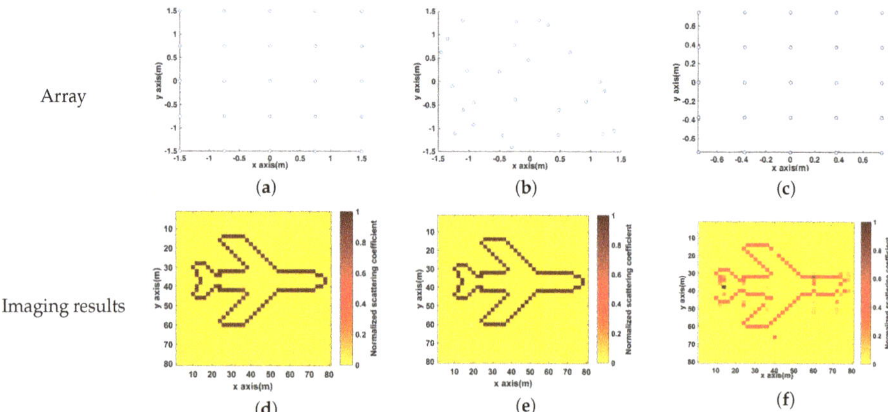

Figure 11. Imaging results under different transmitting array configurations (**a**–**c**) Different transmitting array configurations; (**d**–**f**) Imaging results.

5.4. Performance under Different Center Frequencies

In this subsection, performance under different center frequencies is compared by simulations. I can be seen in Figure 12 that target reconstruction result is not clear when center frequency is 1 GHz, in contrast, when center frequency is 40 GHz, the target reconstruction effect is much better. This is because the resolution of MSCI is related to the center frequency, the higher the center frequency, the better the resolution, and the better the imaging effect under the same grid division.

It can be seen in Figure 12 that the target reconstruction result is not clear when center frequency is 1 GHz; in contrast, when center frequency is 40 GHz, the target reconstruction effect is much better. This is because the resolution of MSCI is related to the center frequency, the higher the center frequency, the better the resolution, and the better the imaging effect under the same grid division.

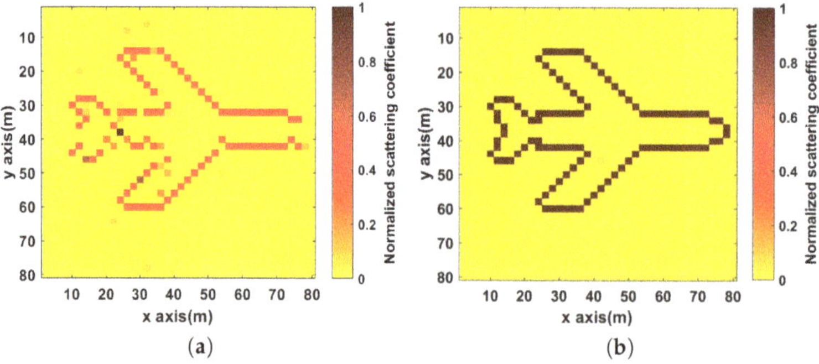

Figure 12. Imaging results under different center frequencies (**a**) 1 GHz; (**b**) 40 GHz.

5.5. Performance under Different Target Scenes

Since Target reconstruction results is obtained by OMP, the reconstruction performance may be affected by the target, more precisely, the sparsity of target. In this subsection, we design simulations to compare the performance under different target scenes.

As shown in Figure 13a–c are three different target scenes. It can be seen that the images become blurred as the complexity of targets increases, which means the less sparse target would make the target reconstruction more difficult and the gain–phase error estimation performance is also affected. Comparing with the results obtained by SACRCI, the spurious scatterers in the bottom three images which are obtained by the proposed method, are much less, and the three targets are identified clearly. It proves that the proposed method can improve the imaging performance by reducing the complexity of correlated imaging processing.

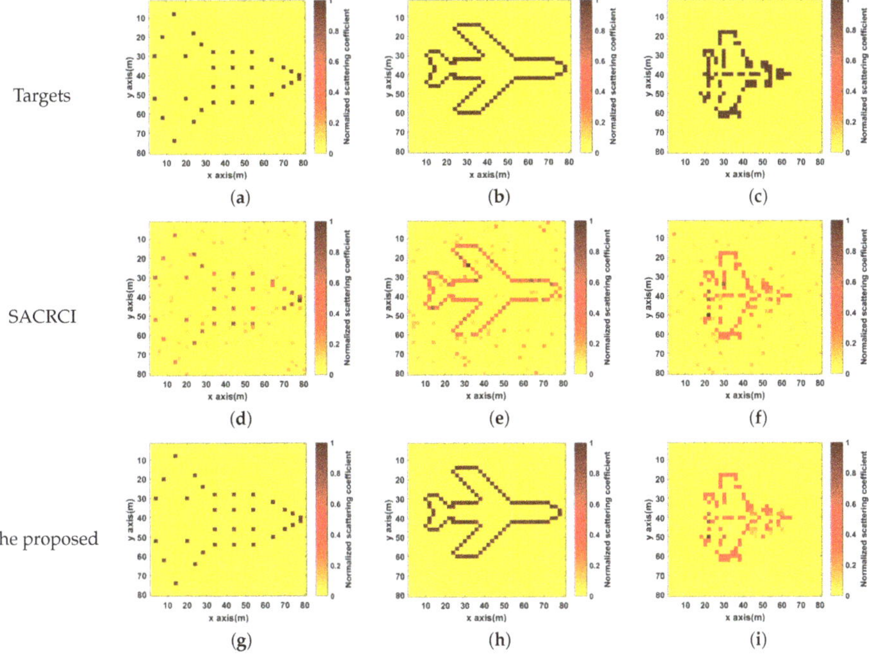

Figure 13. Imaging results for different target scenes (**a**–**c**) Three different target scenes; (**d**–**f**) Imaging results of SACRCI; (**g**–**i**) Imaging results of the proposed method.

5.6. Discussion

Lots of numerical simulations validate potential advantages of the proposed method to shorten the imaging time dramatically and improve the imaging and gain–phase error estimation performance, and show the performance under different SNRs, different targets, different array configurations and different center frequencies. In the actual system, since the proposed method uses the range-gate characteristic of the narrow pulse to divide the imaging area into strips, the transmitting system must have a high rectangular coefficient, and each transmitting element needs a high-precision time-frequency reference. The system must have a high-precision time-frequency synchronization to ensure the accurate separation of the corresponding parts of each strip from the echo. These are great challenges in actual MSCI system.

6. Conclusions

This paper proposes a method of MSCI based on strip-mode self-calibration of gain–phase errors. By dividing the target scene into strips, the target reconstruction and the gain–phase error estimation are solved simultaneously by alternate iteration. By simulations it can be seen that the gain–phase errors calibration and imaging effect have been greatly improved and the time required for the entire imaging process has been greatly shortened. Moreover, to improve imaging and gain–phase error estimation performance furtherly, not only are the gain–phase error estimation results of the previous strip carried into the next strip as the initial value, but also the gain–phase error estimation results of the last strip are the initial value in next round. In conclusion, the proposed method can greatly reduce the time required by the imaging process and improve the imaging quality, so it can rapidly achieve gain–phase errors calibration and target imaging in a large scene.

Author Contributions: All authors contributed extensively to the work presented in this paper. R.X. proposed the original idea, designed the study, performed the simulations and wrote the paper; Y.G. supervised the analysis, edited the manuscript, W.C. and D.W. and provided their valuable suggestions to improve this study.

Funding: This research received no external funding.

Acknowledgments: This work has been supported by the National Natural Science Foundation of China under contact Nos. 61771446, 61431016.

Conflicts of Interest: The authors declare no conflict of interest.

References

1. Ausherman, D.A.; Kozma, A.; Walker, J.L.; Jones, H.M.; Poggio, E.C. Developments in radar imaging. *IEEE Trans. Aerosp. Electron. Syst.* **1984**, *AES-20*, 363–400. [CrossRef]
2. Bao, Z.; Xing, M.; Wang, T. Radar imaging technology. *Beijing Publ. House Electron. Ind.* **2005**, *125*.
3. Lord, R.T.; Inggs, M.R. High resolution SAR processing using stepped-frequencies. In Proceedings of the 1997 IEEE International Geoscience and Remote Sensing Symposium Proceedings. Remote Sensing—A Scientific Vision for Sustainable Development, Singapore, 3–8 August 1997; IEEE: Piscataway, NJ, USA, 1997; Volume 1, pp. 490–492.
4. Chan, Y.K.; Koo, V.C. An introduction to synthetic aperture radar (SAR). *Progress Electromagn. Res.* **2008**, *2*, 27–60. [CrossRef]
5. Li, D.; Li, X.; Qin, Y.; Cheng, Y.; Wang, H. Radar coincidence imaging: an instantaneous imaging technique with stochastic signals. *IEEE Trans. Geosci. Remote Sens.* **2014**, *52*, 2261–2277.
6. Guo, Y.; He, X.; Wang, D. A novel super-resolution imaging method based on stochastic radiation radar array. *Meas. Sci. Technol.* **2013**, *24*, 074013. [CrossRef]
7. Cheng, Y.; Zhou, X.; Xu, X.; Qin, Y.; Wang, H. Radar coincidence imaging with stochastic frequency modulated array. *IEEE J. Sel. Top. Signal Process.* **2017**, *11*, 414–427. [CrossRef]
8. Cao, K.; Zhou, X.; Cheng, Y.; Qin, Y. Improved focal underdetermined system solver method for radar coincidence imaging with model mismatch. *J. Electron. Imaging* **2017**, *26*, 033001. [CrossRef]
9. Cao, K.; Zhou, X.; Cheng, Y.; Fan, B.; Qin, Y. Total variation-based method for radar coincidence imaging with model mismatch for extended target. *J. Electron. Imaging* **2017**, *26*, 063007. [CrossRef]
10. Weiss, A.J.; Friedlander, B. Eigenstructure methods for direction finding with sensor gain and phase uncertainties. *Circuits Syst. Signal Process.* **1990**, *9*, 271–300. [CrossRef]
11. Li, J.; Jin, M.; Zheng, Y.; Liao, G.; Lv, L. Transmit and receive array gain–phase error estimation in bistatic MIMO radar. *IEEE Antennas Wirel. Propag. Lett.* **2015**, *14*, 32–35. [CrossRef]
12. Zhang, L.; Gao, Y.; Wang, K.; Liu, X. A subspace algorithm of calibrating channel gain and phase errors for HRWS SAR imaging. In Proceedings of the 2017 IEEE Radar Conference (RadarConf), Seattle, WA, USA, 8–12 May 2017; pp. 0269–0272.

13. Onhon, N.Ö.; Cetin, M. A sparsity-driven approach for joint SAR imaging and phase error correction. *IEEE Trans. Image Process.* **2012**, *21*, 2075–2088. [CrossRef] [PubMed]
14. Zhou, X.; Wang, H.; Cheng, Y.; Qin, Y. Radar coincidence imaging with phase error using Bayesian hierarchical prior modeling. *J. Electron. Imaging* **2016**, *25*, 013018. [CrossRef]
15. Zhou, X.; Wang, H.; Cheng, Y.; Qin, Y. Sparse auto-calibration for radar coincidence imaging with gain–phase errors. *Sensors* **2015**, *15*, 27611–27624. [CrossRef] [PubMed]
16. Zhou, X.; Wang, H.; Cheng, Y.; Qin, Y. Expansion—compression variance-component-based autofocusing method for joint radar coincidence imaging and gain—phase error calibration. *J. Appl. Remote Sens.* **2017**, *11*, 025002. [CrossRef]
17. Tian, C.; Yuan, B.; Wang, D. Calibration of gain–phase and synchronization errors for microwave staring correlated imaging with frequency-hopping waveforms. In Proceedings of the 2018 IEEE Radar Conference (RadarConf18), Olakhoma City, OK, USA, 23–27 April 2018; pp. 1328–1333.
18. Tian, C.; Jiang, Z.; Chen, W.; Wang, D. Adaptive microwave staring correlated imaging for targets appearing in discrete clusters. *Sensors* **2017**, *17*, 2409. [CrossRef] [PubMed]
19. Guo, Y.; Deng, J.; Liu, F.; Wang, D. Efficient strip-mode microwave correlated imaging method with data fusion. In Proceedings of the 2017 IEEE International Conference on Computer and Information Technology (CIT), Helsinki, Finland, 21–23 August 2017; pp. 16–22.
20. Chen, S.S.; Donoho, D.L.; Saunders, M.A. Atomic decomposition by basis pursuit. *SIAM Rev.* **2001**, *43*, 129–159. [CrossRef]
21. Boyd, S.; Vandenberghe, L. *Convex Optimization*; Cambridge University Press: Cambridge, UK, 2004.
22. Wipf, D.P.; Rao, B.D. Sparse Bayesian learning for basis selection. *IEEE Trans. Signal Process.* **2004**, *52*, 2153–2164. [CrossRef]
23. Fang, J.; Shen, Y.; Li, H.; Wang, P. Pattern-coupled sparse Bayesian learning for recovery of block-sparse signals. *IEEE Trans. Signal Process.* **2015**, *63*, 360–372. [CrossRef]
24. Tropp, J.A.; Gilbert, A.C. Signal recovery from random measurements via orthogonal matching pursuit. *IEEE Trans. Inf. Theory* **2007**, *53*, 4655–4666. [CrossRef]

Article

Geometrical Matching of SAR and Optical Images Utilizing ASIFT Features for SAR-based Navigation Aided Systems

Jakub Markiewicz [1], Karol Abratkiewicz [2,*], Artur Gromek [2], Wojciech Ostrowski [1], Piotr Samczyński [2] and Damian Gromek [2]

1 Department of Photogrammetry, Remote Sensing and Spatial Information Systems, Faculty of Geodesy and Cartography, Warsaw University of Technology, 00-661 Warsaw, Poland; jakub.markiewicz@pw.edu.pl (J.M.); wojciech.ostrowski@pw.edu.pl (W.O.)

2 Institute of Electronic Systems, Faculty of Electronics and Information Technology, Warsaw University of Technology, 00-665 Warsaw, Poland; a.gromek@elka.pw.edu.pl (A.G.); P.Samczynski@elka.pw.edu.pl (P.S.); d.gromek@elka.pw.edu.pl (D.G.)

* Correspondence: k.abratkiewicz@elka.pw.edu.pl

Received: 21 October 2019; Accepted: 6 December 2019; Published: 12 December 2019

Abstract: This article presents a new approach to the estimation of shift and rotation between two images from different kinds of imaging sensors. The first of the image is an orthophotomap that is created using optical sensors with georeference information. The second one is created utilizing a Synthetic Aperture Radar (SAR) sensor.The proposed solution can be mounted on a flying platform, and, during the flight, the obtained SAR images are compared with the reference optical images, and thus it is possible to calculate the shift and rotation between these two images and then the direct georeferencing error. Since both images have georeference information, it is possible to calculate the navigation correction in cases when the drift of the calculated trajectory is expected. The method can be used in platforms where there is no satellite navigation signal and the trajectory is calculated on the basis of an inertial navigation system, which is characterized by a significant error. The proposed method of estimating the navigation error utilizing Affine Scale-Invariant Feature Transform (ASIFT) and Structure from Motion (SfM) is described, and techniques for improving the quality of SAR imaging using despeckling filters are presented. The methodology was tested and verified using real-life SAR images. Differences between the results obtained for a few selected despeckling methods were compared and commented on. Deep investigation of the nature of the SAR imaging technique and noise creation character allows new algorithms to be developed, which can be implemented on flying platforms to support existing navigation systems in which trajectory error occurs.

Keywords: SAR; Synthetic Aperture Radar; ASIFT; Despeckling Filter; Navigation; Structure from Motion; Iterative Closest Point

---

1. Introduction

Over recent years, supporting navigation systems has become particularly important for several reasons. Firstly, the aim is to increase the precision of ammunition and flying objects. Secondly, it is necessary to create systems that can work without the support of GNSS (Global Navigation Satellite System). The lack of a GNSS (GPS, GLONAS, Galileo, or others) signal is a significant limitation whose probability of occurring increases due to potential international conflicts, as well as the possibility of the GNSS signal being jammed or interrupted. For these reasons, it is necessary to create independent systems that allow navigation in the absence of a satellite signal. One of the basic sensors that allows the navigation of objects is the inertial navigation system (INS). However, due to the significant drift and error increasing with time, inertial navigation requires additional systems. Currently, drift is reduced using GNSS systems; however, due to the limitations mentioned above, it is necessary to use additional sensors to support inertial navigation.

In the literature, several solutions have been proposed that allow the navigation of objects in the absence of a GNSS signal. The least effective solutions are purely vision methods, which are ineffective in the case of night missions, cloud cover, fog, smog, or smoke. Other remote sensing methods that do not use sensors working in the visible spectrum are used more often. Such solutions use sensors such as light detection and ranging (LIDAR) and an altimeter, which allows a terrain contour (DEM) to be obtained, and then compares the acquired contour with the data in the database. Such systems, also known as Terrain Contour Matching (TERCOM), are utilized to navigate unmanned aerial vehicles or cruise missiles in cases when a GNSS signal could be unavailable [1–3].

A recently developed technique is the use of Synthetic Aperture Radar (SAR) and Interferometric SAR (InSAR) radar for navigation correction [4–7]. Radio waves used in this technique can be easily used during cloud cover, night, and rain, which makes this method universal and independent of the weather conditions. In this case, a radar sensor and a database with georeferenced images are on board the flying platform. During the flight, the terrain is scanned by the radar and the obtained image is compared to the corresponding one in the database. Thanks to this, the inertial navigation error can be reduced.Such a concept of utilizing SAR and InSAR sensors to support on-board air platform navigational devices was proposed in the SARINA (SAR-based Augmented Integrity Navigation Architecture) project carried out during 2010–2012. The authors of this paper were also involved in this work, and as a result have developed a concept of the system and proved that SAR/InSAR sensors can be successfully used to support navigational devices. The results of this work were published by the authors in [4–6]. Initially, the concept was proved only using simulations at the technical readiness level (TRL) 3–4 in the nine-degree scale, where 9 denotes the system prototype after all the required certifications. In the previous work within the SARINA project, the merging of SAR/InSAR images with an optical image database was based on simple automatic shape recognition of the terrain targets, using target contours extraction. Image processing was then applied using different techniques, such as the Hough transform, to find specific targets and recognize their shape. These techniques were used for SAR image matches. For InSAR, algorithms based on matching 3D SAR interferograms with LIDAR elevation models were developed that were equipped with a simulated on-board database. The very promising results of the previous SARINA project motivated the authors to continue work on the topic, and to start developing a system on a higher TRL level. In the meantime, new algorithms on SAR and optical images have been developed and are widely used for other applications, such as in geodesy. The author intend to test efficiency of these techniques in cases so they could be used in support of air platform navigational devices. One such approach is based on the SIFT technique, and was presented by the authors in [7]. The novelty and usability of this approach is presented in [8]. The SIFT algorithm was used to find and match corresponding keypoints appearing in the SAR and optical images. This approach is extended and investigated in this paper by utilizing the more robust ASIFT (Affine Scale Invariant Feature Transform) method and its limitations. The novelty of the solution proposed and described in this paper is in the utilization of an innovative approach to the shift and rotation estimation between SAR and optical images. By finding characteristic points in both images and applying themodified version of the Structure from Motion (SfM) technique, an error can be estimated, which in turn provides the navigation drift correction in the flying platform. In the authors' opinion, the concept of using the ASIFT technique is interest regarding the checking of the efficiency and precision of mismatched SAR and optical image calculations, which might be further used to correct platform navigation devices according to the algorithms developed by the authors in their previous work [4–6]. To the authors' knowledge, an innovative approach to the shift and rotation estimation between SAR and optical images by also applying SfM techniques has not been used for navigation drift correction on a flying platform, on which the authors of this paper are currently working. In this paper, the authors present an overview of the SIFT and ASIFT techniques and their modifications, as well as the required pre-processing and its limits, which is taken into account when developing systems based on SAR systems for navigation drift corrections. The paper structures is as follows. In Section 2, there is a description of the proposed method presenting an overall problem characterization and solution. In Section 3, the basics of the SfM approach are presented to depict the main keys in this technique. Section 4 presents different types of SAR filters providing speckle noise reduction in images created using SAR radars. The results are presented in Section 5, and the conclusion closes the article. Additionally, the Appendix present the extensives results in a single part of the article. In the corresponding parts of the paper, there are references to the Appendix and images contained to ensure reader clarity.

2. Methodology

2.1. Overview of the Approach

The process of image registration refers to the alignment of two or more images of the same scene which might be obtained with the same sensor, time, and imaging conditions, as well as by different sensors and viewpoints. The process of orientating optical and SAR images is still an open issue, which creates many challenges [9–19]. Many issues can be dealt with in the process of the synergy of optical and SAR data, due to the great differences between passive and active remote sensing techniques. One of the issues is the problem with the speckle noise that influences the detection of robust corresponding/tie points [9,11–13,15,16,18,19]. Another issue is related to the differences in the image geometry acquired from these two devices [9,11–13,15,16,18,19]. Thus, many co-registration approaches have been proposed (e.g., [9–19]), but, in general, these approaches might be divided into two main categories: area- and feature-based methods. Nowadays, the Structure from Motion (SfM) methods, which are mostly based on the feature-based approach, are used for the co-registration of spaceborne SAR and optical images [9,11–13,15,16,18,19]. An extended description of the incremental SfM process is presented in Section 3. The classical SfM approach contains four main steps: (1) feature detection; (2) feature description; (3) descriptor matching; and (4) bundle adjustment. Due to the problems mentioned above, which influence the quality of detected and matched pairs of points, the SfM/co-registration approach has been modified by many authors. The first improvement is in the feature detection—downsampling SAR images or using the despeckling filters in the pre-processing step. Based on this way of pre-processing images, different types of features are detected such as blob, corners, or lines and segments. In the case of the feature description and matching, the modification is in the descriptor, or this step is eliminated and based only on the geometrical relationship. It should be stressed that these presented methods were validated in the spaceborne SAR and optical images.

The proposed methodology of the automation of SAR data registration with orthophotomaps as well as the SAR trajectory improvement, is a multi-stage process. This process is based on the original software and it consists of: (1) SAR data conversion to the raster form with a georeference file; (2) the aligning of orthophotomaps with a SAR raster based on the extended version of the ASIFT algorithm; (3) the relative orientation based on the classical SfM and modifiedIterative Closest Point (ICP) approach; (4) the analysis of the quality of the relative orientation of processed data; and (5) the final bundle adjustment.

In this investigation, the process of optical and SAR images was tested and validated with a high resolution SAR and orthophotomaps from altitude. It should be stressed that the entire investigation was performed on the full resolution of the images, and additionally tested on other pyramid levels. The authors decided to use a well-known SfM approach, but with the following modification: (1) the pre-processing of SAR data with different speckle noise reduction filters in order to increase the possibility of detecting and matching robustness corresponding points; (2) reducing the values of affine angle in the ASIFT algorithm—the values being related to the angle of the antenna; and (3) using a two-stage ICP procedure to eliminate the outliers and compute the correction parameters. Thanks to the use of the ASIFT algorithm, it was possible to detect well-distributed keypoints in the whole area under investigation. The authors decided to eliminate the description and matching step and replace it with the ICP alignment. This allowed keypoints to be treated as a point cloud, and minimize the distance between these two point clouds in an iterative process with the simultaneous correction of translation and rotation of SAR data.

In Figure 1, a diagram of the performed research work and experiments is shown. To perform a complete analysis of the possibility of applying the modified version of the registration algorithms for the detection and matching of correspondence points, a combination of these should be determined to obtain the best results. Verification of the following parameters are required:

1. The keypoints distributions on SAR images and orthophotomaps.
2. The influence of the SAR image correction on the quality and number of the keypoints.
3. The number of detected keypoints used in the final cooregistration process.
4. The orientation accuracy of marked check-points.

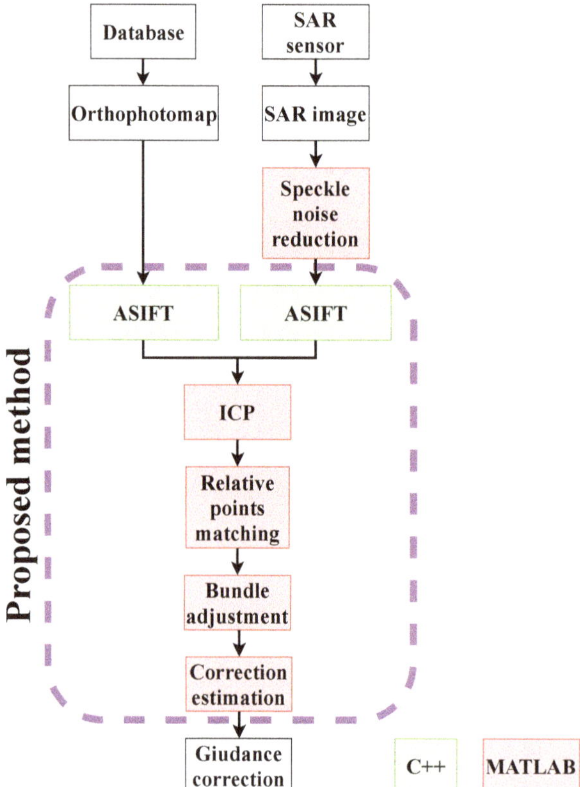

Figure 1. Diagram of the performed research: Processing and orientation of SAR images and orthophotomaps.

The idea of the automatic SAR data registration and trajectory improvement presented in this paper (according to the diagram presented in Figure 1) consists of the following steps:

1. *Generation of the SAR raster with georeferences based on the trajectory information.*
This step is one of the most important parts in the whole automatic registration process. It influences both the computation time and convergence of the ICP process.

2. *Pre-processing of SAR data to reduce the influence of the noise.*

    - RAW data
    - Multilook-2D filter
    - Averaging filter
    - Minimum Mean Square Error filter
    - Enhanced Lee filter
    - Gamma MAP filter
    - SAR Block Matching 3D filter

3. *Selection of the part of the orthophotomap which approximately covers the SAR image, based on the SAR footprint extended by 50 m in each direction.*
4. *Detection and matching of the keypoints using the modified ASIFT algorithms in the SAR images and orthophotomaps:*

    4.1. Detection of keypoints in RAW SAR images on three pyramid levels (full resolutions, as well as **1/2**, 1/4, and 1/16 of the full resolution of the raster)—ASIFT algorithm.
    4.2. Detection of keypoints in SAR images with the speckle noise reduction parameters: 1, 4, 8, and 16 levels (full resolution, and **1/2**, 1/4, and 1/16 of the full resolution of the raster)—ASIFT algorithm.
    4.3. Detection of keypoints on orthophotomap levels (full resolution, as well as **1/2**, 1/4, and 1/16 of the full resolution of the raster)—ASIFT algorithm.
    4.4. Matching keypoints by the incremental ICP method:

        - Approximate registration with 20 iterations and linear threshold deviation 20 m.
        - Removal of orthophotomap keypoints that are outside the SAR area.
        - Registration with 10 iterations and linear threshold deviation 5 m.
        - Removal of the SAR point outliers based on the $RMSE_{xy}$.
        - Final bundle adjustment.

5. *Analysis of the quality of data registration on the marked check-points.*

The presented SAR data registration processing is an original approach, and for this reason original applications were applied (based on the OpenCV library and MATLAB software).

The presented technique was intensively tested, validated, and compared with the methods existing in the literature. Especially two approaches were investigated [9,17]. However, the method described in [17] is inefficient and fails due to the speckle noise on SAR images. The approach presented in [9], in turn, is accurate only on urban areas, and such assumption cannot be applied in the considered case. The authors main goal was to provide universal approach able to work in both, urban and rural areas.

3. The Principles of the Structure from Motion

Modern software packages, application, and function libraries dedicated to raster data orientation, and 3D shape reconstruction utilize algorithms based on a combination of methods commonly applied in Computer Vision (CV) and conventional photogrammetric approaches. These types of algorithms and methods allow the geometry and appearance of an object or an entire scene to be captured, and have been used in video games assets [20], virtual tours [21], virtual and augmented reality [22], and cultural heritage [23–27], among others. One of the most important approaches is the Structure from Motion (SfM) method [20,27–29]. The SfM pipeline allows for the reconstruction of three-dimensional structures based on a series of images (rasters) acquired from different positions (observation points) [20].

Figure 2 shows the overview of the incremental SfM workflow, which contains the following steps: (1) feature extraction; (2) feature matching; (3) geometric verification; (4) reconstruction initialization; (5) image registration; (6) triangulation; and (7) bundle adjustment. To generalize, the SfM approach might be divided into two main parts: the correspondence search phase (1–3) and iterative reconstruction phase (4–6). Based on these two phases, the estimation of the camera position for each image as well as a 3D reconstructed tie point, called a spare point cloud [20], can be done.

In this article, only the correspondence search with the computation of registration parameters is described.

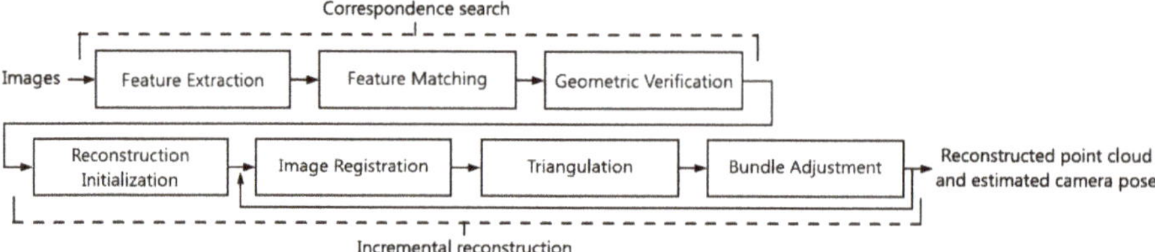

Figure 2. Incremental SfM pipeline [20].

3.1. The Feature Extraction

The feature extraction process is the first step of the SfM pipeline, which is based on detectors. For each image (raster data) given as an input, a group of characteristic points (called keypoints) are detected (excreted) based on the local characteristic of the image intensity. For feature extraction, different methods and algorithms can be used which affect the robustness of the detected features, as well as the efficiency of the matching method. Nowadays, two types of commonly used algorithms are corner (such as FAST, Harris, etc.) and blob detectors (i.e., SIFT and its modification) [30–38]. Brief summaries of studies into data fusion and finding correspondences between SAR and optical images have been recently made by [39] and [14]. Due to the fact that many feature detectors exist, in this section, only SIFT [40] and ASIFT (a modified version of the SIFT) [38] algorithms are focused on, which were used in this investigation.

The SIFT (Scale Invariant Feature Transform) algorithm, which was originally proposed by Lowe [40] for the registration of optical images, has already been adapted for the matching of SAR images. In their studies, [41] focused on the denoising of SAR data with curvelet transformation and the evaluation of SIFT performance on SAR images for various terrain types. The speckle noise, which is characteristic for SAR images, has a vast influence on the SIFT algorithm's performance, and beside denoising (which is further described in Section 4) deeper modification into SIFT feature extraction has also been proposed in the SAR-SIFT approach [19]. However, the noise is not the only difference between SAR and optical images; geometrical differences also have an influence on image registration [14]. In this research, the authors are searching for corresponding points between aerial SAR images where image points are recorded according to the object-to-antenna distance and optical orthophotomaps where bare ground is in ortho projection, but all other objects that are elevated from the ground (e.g., buildings or vegetation) are distorted (shifted) by central projection from the optical image. To overcome these geometric distortions, the authors propose to use the ASIFT (Affine Scale Invariant Feature Transform) [38], which is a modification of the SIFT algorithm. The main idea of the ASIFT algorithm is to simulate a set of sample views of the initial images, obtainable by varying the two camera axis orientation parameters, namely the latitude and the longitude angles, which are not detected by the classical SIFT method [38]. Then, the SIFT method is applied to all the virtual images generated. Thus, ASIFT covers all six parameters of the affine transform and guarantees full affine invariant independence. In the ASIFT algorithm, each image is transformed by simulating all possible affine distortions caused by the change of the initial camera positions. To perform this, the camera model as well as the affine model are utilized (Equations (2) and (3)):

$$u = S_1 G_1 A T u_0, \tag{1}$$

where u is a digital image; u_0 is an (ideal) infinite resolution frontal view of the flat object; T and A are, respectively, a plane translation and a planar projective map due to the camera motion; G_1 is a

Gaussian convolution modeling the optical blur; and S_1 is a standard sampling operator on a regular grid with mesh 1.

$$u(x,y) \to u(ax+by+e, cx_dy+f) \tag{2}$$

$$A = \begin{bmatrix} a & b \\ c & d \end{bmatrix} = H_\lambda R_1(\psi) T_t R_2(\phi) = \lambda \begin{bmatrix} \cos\psi & -\sin\psi \\ \sin\psi & \cos\psi \end{bmatrix} \begin{bmatrix} t & 0 \\ 0 & 1 \end{bmatrix} \begin{bmatrix} \cos\phi & -\sin\phi \\ \sin\phi & \cos\phi \end{bmatrix} \tag{3}$$

where $\lambda > 0$ is the determinant of A, R_i are rotations, $\phi \in [0,\pi)$, and T_t is a tilt, namely a diagonal matrix with first eigenvalue $t > 1$ and the second one equal to 1.

It is possible to prepare the decomposition of the camera motion parameters into the viewing point angels (longitude (ϕ) and latitude ($\theta = \arccos\frac{1}{t}$)), spin of the camera ($\psi$) and zoom factor ($\lambda$). In the ASIFT algorithm, images undergo rotation with the angle ϕ, which is represented by the tilt parameter $t = \frac{1}{\cos\theta}$. In the ASIFT algorithm, the influence of the tilt (latitude rotation) is performed by the t-sampling and the Gaussian convolution with standard deviations$c\sqrt{t^2-1}$ ($c = 0.8$). It is assumed that the latitudes θ are sampled as the geometric series $1, a, a^2, \ldots, a^n$ with $a > 1$ and choosing $a = \sqrt{2}$ is a good compromise between the accuracy and the number of steps. The authors of the ASIFT algorithm proposed the n value equals 5, which results in the tilt being simulated 32 times [38]. In the case of the longitudes ϕ, the arithmetic series $0, \frac{b}{t}, \ldots, \frac{kb}{t}$ with the $b \cong \frac{2\pi}{5}$ and $k\frac{b}{t} < \pi$ is used. After the process of virtual image generation (which includes the skew, tilt, and rotation), any detector, such as SIFT or SURF, might be used. In this study, the SIFT detector was used.

3.2. The Feature Description

After the process of detecting characteristic points, the next step is to describe this by analyzing the nearest points. In the literature, several descriptors are presented such as SIFT, SURF, Daisy, etc. [42], but the authors decided to describe only the SIFT detector, which was used in this study. The main idea of the SIFT descriptor is to compute the local image gradients at the selected scale in the region around the tested keypoint. The full description of the descriptor can be found in Lowe's publication [34], and a further SAR-specific modification of this descriptor can be found in [19]. In the original ASIFT algorithm proposal, the SIFT detector is used. This mathematical computation allows one to determine which detected key-point and its surrounding is highly distinctive yet as invariant as possible to remaining variations, such as changes in illumination or 3D viewpoint. The descriptor calculation is similar to determining the detector as the image gradient magnitude and orientations are sampled around the keypoint localization for each octave and each Gaussian blur.

3.3. The Feature Matching and Images Registration

The detection and description of features for each characteristic point are important components in the process of the detection of conjugate points in digital images. To determine if the keypoint (obtained through the point detection and description process) might be threaded as a tie point, the feature matching process is used. This allows one to take into consideration two points from different images, but which are characterized by the same description as a homologous point. Different strategies can be used for effectively computing matches between images, but two which are usually used are: approximate nearest-neighbor-based point matching [43] and brute-force matching [43]. In the nearest neighbor approach, the points are stored in the k-dimensional space (k-d tree structure). This allows one to compute the nearest neighbors based approximately on the minimal distances between the descriptor values [43]. The brute-force matcher is much simpler because it takes the descriptor of one feature in the first set and matches it with all other features in the second set, using a set of distance calculations. As a result, the closest feature is returned. The feature matching based only on the descriptors is justified in the case of the sensors with a similar wavelength and images obtained from the same optical system. In the case of matching heterogeneous images (e.g., optical and SAR), where a large amount of outliers is expected during feature (descriptors) matching, also adding additional geometrical constraints could be beneficial. [18] used spatial consistent matching, which assumes that the geometrical relationship between matching features should not change too much across images. However, this solution is still largely based on SIFT-like descriptors which are constructed using gradients of image values that can differ significantly between SAR and optical images because of radiometric differences [39]. To overcome this problem, the authors of this paper proposed another way of keypoint matching, where the feature matching is based solely on the geometrical relations between keypoints and utilizes the Iterative Closest Point (ICP) algorithm [44]. The ICP algorithm is a well-known algorithm, implemented in many commercial and open-source software, as well as in programming libraries (such as VTK, open3D, and PCL) [45–47] and is used for oriented point clouds—the minimal distance between two point clouds [44]. There are many variants of the ICP algorithm [44,48–52]; however, in this section, only one of them (thebasic one), proposed by Besl and McKey, is described in more detail.

When considering two datasets, it is possible to determine interrelations between them expressed by the phenomenon:

$$y_i = Rx_i + y_0 \tag{4}$$

where R is a rotation matrix, x_i are the point coordinates in the input point cloud reference system, and y_0 is the translation vector. In the proposed approach, only 2D coordinates are used because orthophotos do not contain height information, and SAR images during pre-processing are projected onto a plane (with mean height). Real-world coordinates can be easily obtained for orthophotomaps because they are georeferenced, and, for SAR images, coordinates are estimated with direct georeferencing using the plane trajectory from GNSS and/or an inertial navigation. The main objective of the ICP method is to align two point clouds, based on shapes or models, by using the Euclidean distance dependence between the nearest point from the initial set of points and the reference. For this purpose, based on the least square method using the distance square minimization (Equation (5)) function, transformation parameters are calculated for points on the areas for which common coverage occurs.

$$e^2 = \sum_i \|Rx_i + y_0 - y_i\|^2 \Rightarrow \min. \tag{5}$$

In each of the iterations of the ICP algorithm, the transformation can be determined using the four main methods: SVD decomposition [53], Hora quaternions [54], Horn's orthogonal matrix [55], and based on Walker's double quater [56]. These algorithms are characterized by similar effectiveness and stability of operation in the case of noisy point clouds [57]. Based on the calculated translation and rotation parameters, the initial point cloud is transformed and the whole process repeats until the minimal distance threshold is not reached. The presented ICP method is commonly used for a 3D point cloud registration. However, in the case of the keypoint matching, only 2D space computation is performed. When using the ICP method, it is important to have a good first approximation of the relative orientation parameters, because without it the solution of the final registration might fail. In the proposed method of the SAR images and orthophotomaps coregistration, this condition is met thanks to the approximate georeference of both data sources. This way of keypoint matching allows one to reduce the problem of the influence of the descriptor, because it is only based on the geometrical relationship between the keypoints detected in SAR and optical images, and could be easily used as long as the approximate georeference of both images are known. One of the most important parts of the image orientation is the geometric verification of the matched keypoints. This correct keypoint matching determines the final correctness of the alignment and quality of the data registration. Depending on the methodology of keypoint matching—descriptor matching or keypoint ICP matching—different methods are used, but overall it should be stressed that the feature matching phase only verifies pairs of points on matched images. Considering the descriptor matching, it is not guaranteed that the matches found actually correspond to 3D points in the scene, and outliers could be included. It is important to find the correct geometric transformation that correctly maps the corresponding point. Using the descriptor method, it is necessary to choose the correct mathematical model for transformation, i.e., homography or similarity transformation. A list of commonly used methods are presented in [20]. To reduce the outliers from the data, it is necessary to implement robust estimation techniques such as RANSAC (Random Sample Consensus [56]) or MLESAC (Maximum Likelihood Estimation SAmple and Consensus), which is a generalization of the RANSAC algorithm [58,59]. In the case of the ICP matching method, the transformation method is predefined. However, for eliminating the outliers, a transformation method such as a similarity transformation can be applied. In this investigation, the registration parameters from the keypoint ICP method are treated as final and applied to the correction of the SAR georeference.

4. SAR Image Preprocessing

Because all kinds of algorithms which detect keypoints such as ASIFT/SIFT/SURF work on intensity images, the presence of speckle noise in SAR images is an obstacle. Speckle noise affects the behavior of keypoint detector algorithms, which is why it has to be reduced. *Speckle noise* is a common phenomenon that accompanies all coherent imaging systems, such as SAR sensors [60]. The source of this "noise" is attributed to random interference between the coherent returns issued from the numerous scatterers present on the surface of a scene, in relation to the wavelength of incident radar wave. The resulting *speckle* has a multiplicative nature, thus SAR imagery is characterized by strong intrinsic noise (hereafter, referred to as *speckle* only). Typically, for a single-look SAR image, **ISNR** (Intrinsic Signal to Noise Ratio) = 0 dB, and we have the same amount of signal and noise power/level. It appears in the image as strong fluctuations in its brightness, hardening the image interpretation.

As described in the literature, utilizing SAR radar images for different purposes often requires additional operations, increasing SNR. If the operations are not performed, the results can be strongly disturbed [7,61]. Some despeckling methods should be taken into account in a processing pipeline, even though they might be computationally complex (for example as shown in Figure 3), because by reducing the influence of noise, better quality results can be obtained, which might be critical in many applications. In the presented approach, speckle noise plays an important or even key role in the final outcome. Despite the resolution reduction caused by the filter usage, the local dynamic of the image is significantly improved, which allows further processing to be carried out. Further processing in this case means keypoint localization, which is the finding of characteristic points in the image. If the speckle noise is present, disturbed pixels may be considered as keypoints even if they are only distorted.

For a fully developed *speckle* (see Figure 4a), the brightness fluctuations in SAR images can be modeled using a gamma distribution (Equation (6)). The gamma distribution is one of the basic types of distributions used in SAR radar imaging (although not the only one [60]):

$$\begin{aligned} p_z(z,n|\sigma,L) &= \frac{n}{\Gamma(L)}\left(\frac{L}{\sigma}\right)^L z^{nL-1}\exp\left(-\frac{Lz^n}{\sigma}\right), \\ n &= 1 \quad \text{for intensity data, } I = A^2 \\ n &= 2 \quad \text{for amplitude data, } A, \end{aligned} \tag{6}$$

with the following statistical moments:

$$E\left[z^m\right] = \frac{\Gamma(L+m/n)}{\Gamma(L)}\left(\frac{\sigma}{L}\right)^{m/n}, \tag{7}$$

where z is the given pixel value; n is the amplitude (A) or intensity (A^2) data format; σ is the expected (true) pixel value; L is the number of (multi)looks/averages; and m is the statistical moment order ($m = 1$ is the mean value; $m = 2$ is the mean square value; etc).

From Equations (6) and (7), providing $n = 1$, the basic maximum likelihood estimators (MLE) can be determined:

$$E[z] = \langle z \rangle, \tag{8}$$

$$ENL\ (\equiv L) = \frac{E^2[z]}{E[z^2]} = \frac{\langle z \rangle^2}{\langle z - \langle z \rangle \rangle^2}, \tag{9}$$

where $E[\bullet]$ is the expected value (operator); $\langle \bullet \rangle$ is the arithmetic averaging; and ENL or $ENIL$ is the equivalent number of (independent) looks.

One of the basic ways to deal with the problem of high *speckle* noise level is the pre-processing of SAR imagery with despeckling filters. Commonly, a classical multilooking technique is applied. Nevertheless, there are plenty of filtration/despeckling algorithms.The authors decided to describe types of filters which were used in experimental processing since the results are different for each filter type.

4.1. ML2D: Multilook–2D Filter

In contrast to the classical multilooking procedure [62], Multilook–2D (which might also be called a non-coherent version of the multilook procedure) works in both image dimensions, i.e. range and cross-range (X and Y) [61]. Thanks to this, a more effective *speckle* reduction can be made with less degradation of the image spatial resolution at the same time.

The idea for the 2D multilooking procedure was taken from optical image processing. The algorithm operates on the entire, already prefocused SAR image, based on Fourier domain processing.

$$\begin{aligned} Z\{\omega_x, \omega_y\} &= \mathbb{F}_{2D}[z(x, y)], \\ Z\{\omega_x, \omega_y\} &\approx \sum_{i=1}^{L} Z_i\{\omega'_x, \omega'_y\}, \\ \tilde{z}(x, y) &= \sum_{i=1}^{L} \mid \mathbb{F}_{2D}^{-1}[\, Z_i\{\omega'_x, \omega'_y\} \cdot W\{\omega'_x, \omega'_y\}\,] \mid, \end{aligned} \tag{10}$$

where $z(x, y)$ is the original noisy SAR image; $\tilde{z}(x, y)$ is the *despeckled* (reconstructed) SAR image; $Z\{\omega_x, \omega_y\}$ is the two-dimensional Fourier spatial frequency spectrum $\{\omega_x, \omega_y\}$; $Z_i\{\omega'_x, \omega'_y\}$ is the partial two-dimensional Fourier spatial frequency spectrum $\{\omega'_x, \omega'_y\}$; $W\{\omega'_x, \omega'_y\}$ is the window (spectrum weighting) function; and L is the number of (multi)looks...

The algorithm transforms the entire radar image into a two-dimensional Fourier space. Then, it divides the spectrum into L sub-bands (with the possibility of overlapping, typically $\leq 50\%$) and filters each sub-band by appropriate weighting. Finally, it reconstructs the image throughout, returning to the spatial domain for each sub-band, generating sub-pictures, and incoherently putting all of them together. The resulting (reconstructed) SAR image is characterized by its reduced L-times speckle level. Undesirable side effects are $\sqrt{L}$-times spatial resolution degradation and visible side lobes resulting from spectrum windowing. The result of filtering a real-life SAR radar image is shown in Figure 4b.

4.2. MEAN: Averaging Filter

One of the simplest noise filtration techniques is averaging image samples over the area around a pixel, combined with a sliding window technique (see Equation (11)). For a gamma distribution [60], the averaging operation also corresponds to the maximum likelihood estimator (MLE) [63].

$$\tilde{z}(k) = \frac{1}{N} \sum_{i=1}^{N} z(k+i), \tag{11}$$

where $z(k)$ is the original noisy SAR image; k is the linear image pel index ($k = x + Width * y$) $Width$ is the image width in pixels/pels; *Height* is the image height in pixels/pels; and x, y is the image pixel(s) indexes (horizontal and vertical); $\tilde{z}(k)$ is the *despeckled* (reconstructed) SAR image; N is the filter window size in pixels/pels, e.g. 3×3, 5×5, etc; and i is the filter window linear index $\in \langle 1, N \rangle$.

As a result of the averaging filtration, the speckle standard deviation is reduced by a factor of $\sqrt{N}$-times, where N is the number of pixels in the filter window. Note that in extreme cases, if the window size is $N = 1$ (only the central pixel without its surroundings will be taken into account), no filtration effect will be noticed. In turn, for a large window size N, the speckle will be reduced at a cost of image detail degradation. An averaging filter that does not include local image statistics results in severe degradation of details (such as: lines, edges and point target blurring). To overcome this issue, only suitably sized windows should be chosen (small windows are usually used e.g. 3×3, 5×5 points in two dimensions). The result of filtering a real-life SAR radar image with an averaging filter is shown in Figure 4c.

4.3. MMSE: Minimum Mean Square Error Filter

The above-mentioned despeckling filters fail when the assumption of constant pixel value within the filter window breaks down. The filter should then adapt to take account of excess fluctuations compared to *speckle* within the analysis window. One approach to such an adaptive filter is to provide a model-free minimum mean-square error (MMSE) filter based on measured local statistics. The solution of minimizing the mean square error for a pixel $\tilde{z}(k)$ is to perform first-order expansion about its local mean value $\bar{z}(k)$ (Equation (11)) so that:

$$\begin{aligned} \tilde{z}(k) &= \bar{z}(k) + \alpha \cdot \left(z(k) - \bar{z}(k)\right), \\ \alpha &= \frac{\overline{V}_{\sigma}}{\overline{V}_{z}} = \frac{\overline{V}_{z} - 1/L}{\overline{V}_{z}\,(1 + 1/L)}, \end{aligned} \tag{12}$$

where $z(k)$ is the original noisy SAR image; k is the linear image pel index ($k = x + Width * y$); $\bar{z}(k)$ is the expected/mean pixel value (see Equation (11)); $\tilde{z}(k)$ is the *despeckled* (reconstructed) SAR image; α is the linear interpolation weight; and $\overline{V}_z$ is the normalized variance.

As can be seen, it is a weighted sum (or linear interpolation) between the mean and given/current pixel value. In cases when there is no fluctuation coming from the image texture, the weighting factor $\alpha \to 0$ and the pixel value is assigned the average value of its surroundings. On the other hand, when the fluctuation of the image texture takes on a significance weighting factor $\alpha \to \frac{L}{L+1}$, the value of the pixel will be scaled by a factor of alpha—which can happen in places where there are lines, edges, or any other texture/spatial features. The result of filtering a real-life SAR radar image with an MMSE filter is shown in Figure 4d. A similar approach was presented by Lee [64,65].

4.4. ELEE: Enhanced Lee Filter

An approach presented by Lee [64] considers an optimal linear filter that is equivalent to a first-order Taylor expansion of the multiplicative noise model $z(k) = \bar{z}(k) \cdot \eta$ about expected $\bar{z}(k)$ and *speckle* component η. Multiplicative noise can be rewritten as an additive one by $z(k) = \bar{z}(k) + (\eta - 1) \cdot \bar{z}(k)$, thus resulting in a similar form to MMSE (Equation (12)), but the weighting factor α is now given a bit differently:

$$\begin{aligned} \tilde{z}(k) &= \bar{z}(k) + \alpha \cdot \left(z(k) - \bar{z}(k)\right), \\ \alpha &= \frac{\overline{V}_{z} - 1/L}{\overline{V}_{z}}, \end{aligned} \tag{13}$$

where $z(k)$ is the original noisy SAR image; k is the linear image pel index ($k = x + Width * y$); $\bar{z}(k)$ is the expected/mean pixel value (see Equation (11)); $\tilde{z}(k)$ is the *despeckled* (reconstructed) SAR image; α is the linear interpolation weight; and $\overline{V}_z$ is the normalized variance.

When there is no image texture variation, it would be expected that the estimate $\overline{V}_z$ is close to pure *speckle*, i.e. $1/L$, so that $\alpha \rightarrow 0$ and $\tilde{z}(k) = \bar{z}(k)$ (same as for the MMSE algorithm). However, texture variability causes $\overline{V}_z$ to be different from the *speckle*. If pixel value $z(k)$ is sufficiently large compared with its surroundings, it yields a large value of $\overline{V}_z$, so that $\alpha \rightarrow 1$ and $\tilde{z}(k) = z(k)$, and the pixel value remains unchanged (no filtration effect). Thus, the response of Lee's filter to strong targets differs from MMSE in that it ignores the *speckle* contribution to the target when making the filtration, corresponding to treating the bright pixel as a point target that would not give rise to *speckle* fluctuations. The result of filtering a real-life SAR radar image with an enhanced Lee filter is shown in Figure 4e.

4.5. GMAP: Gamma MAP Filter

The Gamma filter is a Maximum A Posteriori (MAP) filter based on a Bayesian analysis of the image statistics. It assumes that both the SAR image texture and the *speckle* follows a Gamma distribution (Equation (6)). The imposition of these distributions yields a K-distribution [66], which is recognized to match a large variety of different types of radar clutter, such as land and ocean type cover. The formula of the GMAP filtration is given by:

$$\begin{aligned} \tilde{z}(k) &= \frac{\bar{z}(k) \cdot (\nu - L - 1) + \sqrt{\bar{z}^2(k) \cdot (\nu - L - 1)^2 + 4 \cdot \nu \cdot L \cdot z(k) \cdot \bar{z}(k)}}{2 \cdot \nu}, \\ \nu &= \frac{1}{\overline{V}_\sigma} = \frac{1 + 1/L}{\overline{V}_z - 1/L}, \end{aligned} \tag{14}$$

where $z(k)$ is the original noisy SAR image; k is the linear image pel index ($k = x + Width * y$); $\bar{z}(k)$ is the expected/mean pixel value (see Equation (11)); $\tilde{z}(k)$ is the *despeckled* (reconstructed) SAR image; ν is the texture model order parameter; and $\overline{V}_z$ is the normalized variance.

In the case of pure *speckle*, it would be expected that $\overline{V}_z = 1/L$ so that $\nu \rightarrow \infty$ and $\tilde{z}(k) = \bar{z}(k)$ (as with the MMSE and Lee filters). However, as mentioned above, texture variability causes the estimate $\overline{V}_z$ to be different from the *speckle*. In this case, excess pixel value $z(k)$ is understood as image texture contribution, and the parameter takes small values, thus the Gamma MAP estimator weights the current pixel value by the properly calculated weight $\tilde{z}(k) = z(k)/(1 + 1/L)$. The result of filtering a real-life SAR radar image with Gamma MAP filter is shown in Figure 4f.

4.6. SAR-BM3D: SAR Block Matching 3D Filter

Among different types of despeckling filters [64,67–69], one in particular demonstrates great improvements in speckle filtration performance—SAR-BM3D [70]. The filter has a very complicated structure; however, to outline the processing flow, it can be summarized as shown in Figure 3. The filter is divided into two stages, the first of which makes a coarse estimate of the image content/texture by Non-Local Meanings (NLMs), and the second stage makes a fine estimate of the image texture by Wiener filtration. The algorithm operates in the wavelet transform domain, and the general formula of SAR-BM3D filtration is given by:

$$\begin{aligned} \tilde{Z}(k) &= \bar{Z}(k) + \frac{V_\Sigma^2(k)}{V_\Sigma^2(k) + V_U^2(k)} \cdot \left[Z(k) - \bar{Z}(k)\right], \\ Z(k) &= \Sigma(k) \cdot H = \Sigma(k) + (H - 1) \cdot \Sigma(k) \\ &= \Sigma(k) + U(k), \\ Z(k) &= \mathbb{WT}_{3D}\left[z(k)\right], \quad \tilde{z}(k) = \mathbb{WT}_{3D}^{-1}\left[\tilde{Z}(k)\right], \end{aligned} \tag{15}$$

where $Z(k)$ is the original noisy SAR image in wavelet transform domain, and capitalized letter means a transformed value (e.g., $X = WT[x]$); $\bar{Z}(k)$ is the expected (true/original) image texture in WT transform domain; $\tilde{Z}(k)$ is the *despeckled* (reconstructed) SAR image in WT transform domain; $u(k)$ is the zero mean, additive signal dependent *speckle* noise ($u(k) = (\eta - 1) \cdot \bar{z}(k)$); k is the linear image pel index ($k = x + Width * y$); and η is the fully developed *speckle* noise.

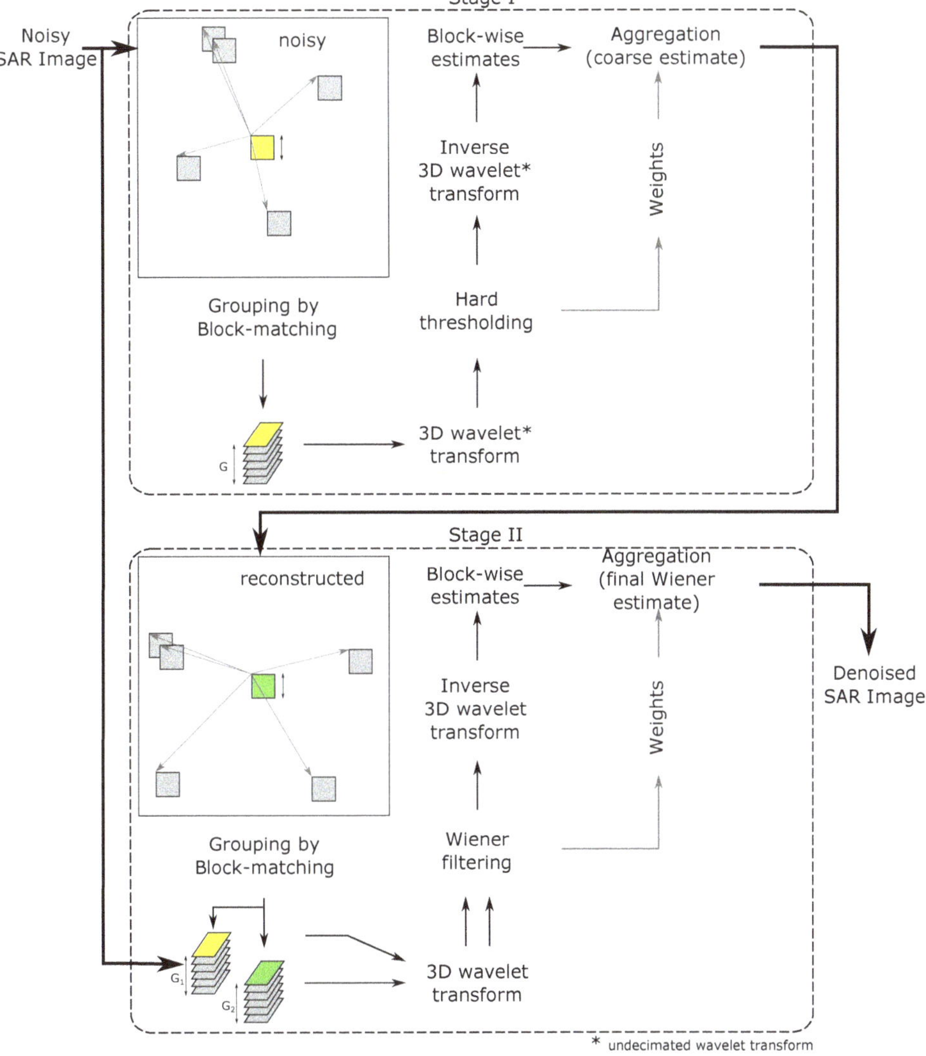

Figure 3. SAR-BM3D despeckling filter block diagram.

The first stage comprises three steps:

- grouping—for each reference block, the most similar blocks are grouped together;
- collaborative filtering—each 3D group undergoes UWT(Undecimated Wavelet Transformation); hard thresholding, and inverse UWT; and
- aggregation—all filtered blocks are suitably weighted, giving a coarse estimate of the image.

The second stage also comprises the same three stages, but with the following differences:

- grouping—blocks are located based on the coarse estimate provided by the first stage;
- collaborative filtering—each 3D group (of noisy blocks) undergoes WT((Decimated) Wavelet Transformation); Wiener filtering, and inverse WT; and

- aggregation—as in Stage 1, giving a final estimate of the image.

The presented filtration algorithm comprises several state-of-the-art techniques, thus giving better performance in terms of signal-to-noise ratio and perceived image quality than any of the other aforementioned filters. The result of filtering a real-life SAR radar image with the SAR-BM3D filter is shown in Figure 4g.

4.7. Filtration Performance

In this subsection, SAR image despeckling filtration results are presented and compared. The image used is shown in Figure 4a; the imaged area is located near Plock city/refinery, Poland. The image was made by WUT's (Warsaw University of Technology) proprietary radar imaging system. The images shown in Figure 4 present results for different filters applied. In Table 1, a summary of the despeckling filtration performance is presented.

It should be emphasized that the implementations of filters were made as so-called MATLAB ® *m*-scripts, and the entire processing chain was made in this computing environment. Thus, the filtration performance in terms of absolute processing time is not meaningful. However, it reflects the computational complexity for each filter.

Table 1. SAR†‡ images filtration performance comparison.

Filter Type	ENL\ENIL	Processing Time
ML2D	23.9	12 s
MEAN	21.7	23 s
MMSE	17.9	43 s
ELEE	19.3	03 m:25 s
GMAP	24.1	03 m:28 s
SAR-BM3D	199.2	04 h:05 m:17 s

† *Original SAR image size* 2540 × 2250 [*px*]; ‡ *Original SAR image ENL* = 1.

From the results presented in Table 1 and Figure 4, it can be seen that SAR-BM3D is the most effective filter. In this case, the equivalent number of looks (ENL) indicator is many times higher than for the others; also the visual assessment is outstanding (see Figure 4g). However, at the same time, it is the slowest one; its execution time is several times longer compared to the other filters, which is a consequence of its sophisticated/complex signal processing pipeline. However, there is a potential to significantly speed up its execution time (e.g., by using CUDA).

(**a**) Original/noisy SAR image

Figure 4. *Cont.*

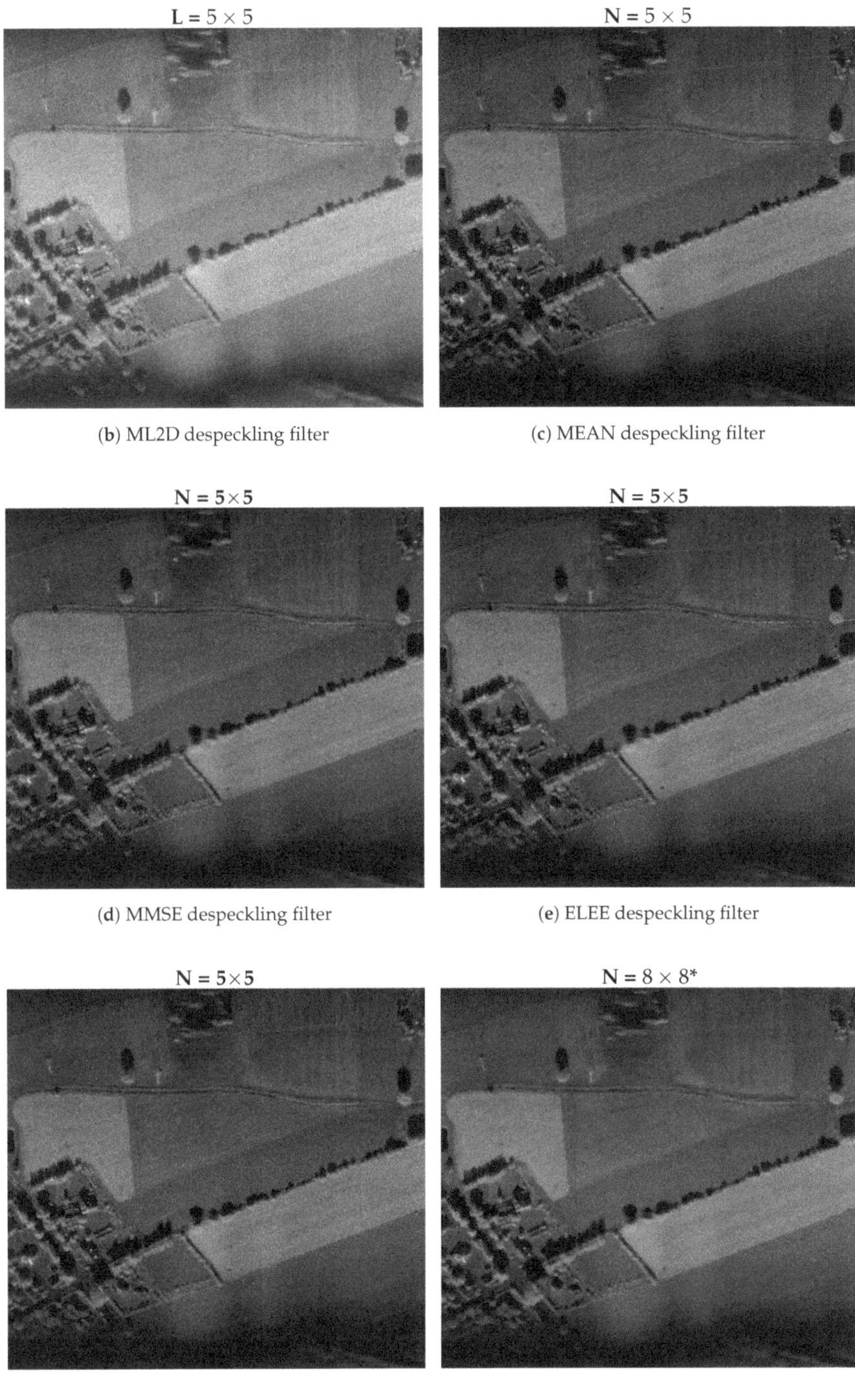

(**b**) ML2D despeckling filter (**c**) MEAN despeckling filter

(**d**) MMSE despeckling filter (**e**) ELEE despeckling filter

(**f**) GMAP despeckling filter (**g**) SAR-BM3D despeckling filter

Figure 4. SAR imagery filtration results: L is the number of (multi)looks and N is the filter window size (equiv. to L). * Even window size forced by UWT transform implementation.

On the other hand, the ML2D despeckling filter provides good performance for both *speckle* reduction and execution time (see Table 1 and Figure 4b). In comparison with the other filters, it performs the fastest, giving very good image quality ($ENL \approx L$ is the desired number of multilooks). Nevertheless, this statement does not disqualify the usability of the other filters.

The ENL indicator is not fully unambiguous in the context of radar to optical image comparison. It is important to preserve details of the unique structural features of the image resulting from the texture of the observed scene. Thus, an assessment based solely on ENL is not conclusive. Therefore, in the following sections, further analysis is carried out for all the filters described above.

5. Results

According to the previously mentioned methods for SAR image despeckling, the ASIFT algorithm was examined to verify its usability in the considered problem. The following scene presented in Figure 5 is considered.The SAR image being considered was gathered during one of the measurement campaigns in which the authors took part. During the raw data acquisition, the GPS data were stored together with the IQ signal samples. The optical image is the reference orthophotomap obtained during the geodetic measurements, and is defined by precise latitude and longitude information. As can be seen, a significant shift is visible in the SAR image, which has to be eliminated using the proposed approach. In theimage, corresponding regions were marked in colors, provingdirect georeference errors between the images. Next, the six described filters were employed to obtain speckle noise reduction, and the results were compared to the original SAR image.

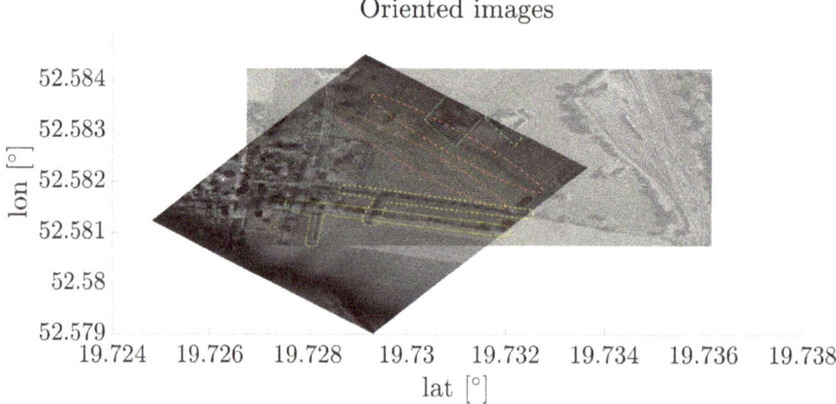

Figure 5. Initially oriented images. The corresponding characteristic areas were marked in colors.

In the process providing point cloud matching, two images are considered. The first is the orthophotomap, which is the same for all cases. The second image was changed to verify the proposed filters. An optical reference image is presented in Figure 6. The total number of keypoints found on the orthophotomap is 126, 164.

Figure 6. Orthophotomap with marked keypoints processed using the ASIFT algorithm.

5.1. Original SAR Image

In the first step, the original SAR image was considered. As mentioned, speckle noise may have significant influence when SAR data are processed in a typical way, ignoring this phenomenon. However, the problem under consideration requires additional processing, because such noise provides unwanted keypoints related to the nature of SAR image creation, not actual artifacts in the image. Figure 7 presents the results of the ASIFT algorithm processing for the initial SAR image.

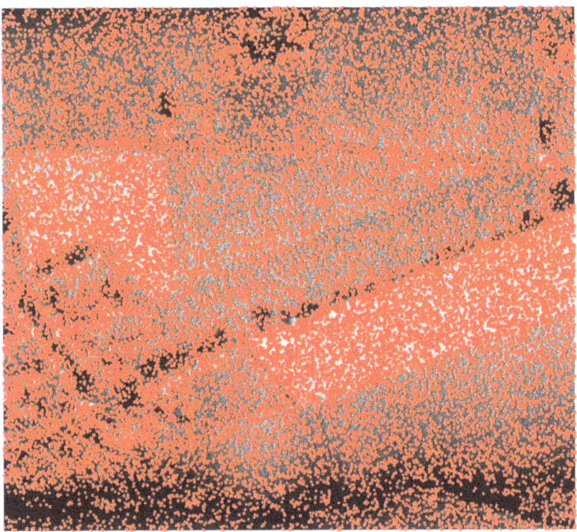

Figure 7. The results of the ASIFT algorithm processing for the initial SAR image.

As can be seen, speckle noise has a significant impact on the number of keypoints as well as their quality. The ASIFT algorithm found 367,584 keypoints, which is almost three times more in comparison to the optical image. Since speckle noise is present in all of the dataset, the distinguishing of the characteristic components or objects is impossible. However, the point clouds assigned to the images were processed, and the results before and after the operation are presented in Figure 8.

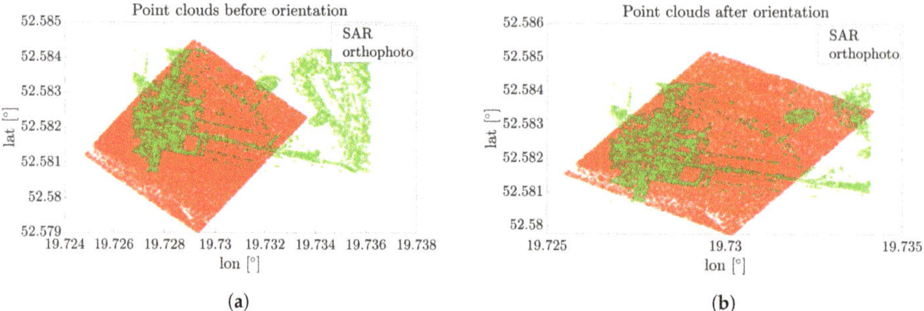

(a) (b)

Figure 8. Point clouds before and after correction for the orthophotomap and original SAR image. (**a**) Point clouds of the SAR and optical images before orientatio, (**b**) Point clouds of the SAR and optical images after orientation.

Because of the distorted data, it is impossible to find orientation correction. Because of the speckle noise, keypoints were found in the entire image, whereas the optical image is covered by keypoints only in the regions with the highest dynamic. As such, point clouds corresponding to keypoints in both images are correlated incorrectly.

For the analyzed point clouds, estimated latitude and longitude correction were calculated as a maximum value of histograms according to each coordinate. The histograms were performed by analyzing the shift of each point from the SAR image to match the optical reference orthophotomap. The results obtained are depicted in Figure 9. The estimated longitude correction is $\Delta_{\Leftrightarrow} = 6.625 \cdot 10^{-4}[^{\circ}]$ and the latitude correction is $\Delta_{\Updownarrow} = 6.45 \cdot 10^{-4}[^{\circ}]$. The results are compared with those obtained for despeckled images below.

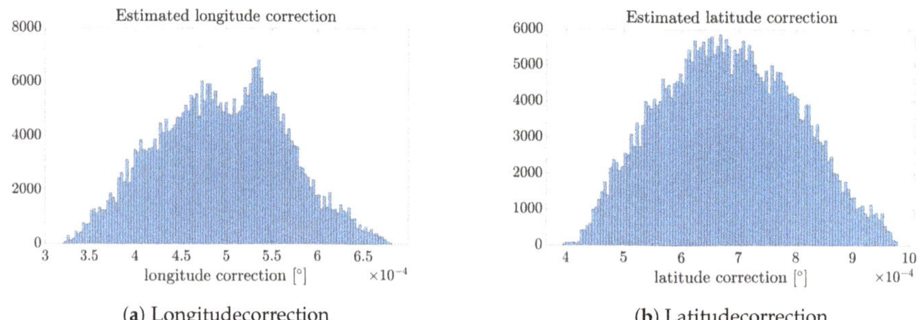

(**a**) Longitudecorrection (**b**) Latitudecorrection

Figure 9. Longitude and latitude correction calculated based on the original SAR image.

5.2. SAR Image Filtered Using ML2D

The ML2D method is presented in Section 4.1. Here, the results of the modified ASIFT algorithm are delivered. First, the keypoints were found in the filtered SAR image. The results obtained are depicted in Figure 10.

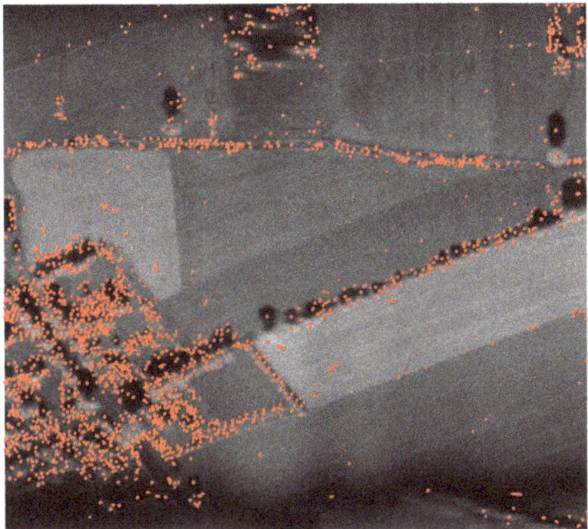

Figure 10. The results of the ASIFT algorithm processing for the SAR image filtered using ML2D method.

As can be noticed, the number of keypoints is significantly reduced by using the despeckling method. After this operation, keypoints were found in characteristic places such as trees, buildings, or roads. The ASIFT algorithm found 49,153 keypoints. The point clouds before and after orientation are presented in Figure 11.

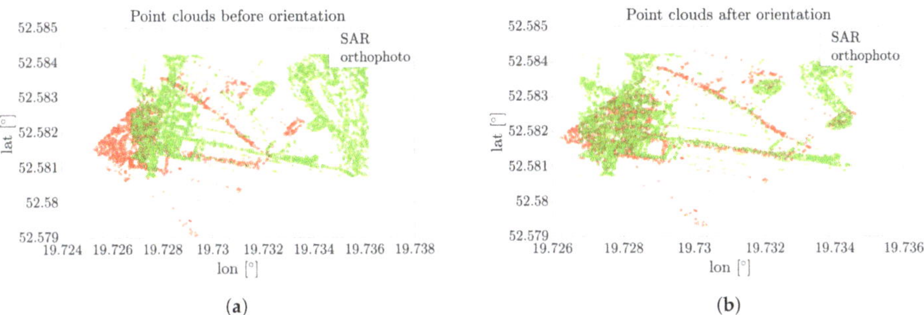

Figure 11. Point clouds before and after correction for the orthophotomap and SAR image filtered using ML2D method. (**a**) Point clouds of the SAR and optical images before orientation, (**b**) Point clouds of the SAR and optical images after orientation

As expected, the filtered image allowed precise results to be obtained. Contours provided by the keypoints in the SAR scene were matched to the same scene, but illustrated using an optical sensor. Thanks to the ML2D method, the correction was estimated. Each coordinate is presented in Figure 12.

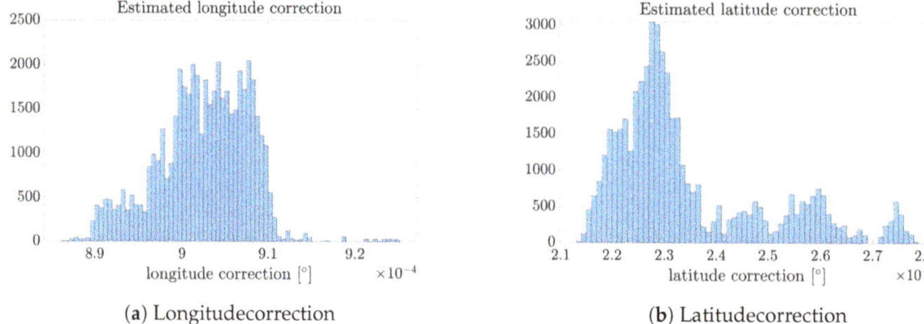

(**a**) Longitudecorrection (**b**) Latitudecorrection

Figure 12. Longitude and latitude correction calculated based on the original SAR image filtered using ML2D method.

The estimated longitude and latitudecorrections are, respectively: $\Delta_{\Leftrightarrow} = 9.775 \cdot 10^{-4}[^\circ]$, $\Delta_{\Updownarrow} = 2.69 \cdot 10^{-4}[^\circ]$. It is worth noting that the character of histograms is different in comparison to the unfiltered data. In the initial case, the error had a Gaussian curve shape. After filtration, the latitude error can be clearly indicated. In the longitude case, the error is ambiguous. This is caused by the character of the SAR and optical image creation. Orthophotomaps are usually made perpendicular (NADIR) to the Earth, whereas SAR radars working in the StripMap mode illuminate scenes from a certain angle (off-NADIR). This dependency provides additional affine transformation, "stretching" an image in the direction of the antenna main lobe. However, by approximating the histogram using a Gaussian curve and interpolating data, the value corresponding to the correction can be estimated in a more precise way than in the original SAR image. Additionally, the correction sets are significantly narrower in comparison to the original data.

The results obtained for the other filters are similar and depicted in the same way. To make the text perspicuous, graphical representations such as keypoint clouds, histograms, and SAR images with marked points are attached in Appendix A.

5.3. SAR Image Filtered Using MEAN Filter

The next considered filter is presented in detail in Section 4.2. This approach is the simplest and probably the most intuitive, and can be quickly implemented to obtain a SAR image with reduced speckle noise. The filtered image with marked keypoints is depicted in Figure A1.

The number of detected keypoints significantly decreased. Despite the simple nature of the filter, the navigation error was estimated. The ASIFT algorithm detected 65,489 points whose distribution focused on characteristic places such as trees, buildings, etc. It enabled keypoints clouds obtained for both optical and SAR images to be comparable, as presented in Figure A2.

The MEAN filter allows improvement of the results to be obtained. According to the histograms presented in Figure A3, the corrected latitude and longitude are, respectively, $\Delta_{\Leftrightarrow} = 9.465 \cdot 10^{-4}[^\circ]$ and $\Delta_{\Updownarrow} = 2.595 \cdot 10^{-4}[^\circ]$. The presented histograms show the correction distribution, based on which the navigation drift is obtainable.

Figures provided for this subsection are attached in Appendix A.1.

5.4. SAR Image Filtered Using MMSE Filter

The MMSE filter, described in Section 4.3, was examined as the next method for speckle noise reduction. As in the previous methods, in the first step, keypoints were detected using the ASIFT method, as presented in Figure A4.

As can be seen, a larger number of points was detected. The algorithm extracted 66,917 keypoints, which is clearly a higher value than in the previous results achieved for the algorithms of speckle noise reduction. By analyzing the result, it can be seen that points were found in places where they were not detected for other noise reduction methods. Keypoints are visible in fields and uniform areas, resulting in poorer point cloud quality. However, the overall point cloud structure has been preserved and correctly covered, as illustrated in Figure A5.

Although the quality of the keypoints is worse, because more points were detected, it was possible to correctly cover the characteristic areas distinguished from the images. A characteristic triangle formed by tree lines and buildings located in the left part of the picture were covered, determining the error being sought.

The calculated longitude and latitude corrections presented in Figure A6 are as follows: $\Delta_{\Leftrightarrow} = 9.355 \cdot 10^{-4}[°]$ and $\Delta_{\Updownarrow} = 2.57 \cdot 10^{-4}[°]$. This is a similar result to the previously used methods. Accuracy is mainly limited by the histogram resolution, which can be improved by interpolation. Despite the detection of a larger number of points, which resulted from a smaller reduction of the speckle noise, it successfully allowed the detected keypoints clouds to be covered, and thus the navigation correction to be recalculated. The results are consistent with the methods presented above.

Figures provided for this subsection are attached in Appendix A.2.

5.5. SAR Image Filtered Using ELEE Filter

The ELEE filter described in detail in Section 4.4 was employed as the next method for the improvement of point clouds covering. As in the case of the MMSE filter, more points were found both in places with increased dynamics and on uniform surfaces. This indicates less noise reduction; however, the overall character of the keypoint cloud was set as in the case of the MMSE filter. The results obtained are presented in Figure A7.

For the discussed method, 58, 165 characteristic points were detected, which is also a significant number compared to the analyzed methods. The keypoint clouds before and after the correction are shown in Figure A8.

Again, the correction was calculated correctly, as evidenced by the coverage of characteristic areas in the images under investigation. It turns out that the method of point cloud shift correction is resistant to such small discrepancies and additional detection of characteristic points resulting from limited filtration of speckle noise. Histograms presenting navigation correction are presented in Figure A9. The estimated longitude and latitude corrections are, respectively: $\Delta_{\Leftrightarrow} = 9.4175 \cdot 10^{-4}[°]$ and $\Delta_{\Updownarrow} = 2.19 \cdot 10^{-4}[°]$.

Figures provided for this subsection are attached in Appendix A.3.

5.6. SAR Image Filtered Using GMAP Filter

The next investigated filter is GMAP, whose extended characteristic is presented in Section 4.5. The result showing the point cloud is illustrated in Figure A10.

The total number of points found in the image is 22, 200, which is the smallest value for all of the considered cases. The algorithm detected characteristic points corresponding to different objects in the scene, which indicates a significant reduction in noise in the image.

As shown in Figure A11, the last filter also provides correct results. The point clouds are similar to those previously presented, making the GMAP filter an effective tool in the proposed method. It is worth noting that, for the considered image, there are a few regions where the resolution is low, such as in the lower part where trees are present, or in the upper part where buildings are depicted. Histograms presenting navigation correction are shown in Figure A12. Comparing the result in Figure A10 to the ones previously obtained, new details are available. In this case, the GMAP method shows a distinguishing object initially "hidden" in noise, which can be used in the point cloud analysis. The same effect is visible in Figure A13 for the SAR-BM3D filter. For the last considered case, the estimated longitude correction is $\Delta_{\Leftrightarrow} = 9.635 \cdot 10^{-4}[°]$, whereas the estimated latitude correction is $\Delta_{\Updownarrow} = 2.445 \cdot 10^{-4}[°]$, which correspond to the previously obtained values.

Figures provided for this subsection are attached in Appendix A.4.

5.7. SAR Image Filtered Using SAR-BM3D Filter

The last examined filter is BM3D, described in detail in Section 4.6. The same high resolution SAR image was processed to obtain a keypoint cloud, allowing navigation the correction to be estimated. The input image with marked keypoints is presented in Figure A13.

The utilized despeckling method significantly improved the image quality (by reducing the noise level), which significantly affected the focusing of the detected keypoints in the most dynamic regions. The keypoints distribution corresponds to the results obtained for the different despeckling methods previously presented (including the number of detected keypoints). However, fewer keypoints were found in comparison to the previously presented outcomes (apart from the GMAP filter). The algorithm detected 24,954 points. In Figure A14, the keypoint clouds before and after orientation are presented.

Thanks to the reduction of speckle noise, negligible improvement was obtained, which affects the unequivocal shift estimation. The correction distribution, presented in Figure A15, is similar to the previous methods. In addition, in this case, the considered set is narrower than it was presented for the original SAR image without despeckling methods. The estimated longitude and latitudecorrections are, respectively: $\Delta_{\Leftrightarrow} = 9.865 \cdot 10^{-4} [^{\circ}]$ $\Delta_{\Updownarrow} = 2.475 \cdot 10^{-4} [^{\circ}]$. This result proves the usability of such a filter for both, despeckling purposes as well as for the proposed modified ASIFT algorithm.

Figures provided for this subsection are attached in Appendix A.5.

5.8. Discussion

Each of the applied filters allowed estimationcorrections to be calculated in accordance with the assumptions. Due to the different nature of the creation of optical images and SAR, some imperfections are visible, but they are negligible in the analyzed problem. The filters processing performance is presented in Section 4.7. Table 2 shows the processing time using the ASIFT algorithm (does not consider filtering time) and the point cloud correlation. Additionally, the results of the correction estimation, as well as the number of points found in each of the images, are summarized.

Table 2. Estimated coordinates correction for different filters.

Filter Type	Longitude Correction [°]	Latitude Correction [°]	Rotation [°]	Processing Time [s]	Amount of Keypoints
No filter	6.6250×10^{-4}	6.4500×10^{-4}	4.0919	111.7241	367,584
ML2D	9.7750×10^{-4}	2.6900×10^{-4}	0.9416	79.9240	49,153
MEAN	9.4650×10^{-4}	2.5950×10^{-4}	0.5612	71.8246	65,489
MMSE	9.3550×10^{-4}	2.5700×10^{-4}	0.6804	77.3086	66,917
ELEE	9.4175×10^{-4}	2.1900×10^{-4}	1.2547	71.1760	58,165
GMAP	9.6350×10^{-4}	2.4450×10^{-4}	0.0051	66.2229	22,200
BM3D	9.8650×10^{-4}	2.4750×10^{-4}	0.3204	66.9932	24,954

As can be seen, similar results were obtained for all filters tested. The outcomes of the rotation estimation are characterized by the following relationship: the greater is the number of points found, the greater is the rotation correction determined. Interestingly, the calculation time is not strongly dependent on the number of keypoints detected. However, storing points requires more memory, and, even if the calculation time is similar, the memory required is larger for the method detecting a greater number of points. It should be noted that all computation was performed on the CPU without any parallelization. To decrease the computation time and boost the computation speed, a GPU should be used.

The experiments proved that the proposed method for the geometrical matching of SAR and optical images utilizing ASIFT features for SAR-based Navigation Aided Systems is suitable for compute the corrections to the SAR images' direct georeferencing. In the literature, there are many methods and algorithms for this type of data co-registration [9,11–13,15,16,18,19]. All of these algorithms were tested on SAR images and optical images acquired from space. The proposed methodology of data integration is based on high resolution (in full resolution) SAR images and orthophotomaps obtained from altitude. Thus, it is hard to compare the presented method with the method described in [9,11–13,15,16,18,19], because of the spatial resolution of satellite optical images and the size of overlapping areas. For this reason, it required guaranteeing well-distributed (in the whole investigation area), robust corresponding points. To achieve this, a modification of the SfM approach based on the ASIFT detector and he elimination of the keypoints description, as well as description matching step and using the two-step ICP method, was performed. To compute the transformation parameters, methods based on finding the pairs of points are used. This way of determining corresponding points is useful when the points are well distributed throughout the overlapping areas. Unfortunately, when data from attitude are processed, this relationship may not be met and the matching pairs of points will not be evenly distributed throughout the study area. This might cause wrong or inaccurate determination of correction parameters. Therefore, the use of the ASIFT algorithm allows the detection of more points evenly distributed throughout the entire area of work. In connection with the ICP method, the keypoints will not be aligned in pairs, but will form a rigid body that reduces the influence of the outliers on the final determining transformation elements.

In Figure 13, the oriented SAR and optical images before and after correction are presented. In this case, correction calculated by the ELEE method was utilized, however the results are comparable for all cases when despeckling filters are used. As can be seen, characteristic objects such as trees, roads, and buildings are covered.

On the basis of the results obtained, the calculated correction value per kilometer can also be estimated. Assuming that $1^\circ \approx 111.1$ km, the longitude correction is about 105 m, while for latitude the correction was about 25 [m]. This is a significant value, especially in the case of military systems for which the required precision should be as high as possible. For the case where the despeckling filter was not applied, the navigation correction is inappropriate and amounts to 73.6 m for longitude and 71.6 m for latitude, which are the mean values of the differences between the original coordinates of the SAR and optical images taken into consideration. This shows how important it is to combine the presented methods and implement the processing pipeline. The juxtaposition also shows that, from the point of view of the implementation of navigation correction on a flying platform, it is sufficient to use simple despeckling filters, which can significantly reduce computational effort and thus accelerate processing, which is particularly important for fast flying objects. It should be noted that the final resolution of the correction is limited by the radar bandwidth, for which the range resolution is expressed by the relationship $\delta R = \frac{c}{2B}$, where c is the speed of light and B is radar signal bandwidth. For the proposed method to make sense, high resolution radars should be used, which, unfortunately, are associated with a greater financial outlay and a requirement for high processing power of the computing unit. In addition, it should be taken into account that the obtained high resolution radar images must be filtered to reduce speckle noise, which limits the resolution of the entire imaging. Moreover, phenomena such as heterogeneous movement of the radar carrier platform or changing its speed during the flight in an unpredictable way also degrade the quality of the image. Nevertheless, taking into account and minimizing such phenomena, it is possible to use the method proposed by the authors, which was proved experimentally.

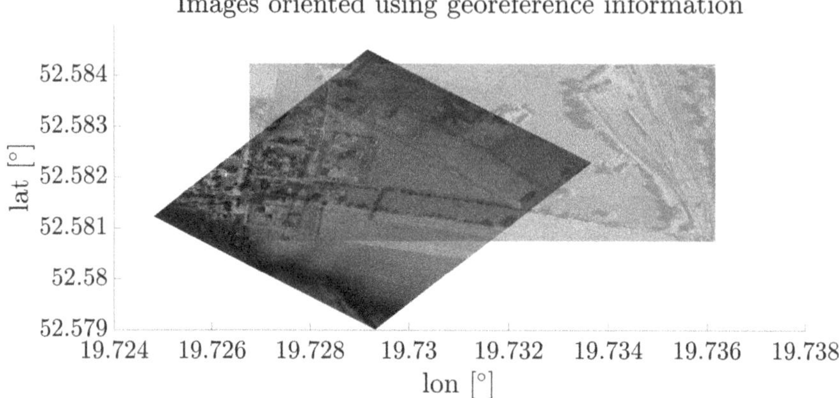

(**a**) Initially oriented images

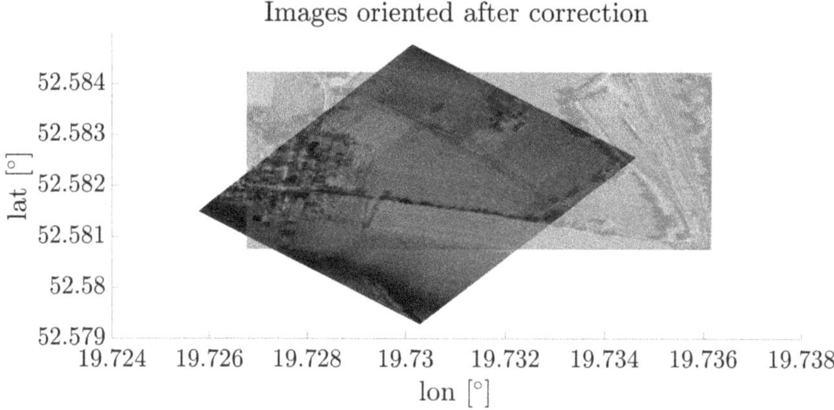

(**b**) Images oriented after correction

Figure 13. Oriented images before and after correction.

6. Conclusions

This paper presents a novel algorithm for navigation correction estimation dedicated to flying platforms (e.g., drones, airplanes, cruise missiles, etc.) in cases when satellite navigation systems are unavailable. This method utilizes various techniques to:

- filter SAR images;
- find keypoints on SAR and optical images;
- find shift and rotation correction between images; and
- calculate navigation correction between an acquired SAR image and a reference optical image.

Combining such methods in a single solution is a novel approach proposed by the authors. Taking advantage of several techniques, an innovative algorithm was proposed, tested and validated to confirm its usability for SAR images obtained during a measurement campaign. This solution may be useful in military and civilian applications, when the lack of a GNSS signal is a critical problem which makes flying missions impossible. Merged techniques such as ASIFT-based keypoint extraction and SfM-based keypoints matching make this method robust and resistant to noise and interference. Thus, the presented methodology can be successfully integrated with existing systems to enhance their precision and dependability. Additionally, the authors provide a comparison of several filters, including their computational complexity and performances. This presents a wide variety of uses for this technique depending on the solution. In the future, the authors intend to implement the method on a real-time platform and test and verify the proposed methodology in real conditions, which should confirm its usability. The target platform is the GPU (Graphics Processing Unit), allowing fast computing to be obtained (both SAR processing and correction estimation). Another crucial issue is the implementation of the functionality that would be able to cope with the situation where the imaged scene has been significantly changed compared to the one that was saved in the database (as an optical image). Such a situation may arise when, for example, the imaged area has been damaged as a result of a disaster or war, and the database on board the flying object does not have current pictures. The estimation may then be ineffective, which is undesirable. This is currently the main problem the authors are working on.

Author Contributions: project administration, D.G.; formal analysis, J.M., W.O., A.G., and K.A.; funding acquisition, D.G. and P.S.; investigation, J.M., W.O., A.G., and K.A.; methodology, J.M., W.O., and A.G.; resources, D.G.; software, J.M. and K.A.; supervision, P.S.; validation, D.G.; visualization, A.G. and K.A., writing—original draft, K.A., A.G., J.M., and W.O.; and writing—review and editing, P.S.

Funding: This work was done within the frame of project No. DOB-2P/03/05/2018 funded by the Polish National Center for Research and Development (NCBiR).

Conflicts of Interest: The authors declare no conflict of interest.

Appendix A

In this appendix, the results obtained in Section 5 are presented.

Appendix A.1. Figures for SAR Image Filtered Using MEAN filter

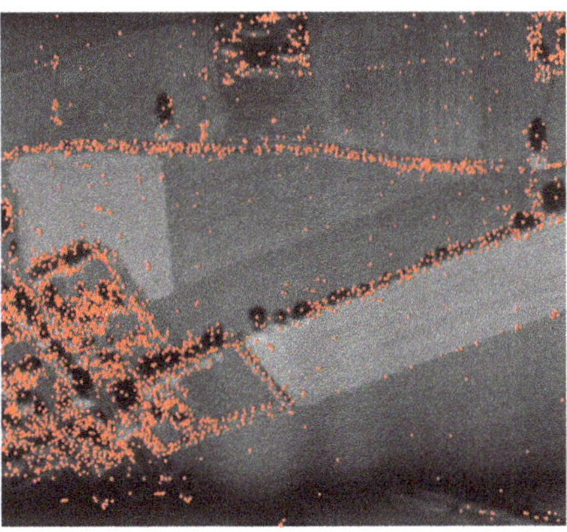

Figure A1. The results of the ASIFT algorithm processing for the SAR image filtered using MEAN method.

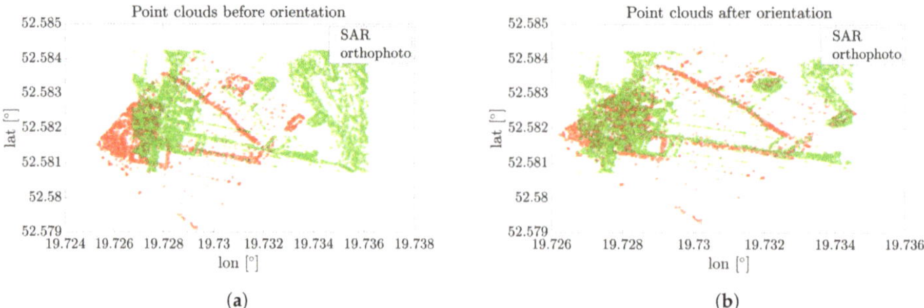

(**a**) (**b**)

Figure A2. Point clouds before and after correction for the orthophotomap and SAR image filtered using MEAN filter (**a**) Point clouds of the SAR and optical images before orientation, (**b**) Point clouds of the SAR and optical images after orientation.

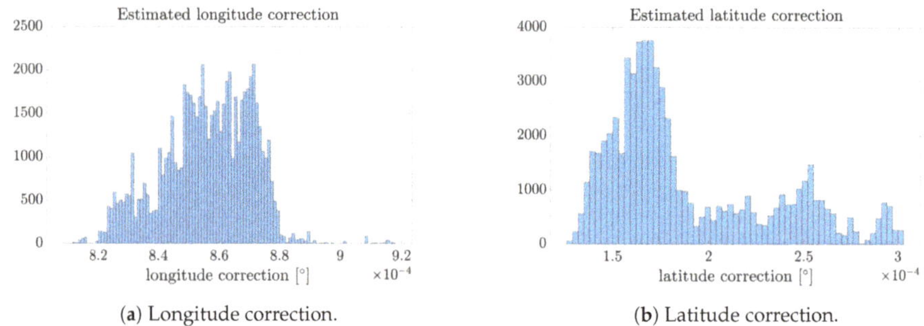

(**a**) Longitude correction. (**b**) Latitude correction.

Figure A3. Longitude and latitude correction calculated based on the original SAR image filtered using ML2D method.

Appendix A.2. Figures for SAR Image Filtered Using MMSE Filter

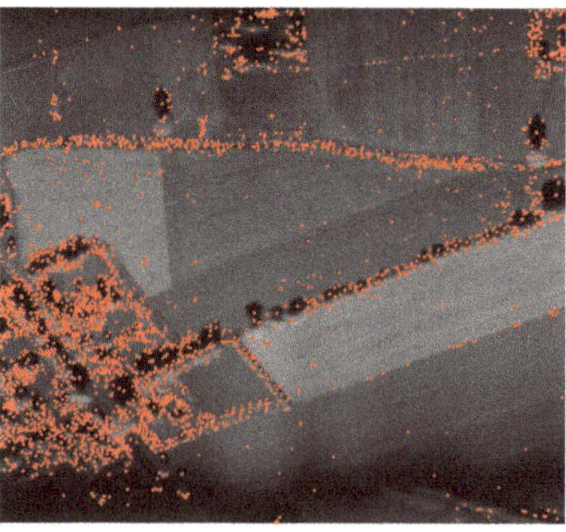

Figure A4. The results of the ASIFT algorithm processing for the SAR image filtered using MMSE filter.

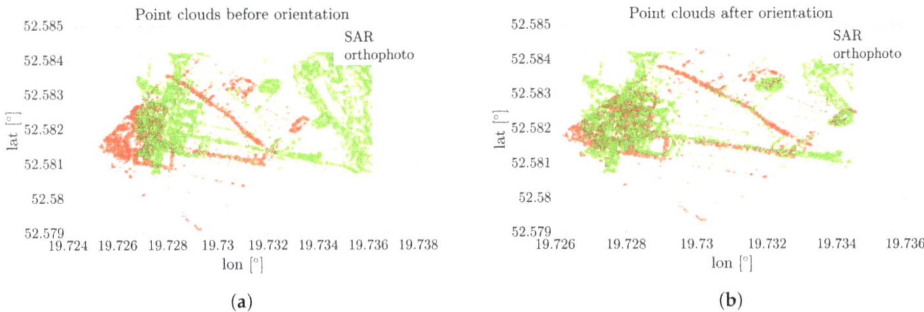

(**a**) (**b**)

Figure A5. Point clouds before and after correction for the orthophotomap and SAR image filtered using MMSE filter. (**a**) Point clouds of the SAR and optical images before orientation, (**b**) Point clouds of the SAR and optical images after orientation.

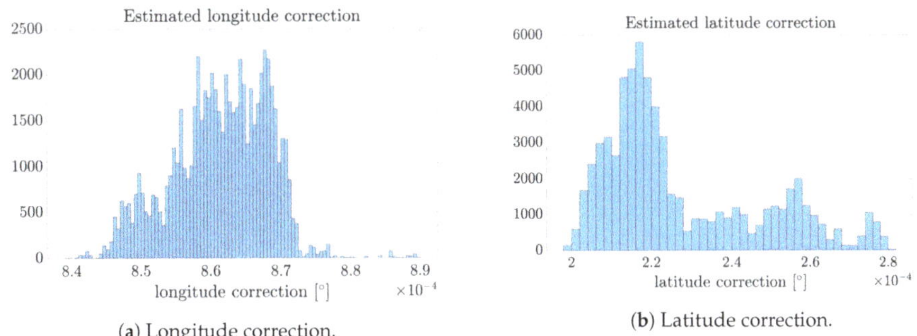

(**a**) Longitude correction. (**b**) Latitude correction.

Figure A6. Longitude and latitude correction calculated based on the original SAR image filtered using MMSE filter.

Appendix A.3. Figures for SAR Image Filtered Using ELEE Filter

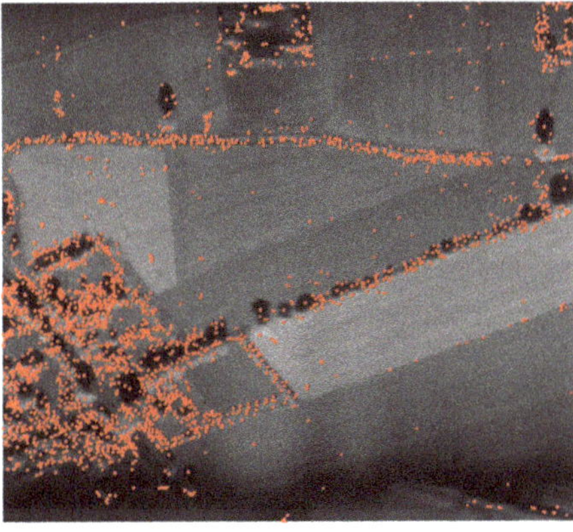

Figure A7. The results of the ASIFT algorithm processing for the SAR image filtered using ELEE filter.

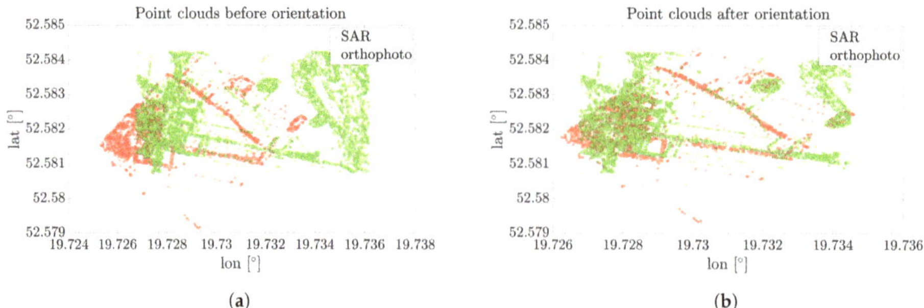

(a) (b)

Figure A8. Point clouds before and after correction for the orthophotomap and SAR image filtered using ELEE filter. (**a**) Point clouds of the SAR and optical images before orientation, (**b**) Point clouds of the SAR and optical images after orientation.

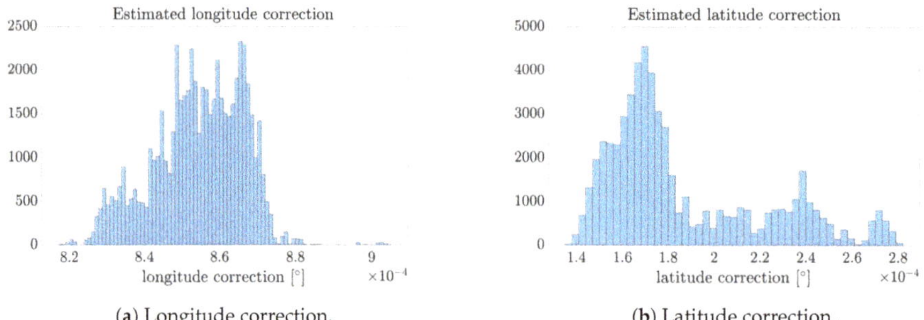

(**a**) Longitude correction. (**b**) Latitude correction.

Figure A9. Longitude and latitude error calculated based on the original SAR image filtered using ELEE filter.

Appendix A.4. Figures for SAR Image Filtered Using GMAP Filter

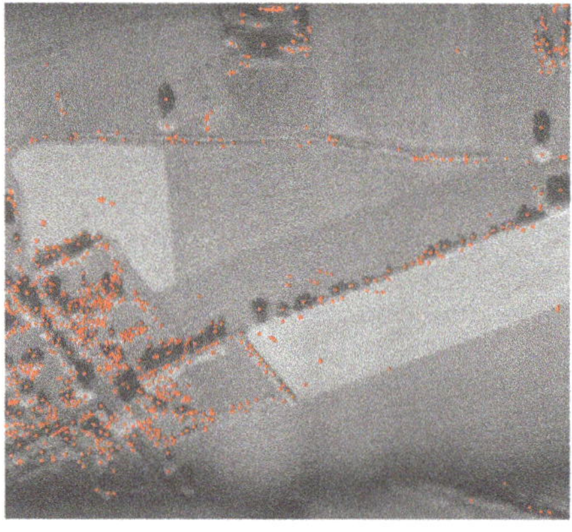

Figure A10. The results of the ASIFT algorithm processing for the SAR image filtered using GMAP filter.

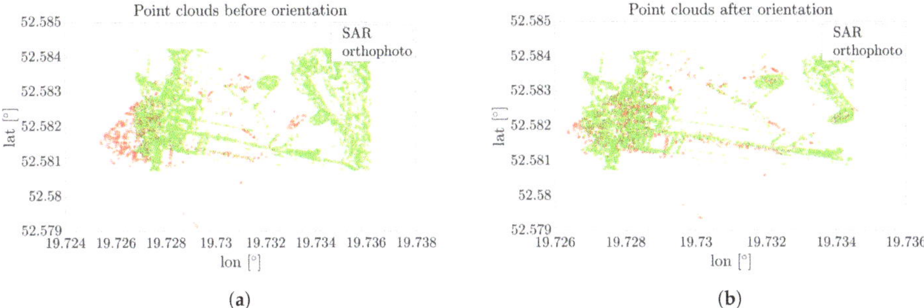

(**a**) (**b**)

Figure A11. Point clouds before and after correction for the orthophotomap and SAR image filtered using GMAP filter. (**a**) Point clouds of the SAR and optical images before orientation, (**b**) Point clouds of the SAR and optical images after orientation.

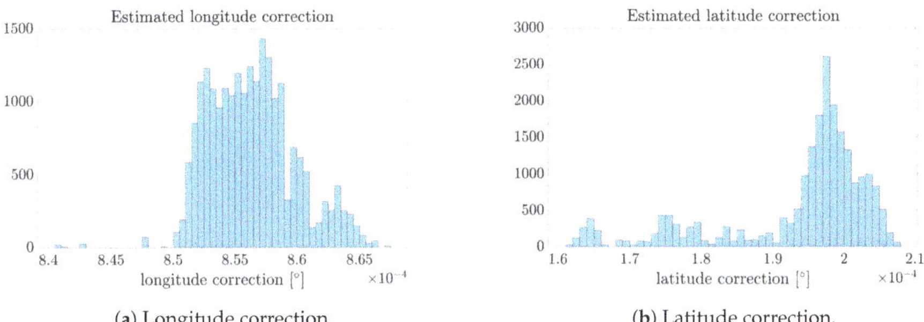

(**a**) Longitude correction. (**b**) Latitude correction.

Figure A12. Longitude and latitude error calculated basing on the original SAR image filtered using GMAP filter.

Appendix A.5. Figures for SAR Image Filtered Using SAR-BM3D Filter

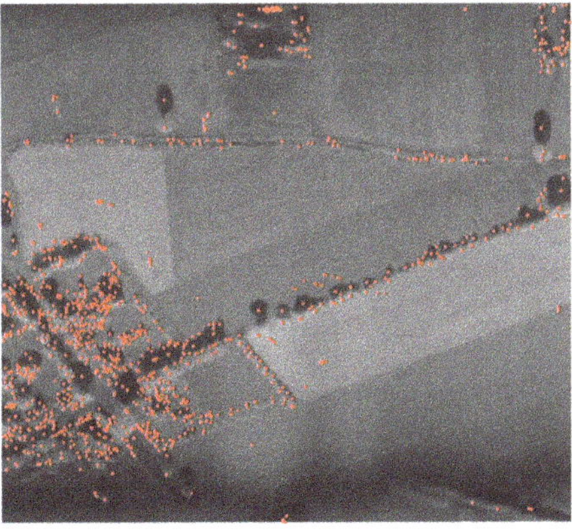

Figure A13. The results of the ASIFT algorithm processing for the SAR image filtered using SAR-BM3D filter.

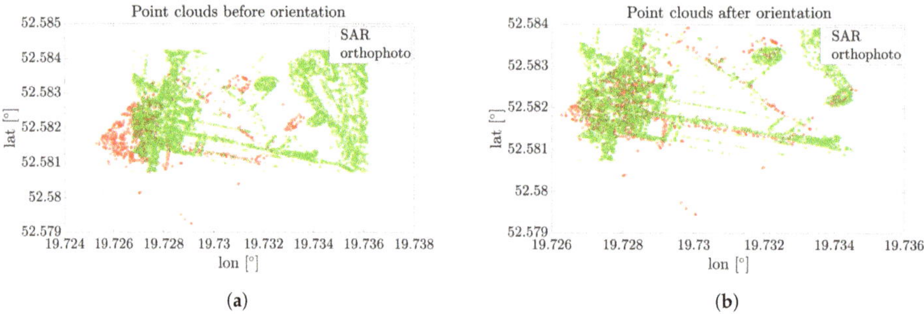

(**a**) (**b**)

Figure A14. Point clouds before and after correction for the orthophotomap and SAR image filtered using SAR-BM3D filter. (**a**) Point clouds of the SAR and optical images before orientation, (**b**) Point clouds of the SAR and optical images after orientation.

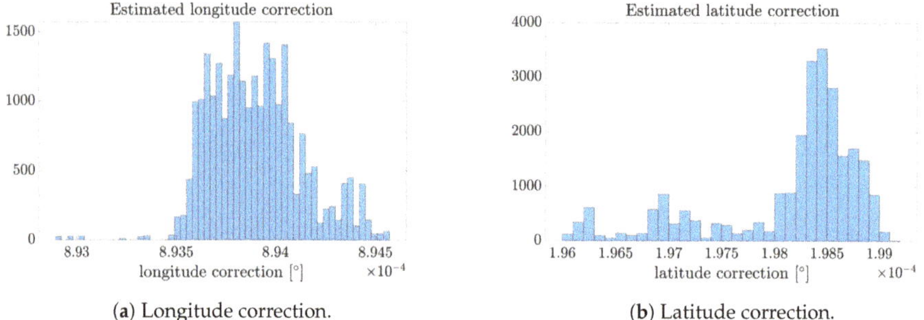

(**a**) Longitude correction. (**b**) Latitude correction.

Figure A15. Longitude and latitude correction calculated based on the original SAR image filtered using SAR-BM3D filter.

References

1. Golden, P. J. Terrain contour matching (TERCOM): A cruise missile guidance aid. *Image Process. Missile Guid.* **1980**, *238*, 10–18. [CrossRef]
2. Vaman, D. TERRAIN REFERENCED NAVIGATION: History, trends and the unused potential. In Proceedings of the 31st Digital Avionics Systems, Williamsburg, VA, USA, 14–18 October 2012; pp. 1–28.
3. Hwang, Y.; Tahk, M.J. Terrain Referenced UAV Navigation With LIDAR—A Comparison of Sequential Processing and Batch Processing Algorithms. In Proceedings of the 28th International Congress of the Aeronautical Sciences, ISAC, Brisbane, Australia, 23–28 September 2012; pp. 1–7.
4. Greco, M.; Pineli, G.; Kulpa, K.; Samczyński, P.; Querry, B.; Querry, S. The study on SAR images exploitation for air platform navigation purposes. In Proceedings of the International Radar Symposium (IRS 2011), Leipzig, Germany, 7–9 September 2011; pp. 347–352.
5. Greco, M.; Kulpa, K.; Pinelli, G.; Samczyński, P. SAR and InSAR georeferencing algorithms for Inertial Navigations Systems. In Proceedings of the SPIE, Bellingham, WA, USA, 7 October 2011.
6. Greco, M.; Querry, S.; Pinelli, G.; Kulpa, K.; Samczyński, P.; Gromek, D.; Gromek, A.; Malanowski, M.; Querry, B.; Bonsignore, A. SAR-based Augmented Integrity Navigation Architecture. In Proceedings of the IRS 2012, Warsaw, Poland, 23–25 May 2012; pp. 225–229.
7. Abratkiewicz, K.; Gromek, D; Samczyński, P.; Markiewicz J.; Ostrowski, W. The Concept of Applying a SIFT Algorithm and Orthophotomaps in SAR-based Augmented Integrity Navigation Systems. In Proceedings of the 2019 20th International Radar Symposium (IRS), Ulm, Germany, 26–28 June 2019; pp. 1–12.
8. Gromek, D.; Samczyński, P.; Abratkiewicz, K.; Drozdowicz, J.; Radecki, K.; Stasiak, K. System Architecture for a Miniaturized X-band SAR-based Navigation Aid System. In Proceedings of the 2019 Signal Processing Symposium (SPSympo), Krakow, Poland, 17–19 September 2019; pp. 271–276. [CrossRef]

9. Salehpour, M.; Behrad, A. Hierarchical approach for synthetic aperture radar and optical image coregistration using local and global geometric relationship of invariant features. *J. Appl. Remote Sens.* **2017**, *11*, 015002. [CrossRef]
10. Schmitt, M.; Tupin, F.; Zhu, X.X.; Saclay, P. *FUSION OF SAR AND OPTICAL REMOTE SENSING DATA—CHALLENGES AND RECENT TRENDS Signal Processing in Earth Observation*; Technical University of Munich (TUM): Munich, Germany, 2017; pp. 5458–5461.
11. He, C.; Fang, P.; Xiong, D.; Wang, W.; Liao, M. A point pattern chamfer registration of optical and SAR images based on mesh grids. *Remote Sens.* **2018**, *10*, 1837. [CrossRef]
12. Wang, B.; Zhang, J.; Lu, L.; Huang, G.; Zhao, Z. A Uniform SIFT-Like Algorithm for SAR Image Registration. *IEEE Geosci. Remote Sens. Lett.* **2015**, *12*, 1426–1430. [CrossRef]
13. Costantini, M.; Zavagli, M.; Martin, J.; Medina, A.; Barghini, A.; Naya, J.; Hernando, C.; Macina, F.; Ruiz, I.; Nicolas, E.; et al. Automatic coregistration of SAR and optical images exploiting complementary geometry and mutual information. *Int. Geosci. Remote Sens. Symp.* **2018**, *2018*, 8877–8880.
14. Kai, L.; Xueqing, Z. Review of research on registration of sar and optical remote sensing image based on feature. In Proceedings of the 2018 IEEE 3rd International Conference on Signal and Image Processing (ICSIP), Shenzhen, China,13–15 July 2018; pp. 111–115.
15. Liu, F.; Bi, F.; Chen, L.; Shi, H.; Liu, W. Feature-Area Optimization: A Novel SAR Image Registration Method. *IEEE Geosci. Remote Sens. Lett.* **2016**, *13*, 242–246. [CrossRef]
16. Zhang, H.; Li, G.; Lin, H. An automatic co-registration approach for optical and SAR data in urban areas. *Ann. GIS* **2016**, *22*, 235–243. [CrossRef]
17. Hellwich, O.; Wefelscheid, C.; Lukaszewicz, J.; Hänsch, R.; Siddique, M.A.; Stanski, A. Integrated matching and geocoding of SAR and optical satellite images. *Lect. Notes Comput. Sci.* **2013**, *7887*, 798–807.
18. Fan, B.; Huo, C.; Pan, C.; Kong, Q. Registration of optical and sar satellite images by exploring the spatial relationship of the improved SIFT. *IEEE Geosci. Remote Sens. Lett.* **2013**, 10, 657–661. [CrossRef]
19. Dellinger, F.; Delon, J.; Gousseau, Y.; Michel, J.; Tupin, F. SAR-SIFT: A SIFT-like algorithm for applications on SAR images. *IEEE Trans. Geosci. Remote Sens.* **2012**, 3478–3481. [CrossRef]
20. Bianco, S.; Ciocca, G.; Marelli, D. Evaluating the Performance of Structure from Motion Pipelines. *J. Imaging* **2018**, *4*, 98. [CrossRef]
21. Saurer, O.; Fraundorfer, F.; Pollefeys, M. OmniTour: Semi-automatic generation of interactive virtual tours from omnidirectional video. In Proceedings of the 3DPVT (Int. Symp. on 3D Data Processing, Visualization and Transmission) Espace Saint Martin, Paris, France, 17–20 May 2010.
22. Noh, Z.; Sunar, M.S.; Pan, Z. A review on augmented reality for virtual heritage system. In *Lecture Notes in Computer Science (Including Subseries Lecture Notes in Artificial Intelligence and Lecture Notes in Bioinformatics)*; Springer: Berlin/Heidelberg, Germany, 2009.
23. Hatzopoulos, J.N.; Stefanakis, D.; Georgopoulos, A.; Tapinaki, S.; Pantelis, V.; Liritzis, I. Use of various surveying technologies to 3D digital mapping and modelling of cultural heritage structures for maintenance and restoration purposes: The Tholos in Delphi. *Greece Mediterr. Archaeol. Archaeom.* **2017**, *17*, 311–336. [CrossRef]
24. Luhmann, T.; Robson, S.; Kyle, S.; Boehm, J. *Close-Range Photogrammetry and 3D Imaging*; De Gruyter: Berlin, Germany; Boston, MA, USA, November 2013; ISBN 9783110302783.
25. Remondino, F. IMAGE-BASED 3D MODELLING: A REVIEW. *Photogramm. Rec.* **2006**, *21*, 269–291. [CrossRef]
26. Schindler, K. Developments in Computer Vision and its Relation to Photogrammetry. In Proceedings of the XXII ISPRS Congress, Melbourne, Australia, 25 August–1 September 2012.
27. Friedt, J. Photogrammetric 3D Structure Reconstruction Using Micmac Basics of SfM Getting Started: Picture Acquisition Methods. 2014; pp. 1–28. Avalible online: http://jmfriedt.free.fr/lm_sfm_eng.pdf (accessed on 9 December 2019)
28. Apollonio, F.I.; Ballabeni, A.; Gaiani, M.; Remondino, F. Evaluation of feature-based methods for automated network orientation. *Int. Arch. Photogramm. Remote Sens. Spat. Inf. Sci.—ISPRS Arch.* **2014**, *40*, 47–54. [CrossRef]
29. Westoby, M.J.; Brasington, J.; Glasser, N.F.; Hambrey, M.J.; Reynolds, J.M. "Structure-from-Motion" photogrammetry: A low-cost, effective tool for geoscience applications. *Geomorphology* **2012**. [CrossRef]
30. Markiewicz, J. The use of computer vision algorithms for automatic orientation of terrestrial laser scanning data. *Int. Arch. Photogramm. Remote Sens. Spat. Inf. Sci.—ISPRS Arch.* **2016**, *41*, 315–322. [CrossRef]

31. Moussa, W.; Abdel-Wahab, M.; Fritsch, D. an Automatic Procedure for Combining Digital Images and Laser Scanner Data. *ISPRS—Int. Arch. Photogramm. Remote Sens. Spat. Inf. Sci.* **2012**, *XXXIX-B5*, 229–234. [CrossRef]
32. Moussa, W. Integration of Digital Photogrammetry and Terrestrial Laser Scanning for Cultural Heritage Data Recording. Ph.D. Thesis, University of Stuttgart, Stuttgart, Germany, 2014.
33. Harris, C.; Stephens, M. A Combined Corner and Edge Detector. *Proc. Alvey Vis. Conf.* **1988**, *1988*, 23.1–23.6. [CrossRef]
34. Lowe, D.G. Distinctive image features from scale-invariant keypoints. *Int. J. Comput. Vis.* **2004**, *60*, 91–110.[CrossRef]
35. Bay, H.; Tuytelaars, T.; Van Gool, L. SURF: Speeded up robust features. *Lect. Notes Comput. Sci.* **2006**, *3951*, 404–417. [CrossRef]
36. Cyganek, B.; Siebert, J.P. *An Introduction to 3D Computer Vision Techniques and Algorithms*; Wiley and Sons Ltd.: West Sussex, UK, 2009; ISBN 9780470017043.
37. Bay, H.; Ess, A. Speeded-Up Robust Features (SURF). *Comput. Vis. Image Understanding* **2008**, *110*, 346–359. [CrossRef]
38. Yu, G.; Morel, J.-M. ASIFT: An Algorithm for Fully Affine Invariant Comparison. *Image Process. Line* **2011**, *1*, 11–38. [CrossRef]
39. Schmitt, M.; Tupin, F.; Zhu, X.X. Fusion of SAR and optical remote sensing data—Challenges and recent trends. In Proceedings of the International Geoscience and Remote Sensing Symposium (IGARSS), Fort Worth, TX, USA, 23–28 July 2017.
40. Lowe, D.G. Object recognition from local scale-invariant features. *Proc. Seventh IEEE Int. Conf. Comput. Vis.* **1999**, *2*, 1150–1157. [CrossRef]
41. Liu, J.; Yu, X. Research on SAR image matching technology based on SIFT. Available online: https://www.isprs.org/proceedings/XXXVII/congress/1_pdf/67.pdf (accessed on 9 December 2019).
42. Tola, E.; Lepetit, V.; Fua, P. DAISY: An efficient dense descriptor applied to wide-baseline stereo. *IEEE Trans. Pattern Anal. Mach. Intell.* **2010**, *32*, 815–830. [CrossRef] [PubMed]
43. Tran, T.T.H.; Marchand, E. Real-time keypoints matching: Application to visual serving. *Proc.—IEEE Int. Conf. Robot. Autom.* **2007**, 3787–3792. [CrossRef]
44. Besl, P.; McKay, N. A Method for Registration of 3-D Shapes. *IEEE Trans. Pattern Anal. Mach. Intell.* **1992**, *14*, 239–256. [CrossRef]
45. PCL PCL. Available online: http://www.pointclouds.org/downloads/ (accessed on 24 July 2019).
46. VTK VTK. Available online: https://vtk.org/ (accessed on 24 July 2019).
47. Open3D Open3D. Available online: http://www.open3d.org/ (accessed on 24 July 2019).
48. Theiler, P.W.; Schindler, K. Automatic Registration of Terrestrial Laser Scanner Point Clouds Using Natural Planar Surfaces. *ISPRS Ann. Photogr. Remote Sens. Spat. Inf. Sci.* **2012**, *3*, 173-178. [CrossRef]
49. Liu, Y. Automatic registration of overlapping 3D point clouds using closest points. *Image Vis. Comput.* **2006**, *24*, 762–781. [CrossRef]
50. Bae, K.H.; Lichti, D.D. A method for automated registration of unorganised point clouds. *ISPRS J. Photogramm. Remote Sens.* **2008**, *63*, 36–54. [CrossRef]
51. Sprickerhof, J.; Nüchter, A.; Lingemann, K.; Hertzberg, J. An Explicit Loop Closing Technique for 6D SLAM. In Proceedings of the 4th European Conference on Mobile Robots, Dubrovnik, Croatia, 23–25 September 2009; pp. 1–6.
52. Vosselman, G.; Maas, H.-G. Airborne and Terrestrial Laser Scanning. *Current* **2010**, *XXXVI*, 318.
53. Arun, K.S.; Huang, T.S.; Blostein, S.D. Least-Squares Fitting of Two 3-D Point Sets. *IEEE Trans. Pattern Anal. Mach. Intell.* **1987**, *9*, 698–700. [CrossRef] [PubMed]
54. Horn, B.K.P. Closed-form solution of absolute orientation using unit quaternions. *J. Opt. Soc. Am. A* **1987**, *4*, 629. [CrossRef]
55. Horn, B.; Hilden, P.K.; Hugh, M.; Shahriar, N. Closed-form solution of absolute orientation using orthonormal matrices. *J. Opt. Soc. Am. A* **1988**, *5*, 1127. [CrossRef]
56. Walker, M.W.; Shao, L.; Volz, R.A. Estimating 3-D location parameters using dual number quaternions. *CVGIP Image Underst.* **1991**, *54*, 358–367. [CrossRef]
57. Eggert, D.W.; Lorusso, A.; Fisher, R.B. Estimating 3-D rigid body transformations: A comparison of four major algorithms. *Mach. Vis. Appl.* **1997**, *9*, 272–290. [CrossRef]

58. Fischler, M.A.; Bolles, R.C. Random sample consensus: A paradigm for model fitting with applications to image analysis and automated cartography. *Commun. ACM* **1981**. [CrossRef]
59. Torr, P.H.S.; Zisserman, A. MLESAC: A new robust estimator with application to estimating image geometry. *Comput. Vis. Image Underst.* **2000**, *78*, 138–156. [CrossRef]
60. Oliver, C.; Quegan, S. *Understanding Synthetic Aperture Radar Images*; SciTech Radar and Defense Series; SciTech Publ.: Raleigh, NC, USA, 2004; ISBN 1891121316.
61. Gromek, A. SAR Imagery Non-Coherent Change Detection Methods. Ph.D. Dissertation, Warsaw Univ. of Tech. Publ., Warsaw, PL, USA, July 2019.
62. Jakowatz, C.V.; Wahl, D.E.; Eichel, P.H.; Ghiglia, D.C.; Thompson, P.A. *Spotlight-Mode Synthetic Aperture Radar: A Signal Processing Approach*; Kluwer Academic Publ.: Norwell, MA, USA, 1996; pp. 105–218; ISBN 0792396774.
63. Scott R. *Maximum Likelihood Estimation: Logic and Practice*; SAGE Publ.: Newbury Park, CA, USA, 1993; pp. 46–56; ISBN 0803941072..
64. Lee, J.S. Refined filtering of image noise using local statistics. *Comp. Graph. Image Proc.* **1981**, *15*, 380–390. [CrossRef]
65. Lee, J.S. A simple speckle smoothing algorithm for synthetic aperture radar images. *I3E Trans. Syst. Man Cybern.* **1983**, *13*, 85–89. [CrossRef]
66. Jakeman, E. On the statistics of K-distributed noise. *J. Phys. A Math. Gen.* **1980**, *13*, 31. [CrossRef]
67. Frost, V.S. A model for radar images and its application to adaptive filtering of multiplicative noise. *I3E Trans. Pattern Anal. Machine Intell.* **1982**, *4*, 157–166. [CrossRef] [PubMed]
68. Kuan, D.T. Adaptive restoration of images with speckle. *I3E Trans. Acoust. Speech Signal Process.* **1987**, *35*, 373–383. [CrossRef]
69. Lopes, A. Structure detection and adaptive speckle filtering in SAR images. *Int. J. Remote Sens.* **1993**, *14*. [CrossRef]
70. Poderico, M. Denoising of SAR Image. Ph.D. Thesis, University of Naples Federico II, Naples, IT, USA, November 2011.

Article

Wavelength-Resolution SAR Ground Scene Prediction Based on Image Stack

Bruna G. Palm [1,*], Dimas I. Alves [2,3], Mats I. Pettersson [4], Viet T. Vu [4], Renato Machado [5], Renato J. Cintra [6,7,8], Fábio M. Bayer [9], Patrik Dammert [10] and Hans Hellsten [10]

1 Programa de Pós-Graduação em Estatística, Universidade Federal de Pernambuco, Recife 50670-901, Brazil
2 Departamento de Engenharia de Telecomunicações, Universidade Federal do Pampa, Alegrete 97546-550, Brazil; dimasalves@unipampa.edu.br
3 Departamento de Engenharia Elétrica, Universidade Federal de Santa Catarina, Florianópolis 88040-900, Brazil
4 Department of Mathematics and Natural Sciences, Blekinge Institute of Technology, 371 79 Karlskrona, Sweden; mats.pettersson@bth.se (M.I.P.); viet.thuy.vu@bth.se (V.T.V.)
5 Department of Telecommunications, Aeronautics Institute of Technology (ITA), São José dos Campos, SP 12228-900, Brazil; rmachado@ita.br
6 Signal Processing Group and Departamento de Estatística, Universidade Federal Pernambuco, Recife 50670-901, Brazil; rjdsc@de.ufpe.br
7 Department of Electrical and Computer Engineering, University of Calgary, Calgary, AB T2N1N4, Canada
8 Department of Electrical and Computer Engineering, Florida International University, Miami, FL 33199, USA
9 Departamento de Estatística and LACESM, Universidade Federal de Santa Maria, Santa Maria 97105-900, Brazil; bayer@ufsm.br
10 Saab Electronic Defence Systems, 412 76 Gothenburg, Sweden; patrik.dammert@saabgroup.com (P.D.); hans.hellsten@saabgroup.com (H.H.)
* Correspondence: brunagpalm@gmail.com

Received: 13 February 2020; Accepted: 31 March 2020; Published: 3 April 2020

Abstract: This paper presents five different statistical methods for ground scene prediction (GSP) in wavelength-resolution synthetic aperture radar (SAR) images. The GSP image can be used as a reference image in a change detection algorithm yielding a high probability of detection and low false alarm rate. The predictions are based on image stacks, which are composed of images from the same scene acquired at different instants with the same flight geometry. The considered methods for obtaining the ground scene prediction include (i) autoregressive models; (ii) trimmed mean; (iii) median; (iv) intensity mean; and (v) mean. It is expected that the predicted image presents the true ground scene without change and preserves the ground backscattering pattern. The study indicates that the the median method provided the most accurate representation of the true ground. To show the applicability of the GSP, a change detection algorithm was considered using the median ground scene as a reference image. As a result, the median method displayed the probability of detection of 97% and a false alarm rate of $0.11/\text{km}^2$, when considering military vehicles concealed in a forest.

Keywords: CARABAS II; ground scene prediction; image stack; multi-pass; SAR images

1. Introduction

Common tasks in synthetic aperture radar (SAR) statistical image processing include the identification and classification of distinct ground type [1–5], modeling [6–9], and change detection [10–13]. In special, wavelength-resolution low-frequency SAR systems are useful for natural disasters monitoring, foliage-penetrating applications, and detection of concealed targets [14].

The wavelength-resolution SAR system is usually associated with ultrawideband (UWB) radar signal and ultrawidebeam antenna [15]. With such, the maximum resolution is achieved and it is in the order of radar signal wavelength. Additionally, available UWB SAR systems only operate at low frequencies. One essential feature of wavelength-resolution SAR systems is that the speckle noise does not influence the acquired images since it is likely that only a single scatter is present in the resolution cell. Additionally, small scatterers present in the ground area of interest do not contribute to the backscattering for low-frequency radar systems. Thus, small structures, such as tree branches and leaves, are not shown in SAR images [16]. Because large scatterers are associated with low-frequency components, they tend to be less influenced by environmental effects and are stable in time. Hence, by using multi-passes with identical heading and incidence angle of the illuminating platform at a given ground area, an image package with similar statistics can be obtained [17]. In [18], clutter statistical models for stacks of very-high-frequency (VHF) wavelength-resolution SAR images are discussed. The SAR image stacks are a frequent topic of study for SAR systems with high resolution [19–21]. However, the literature lacks the use of large image stacks for wavelength-resolution SAR for change detection applications.

Change detection algorithms (CDA) have been widely considered over the years in the detection of distinct targets in SAR images [22–24]. In particular, the wavelength-resolution SAR change detection is an important topic of research and has been studied for more than a decade [17]. Wavelength-resolution systems have also shown unique results with high detectability rate on a low false alarm rate per square km, as presented, for example, in [17,24]. The nature of the wavelength-resolution SAR imagery can be exploited to facilitate the design of CDAs, since (i) the contribution of small scatterers to radar echoes is not significant for the wavelength of several meters; (ii) scatter from large objects are the main contribution; (iii) large scatterers are usually stable in time and less sensitive to environmental effects; and (iv) the wavelength-resolution almost totally cancel the speckle noise [16] in the SAR image given a very stable backscattering between measurements.

A CDA is used to detect changes in a ground scene between distinct measurements in time, such as natural disasters like floods and wildfires or human-made interferences [14]. Generally, in wavelength-resolution systems, a CDA can be simply obtained by the subtraction of two single-look images (reference and surveillance), followed by a thresholding operation. However, an image stack can be considered instead of just two images in a CDA; such a collection of images leads to improved detection performance, as discussed in [17]. This information is used to eliminate clutter and noise in the surveillance image [17], and consequently, enhancing CDA results. Recently, a study using a small stack of multi-pass wavelength-resolution SAR images for change detection was introduced in [17].

In [25], the autoregressive (AR) model was employed as a preliminary study considering a ground scene prediction (GSP) based on a single wavelength-resolution SAR image stack. The resulting predicted image was submitted as input data to a change detection algorithm, based only on subtraction, thresholding, and morphological operations. The CDA in [25] corresponds to the detection analysis step of the CDA used in [26]. Despite its simplicity, the change detection results in [25] were competitive when compared with the ones recently presented in [17,27].

Multi-pass SAR images cannot be exactly equidistantly observed over time since the noise across the image stack is not related to the time order. As a consequence, the use of a time series model, commonly employed in statistical signal processing [28–31], may not be the most suitable approach to obtain a GSP, and, consequently, resulting in lower performance in a CDA. Additionally, the backscattering of the images in the stack is stable in time, i.e., a sequence of pixels for each position follows a similar pattern, and changes in such behavior are understood as outliers. Thus, an image filtering considering robust statistical methods, such as trimmed mean and median [32,33], might be better candidates to obtain a ground scene prediction. These approaches can provide an accurate prediction of the ground scene, avoid the time order problem, and exclude the pixels that do not follow the sequence pattern. Indeed, the median and the trimmed mean filters are traditionally used to remove impulse noise from an image [34–41].

To the best of our knowledge, the study in [25] is the only work related to the ground scene prediction for wavelength-resolution SAR image stacks. Our paper extends the results presented in [25] with four other statistical methods to predict a ground scene for three SAR image stacks, since statistical methods are commonly employed in SAR image processing [1,2,5–11,13]. The selected statistical methods to obtain the prediction image are (i) autoregressive models; (ii) trimmed mean; (iii) median; (iv) intensity mean; and (v) mean. The predicted ground scene methods are sought to preserve the ground backscattering statistical characteristics of the images in the stack and presents predicted pixel values closer to the original images. It is expected that the predicted images represent the true ground scenes, allowing applications, such as monitoring of forested areas and natural disasters. In this paper, our goal is twofold. First, we propose the use of statistical methods to obtain a ground scene prediction image based on a wavelength-resolution SAR image stacks. Second, we consider this new image as a reference image in a change detection algorithm. In particular, we employed the median GSP image obtained based on stack statistics as a reference image in a CDA based on the detection analysis step of the CDA presented in [26], which was evaluated in terms of target detection probability and false alarm rate. The results reported in [12,17,24] were adopted as the reference model for comparison.

The paper is organized as follows. In Section 2, we describe the considered change detection method and a suite of selected statistical methods for ground scene prediction. Section 3 presents experimental results, including a description of the considered data set, the ground scene prediction results, and the change detection results. Then, a change detection method based on the discussed GSP approaches is introduced. Finally, Section 4 concludes the paper.

2. Change Detection Method

The change detection method used in this paper applied the processing scheme given in Figure 1. An image stack is processed by a desirable GSP method furnishing the GSP image. The changes are simply obtained with the subtraction of the image of interest (surveillance image) from the GSP image (reference image). For change detection, we applied thresholding to the difference image and then used morphological operations for false alarm minimization. The methods employed to obtain the GSP images are described in the next section.

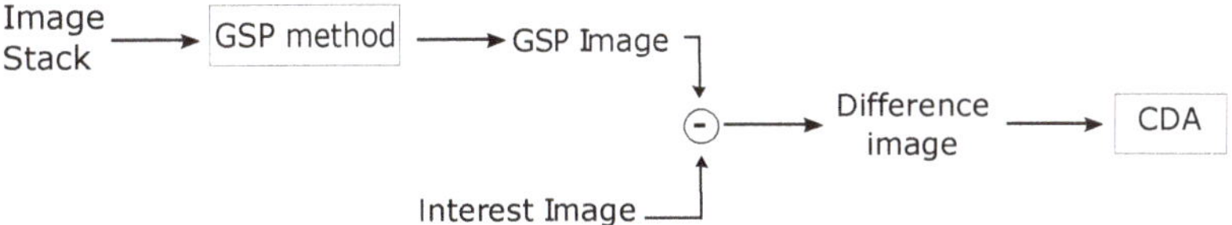

Figure 1. Processing scheme for change detection. The ground scene prediction (GSP) image is the reference image and the interest image is the surveillance image. The change detection algorithm (CDA) is performed applying thresholding and morphological operations in the difference image. Note that the difference image is based on the subtraction between single-look image pixels as a consequence of the stability in backscattering using a wavelength-resolution synthetic aperture radar (SAR) system.

The employed CDA consists of two mathematical morphology steps. First, an opening operation [42] aimed at removing small pixel values, which are regarded as noise. The second step is a dilation that prevents the splitting of the interest targets in multiple substructures. The first step uses a 3×3 pixel square structuring element, whose size is determined by the system resolution; the second step considers a 7×7 pixel structuring element, which is linked to the approximate size of the targets (about 10×10 pixels).

2.1. Ground Scene Prediction

As discussed in [18], an image stack is composed of images with similar heading and incidence angle of the same illuminating platform. As a consequence of this similarity, the SAR images in the stack are very similar and stable in time. Thus, a sequence of each pixel position can be extracted from the stack, as illustrated in Figure 2.

The data set considered in this paper is composed of wavelength-resolution SAR images, i.e., the resolution of the SAR image is in the order of the radar signal wavelength [16]. Therefore, there may only be a single scatter in the resolution cell. As a consequence, the considered images are not affected by speckle noise, which is typically a strong source of noise in SAR images in higher frequency bands. Thus, the backscattering from the image stack is stable in time, allowing an accurate GSP.

Figure 2. Stack of images to be considered in GSP. The methods should be applied for each pixel position, as evidenced by the vertical line.

We consider five statistical methods to obtain ground scene predictions. The techniques are applied in a sequence of pixels, as described in the following.

2.2. AR Model

The AR model was adopted to compute the GSP, which can be defined as [43]

$$y[n] = -\sum_{k=1}^{p} a[k]y[n-k] + u[n], \quad n = 1, 2, \ldots, N, \tag{1}$$

where $y[n]$ is the value of each pixel in one image, N is the number of images in the stack, $a[k]$ are the autoregressive terms, $u[n]$ is white noise, and p is the order of the model [43]. The autoregressive terms $a[k]$ in Equation (1) can be estimated by the Yule–Walker method [43,44].

Hence, the estimated autoregressive terms $\hat{a}[k]$ are the solutions of the equation system, given by [43]

$$\begin{bmatrix} r_{yy}[0] & r_{yy}[1] & \ldots & r_{yy}[p-1] \\ r_{yy}[1] & r_{yy}[0] & \ldots & r_{yy}[p-2] \\ \vdots & \vdots & \ddots & \vdots \\ r_{yy}[p-1] & r_{yy}[p-2] & \ldots & r_{yy}[0] \end{bmatrix} \begin{bmatrix} a[1] \\ a[2] \\ \vdots \\ a[p] \end{bmatrix} = - \begin{bmatrix} r_{yy}[1] \\ r_{yy}[2] \\ \vdots \\ r_{yy}[p] \end{bmatrix}, \tag{2}$$

where $r_{yy}[\cdot]$ is the sample autocorrelation function. Information about large sample distributions of the Yule–Walker estimator, order selection, and confidence regions for the coefficients can be found in [45]. Considering the estimated autoregressive terms $\hat{a}[k]$, it is possible to forecast h steps ahead with the AR model as [44]

$$\hat{y}[N+h] = -\sum_{k=1}^{p} \hat{a}[k]y[N+h-k]. \tag{3}$$

The ground scene prediction image is obtained by forecasting the one-step ahead ($h = 1$) pixel value for each pixel in the image.

2.3. Trimmed Mean, Median, and Mean

For SAR images whose backscattering is stable in time, robust methods can be applied to obtain a GSP. We consider the trimmed mean to obtain a GSP, which is given by

$$\bar{y}_{\text{tm}} = \frac{2}{N - 2m} \sum_{n=m+1}^{N-m} y^{\star}[n], \tag{4}$$

where $y^{\star}[n]$ is the ordered sequence of $y[n]$, $m = (N-1)\alpha$, and $\alpha \in [0, 1/2)$ [32,33]. If $\alpha = 0$ or $\alpha \to 0.5$, then the trimmed mean corresponds to the sample mean and median, respectively [32], which are considered as methods for GSP derivation.

2.4. Intensity Mean

We also use the intensity mean for obtaining ground scene predictions, given by

$$\bar{y}_{\text{im}} = \sqrt{\frac{1}{N} \sum_{n=1}^{N} y[n]^2}. \tag{5}$$

Compared to other statistical methods, the intensity mean has the advantage of providing physical interpretation about the image reflection. However, the intensities' values contribute evenly to the prediction results, which can be strongly affected by the changes in the ground scene [32].

3. Experimental Results

In this section, we present the results obtained from the discussed ground scene prediction methods and describe an approach for change detection based on such methods.

3.1. Data Description

In this study, we considered a data set obtained from CARABAS II, a Swedish UWB VHF SAR system whose images are available in [46]. The system is a low-frequency wavelength-resolution system which means that the images have almost no speckle noise. The data set was divided into three stacks with eight images each, i.e., two out of six passes have identical flight headings. Two passes have a flight heading of 255°, two of 135°, and two of 230°, and the heading is defined as 0° pointing towards the north with clockwise increasing heading. The images in the stacks have the same flight geometry but are associated with four different targets' deployments (missions 1 to 4) in the ground scene. Hence, with four missions and six passes for each mission, there are 24 magnitude single-look SAR images. The images cover a scene of size 2 km $\times$ 3 km and are georeferenced to the Swedish reference system RR92, which can easily be transformed to WGS84 [12,26].

The first stack is composed of images corresponding to flight passes 1 and 3; the second stack, with passes 2 and 4; and the last stack is composed of images associated with passes 5 and 6. In all images, the backscattering was stable in time, and only target changes are expected within the image stacks.

Each image is represented as a matrix of 3000 $\times$ 2000 pixels, corresponding to an area of 6 km^2. As reported in [12], the spatial resolution of CARABAS II is 2.5 m in azimuth and 2.5 m in range. The ground scene is dominated by boreal forest with pine trees. Fences, power lines, and roads were also present in the scene. Military vehicles were deployed in the SAR scene and placed uniformly, in a manner to facilitate their identifications in the tests [26]. Each image has 25 targets with three different sizes and the spacing between the vehicles was about 50 m. For illustration, one image of Stack 1 is shown in Figure 3. In this image, the vehicles were (i) obscured by foliage; (ii) deployed in the top left of the scene; and (iii) oriented in a southwestern heading. This deployment corresponds to mission 1. In missions 2, 3 and 4, these vehicles were deployed in other locations and were oriented in a northwestern, southwestern, and western heading, respectively [12,26].

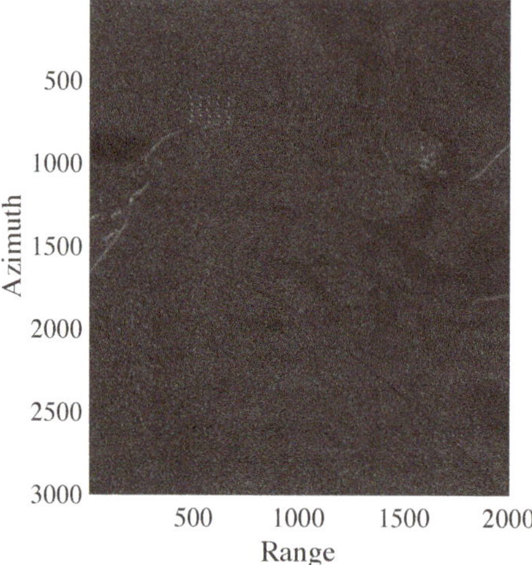

Figure 3. Sample image from CARABAS II data set—Stack 1: mission 1 and pass 1.

3.2. Ground Scene Prediction Evaluation

The AR model parameter estimation requires (i) fitting 6,000,000 models (one fit for each pixel) in each stack and (ii) evaluating the best model for each pixel sequence. Such demands lead to a significant computational burden. For simplicity, we considered $p = 1$ in the AR model. Within the image stack, the two images related to the targets have the highest pixel values in the areas where the targets were deployed. Thus, to compute the trimmed mean, we considered $m = 2$ ($\alpha \approx 0.3$), expecting to remove the pixels related to the targets, since it is desired that the predicted image presents the true ground scene without change.

Figures 4 and 5 show the ground scene prediction for Stack 1, considering the discussed methods and a zoomed image in the region where the targets were deployed. In Figure 4, the deployed targets are visually present. However, the targets are absent in the images predicted with the trimmed mean and median, as shown in Figure 5. The areas highlighted by rectangles and circles in the images in Figure 4 indicate the regions where the targets were deployed during the measurement campaign. The circles show selected military vehicles that can be viewed. With such visual analysis, the trimmed mean and median show better performance, i.e., better prediction of the ground scene. For brevity, we limited our presentation to the GSP images from Stack 1, which is representative of all considered stacks.

Table 1 displays descriptive statistics of the employed images, such as average, standard deviation, skewness, and kurtosis. It is desirable that a GSP presents not only a good visual representation of the true ground, but also preserves the statistical characteristics of the image of interest. In Table 1, we highlighted the two best methods according to each considered measure. In the majority of the scenarios, the AR model and median methods outperformed the remaining methods.

To evaluate the difference between the ground scene prediction methods, we computed some standard quality adjustment measures. The criteria are the mean square error (MSE), mean absolute percentage error (MAPE), and median absolute error (MdAE), which can be defined as follows [47].

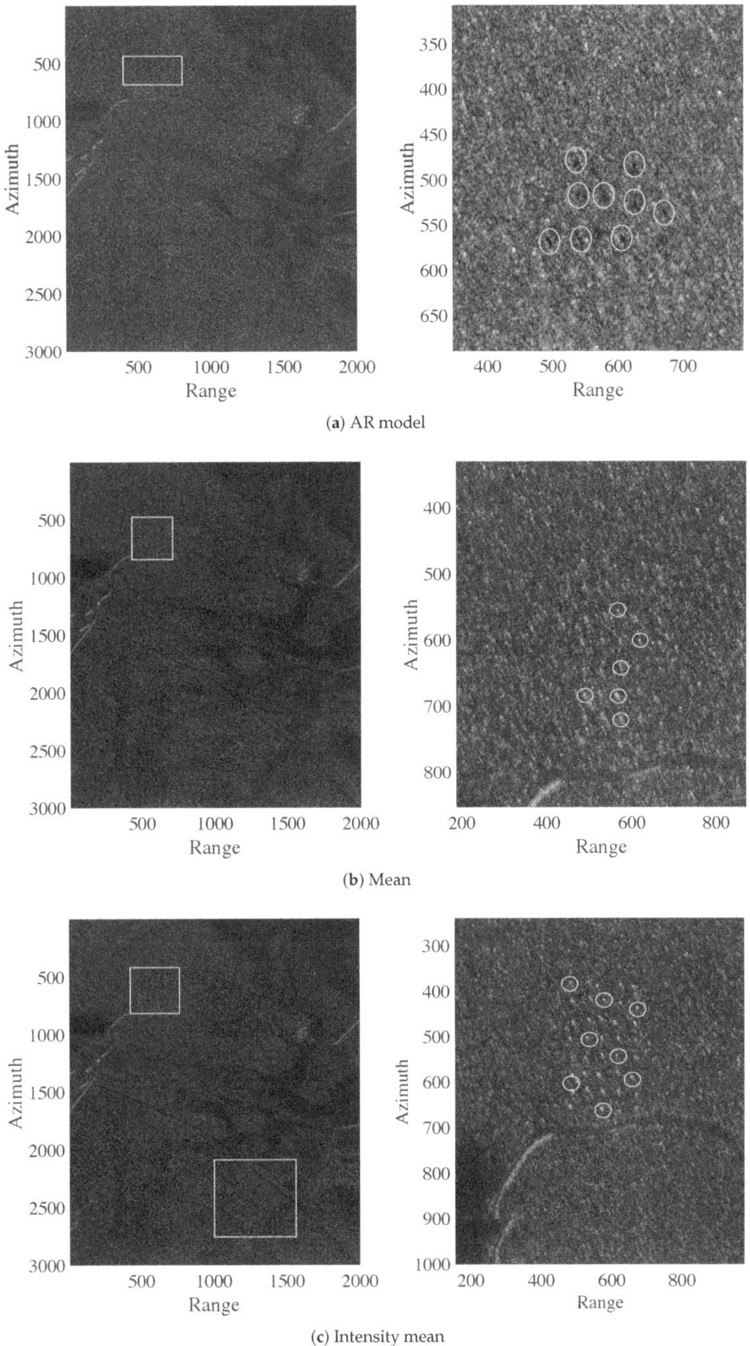

Figure 4. Ground scene prediction images for Stack 1 based on the autoregressive (AR) model, mean, and intensity mean methods. The areas highlighted by rectangles in the images represent the regions where the targets are deployed. The circles show selected military vehicles that can be viewed.

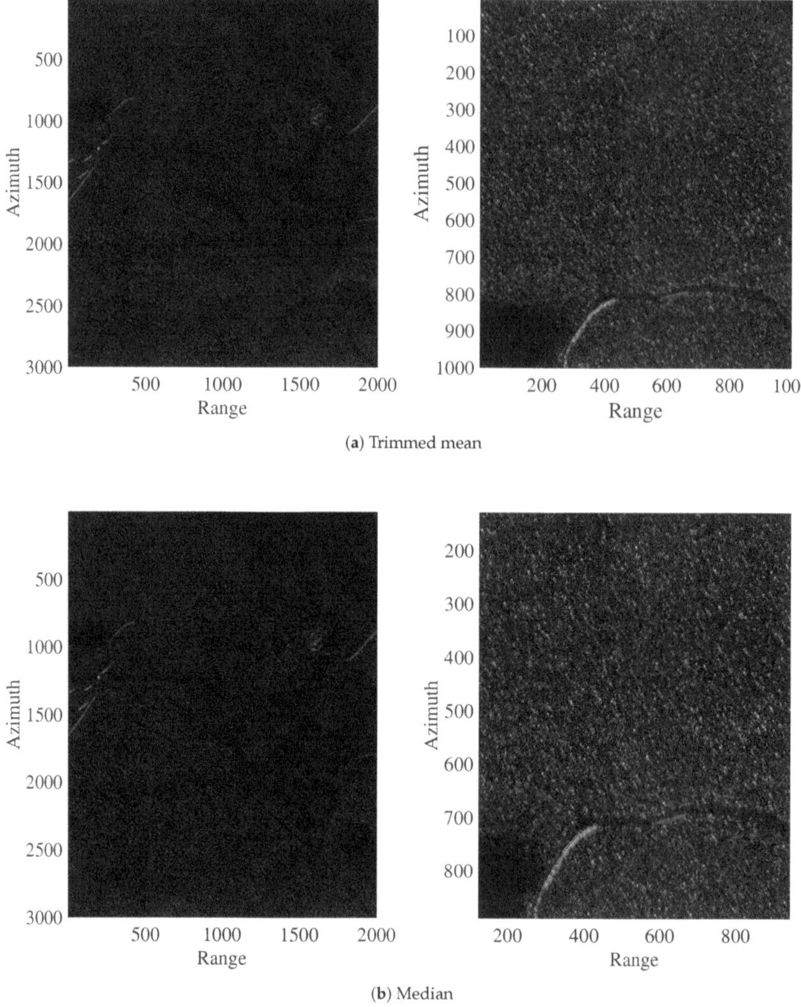

Figure 5. Ground scene prediction images for Stack 1 based on trimmed mean and median methods.

$$\text{MSE} = \frac{1}{Q} \sum_{q=1}^{Q} (x[q] - \widehat{x}[q])^2, \tag{6}$$

$$\text{MAPE} = \frac{1}{Q} \sum_{q=1}^{Q} \frac{|x[q] - \widehat{x}[q]|}{|x[q]|}, \tag{7}$$

$$\text{MdAE} = \text{Median}\left(|x[q] - \widehat{x}[q]|\right), \quad q = 1, 2, \ldots, Q, \tag{8}$$

where $x[q]$ and $\widehat{x}[q]$ are the pixel values of the interest and predicted images, respectively, Q is the number of pixels, and Median$(\cdot)$ is the median value of $|x[q] - \widehat{x}[q]|$, for $q = 1, 2, \ldots, Q$. These goodness-of-fit measures are usually considered to compare different methods applied to the same data set [47]. They are expected to be as close to zero as possible.

Table 1. Average, standard deviation, skewness, and kurtosis of one interest image and the ground scene prediction. The interest image in Stacks 1, 2 and 3, is the image of mission 1 and passes 1, 2 and 5, respectively. The two values of each measure that yielded the closest values with the interest image are **highlighted**.

	Average	Standard Deviation	Skewness	Kurtosis
		Stack 1		
Interest image	0.1442	0.0894	1.8597	14.1740
AR model	0.1101	**0.0725**	**2.1120**	**13.5190**
Trimmed mean	**0.1430**	0.0680	2.9051	21.2919
Median	**0.1424**	**0.0688**	**2.8231**	**20.4990**
Mean	0.1467	0.0663	3.0516	22.8448
Intensity mean	0.1592	0.0667	3.0090	22.8725
		Stack 2		
Interest image	0.1373	0.0968	2.9345	30.5666
AR model	0.0997	0.0784	**3.6398**	**40.9991**
Trimmed mean	**0.1344**	**0.0806**	4.4488	55.4260
Median	0.1339	**0.0812**	**4.3664**	**53.9367**
Mean	**0.1376**	0.0792	4.6022	58.3558
Intensity mean	0.1485	0.0792	4.5487	57.8894
		Stack 3		
Interest image	0.1451	0.0905	1.8583	14.0932
AR model	0.0997	**0.0683**	**2.2034**	**14.6539**
Trimmed mean	**0.1372**	0.0665	2.8811	22.0954
Median	0.1366	**0.0674**	**2.8090**	**21.3242**
Mean	**0.1410**	0.0646	2.9582	22.9540
Intensity mean	0.1534	0.0655	2.9170	22.9794

For the quality adjustment measures, the target regions in the image were excluded since we expect to obtain an accurate ground scene prediction, and no target deployment should influence the measurements. Table 2 summarizes the results of the quality adjustment measures for the five considered statistical methods, and the best measurements are highlighted. The mean method presents the best performance according to MSE measurements, while the median method excels in terms of MAPE and MdAE measures in all the stacks. However, the MSE values obtained with the mean and median methods are similar. The results provided in Tables 1 and 2 consider the same reference image of each stack. Regardless of the selected image, the median method presented good performance according to MAPE, MdAE, and statistics measures.

Based on visual inspection, statistical characteristics, and quality adjustment measures, the median method yields the most reliable prediction among the considered methods. Therefore, we separate the predicted images from the median method as reference images in the change detection algorithm detailed in the next section.

Table 2. Measures of quality of the ground scene prediction image. The interest image in Stacks 1, 2 and 3 is the image of mission 1 and passes 1, 2 and 5, respectively. We **highlighted** the values of each quality adjustment measure that yielded the smallest values.

		MSE	MAPE	MdAE
Stack 1	AR model	0.0077	0.6756	0.0548
	Trimmed mean	**0.0036**	0.6187	0.0364
	Median	0.0037	**0.6125**	**0.0351**
	Mean	**0.0036**	0.6489	0.0401
	Intensity mean	0.0039	0.7505	0.0426
Stack 2	AR model	0.0068	0.6450	0.0502
	Trimmed mean	**0.0030**	0.5971	0.0326
	Median	0.0031	**0.5912**	**0.0315**
	Mean	**0.0030**	0.6254	0.0359
	Intensity mean	0.0032	0.7204	0.0378
Stack 3	AR model	0.0083	0.6337	0.0557
	Trimmed mean	0.0037	0.5809	0.0357
	Median	0.0038	**0.5751**	**0.0346**
	Mean	**0.0036**	0.6104	0.0392
	Intensity mean	0.0037	0.7011	0.0410

3.3. Change Detection Results

As indicated in Figure 1, we use the obtained GSP image and the interest image for change detection based on image subtraction. Two examples of subtraction images are shown in Figure 6. Figure 6a highlights the deployed targets, while Figure 6b focuses on the targets and the back-lobe structures. A comparison between the difference image shown in Figure 6b to the related GSP image suggests that the back-lobe structures are related to issues in the SAR system and the image formation algorithm.

Figure 7 shows the pixels' values of the image given in Figure 6a in a vectorized form. In general, the subtracted image pixels values are randomly distributed in $(-0.4, 0.4)$. As discussed in [16], the distribution of the values of the CARABAS II subtracted image approximately follows the Gaussian distribution and the regions where no change occurs are stable. Thus, the threshold (λ) can be simply chosen as

$$C = \frac{\lambda - \widehat{\mu}}{\widehat{\sigma}}, \tag{9}$$

where C is a constant, $\widehat{\mu}$ is the estimated mean, and $\widehat{\sigma}$ is the estimated standard deviation of the considered amplitude pixels in the image. For evaluation, we set $C \in \{2, 3, 4, 5, 6\}$, resulting in different false alarm rates (FAR), which range from full detection to almost null false alarm rate.

Table 3 summarizes the change detection results corresponding to a single constant $C = 5$. Among 600 deployed vehicles in the missions, 579 were correctly detected. There are 22 detected objects that can not be related to any vehicle and are considered to be false alarms. Thus, the detection probability is about 97%, while the false alarm rate is $0.15/\text{km}^2$ (total of $144/\text{km}^2$). Ten of the 22 false alarms are related to the back-lobe structures, i.e., they are not actually false alarms and may stem from system and image formation issues. Additionally, in general, the undetected targets are related to missions 2 and 4. These undetected military vehicles are more difficult to detect since they have the smaller sizes and magnitude values, and, consequently, pixel values closer to the forest ones.

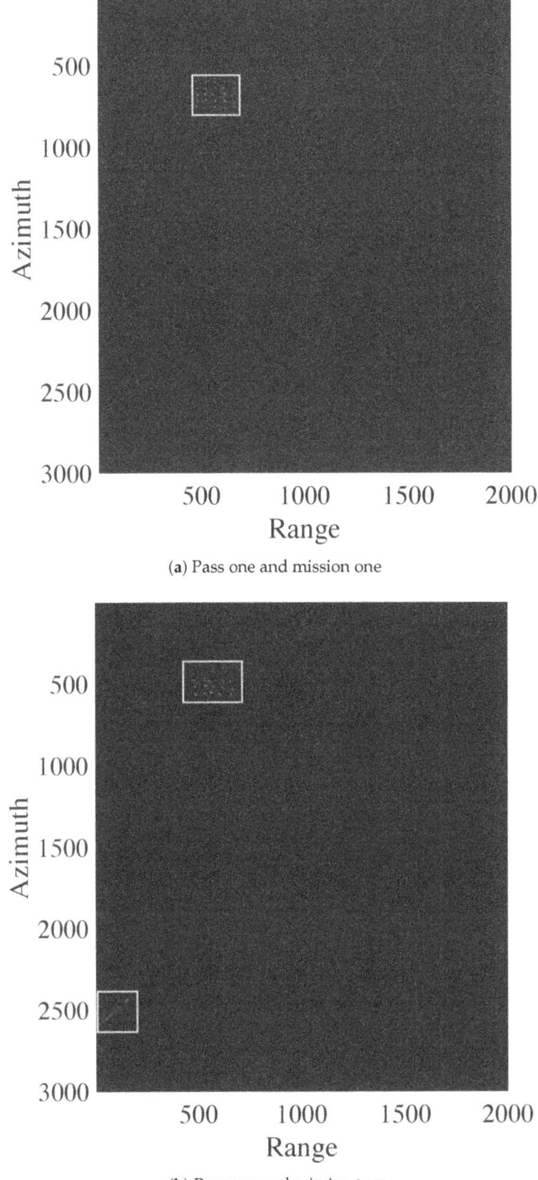

(**a**) Pass one and mission one

(**b**) Pass one and mission two

Figure 6. Subtraction of an interest image from the median ground scene prediction image. The areas highlighted by rectangles in the images represent the region with higher pixel values.

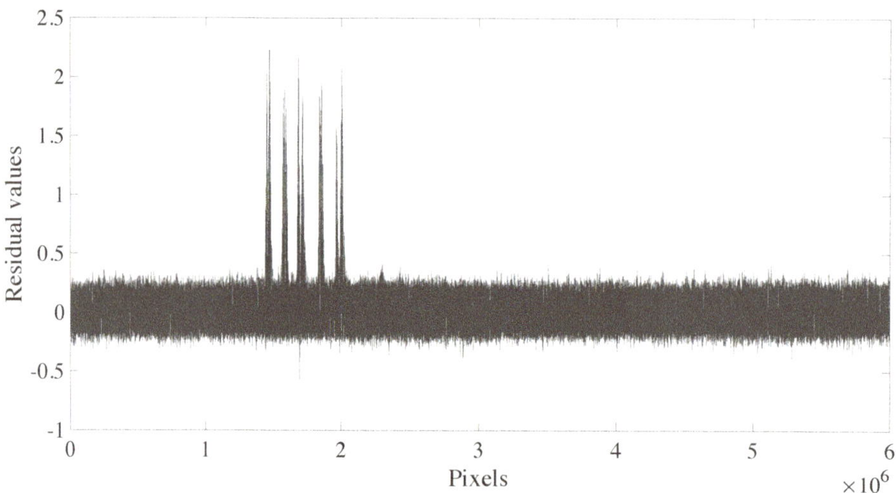

Figure 7. Result of the subtraction of the ground scene prediction image from the image obtained from mission 1 and pass 1.

Table 3. Change detection results obtained with $C = 5$.

Case of Interest Mission	Pass	Number of Known Targets	Detected Targets	P_d	Number of False Alarms
1	1	25	25	1.00	0
2	1	25	25	1.00	3
3	1	25	25	1.00	0
4	1	25	23	0.92	2
1	2	25	25	1.00	0
2	2	25	25	1.00	1
3	2	25	25	1.00	2
4	2	25	23	0.92	1
1	3	25	25	1.00	2
2	3	25	23	0.92	0
3	3	25	25	1.00	3
4	3	25	23	0.92	0
1	4	25	25	1.00	0
2	4	25	25	1.00	0
3	4	25	25	1.00	1
4	4	25	23	0.92	0
1	5	25	25	1.00	0
2	5	25	15	0.60	6
3	5	25	25	1.00	0
4	5	25	24	0.96	0
1	6	25	25	1.00	0
2	6	25	25	1.00	1
3	6	25	25	1.00	0
4	6	25	25	1.00	0
Total		600	579	0.97	22

3.4. Evaluation

The performance of change detection was evaluated by the probability of detection (P_d) and FAR. The quantity P_d was obtained from the ratio between the number of detected targets and the total numbers of known targets, while FAR is defined by the number of false alarms detected per square kilometer [26]. Figure 8 presents the receiver operating characteristic (ROC) curves [48] of the change detection results, showing the probability of detection versus the false alarm rates for the different evaluated values of C. We compared the change detection results obtained from the proposed method with the results described in [12,17,24]. The proposed method excels in terms of probability of detection and false alarm rate in comparison to [12,17,24].

For example, for a detection probability of 98%, our proposed change detection method presents $\log_{10}(\text{FAR})$ about -0.5, while [12,17,24] have $\log_{10}(\text{FAR})$ about 1.4, -0.3 and 0.14, respectively. For $\log_{10}(\text{FAR}) = -0.9$, i.e., a very low FAR, the probability detection given by [12] drops to 60%, while our proposal still maintains the probability of detection more than 90%. The detection probability of our proposed method and [17] reach 100% with $\log_{10}(\text{FAR}) \approx 1$, while [12,24] have full detection for $\log_{10}(\text{FAR}) \approx 1.5$ and $\log_{10}(\text{FAR}) \approx 2$, respectively. Additionally, detection probability improvements of our method compared to [17] are found in the range of $(0.93, 0.98)$. For example, for a probability of detection of 0.97%, our proposed change detection method presents $\log_{10}(\text{FAR})$ about -0.8, while [17] has $\log_{10}(\text{FAR}) \approx -0.2$.

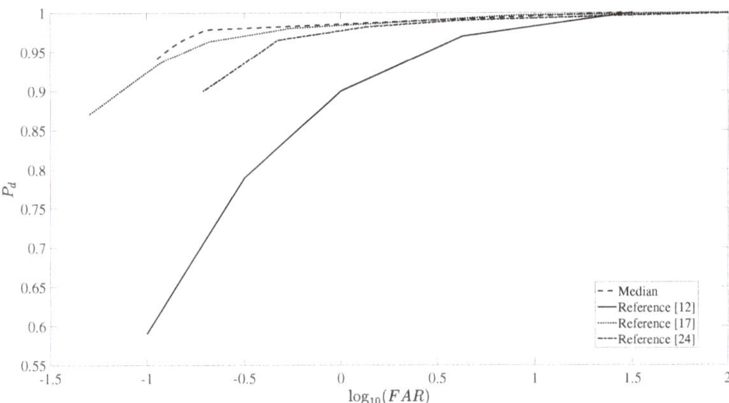

Figure 8. The receiver operating characteristic (ROC) curves obtained with the CDA with the background predicted scene as the reference image compared with the best ROC curves extracted from [12,17,24].

4. Conclusions

In this paper, we presented five methods to obtain ground scene prediction of SAR images based on image stack. The experimental results revealed that, among the considered techniques, the median method yielded the most accurate ground prediction. The statistical characteristics of the obtained GSP image were similar to the image of interest. Moreover, the median method excels in terms of quality adjustment measures, and the changes in the image stack were not visually presented in the predicted image. The GSP image based on the method was used as a reference image in a CDA, presenting competitive performance when compared with recently published results.

Author Contributions: Writing—original draft, B.G.P. and D.I.A.; Writing—review and editing, M.I.P., V.T.V., R.M., R.J.C., and F.M.B.; Resources and supervision, P.D. and H.H. All authors have read and agreed to the published version of the manuscript.

Funding: This research received no external funding.

Acknowledgments: This study was financed in part by the Conselho Nacional de Desenvolvimento Científico e Tecnológico, CNPq, Brazil, Coordenação de Aperfeiçoamento de Pessoal de Nível Superior, CAPES, Brazil, Swedish-Brazilian Research and Innovation Centre (CISB), and Saab AB.

Conflicts of Interest: The authors declare no conflict of interest.

References

1. Hoekman, D.H.; Quiriones, M.J. Land cover type and biomass classification using AirSAR data for evaluation of monitoring scenarios in the Colombian Amazon. *IEEE Trans. Geosci. Remote Sens.* **2000**, *38*, 685–696. [CrossRef]
2. Tison, C.; Nicolas, J.M.; Tupin, F.; Maître, H. A new statistical model for Markovian classification of urban areas in high-resolution SAR images. *IEEE Trans. Geosci. Remote Sens.* **2004**, *42*, 2046–2057. [CrossRef]
3. Cintra, R.J.; Frery, A.C.; Nascimento, A.D. Parametric and nonparametric tests for speckled imagery. *Pattern Anal. Appl.* **2013**, *16*, 141–161. [CrossRef]
4. Inglada, J.; Mercier, G. A new statistical similarity measure for change detection in multitemporal SAR images and its extension to multiscale change analysis. *IEEE Trans. Geosci. Remote Sens.* **2007**, *45*, 1432–1445. [CrossRef]
5. Palm, B.G.; Bayer, F.M.; Cintra, R.J.; Pettersson, M.I.; Machado, R. Rayleigh Regression Model for Ground Type Detection in SAR Imagery. *IEEE Geosci. Remote Sens. Lett.* **2019**, *16*, 1660–1664. [CrossRef]
6. Sportouche, H.; Nicolas, J.M.; Tupin, F. Mimic capacity of Fisher and generalized Gamma distributions for high-resolution SAR image statistical modeling. *IEEE J. Sel. Top. Appl. Earth Obs. Remote Sens.* **2017**, *10*, 5695–5711. [CrossRef]
7. Eltoft, T.; Hogda, K.A. Non-Gaussian signal statistics in ocean SAR imagery. *IEEE Trans. Geosci. Remote Sens.* **1998**, *36*, 562–575. [CrossRef]
8. Amirmazlaghani, M.; Amindavar, H.; Moghaddamjoo, A. Speckle suppression in SAR images using the 2-D GARCH model. *IEEE Trans. Image Process.* **2009**, *18*, 250–259. [CrossRef]
9. Belcher, D.P.; Cannon, P.S. Ionospheric effects on synthetic aperture radar (SAR) clutter statistics. *IET Radar Sonar Navig.* **2013**, *7*, 1004–1011. [CrossRef]
10. Gudnason, J.; Cui, J.; Brookes, M. HRR automatic target recognition from superresolution scattering center features. *IEEE Trans. Aerosp. Electron. Syst.* **2009**, *45*, 1512–1524. [CrossRef]
11. Zheng, Y.; Zhang, X.; Hou, B.; Liu, G. Using Combined Difference Image and *k*-Means Clustering for SAR Image Change Detection. *IEEE Geosci. Remote Sens. Lett.* **2014**, *11*, 691–695. [CrossRef]
12. Ulander, L.M.; Lundberg, M.; Pierson, W.; Gustavsson, A. Change detection for low-frequency SAR ground surveillance. *IEEE Proc. Radar Sonar Navig.* **2005**, *152*, 413–420. [CrossRef]
13. Mercier, G.; Moser, G.; Serpico, S.B. Conditional copulas for change detection in heterogeneous remote sensing images. *IEEE Trans. Geosci. Remote Sens.* **2008**, *46*, 1428–1441. [CrossRef]
14. Ulander, L.M.; Pierson, W.E.; Lundberg, M.; Follo, P.; Frolind, P.O.; Gustavsson, A. Performance of VHF-band SAR change detection for wide-area surveillance of concealed ground targets. In Proceedings of the Algorithms for Synthetic Aperture Radar Imagery XI, International Society for Optics and Photonics, Orlando, FL, USA, 2 September 2004; Volume 5427, pp. 259–271.
15. Hellsten, H.; Ulander, L.M.; Gustavsson, A.; Larsson, B. Development of VHF CARABAS II SAR. *Radar Sens. Technol.* **1996**, *2747*, 48–61.
16. Machado, R.; Vu, V.T.; Pettersson, M.I.; Dammert, P.; Hellsten, H. The stability of UWB low-frequency SAR images. *IEEE Geosci. Remote Sens. Lett.* **2016**, *13*, 1114–1118. [CrossRef]
17. Vu, V.T. Wavelength-resolution SAR incoherent change detection based on image stack. *IEEE Geosci. Remote Sens. Lett.* **2017**, *14*, 1012–1016. [CrossRef]
18. Alves, D.I.; Palm, B.G.; Pettersson, M.I.; Vu, V.T.; Machado, R.; Uchoa-Filho, B.F.; Dammert, P.; Hellsten, H. A Statistical Analysis for Wavelength-Resolution SAR Image Stacks. *IEEE Geosci. Remote Sens. Lett.* **2019**, *17*, 227–231. [CrossRef]
19. Baselice, F.; Ferraioli, G.; Pascazio, V. Markovian change detection of urban areas using very high resolution complex SAR images. *IEEE Geosci. Remote Sens. Lett.* **2013**, *11*, 995–999. [CrossRef]
20. Montazeri, S.; Zhu, X.X.; Eineder, M.; Bamler, R. Three-dimensional deformation monitoring of urban infrastructure by tomographic SAR using multitrack TerraSAR-X data stacks. *IEEE Trans. Geosci. Remote Sens.* **2016**, *54*, 6868–6878. [CrossRef]
21. Wang, Y.; Zhu, X.X.; Bamler, R. An efficient tomographic inversion approach for urban mapping using meter resolution SAR image stacks. *IEEE Geosci. Remote Sens. Lett.* **2014**, *11*, 1250–1254. [CrossRef]
22. White, R.G. Change detection in SAR imagery. *Int. J. Remote Sens.* **1991**, *12*, 339–360. [CrossRef]

23. Ulander, L.; Frolind, P.O.; Gustavsson, A.; Hellsten, H.; Jonsson, T.; Larsson, B.; Stenstrom, G. Performance of the CARABAS-II VHF-band synthetic aperture radar. In Proceedings of the Geoscience and Remote Sensing Symposium (IGARSS'01), Sydney, NSW, Australia, 9–13 July 2001; Volume 1, pp. 129–131.
24. Vu, V.T.; Gomes, N.R.; Pettersson, M.I.; Dammert, P.; Hellsten, H. Bivariate Gamma Distribution for Wavelength-Resolution SAR Change Detection. *IEEE Trans. Geosci. Remote Sens.* **2018**, *57*, 473–481. [CrossRef]
25. Palm, B.G.; Alves, D.I.; Vu, V.T.; Pettersson, M.I.; Bayer, F.M.; Cintra, R.J.; Machado, R.; Dammert, P.; Hellsten, H. Autoregressive model for multi-pass SAR change detection based on image stacks. In Proceedings of the Image and Signal Processing for Remote Sensing XXIV, International Society for Optics and Photonics, Berlin, Germany, 10–12 September 2018; Volume 10789.
26. Lundberg, M.; Ulander, L.M.; Pierson, W.E.; Gustavsson, A. A challenge problem for detection of targets in foliage. In Proceedings of the Algorithms for Synthetic Aperture Radar Imagery XIII, Orlando, FL, USA, 17–21 May 2006; Volume 6237.
27. Vu, V.T.; Pettersson, M.I.; Machado, R.; Dammert, P.; Hellsten, H. False alarm reduction in wavelength-resolution SAR change detection using adaptive noise canceler. *IEEE Trans. Geosci. Remote Sens.* **2017**, *55*, 591–599. [CrossRef]
28. Ghirmai, T. Representing a cascade of complex Gaussian AR models by a single Laplace AR model. *IEEE Signal Process. Lett.* **2015**, *22*, 110–114. [CrossRef]
29. Liu, B.; Reju, V.G.; Khong, A.W. A linear source recovery method for underdetermined mixtures of uncorrelated AR-model signals without sparseness. *IEEE Trans. Signal Process.* **2014**, *62*, 4947–4958. [CrossRef]
30. Biscainho, L.W. AR model estimation from quantized signals. *IEEE Signal Process. Lett.* **2004**, *11*, 183–185. [CrossRef]
31. Milenkovic, P. Glottal inverse filtering by joint estimation of an AR system with a linear input model. *IEEE Trans. Acoust. Speech Signal Process.* **1986**, *34*, 28–42. [CrossRef]
32. Maronna, R.A.; Martin, R.D.; Yohai, V.J.; Salibián-Barrera, M. *Robust Statistics: Theory and Methods (with R)*; Wiley: Hoboken, NJ, USA, 2018.
33. Hampel, F.R.; Ronchetti, E.M.; Rousseeuw, P.J.; Stahel, W.A. *Robust statistics: The Approach Based on Influence Functions*; Wiley Online Library: Hoboken, NJ, USA, 2011.
34. Bustos, O.; Ojeda, S.; Vallejos, R. Spatial ARMA models and its applications to image filtering. *Braz. J. Probab. Stat.* **2009**, *23*, 141–165. [CrossRef]
35. Wang, Z.; Zhang, D. Progressive switching median filter for the removal of impulse noise from highly corrupted images. *IEEE Trans. Circuits Syst. II Analog Digit. Signal Process.* **1999**, *46*, 78–80. [CrossRef]
36. Kirchner, M.; Fridrich, J. On detection of median filtering in digital images. In Proceedings of the Media Forensics and Security II. International Society for Optics and Photonics, San Jose, CA, USA, 18–20 January 2010; Volume 7541, p. 754110.
37. Chen, C.; Ni, J.; Huang, J. Blind detection of median filtering in digital images: A difference domain based approach. *IEEE Trans. Image Process.* **2013**, *22*, 4699–4710. [CrossRef]
38. Zhang, Y.; Li, S.; Wang, S.; Shi, Y.Q. Revealing the traces of median filtering using high-order local ternary patterns. *IEEE Signal Process. Lett.* **2014**, *21*, 275–279. [CrossRef]
39. Chen, J.; Kang, X.; Liu, Y.; Wang, Z.J. Median filtering forensics based on convolutional neural networks. *IEEE Signal Process. Lett.* **2015**, *22*, 1849–1853. [CrossRef]
40. Oten, R.; de Figueiredo, R.J. Adaptive alpha-trimmed mean filters under deviations from assumed noise model. *IEEE Trans. Image Process.* **2004**, *13*, 627–639. [CrossRef] [PubMed]
41. Ahmed, F.; Das, S. Removal of high-density salt-and-pepper noise in images with an iterative adaptive fuzzy filter using alpha-trimmed mean. *IEEE Trans. Fuzzy Syst.* **2013**, *22*, 1352–1358. [CrossRef]
42. Gonzalez, R.C.; Woods, R. *Digital Image Processing*; Prentice Hall: Upper Saddle River, NJ, USA, 2008.
43. Kay, S.M. *Fundamentals of Statistical Signal Processing: Detection Theory*; Prentice Hall: Upper Saddle River, NJ, USA, 1998; Volume II.
44. Brockwell, P.J.; Davis, R.A. *Introduction to Time Series and Forecasting*; Springer: Berlin, Germany, 2016.
45. Brockwell, P.J.; Davis, R.A. *Time Series: Theory and Methods*; Springer: Berlin, Germany, 2013.
46. SDMS. Sensor Data Management System Public Web Site. 2018. Available online: https://www.sdms.afrl.af.mil/index.php (accessed on 28 February 2020).

47. Hyndman, R.J.; Koehler, A.B. Another look at measures of forecast accuracy. *Int. J. Forecast.* **2006**, *22*, 679–688. [CrossRef]
48. Metz, C.E. Basic principles of ROC analysis. In *Seminars in Nuclear Medicine*; Elsevier: Amsterdam, The Netherlands, 1978; Volume 8, pp. 283–298.

Article

A Multi-Scale U-Shaped Convolution Auto-Encoder Based on Pyramid Pooling Module for Object Recognition in Synthetic Aperture Radar Images

Sirui Tian [1,*], Yiyu Lin [2], Wenyun Gao [3], Hong Zhang [4] and Chao Wang [4,5]

1 Department of Electronic Engineering, School of Electronic and Optical Engineering, Nanjing University of Science and Technology, Nanjing 210094, China
2 Department of Electrical and Computer Engineering, University of California, Riverside, Riversidem, CA 92521, USA; ylin170@ucr.edu
3 College of Computer and Information, Hohai University, Nanjing 211100, China; gao_wy@les.cn
4 Key Laboratory of Digital Earth Science, Institute of Remote Sensing and Digital Earth, Chinese Academy of Sciences, Beijing 100094, China; zhanghong@radi.ac.cn (H.Z.); wangchao@radi.ac.cn (C.W.)
5 College of Resources and Environment, University of Chinese Academy of Sciences, Beijing 100049, China
* Correspondence: tiansirui@njust.edu.cn

Received: 5 February 2020; Accepted: 7 March 2020; Published: 10 March 2020

Abstract: Although unsupervised representation learning (RL) can tackle the performance deterioration caused by limited labeled data in synthetic aperture radar (SAR) object classification, the neglected discriminative detailed information and the ignored distinctive characteristics of SAR images can lead to performance degradation. In this paper, an unsupervised multi-scale convolution auto-encoder (MSCAE) was proposed which can simultaneously obtain the global features and local characteristics of targets with its U-shaped architecture and pyramid pooling modules (PPMs). The compact depth-wise separable convolution and the deconvolution counterpart were devised to decrease the trainable parameters. The PPM and the multi-scale feature learning scheme were designed to learn multi-scale features. Prior knowledge of SAR speckle was also embedded in the model. The reconstruction loss of the MSCAE was measured by the structural similarity index metric (SSIM) of the reconstructed data and the images filtered by the improved Lee sigma filter. A speckle suppression restriction was also added in the objective function to guarantee that the speckle suppression procedure would take place in the feature learning stage. Experimental results with the MSTAR dataset under the standard operating condition and several extended operating conditions demonstrated the effectiveness of the proposed model in SAR object classification tasks.

Keywords: multi-scale representation learning (MSRL); pyramid pooling module (PPM); compact depth-wise separable convolution (CSeConv); convolution auto-encoder (CAE); object classification; synthetic aperture radar (SAR)

---

1. Introduction

As the vital task of object classification with synthetic aperture radar (SAR) images, feature engineering intends to obtain robust representations of intrinsic properties to distinguish various targets in high-resolution radar images. Although numerous hand-designed features have been proposed to represent both the spatial and electromagnetic characteristics of targets over the past decades, feature learning is still a challenging task for SAR-based automatic target recognition (SAR ATR) applications.

In general, the traditional hand-designed features include two categories: the generalized features [1–3] and the SAR-specialized features [4–7]. The former ones involve features from other domains considering little of the characteristics of SAR imagery, while the latter ones refer to those designed for specific SAR ATR tasks. Despite their high accuracy while dealing with the benchmark or specific SAR dataset, all these handcrafted features have certain obstacles. A major drawback is the requirement of detailed prior knowledge about the potential applications that are sometimes unavailable. Another obstacle is that many features, especially those based on scattering models, hold a series of assumptions for operation conditions (OCs), leading to performance degradation when the assumptions are inconsistent with the OCs. Accordingly, it is necessary to devise new feature learning algorithms which can adaptively learn representations from various data, considering complicated situations.

With the theoretical progress of machine learning, the deep learning (DL) model, which has turned out to be adept at automatically discovering intricate information in high-dimensional raw data [8], has been employed to tackle SAR ATR tasks and achieved the superior performance than hand-designed features. Although the supervised DL models have obtained state-of-the-art results, their requirement of a great many labelled data is the major obstacle in SAR ATR. The labelled benchmarks are too small to train a supervised deep network effectively, and overfitting caused by limited labelled samples is often one of the main causes of performance degradation of the supervised model. To handle this problem, various unsupervised DL models are employed and developed, including the autoencoder (AE) [9,10], the generative adversarial network (GAN) [11,12], and the restricted Boltzmann machine (RBM) [13]. Due to the fact of its simple implementation and attractive computational cost, the AE has widely been used in SAR ATR which minimizes the distortion between the inputs and the reconstructions to guarantee that the mapping process preserves the information of the inputs.

In earlier works, the autoencoder was utilized to derive refined representations from the predefined features or preprocessed images before feeding them into a traditional classifier such as the softmax or the support vector machine (SVM) [14–20]. In Reference [14], Geng et al. presented a deep convolution AE (CAE) for SAR image classification. Two kinds of handcrafted features, a gray-level co-occurrence matrix (GLCM) and the Gabor filter banks, were jointly fed into the CAE. The learned representation was subsequently fed to a softmax classifier for land-cover classification. In Reference [15], the geometric parameters and the local texture features were combined to train a stacked AE (SAE) for vehicle classification in SAR images. Gleich and Planinšic [16] estimated the log commulants of SAR data patches via the dual-tree oriented wavelet transform and input them into an SAE to derive representations for scene patch categorization. Zhang et al. [17] devised a framework to learn the robust representation of polarimetric SAR (PolSAR) data based on the spatial information, which was characterized by the spatial distance to the central pixel. Their framework was subsequently improved in [18] with a multi-scale strategy in which the spatial information was obtained by taking neighborhood windows of different scales before the stacked sparse AE (SSAE) was applied to extract features at different scales for land cover classification in PolSAR images. Hou et al. [19] devised a PolSAR image classification method based on both the multilayer AE (MLAE) and the superpixel trick. The superpixels produced by the Pauli decomposition to integrate contextual information of the neighborhood was refined by an MLAE to generate a robust representation. Chen and Jiao [20] fed the discriminative feature extracted by the multilayer projective dictionary into an SAE to realize the nonlinear relationship between the elements of feature vectors in an adaptive way.

To further promote the performance in representation learning, various models based on the AE framework have also been utilized [21–24]. In Reference [21], the stacked contractive AE (SCAE) was utilized to extract temporal characteristics from superpixels for change detection in SAR images which was restricted by a contractive penalty with the Frobenius norm of the Jacobian. Xu et al. [22] developed an improved variational AE (VAE) based on the residual network to draw latent representations for vehicle classification in SAR images. Song et al. [23] devised an adversarial autoencoder neural network (AAN) to learn intrinsic characteristics and generate new samples at different azimuth angles by adversarial training. Kim and Hirose [24] proposed a quaternion autoencoder and a quaternion self-organizing map (SOM) for PolSAR image classification. The quaternion AE was introduced to extract representations based on the natural distribution of PolSAR features. The extracted features were classified by the quaternion SOM in an unsupervised manner, by which new and more detailed land categories could be discovered. In Reference [25], a deep bimodal AE was proposed for land cover classification by fusing the SAR and the multispectral images. The bimodal AE provided independent encoding modalities in the front part to learn the features of SAR data and fused the feature of each modality with shared representation layers to obtain the representations for classification.

Despite the explosive growth of unlabeled SAR images with the development of high-resolution SAR systems, the training dataset (even the unlabeled benchmarks) available for specific tasks or targets are limited and incomplete. To handle this problem, model transferring is recommended as another solution to improve the representation learning capability with small sample size and limited training resources. Huang et al. [26] devised an assembled CNN model that combines a CAE with a CNN, sharing the encoder part of the CAE. The CAE was pre-trained with a large number of unlabeled SAR images, and its encoder part that connected with a fully connected layer was fine-tuned with the limited target patches. Mohammad et al. [27] proposed a domain adaptation algorithm to transfer knowledge from the earth observation (EO) domain to the SAR domain. They trained two deep encoders coupled through their last layer to map data points from the EO and the SAR domains to the shared embedding space, such that the distance between the distributions of the two domains was minimized in the latent embedding space. In Reference [28], a DL-based workflow was proposed to map forest above-ground biomass by integrating Landsat 8 and Sentinel-1A images with airborne light detection and ranging (LiDAR) data. They demonstrated the advantage of a stacked sparse autoencoder network in comparison to other prediction techniques. De et al. [29] proposed an AE-based technique for urban area classification in PolSAR images which leveraged a synthetic target database for data augmentation (DA). The synthetic dataset obtained by rotation and collation was fed to an SAE to generate a compact representation of the information in the augmented dataset. Although these model transferring methods have alleviated overfitting of DL models caused by small datasets and achieved the state-of-the-art performance, it is quite difficult to design the transferring schemes for specific SAR ATR tasks in various extended operation conditions (EOCs). Besides, the selection of the pre-trained model and the natural image dataset for information transferring will also greatly affect the performance of these approaches. If there is a great difference between the natural images and the objective SAR dataset, the representation learning capability of the transferred models will suffer serious degradation.

Another way to promote the performance of the AE models with a limited training dataset is to incorporate prior knowledge in the model with certain regularization terms and task-specific cost functions. The training process of the AE refers to estimating the trainable parameters of the model and can be achieved by optimizing the objective function consisting of a reconstruction loss and certain regularization terms [9]. In References [30,31], the supervised information was embedded in the cost function by designing label-related regularization terms. Deng et al. [30] devised a Euclidean distance restriction in the cost function which encouraged the intra-class distance of features to be a small value near zero and the inter-class distance to be close to a constant. A similar idea was applied in Reference [31], where the objective function was tuned according to the SAR ATR task. The authors devised a regularization term based on the modified triplet loss that combines the semi-hard triplet loss with the intra-class distance penalty to learn discriminative features with a small intra-class divergence and a large inter-class divergence. In References [32–34], the task-specialized prior knowledge was embedded in the objective function of the AE-based model. Xie et al. [32] proposed a new type of AE and CAE with a modified objective function according to the task of PolSAR image classification, where the distortion of the reconstructed data to the inputs was measured by the Wishart distance instead of the ordinary mean square error (MSE) or cross-entropy. Similarly, Wang et al. [33] devised a hybrid AE for land cover classification, where the Wishart distance and Euclidean distance were jointly applied to evaluate the reconstruction error between the input and the output according to the distribution of PolSAR data matrix. In Reference [34], Li et al. proposed a stacked fisher AE for change detection, where the ratio difference image (RDI) of multi-temporal SAR images was used as the input and the distribution of the RDI was introduced to construct the objective function with sparsity regularization.

Although these AE-based models have developed an effective way to learn the robust representation via an unlabeled SAR dataset and achieved competitive results, the performance of most of these models is still slightly inferior to their supervised counterparts [35–39] and some handcrafted features [4,5,7] that are based on the electromagnetic scattering models. The major reasons include the following:

(1) Most of these models learn representations at a large single scale with the hierarchical structure. However, without using local and detailed discriminative information at multiple scales, the classification performance of the features learned by these methods is limited;
(2) Most of the models are optimized according to the minimum reconstruction deviation criterion, importing useless information of speckle in the learned feature and diluting the discriminative features that could benefit classification and ATR tasks;
(3) Small incomplete training benchmarks in SAR ATR limit the application of complicated and deeper DL model due to the large number of trainable parameters and overfitting arising therefrom.

In this paper, a novel unsupervised multi-scale CAE (MSCAE) is proposed which can extract features at different scales and discard useless information of speckle and background clutter. The proposed model provides a framework to learn multi-scale features at two levels: the modality level feature learning achieved by the U-shaped structure and the branch level feature extracted by the pyramid pooling module (PPM). A modified objective function was devised to tackle the performance degradation caused by speckle. The reconstruction loss of the MSCAE was measured between the output and the input filtered by the improved Lee sigma filter (ILSF) [40] to alleviate the influence of serious speckle in SAR images. The structural similarity index metric (SSIM) was employed as the measurement of the reconstruction deviation, taking full advantage of the targets' characteristics such as the structure and the variation of backscattering intensity. An additional filter regularization term was also incorporated in the objective function that measures the dissimilarity of the encoded features of the raw data and the ILSF filtered inputs, guaranteeing that the speckle suppression procedure occurred during the encoding stage. Moreover, to handle the performance degradation caused by the limited training dataset, a new convolution layer, named compact depth-wise separable convolution (CSeConv) layer, and its deconvolution counterpart (CSeDeConv layer) were also developed to reduce the number of the trainable parameters in the model, alleviating overfitting caused by limited training samples.

The rest of this paper is organized as follows: Section 2 illustrates the key technologies used to build our MSCAE model, including the CSeConv and CSeDeConv, the PPM processing, and the specially designed objective function. Furthermore, the technical details of network topology are also given. Section 3 conducts a series of comparative experiments based on the moving and stationary target acquisition and recognition (MSTAR) dataset [41,42]. The experimental results of the proposed network with SOC and various EOCs are presented. Section 4 concludes our work.

2. The Multi-Scale Convolution Auto-Encoder

2.1. Overall Structure of the MSCAE

In this part, we discuss the characteristics and general layout of the proposed MSCAE. As shown in Figure 1, the MSCAE consists of a series of uniform modalities to learn representations and generate the reconstructed feature map at different modality levels. Each modality includes an encoder part and the corresponding decoder part.

In the encoder part of a modality, the CSeConv and the PPM module are applied to learn multi-scale features at branch level. The input feature map will be convolved with a 5×5 CSeConv layer with a stride 2 which means that both the width and the height of the output feature map will be half the size of the input. Subsequently, the batch normalization (BN) and the rectified linear unit

(ReLU) activation function will be applied to the convolved feature map. The output feature map will be processed in two branches: one for multi-scale representation learning with the PPM module and the other for the processes in next modality after downsampled by a 2×2 max-pooling layer. It should be noted that at the coarsest modality level, the PPM module is neglected, and the convolved feature map is optional. If the input feature map is larger than 5×5, the convolution layer will be applied. Otherwise, it will also be neglected. The processed feature map will be directly vectorized to form the feature vector of the coarsest level.

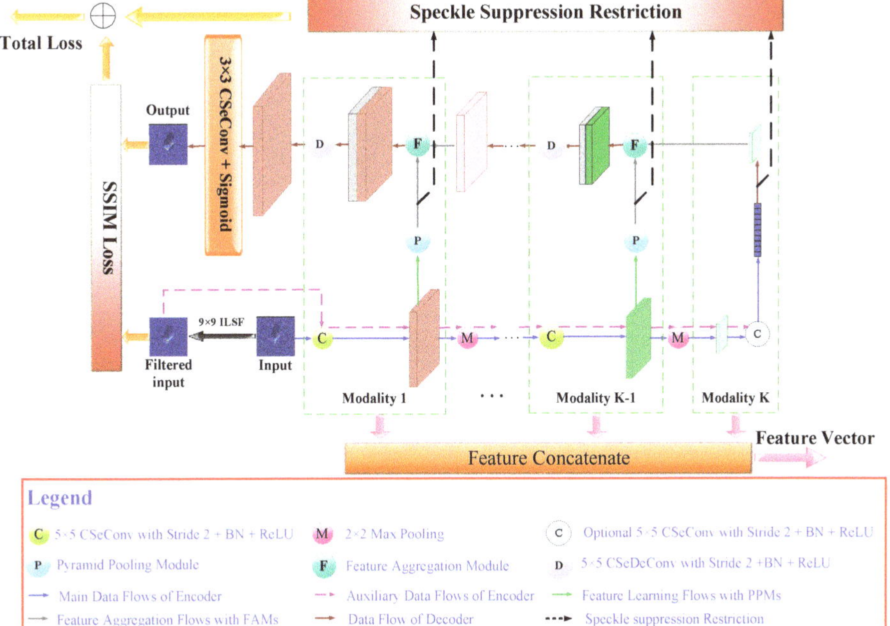

Figure 1. The overall architecture of the U-shaped multi-scale convolution auto-encoder. The feature vector learnt at each modality level will be converted into vector and concatenated to form the feature vector for SAR ATR. CSeConv stands for the compact depth-wise separable convolution layer; BN stands for batch normalization; ReLU stands for the rectified linear unit activation function; CSeDeConv denotes the compact separable deconvolution layer; PPM stands for the pyramid pooling modul; FAM stands for the feature aggregation module; ILSF refers to the improved Lee sigma filter; SSIM refers to the structural similarity index metric.

In the decoder part of each modality, the feature aggregation module (FAM) is adopted to combine the feature vector learned by the PPM with the feature map reconstructed from the coarser modality level. The combined feature map is convolved by a 5×5 CSeDeConv layer with a stride 2 to upsample the feature map as well as reduce the channel number. At the first modality level, the reconstructed feature map is convolved with a 3×3 CSeConv layer followed by a sigmoid activation function, and the reconstructed image is generated.

The SSIM loss is measured between the reconstructed image and the image filtered by the 9×9 ILSF to diminish the influence of the speckle. Besides, the speckle suppression restriction is also computed to force the speckle suppression taking place at the feature learning procedure. Therefore, an additional auxiliary data flow is fed to the encoder part of the proposed model, where the image filtered by the 9×9 ILSF is encoded with similar modules and parameters at each modality level. The similarity between the feature vectors learned from the unfiltered image and those learned from the filtered one will be compared and summarized to construct the speckle suppression restriction.

The weighted sum of the SSIM loss and the speckle suppression restriction forms the objective function of the proposed model. Once the model is trained with the dataset, the encoder part will be utilized to learn representations and the feature vector learned at each modality level will be concatenated to generate the final feature vector for SAR ATR.

2.2. Compact Depth-Wise Separable Convolution and the Corresponding Deconvolution

The convolution layer, which is the basic structure in CNNs and CAEs, has the capability of capturing local patterns of input data and generating new representations of jointly encoding space and channel information. As presented in Figure 2, the standard convolution layer [43] creates C_{out} trainable convolution kernels that are convolved with the $W_{in} \times H_{in} \times C_{in}$ input F_{in} to produce a $W_{out} \times H_{out} \times C_{out}$ feature map F_{out}. Here, $W_{in} \times H_{in}$ and $W_{out} \times H_{out}$ are the spatial size of the input and the output feature maps, respectively; C_{in} and C_{out} are the channels of F_{in} and F_{out}, respectively. The size of each trainable convolution kernel is $N_k \times N_k \times C_{in}$ with N_k being the size of the sliding filter.

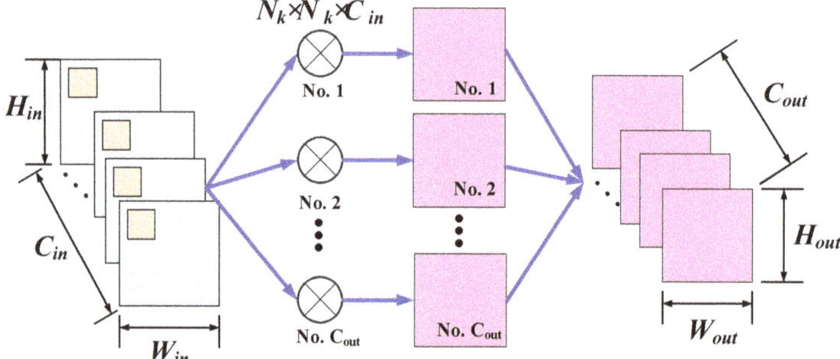

Figure 2. The convolution procedure of a standard convolution layer. Given a $W_{in} \times H_{in} \times C_{in}$ feature map F_{in} with $W_{in} \times H_{in}$ being the spatial size and C_{in} being the number of channels of the input feature map, respectively, the standard convolution layer utilized C_{out} trainable convolution kernels whose size is $N_k \times N_k \times C_{in}$ to produce the $W_{out} \times H_{out} \times C_{out}$ feature map F_{out} with $W_{out} \times H_{out}$ being the spatial size and C_{out} being the number of channels of F_{out}, respectively.

In comparison with the fully connected layer, the number of trainable parameters in a convolution layer is much fewer due to the shared convolution kernels, substantially decreasing the computational cost and improving the performance with a small dataset. However, in large-scale deep networks, where the size of the convolution kernels is quite large and the channel number rises rapidly as the depth of the network increases, the high computational cost and overfitting caused by massive trainable parameters are still major causes of performance deterioration. To diminish these problems, various factorized convolution operators are utilized.

The depth-wise (DW) separable convolution (SeConv) [44], presented in Figure 3, is a typical factorized convolution operator in channel level which factorizes the standard convolution into two steps via the DW convolution and the pointwise (PW) convolution. In the DW convolution step, C_{in} filters with the size of $N_k \times N_k \times 1$ are applied to every input channel of the $W_{in} \times H_{in} \times C_{in}$ feature map F_{in} and produce the intermediate feature map that has the same number of channels as that of the inputs. Subsequently, the PW convolution utilizes C_{out} filters with the size of $1 \times 1 \times C_{in}$ to combine the output of the depth-wise layer and produce the final output feature map F_{out}.

Another factorized convolution layer is the kernel decomposition convolution (DeCConv) layer proposed by Simonyan and Zisserman [45] which decomposes the large convolution kernel into a series of 3×3 small kernels as depicted in Figure 4. Specifically, a large convolution kernel with the size $N_k \times N_k$ is approximated by M cascaded 3×3 filters, where the number of the 3×3 filters is determined by:

$$M = (N_k - 1)/2 \tag{1}$$

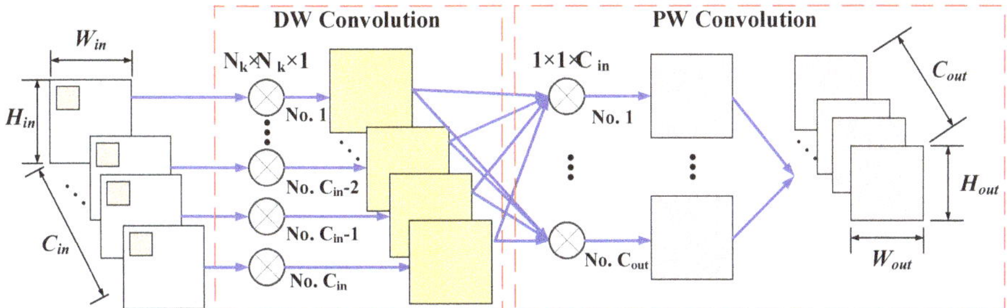

Figure 3. The convolution procedure of the depth-wise (DW) separable convolution (SeConv) layer that decomposes the standard convolution into two steps: the DW convolution and the point-wise (PW) convolution. During the DW convolution process, each channel of the $W_{in} \times H_{in} \times C_{in}$ feature map, F_{in}, is convolved with a filter the size of $N_k \times N_k \times 1$ to generate an intermediate feature map that has C_{in} channels. Subsequently, C_{out} PW filters with the size $1 \times 1 \times C_{in}$ are adopted to combine the output of the depth-wise layer and produce the final output feature map, F_{out}, with the size of $W_{out} \times H_{out} \times C_{out}$

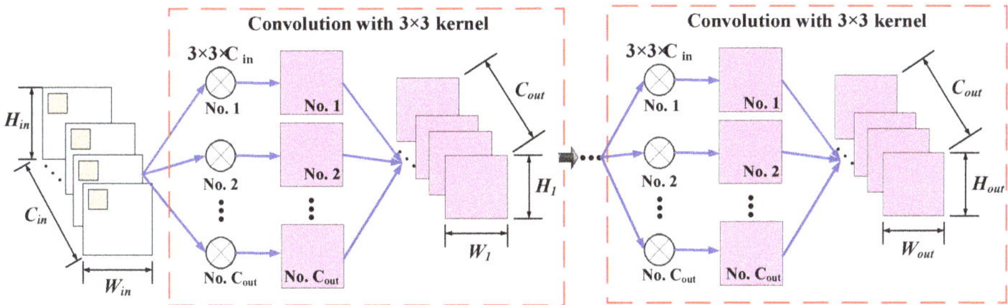

Figure 4. The procedure of the kernel decomposition convolution layer that decomposes the large convolution kernels into M stacked 3×3 filters. Given the $W_{in} \times H_{in} \times C_{in}$ feature map, F_{in}, C_{out} small convolution kernels with the size of $3 \times 3 \times C_{in}$ are applied to generate the first intermediate feature map with the size of $W_1 \times H_1 \times C_{out}$. Subsequently, the first intermediate feature map is convolved successively with C_{out} small convolution kernels with the size of $3 \times 3 \times C_{out}$ for $M-1$ times, and the output feature map F_{out} with the size of $W_{out} \times H_{out} \times C_{out}$ is produced.

During the convolution procedure with stacked 3×3 filters, activation functions can be employed after each convolution operator. Given the $W_{in} \times H_{in} \times C_{in}$ feature map, F_{in}, C_{out} small convolution kernels with the size of $3 \times 3 \times C_{in}$ are applied to generate the first intermediate feature map with the size of $W_1 \times H_1 \times C_{out}$. Subsequently, the first intermediate feature map is convolved successively with C_{out} small convolution kernels with the size of $3 \times 3 \times C_{out}$ for $M-1$ times and the output feature map F_{out} with the size of $W_{out} \times H_{out} \times C_{out}$ is produced. It is reported that this scheme could not only significantly decrease the trainable parameters and computational cost but also improve the representation learning capability of the convolution layer due to the increasing nonlinearity induced by the activation function of the cascaded 3×3 convolution layers.

Although these schemes significantly decrease the trainable parameters and computational cost, the small benchmark in SAR ATR still limit the application of deeper and complicated models. In this paper, a more compact convolution layer and its deconvolution counterpart are proposed. The proposed layers combine the kernel decomposition scheme and the DW SeConv scheme, thereby requiring a smaller number of trainable parameters as well as introducing more nonlinearity for better representation learning. The proposed compact DW separable convolution (CSeConv) process is depicted in Figure 5a. Similar to the DW SeConv layer, the standard convolution is split into two steps: the DW convolution for separable convolution at the channel level and the PW convolution to combine the filtered features of all channels. Besides, the kernel decomposition scheme is also adopted in the

DW convolution step, as each DW convolution can be considered as an input with single-channel convolving with only one kernel. Each large DW kernel is decomposed into a bunch of 3×3 filters, each of which is followed by a nonlinear activation function to provide additional nonlinearity [45]. Accordingly, the trainable parameters can be further decreased by the combined scheme.

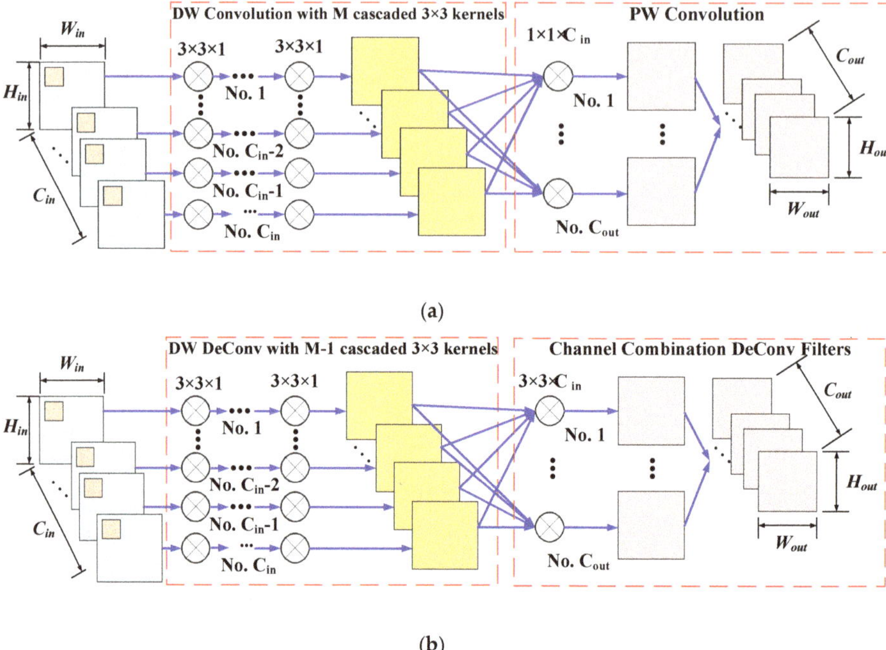

Figure 5. The proposed compact depth-wise separable convolution (CSeConv) and the corresponding compact separable deconvolution (CSeDeConv)layer which combines the depth-wise (DW) separable convolution/deconvolution scheme and the kernel decomposition scheme to reduce the trainable parameters. (**a**) The procedure of the CSeConv layer; (**b**) the details of the CSeDeConv. The size of the input feature map and output feature map in (**a**) and (**b**) are $W_{in} \times H_{in} \times C_{in}$ and $W_{out} \times H_{out} \times C_{out}$, respectively.

Its deconvolution counterpart, i.e., the compact separable deconvolution (CSeDeConv) layer presented in Figure 5b, is devised in the same manner, composed of two steps: the DW separable deconvolution and the channel-level combination DeConv. In the first step, the deconvolution operator was applied to each channel of the $W_{in} \times H_{in} \times C_{in}$ input feature map, F_{in}. The kernel decomposition scheme is also employed to split the $N_k \times N_k$ deconvolution kernel into $M - 1$ concatenated 3×3 filters at the channel level, where M is determined according to (1). Subsequently, C_{out} deconvolution filters with the size of $3 \times 3 \times C_{in}$ are utilized to combine the output of the DW separable deconvolution step and generate the output feature map F_{out} with the size of $W_{out} \times H_{out} \times C_{out}$. It should be noted that if the stride of either the CSeConv or the CSeDeConv is larger than 1, the convolution/deconvolution operator with the given stride will be implemented in the last channel level combination step.

To demonstrate the validation of the CSeConv and the CSeDeConv, the mixed national Institute of standards and technology database (MNIST) of handwritten digits was utilized for evaluation. A three-layer CAE model with one standard convolution layer and one deconvolution layer was employed as the baseline model for comparison. Both the convolution layer and the deconvolution layer had four 5×5 filters, and the strides of both the convolution layer and the deconvolution layer were 4. The trainable parameters were initialized with the He initialization [46], while the activation functions of both the convolution and deconvolution layer were ReLU. In the experiment, the convolution and deconvolution layers were replaced by the proposed CSeConv and the CSeDeConv, respectively. Accordingly, four CAEs could be generated for comparison: the baseline CAE, the CAE with the CSeConv layer (CCAE), the CAE with the CSeDeConv layer (CDCAE), and the compact CAE with the CSeConv layer and CSeDeConv layer (CompactCAE). The original images and the reconstructed results of the four CAE models are shown in Figure 6a to illustrate the validation of the proposed layer. Moreover, the training losses of the four models are compared in Figure 6b. As shown in Figure 6, both the reconstruction results and the training losses of the four CAEs were approximately the same which demonstrate the validation of the proposed CSeConv layer and CSeDeConv layer.

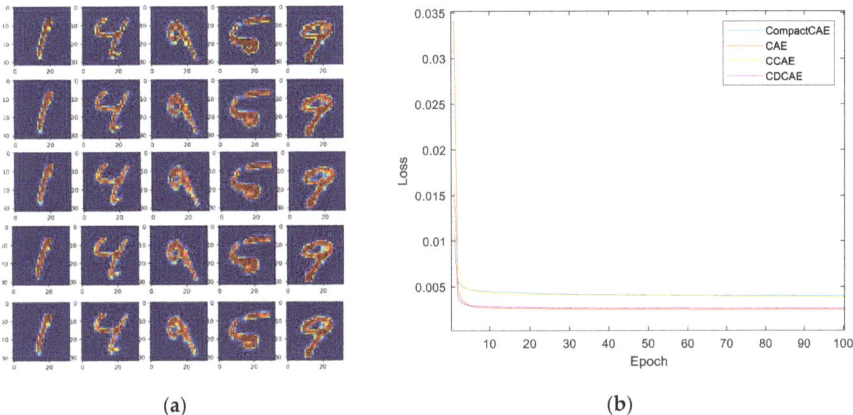

Figure 6. Validation Experiments with the mixed national Institute of standards and technology database (MNIST) **dataset.** (**a**) The reconstruction results of the baseline convolution auto-encoder (CAE), the CAE with the compact depth-wise separable convolution (CSeConv) layer (CCAE), the CAE with the compact separable deconvolution (CSeDeConv) layer (CDCAE), and the compact CAE with the CSeConv layer and CSeDeConv layer (CompactCAE) from the top row to the bottom row. (**b**) The training losses of the four models.

A brief analysis of the trainable parameters and computational consumption of various convolution layers are made and compared in Table 1. In our comparison, the size of the input feature map is supposed to be $W_{in} \times H_{in}$ with C_{in} channels, and the size of the convolution kernel is $N_k \times N_k \times C_{in}$. In order to simplify the analysis, the stride of the convolution is assumed to be 1, and the padding mode of the convolution is set to unify the input and output feature maps in size. Consequently, the size of the output feature map is $W_{in} \times H_{in}$ with C_{out} channels. Besides, the addition of feature aggregation is also ignored as in Reference [36] when the computational consumption with different convolution layers is compared. The number of trainable parameters K_{param}, the computation consumption L_{comp}, and the ratio of calculation consumption between the improved convolution layer and the standard convolution $R_{opt} = L_{comp}^{Other} / L_{comp}^{Standard}$ are all listed in Table 1. It can easily be found that the number of trainable parameters and the calculation consumption have been effectively reduced compared with the standard convolution and other mainstream convolution layers. Moreover, the reduction of the trainable parameters and the ratio of calculation consumption is only related to the number and size of the convolution kernel.

Table 1. Analysis of the trainable parameters and computational consumption of the proposed convolution layer and some mainstream convolution layers.

Layer Name	In	Out	Kernel Size			K_{param}	L_{comp}	R_{opt}
			Ordinary	DW	PW			
Standard			$N_k \times N_k \times C_{in}$	/	/	$C_{out} \times N_k \times N_k \times C_{in}$	$W_{in} \times H_{in} \times N_k \times N_k \times C_{in} \times C_{out}$	1
DW SeConv			/	$N_k \times N_k \times 1$	$1 \times 1 \times C_{in}$	$N_k \times N_k \times C_{in} + C_{in} \times C_{out}$	W_in × H_in × N_k × N_k × C_in + W_in × H_in × C_in × C_ou	$\frac{1}{C_{out}} + \frac{1}{N_k^2}$
DeCConv	$W_{in} \times H_{in} \times C_{in}$	$W_{in} \times H_{in} \times C_{out}$	1st: $3 \times 3 \times C_{in}$; Other:$3 \times 3 \times C_{out}$	/	/	$3 \times 3 \times C_{in} \times C_{out} + 3 \times 3 \times C_{out} \times C_{out} \times (N_k - 3)/2$	W_in × H_in × 3 × 3 × C_in × C_out + W_in × H_in × 3 × 3 × C_out × C_out × ((N_k − 3))/2	$\frac{9}{N_k^2} + \frac{9C_{out}(N_k-3)}{2C_{in} \times N_k^2}$
CSeConv			/	$3 \times 3 \times 1$	$1 \times 1 \times C_{in}$	$3 \times 3 \times C_{in} \times (N_k - 1)/2 + 1 \times 1 \times C_{in} \times C_{out}$	W_in × H_in × 3 × 3 × C_in × ((N_k − 1))/2 + W_in × H_in × C_in × C_out	$\frac{9 \times (N_k-1)}{2N_k^2 \times C_{out}} + \frac{1}{N_k^2}$
CSeDeConv			/	$3 \times 3 \times 1$	$3 \times 3 \times C_{in}$	$3 \times 3 \times C_{in} \times (N_k - 3)/2 + 3 \times 3 \times C_{in} \times C_{out}$	W_in × H_in × 3 × 3 × C_in × ((N_k − 3))/2 + W_in × H_in × 3 × 3 × C_in × C_out	$\frac{9 \times (N_k-3)}{2N_k^2 \times C_{out}} + \frac{9}{N_k^2}$

2.3. *Multi-Scale Representation Learning with Pyramid Pooling Module and Feature Aggregation Module*

2.3.1. Pyramid Pooling Module for Multi-Scale Feature Extraction

In most convolution-based deep networks, the spatial pooling operator is utilized as a crucial element to fuse characteristics of nearby feature bins into a compact representation. The objective of the spatial pooling process is to transform the joint feature representation into a new compressed, more effective one that preserves discriminative information while discarding irrelevant detail, the crux of which is to determine what can benefit the classification performance. Various pooling operators have been devised based on the sum, the average, the maximum, or some other combination rules and achieved significant success in computer vision and SAR ATR tasks [47]. However, most of the spatial pooling operators usually obtain the compact representation at a fixed-size receptive field which is possibly improper to the structure of the intrinsic characteristics and will lead to either information loss with too large of a size or feature dilution with too small of a size. Besides, for targets with a complicated characteristic structure, the fixed-size pooling operators that can only learn features at a fixed scale is also the major cause for incomplete representation learning and the consequent performance degradation.

To tackle the problem caused by the fixed-size pooling operators, the pyramid pooling module (PPM) was devised which was first adopted to generate fixed-length representations from inputs with varying sizes for deep visual recognition [48]. The PPM provides an effective way to obtain intrinsic characteristics of complicated targets from the view of multiple scales. In this paper, a modified version of PPM was devised and adopted in the proposed model to obtain a multi-scale representation at each modality level. As depicted in Figure 7, a typical PPM in the proposed model consists of four sub-branches with varying local reception field for pooling to capture the context information of the input feature maps. The first and the last sub-branches are the global max-pooling layer in the channel level and feature map level, respectively. For the two middle sub-branches, each of them consists of an adaptive max-pooling layer and a 3×3 CSeConv layer followed by a BN operator and a ReLU activation function. To be more specific, let us suppose the size of the input feature map is $W_{in} \times H_{in} \times C_{in}$. Accordingly, the size of the output feature maps of the first and the last global max-pooling layers are $1 \times 1 \times C_{in}$ and $W_{in} \times H_{in} \times 1$, respectively. The size of the output feature maps of the adaptive max-pooling layers in the two middle branches will be $\frac{W_{in}}{2} \times \frac{H_{in}}{2} \times C_{in}$ and $\frac{W_{in}}{4} \times \frac{H_{in}}{4} \times C_{in}$. The following CSeConv layers in each branch is employed to compress the pooled multi-channel feature map into a single-channel feature map, i.e., the sizes of the output feature maps of the CSeConv layers in the two middle sub-branches are $\frac{W_{in}}{2} \times \frac{H_{in}}{2} \times 1$ and $\frac{W_{in}}{4} \times \frac{H_{in}}{4} \times 1$. Finally, the output feature maps of the four sub-branches are converted to a single column vector and concatenated to construct the representation at the current modality level. It should be noted that if the size of the input feature map is too small to obtain the feature maps in the overall four sub-branches, part of the branches can be removed from the typical PPM and the corresponding feature maps can be neglected according to the image size.

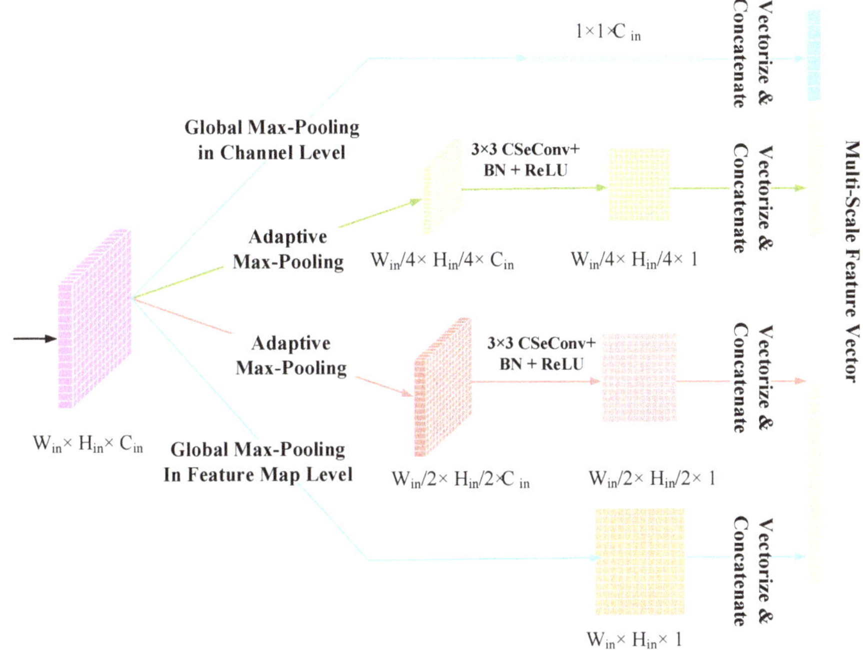

Figure 7. The architecture of the proposed pyramid pooling module (PPM) for multi-scale representation learning at each modality level. Given an input feature map with the size of $W_{in} \times H_{in} \times C_{in}$, the output feature maps of the four sub-branches of the PPM are with the size of $1 \times 1 \times C_{in}$, $\frac{W_{in}}{4} \times \frac{H_{in}}{4} \times C_{in}$, $\frac{W_{in}}{2} \times \frac{H_{in}}{2} \times C_{in}$, and $W_{in} \times H_{in} \times 1$.

2.3.2. Feature Aggregation Module (FAM) for Feature Map Reconstruction

The utilization of our PPMs in the encoder stage allows the model to learn multi-scale representations from the input SAR image at different modality levels. However, a new problem that deserves to be solved is how to seamlessly merge the feature maps from PPMs at different modality levels and obtain the reconstructed image that is the essential objective of an AE model. To this end, a series of FAMs are developed each of which contains two parts as illustrated in Figure 8.

In the first part, the feature maps at different scales of the PPM are combined to produce a new feature map at the current modality level. The multi-scale feature vector at one modality level obtained by a PPM is first decomposed and reshaped into feature maps of different scales according to the PPM at the same modality level. Subsequently, the feature maps of different scales are processed in separate sub-branches. For two middle sub-branches, an upsampling operator and a smooth operator consisting of a 3×3 CSeDeConv layer, a BN operator and a ReLU activation function are executed. For the first and the last sub-branches, only the upsampling operator and the smooth operator are applied, respectively. Finally, the processed feature maps of different scales are weighted and summed to generate the feature map of the current modality level. In the second part, the feature map from the coarser level is merged with the combined multi-scale feature map at the current level. The upsampling process followed by a 3×3 CSeDeConv layer, a BN operator and a ReLU activation function is applied to obtain the feature map of a coarser level that is the same size as the feature map at the current level.

Subsequently, feature maps from different levels are concatenated together to generate the merged feature map.

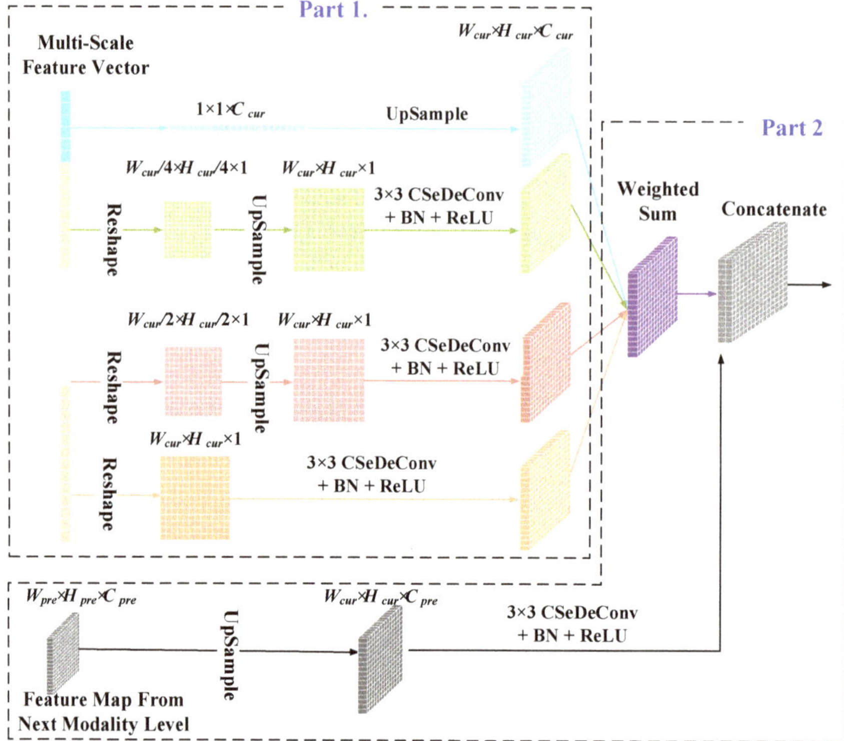

Figure 8. The architecture of the proposed feature aggregation module (FAM) for merging multi-scale representation of different modality levels which consists of two parts. The first part of the proposed module combines multi-scale feature maps and generates a new feature map at the current modality level. The input feature vector with the size of $\frac{21}{16}W_{cur} \times H_{cur} + C_{cur}$ is decomposed and reshaped to produces the feature maps in the four sub-branches with the size of $1 \times 1 \times C_{cur}$, $\frac{W_{cur}}{2} \times \frac{H_{cur}}{2} \times 1$, $\frac{W_{cur}}{4} \times \frac{H_{cur}}{4} \times 1$ and $W_{cur} \times H_{cur} \times 1$ Subsequently, the upsampling operator and the following smoothing operator are applied to each sub-branch. The output feature maps of the four sub-branches are weighted and added to generate the feature map of the first part with the size of $W_{cur} \times H_{cur} \times C_{cur}$. The second part of the FAM merges the feature map with the size of $W_{pre} \times H_{pre} \times C_{pre}$ from the coarser modality level with the combined feature map with the size $W_{cur} \times H_{cur} \times C_{cur}$ at the current modality level. The input feature map at the coarser level is upsampled and smoothed to generate the upsampled feature map of the coarser level with the size of $W_{cur} \times H_{cur} \times C_{pre}$ The upsampled feature map is concatenated with the feature map at the current modality level, generating the new feature map with the size of $W_{cur} \times H_{cur} \times \left(C_{pre} + C_{cur}\right)$.

To be more specific, in the first step, suppose the input feature vector at the current modality level has the size of $\frac{21}{16}W_{cur} \times H_{cur} + C_{cur}$. The input multi-scale feature vector will be decomposed and reshaped the feature maps with the size of $1 \times 1 \times C_{cur}$, $\frac{W_{cur}}{2} \times \frac{H_{cur}}{2} \times 1$, $\frac{W_{cur}}{4} \times \frac{H_{cur}}{4} \times 1$, and $W_{cur} \times H_{cur} \times 1$, respectively. Subsequently, the upsampling operator and the following smoothing operator are applied. Accordingly, the output feature maps of the four sub-branches will have the same size of $W_{cur} \times H_{cur} \times C_{cur}$. Finally, the four feature maps are weighted and added to generate the feature map of the current modality level with the size of $W_{cur} \times H_{cur} \times C_{cur}$. In the second step, let the input feature map at the coarser level have the size of $W_{pre} \times H_{pre} \times C_{pre}$ with $W_{pre} \times H_{pre}$ being the spatial size of the input feature map and C_{pre} being the number of channels at the coarser level. The feature map from the coarser level is upsampled and smoothed by a 3×3 CSeDeConv layer, a BN operator, and a ReLU activation function. Finally, the upsampled feature map of the coarser level with the size of $W_{cur} \times H_{cur} \times C_{pre}$ is concatenated with the feature map at the current modality level, generating the new feature map with the size of $W_{cur} \times H_{cur} \times \left(C_{pre} + C_{cur}\right)$.

2.4. Loss Function Based on the Modified Reconstruction Loss and Speckle Filtering Restriction

Typically, an AE-based model provides a symmetrical frame on learning latent representation of candidate targets by mapping the inputs into a low dimensional feature space at the encoder stage and approximately reconstruct the inputs from the learned features at the decoder stage. The objective of the AE-based model is to minimize the loss function that measures the distortion between the inputs and the outputs to guarantee that the mapping process preserves the information of the inputs. The commonly used loss functions, such as the MSE, cross-entropy, and the Minkowski distance in the field of deep learning, concern the total bias of pixel values or distributions and neglect the structural information of the candidate targets, leading to performance degradation in SAR ATR. Therefore, the SSIM loss function [49] is employed in the proposed model which simultaneously compares the similarity of two images over the structure, the luminance, and the contrast, gaining significant success in the computer vision domain. Suppose the input image of an AE-based model is x and the output of the model is $\hat{x}$, the SSIM loss function can be:

$$L_{SSIM}(x, \hat{x}) = E\left(\frac{(2\mu_x\mu_{\hat{x}} + c_1)(2\sigma_{x\hat{x}} + c_2)}{\left(\mu_x^2 + \mu_{\hat{x}}^2 + c_1\right)\left(\sigma_x^2 + \sigma_{\hat{x}}^2 + c_2\right)}\right), \tag{2}$$

where μ_x and $\mu_{\hat{x}}$ are the local average in a 11×11 sliding window of x and $\hat{x}$, respectively; $\sigma_{\hat{x}}^2$ and $\sigma_{\hat{x}}^2$ are the local variance in a 11×11 sliding window of x and $\hat{x}$, respectively; $\sigma_{x\hat{x}}$ is the correlation coefficient in a 11×11 sliding window; $c_1 = (K_1L)^2$ and $c_2 = (K_2L)^2$ are two constants with $K_1 = 0.01$ and $K_2 = 0.03$; L is the dynamic range of the pixel values (1.0 for normalized SAR images); $E(\cdot)$ is the expectation operator. While calculating the SSIM loss of two images, the sliding window will be moved pixel by pixel over the entire image. At each step, the local statistics and the local SSIM loss are computed in the window. Finally, the SSIM of the entire image is computed by averaging the local SSIM of each step.

Another problem is that in most conditions there is serious speckle in the target patches which not only have little information about the target but affect the ATR capability of the learned features. To alleviate their influence, the reconstruction loss is modified by measuring the distortion between the outputs and the speckle filtered images instead of the original input data. Consequently, the model will be forced to learn the characteristics of the targets rather than the background clutter, and the prior knowledge of speckle suppression can be embedded in the MSCAE during the model training procedure. In the proposed model, the ILSF is employed to generate the speckle suppressed image due to the fact of its excellent capability in maintaining detailed structures, strongly reflecting and scattering targets, and smoothing undesired background clutter [50]. Moreover, an additional restriction is devised to guarantee that the speckle suppression process is taken place in the encoder stage and little information on the speckle will be learned. The restriction is implemented by comparing the difference between the features learned from the original inputs and those learned from the speckle suppressed images. Accordingly, the loss function of the proposed model is

$$L_{MSCAE} = \frac{1}{N}\sum_{i=1}^{N} L_{SSIM}\left(\hat{x}_i, x_i^{ILSF}\right) + \alpha \sum_{j=1}^{C} \|h_{ij} - h_{ij}^{ILSF}\|_2 \tag{3}$$

where $D_{train} = \{x_i\}_{i=1}^{N}$ is the training dataset with x_i being the ith target patch and N being the number of samples; $\hat{x}_i$ and x_i^{ILSF} are the output image of the proposed model and the speckle suppressed version of x_i, respectively; $\|\cdot\|_2$ is the $l-2$ norm; h_{ij} and h_{ij}^{ILSF} are the encoded feature vectors of x_i and x_i^{ILSF} at scale j, respectively; C is the number of modality levels; α is the coefficient of the speckle suppression restriction, which can be $0.01/C$ in most SAR ATR tasks.

3. Experiments and Discussion

3.1. Experimental Data Sets

In this study, the representation learning capability of the proposed model was evaluated by the MSTAR dataset [42] which is jointly sponsored by the US Defense Advanced Research Projects Agency (DARPA) and Air Force Research Laboratory (AFRL). There were a total of ten distinctive types of vehicles in the dataset as shown in Figure 9, including the armored personnel carrier BMP-2, BRDM-2, BTR-60, and BTR-70; the tank T-62 and T-72; the rocket launcher 2S1; the air defense unit ZSU-234; the truck ZIL-131; and the bulldozer D7. The images were collected by an X-band SAR in spotlight mode with the resolution of 0.3 m × 0.3 m and split into tens of thousands of small patches centered on the candidate targets and surrounded by varying background clutter. These small patches provide full-aspect coverage from 0° to 360° and different views at various depression angles for each type of the ten vehicles. Detailed information including the type, the serial number (Serial No.), the depression angle, and the number of samples are all listed in Table 2.

Figure 9. Photographs (the first row) and SAR imagery examples (the second row) of the moving and stationary target acquisition and recognition (MSTAR) dataset for model evaluation. From left to right, the types of vehicles are 2S1, BMP-2, BRDM-2, BTR-60, BTR-70, D7, T-62, T-72, ZIL-131, and ZSU-234.

Table 2. Detailed information of the MSTAR dataset.

Type	Serial Number	Number of Samples	
		17° Depression	15° Depression
2S1	b01	299	274
	9563	233	195
BMP-2	9566	232	196
	c21	233	196
BRDM-2	E-71	298	274
BTR-60	K10yt7532	256	195
BTR-70	c71	233	196
D7	92v13015	299	274
T-62	A51	299	273
	132	232	196
T-72	812	231	195
	s7	233	191
ZIL-131	E12	299	274
ZSU-234	d08	299	274

In order to ensure comprehensive access to the performance, the proposed MSCAE was tested under standard operating condition (SOC) and various extended operating conditions (EOCs) including substantial variations in the signal-to-noise ratio (SNR), resolution, and version. In our experiments, the proposed model was first validated on three similar targets, namely, BMP-2, BTR-70, and T-72, to validate its performance under SOC and version variants. Subsequently, validation of the SSIM loss, the PPM, the CseConv, and CSeDeConv and the speckle suppression scheme are all discussed based on the three-target dataset. Experiments on 10 class MSTAR data were also conducted to evaluate the performance under the extension of the target type. Finally, the robustness of the proposed model under various conditions, including noise corruption and resolution variance, was also evaluated with the ten-target dataset. For both the three-target dataset and the ten-target dataset, the patches acquired at the 17° depression angle were utilized as training samples, while those obtained at the 15° depression angle constructed the test set. Similar to the experimental setting in References [30,31], only the data from BMP2-9563 and T72-132 were used, as the samples of the BMP-2 and T-72 were used to construct the training dataset. But in the test dataset, images of all serial numbers (i.e., version variants) were used to test the performance of the proposed method.

3.2. Experiment Configuration

3.2.1. Data Preprocessing

In most deep networks, including the proposed model, the size of the input images is required to be the same. Meanwhile, the size of target patches in the MSTAR dataset can vary from 128×128 to 158×158. Consequently, the input target patches should be resized to the same shape, which is 128×128 in this study, before being used for model training and performance validation. In this study, the image crop processing based on the centroid of the target region is adopted. The ILSF is firstly applied to suppress the speckle and background clutter in the small patches. Subsequently, a two-parameter constant false alarm rate (CFAR) detector is executed to obtain the target region of each patch. The centroid is calculated by averaging the coordinates of the target pixels weighted by their pixel values. Finally, only the 128×128 region surrounding the centroid will remain, while other regions will be removed.

Another preprocessing step is the normalization process. It can be found that in many target patches, the intensity of targets seriously varies which possibly conceals the differences among targets and, thus, affect the performance of the learned features. Accordingly, intensity normalization was adopted to alleviate the amplitude variation in target patches, mapping the pixel intensities onto the range $[0, 1]$. Except for image resizing and normalization, no other preprocessing, such as data augmentation (DA) or target segmentation, were applied.

3.2.2. Model Configuration and Experiment Design

In our experiments, an MSCAE with four modality levels was utilized to obtain multi-scale representations of the MSTAR data. The model parameters and the fan-ins and fan-outs of each level are listed in Table 3. As depicted in the table, there were some changes in the third and fourth levels. In the third level, the size of the feature map was downsampled from $16 \times 16 \times 16$ to $4 \times 4 \times 32$ after the 2×2 max-pooling and the 5×5 CseConv with a stride of 2. While feeding the feature map to the PPM block, the output feature maps of the four sub-branches of the typical PPM should have the size of $4 \times 4 \times 1$, $2 \times 2 \times 32$, $1 \times 1 \times 32$ and $1 \times 1 \times 32$, respectively. The outputs of the third and the fourth sub-branch were the same, bringing in redundant information and had little contribution to the target discrimination. Accordingly, the fourth sub-branch, which provides a feature map with global max-pooling in channel level, was removed, and the corresponding feature map was neglected. The FAM in the same modality level was also changed by removing the corresponding sub-branches while combining the feature maps obtained by the PPM. In the fourth modality level, since the size of the input feature map was smaller than 5×5, the optional CseConv layer was removed, and only the 2×2 max-pooling layer was applied before drawing the feature vector of the coarsest level.

Table 3. The main structure of the MSCAE with four modality levels utilized in the experiments.

Stage	Level	Input Size	Processes	Output Size	Feature Size
Encoder	1	$128 \times 128 \times 1$	*CSeConv*, $5 \times 5 \times 1$, 8, *stride* 2 *BN + ReLU* *PPM Block*	$64 \times 64 \times 8$	$64 \times 64 \times 1$ $32 \times 32 \times 8$ $16 \times 16 \times 8$ $1 \times 1 \times 8$
	2	$64 \times 64 \times 8$	*Max Pooling*, 2×2 *CSeConv*, $5 \times 5 \times 8$, 16, *stride* 2 *BN + ReLU* *PPM Block*	$16 \times 16 \times 16$	$16 \times 16 \times 1$ $8 \times 8 \times 16$ $4 \times 4 \times 16$ $1 \times 1 \times 16$
	3	$16 \times 16 \times 16$	*Max Pooling*, 2×2 *CSeConv*, $5 \times 5 \times 16$, 32, *stride* 2 *BN + ReLU* *PPM Block*	$4 \times 4 \times 32$	$4 \times 4 \times 1$ $2 \times 2 \times 32$ $1 \times 1 \times 32$
	4	$4 \times 4 \times 32$	*Max Pooling*, 2×2 *Vectorization*	$2 \times 2 \times 32$	$128 \times 1 \times 1$
Decoder	4	$128 \times 1 \times 1$	Reshape	$2 \times 2 \times 32$	-
	3	$2 \times 2 \times 32$ $176 \times 1 \times 1$	*FAM Block* *CSeDeConv*, $5 \times 5 \times 64$, 16, *Stride* 2 *BN + ReLU*	$8 \times 8 \times 16$	-
	2	$8 \times 8 \times 16$ $1152 \times 1 \times 1$	*FAM Block* *CSeDeConv*, $5 \times 5 \times 32$, 8, *Stride* 2 *BN + ReLU*	$32 \times 32 \times 8$	-
	1	$32 \times 32 \times 8$ $14344 \times 1 \times 1$	*FAM Block* *CSeDeConv*, $5 \times 5 \times 16$, 4, *Stride* 2 *CSeDeConv*, $3 \times 3 \times 4$, 1, *Stride* 1 *Sigmoid*	$128 \times 128 \times 1$	-

All the convolution kernels of the MSCAE were also initialized with the He initialization [46]. After initialization, the model was trained with the preprocessed training dataset, and the Adam optimizer [51] was utilized to optimize the model with an initial learning rate of 0.001. The exponential decay rates β_1 and β_2 for the moment estimates were 0.9 and 0.999, respectively. The batch of the training samples was 32. The maximum number of iterations was 500, and the early-stopping scheme was enabled to terminate the training if the improvement of the training loss was less than the threshold. In our experiments, the nonlinear SVM (NSVM) was employed for classification after extracting features from the target patches. Moreover, to avoid fluctuations in the results caused by random steps in the model initialization and optimization, each experiment was repeated ten times, and the average of the results were utilized for performance evaluation.

Experiments were carried out in a 64 bit Windows 10 system. The proposed model was mainly built on the Google Tensorflow v1.5.0 deep learning library in the Python development environment PyCharm. The hardware platform was a specially adapted DELL T5810 workstation with an Intel Xeon E5-1607 v3 @ 3.10 GHz CPU, 32 GB DDR4 RAM and an NVIDIA K40c (12G memory) GPU with CUDA8.0 accelerating calculation.

3.3. Evaluation on Three-target Classification

The average results of the ten experiments with the three-target dataset are depicted in Table 4. The performance was measured by the probability of correct classification (P_{cc}) which is calculated through the number of targets recognized correctly divided by the number of all the targets. The results with and without version variants are listed in the sixth and fifth columns of the table, respectively. A comparison experiment was also conducted to demonstrate the performance improvement induced by the multi-scale feature learning architecture. Classification rates with various feature combinations are evaluated and compared.

As shown in Table 4, for the experiment without version variants (i.e., under SOC), the proposed model had the highest accuracy (i.e., 99.73%) when features from all modality levels were utilized. Meanwhile, in the case with variants only, the accuracy of the proposed model with various feature combinations suffered a slight degradation due to the differences in local structure and small equipment of varied serial numbers. However, the accuracy obtained by the model with features from all levels was still higher than 98%, indicating a good generalization performance. The average accuracies of these methods with all the test data are listed in the seventh column of Table 4. It can be found that when features from multiple scales are combined, the average P_{cc} of 99.14% is competitive to the state-of-the-art results provided by the supervised neural networks. The major reason is that the proposed two-level multi-scale feature extraction structure guarantees that the MSCAE can learn the high-level abstracted properties while preserving the detailed information that is neglected by most DL networks. Besides, the specifically designed objective function with speckle suppression and SSIM can diminish the influence of serious speckle in SAR images and take full advantage of the target structure caused by the backscattering. In addition, the proposed compact convolution and deconvolution processes greatly decrease the number of trainable parameters and introduce more nonlinearity that slightly benefits the model capability.

Table 4. Classification results on the three-target dataset.

Scheme	BMP-2	BTR-70	T-72	Without Variants	Variants Only	Average P_{cc}
L 1	89.05%	99.39%	99.79%	96.15%	94.34%	95.12%
L 2	87.01%	98.88%	96.02%	95.13%	92.28%	92.56%
L 3	82.35%	98.06%	96.01%	93.08%	90.37%	90.44%
L 4	77.92%	93.88%	96.80%	90.23%	86.78%	88.26%
L 1+2	95.33%	99.39%	99.14%	98.47%	97.35%	97.54%
L 1+3	94.52%	98.98%	99.38%	98.30%	96.95%	97.24%
L 1+4	93.94%	99.39%	99.62%	98.26%	96.92%	97.14%
L 2+3	96.13%	97.76%	96.70%	97.55%	96.27%	96.61%
L 2+4	93.52%	98.98%	99.48%	97.75%	96.74%	96.85%
L 3+4	92.22%	98.67%	99.48%	97.38%	96.05%	96.25%
L 1+2+3	96.70%	99.39%	99.90%	99.25%	98.15%	98.45%
L 1+2+4	95.57%	99.39%	99.79%	98.91%	97.62%	97.92%
L 1+3+4	95.30%	99.39%	99.90%	98.26%	97.53%	97.85%
L 2+3+4	95.06%	99.18%	99.90%	98.50%	97.54%	97.71%
All	99.05%	99.69%	99.04%	99.73%	98.69%	99.14%

Other features extraction methods were also compared with the proposed method for further evaluation including the baseline handcrafted methods and the DL networks which were obtained from the state-of-the-art results. The baseline handcrafted methods include the PCA-kernel SVM (PCA-KSVM) [52], the joint sparse representation based method (JSRC) [53], the particle swarm optimization with Hausdorff distance (PSO-HD) [54], the non-negative matrix factorization (NMF) method [55], the attributed scattering center matching method (ASCM) [7], and the 3D scattering center model reconstruction method (3D-SCM) [5]. Among these methods, the PCA-KSVM method employs the nonlinear PCA to extract discriminative feature and, subsequently, feeds the features into the SVM classifier. The JSRC method exploits the inter-correlations among the multiple views using joint sparse representation over a training dictionary. The PSO-HD is a pattern matching method that minimizes the Hausdorff distance over rigid transformations. The NMF method utilized the NMF with an $L_{1/2}$ norm constraint to extract features in SAR images. The ASCM and the 3D-SCM were devised based on the backscattering model of the SAR image and achieved state-of-the-art results in SAR ATR. The ASCM proposed a SAR ATR method, where the ASCs were utilized for target reconstruction and similarity measurement. In the 3D-SCM method, the 3D scattering center model, established offline from the CAD model of the target, was employed to predict the 2D scattering centers for template matching. The DL networks for comparison were composed of the restricted RBM (RRBM) [56], the CNN with DA (DA-CNN) [57] and additional data generated by image processing methods, the CNN with SVM (CNN+SVM) [37], the A-Convnet [57] that replaced the fully connected layers with a convolution layer in a CNN, the sparse AE pre-trained CNN (AE-CNN) [58] where the convolution kernel was trained on randomly sampled image patches using unsupervised sparse auto-encoder, the ED-AE [30], and the

Triplet-DAE [31]. Among these methods, the CNN with SVM and the A-Convnet were implemented in our codes with Python. In our implementation, preprocessing included image cropping, speckle filtering with ILSF, and normalization was applied to these methods. Besides, the additional DA scheme was also executed to generate sufficient training samples for the CNN with SVM model and the A-Convnet according to References [37,38]. The configurations of the CNN with SVM and the A-Convnet were determined according to References [37,38]. The accuracies of all the methods are shown in Figure 10. The features learned by the proposed method had a better classification capability than most handcrafted methods, even comparing them with ASCM and 3D-SCM which achieved state-of-the-art results for handcrafted features, because of the multi-scale feature learning scheme and the specifically designed objective function. Comparison with deep networks, including the DA-CNN, the RRBM, the AE-CNN, and the ED-AE, also indicates that the proposed model outperformed most of the DL models which have specialized restrictions for finding discriminative features. Even compared with the CNN+SVM and the A-Convnet that achieved state-of-the-art results, the proposed method obtained a comparable result.

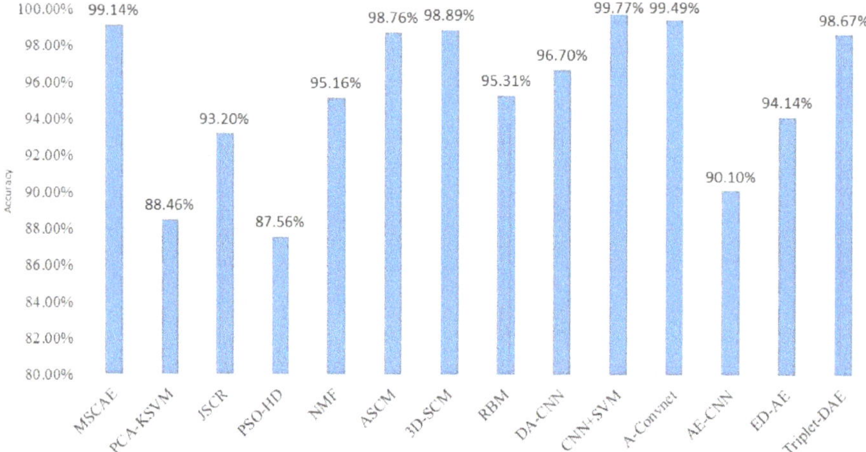

Figure 10. Performance comparison with baseline handcrafted feature extraction methods and deep representation models via the three-target dataset.

3.4. Validation of the Model Component

In order to investigate the contribution of each proposed component in the MSCAE, including the PPM, the CSeConv, the SSIM measurement, the ILSF, and the speckle suppression restriction, validation experiments were conducted. Each component was removed from the MSCAE to reveal the performance improvement induced by it. Accordingly, we obtained five models for performance validation, marked as MSCAE no. 1 to no. 5 in Table 5. In model no. 1, the PPMs and the corresponding FAMs at each modality were removed from the MSCAE, and the multi-scale features were only generated by the U-shaped architecture. In model no. 2, the CSeConv and the corresponding CSeDeConv layers were all replaced by the standard convolution layers. In model no. 3, the SSIM measurement was replaced by the MSE loss. In model no. 4, all the data flows and restrictions that related to the ILSF were removed from the MSCAE model, while in model no. 5 only the restriction term in the objective function was removed. The three-target MSTAR dataset was utilized to evaluate their performance and each experiment was conducted ten times. The average accuracy of the five models and the proposed MSCAE models are listed in Table 5. As shown in the table, the PPM and the SSIM measurement contributed the most to improving the accuracy, 2.85% and 1.83% respectively, while CSeConv and the CSeDeConv only improved the accuracy approximately 0.41%. However, the proposed CSeConv and CSeDeConv can remarkably reduce the number of trainable parameters in the proposed model that can greatly benefit its performance with a small training dataset. To further illustrate their contribution, an experiment which evaluated the model performance with a limited training sample was conducted by randomly removing a part of the sample in the training set. In this experiment, only $1/n$ images were randomly selected from the dataset as training samples with n varying from one to ten. The average accuracies and their standard deviations with the proposed MSCAE and model no. 2 are presented in Figure 11. As shown in the figure, the proposed model achieved an accuracy higher than 90% when only 20% of the sample was utilized to train the model, while the P_{cc} of the MSCAE no. 2 that had much more trainable parameters than the proposed model fell below 85%. Moreover, when the size of the training dataset was only 1/10 of the original one, the P_{cc} of the MSCAE no. 2 fell below 60% and the proposed MSCAE still had an accuracy higher than 70%.

Table 5. Validation of the components in the proposed MSCAE model.

Model	Description	BMP-2	BTR-70	T-72	P_{cc}
No. 1	No PPM and FAM	93.41%	98.81%	99.66%	96.85%
No. 2	No CSeConv and CSeDeConv	97.33%	99.66%	99.83%	98.73%
No. 3	MSE Measurement	93.98%	99.83%	99.83%	97.31%
No. 4	No ILSF	95.17%	99.15%	99.71%	97.68%
No. 5	No Speckle Suppression Restriction	96.15%	99.49%	99.60%	98.10%
Baseline	The proposed MSCAE	99.05%	99.69%	99.04%	99.14%

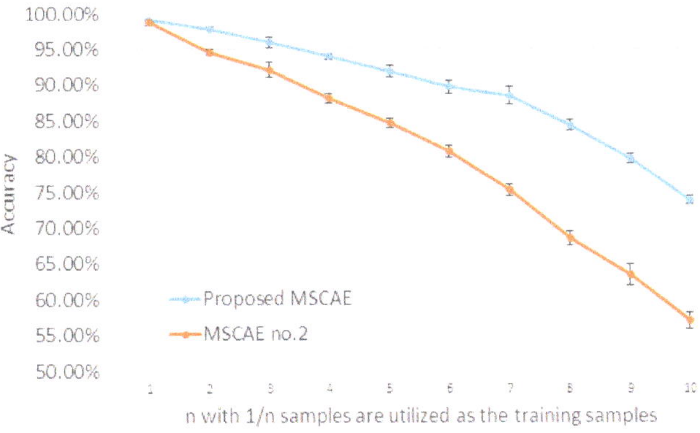

Figure 11. Validation of the proposed model with a small dataset when $1/n$ samples were randomly selected to train the model.

3.5. Evaluation on Ten-Target Classification

The average results of the ten experiments with the ten-target MSTAR dataset are depicted in Table 6. It can be found that by combining the features of all the four levels, the classification accuracy obtained an improvement that increased from 98.5% to 98.9%, in comparison with the highest P_{cc} achieved by the feature vector that combined the learned representations of the first three modality levels. Many other feature extraction methods and representation learning models were also compared with the MSCAE for further evaluation, including the baseline handcrafted features, the unsupervised DL models, and the supervised models. The baseline handcrafted features for evaluation includes the NMF method [55], the sparse representation of monogenic signal via Riemannian manifolds (SRRMs) [59], the weighted multi-task kernel sparse representation (WMTKSR) [60], and the ASCM [7]. Among these methods, the SRRM utilizes the covariance descriptor of the monogenic signal as the features, and classifies the targets with the Riemannian manifold embedded in an implicit reproduction of the kernel Hilbert space (RKHS). The WMTKSR maps the multi-scale monogenic features into a high-dimensional kernel feature space using the nonlinear mapping associated with a kernel function, and the classification process is formulated as a joint covariate selection problem across a group of related tasks. The unsupervised DL models comprise the multi-discriminator generative adversarial network (MGAN-CNN) that generates unlabeled images with GAN and sets them as the input of CNN together with original labeled images [61], the feature fusion SAE (FFAE) [15] that extracts 23 baseline features and three-patch local binary pattern (TPLBP) features and, subsequently, feeds them into an SAE for feature fusion and the variational AE based on residual network (ResVAE) [22]. The supervised models for performance evaluation are the ED-AE [30], the Triplet-DAE [31], the CNN with SVM [37], the A-Convnet [38], the ESENet that based on a new enhanced squeeze and excitation (enhanced-SE) module [35], and the hierarchical fusion of CNN and ASC (ASC-CNN) that provide a complicated scheme to fuse the decision of the ASC model and the CNN [39]. Among these methods, the CNN with SVM and the A-Convnet are implemented in our codes with Python. In our implementation, preprocessing included image cropping, speckle filtering with ILSF, and normalization was applied to the two methods. Besides, an additional DA scheme was also executed to generate sufficient training samples for the CNN with an SVM model and the A-Convnet according to References [37,38]. Therefore, both the results of the two models with and without DA processes were compared in our experiment to make a comprehensive and equal analysis. The configurations of the CNN with SVM and the A-Convnet were determined according to References [37,38]. In the experiments, each of the CNN with SVM and the A-Convnet was executed and tested ten times, and the average classification accuracy was utilized for performance evaluation.

Table 6. Classification results of the ten-target MSTAR dataset.

Scheme	2S1	BMP-2	BRDM-2	BTR-60	BTR-70	D7	T-62	T-72	ZIL-131	ZSU-234	P_{cc}
L. 1	95.3%	89.8%	93.1%	96.9%	99.0%	98.2%	95.6%	99.8%	98.5%	99.3%	96.1%
L. 2	88.7%	86.2%	88.3%	93.3%	94.4%	93.8%	89.7%	97.8%	98.2%	98.9%	92.7%
L. 3	83.2%	91.8%	87.6%	88.2%	93.4%	94.2%	85.7%	95.4%	94.9%	96.7%	91.6%
L. 4	75.5%	85.2%	81.8%	81.5%	91.8%	94.5%	83.2%	92.8%	79.9%	97.8%	86.9%
L. 1+2	94.9%	95.7%	93.8%	97.9%	98.5%	99.3%	96.3%	99.8%	99.3%	100.0%	97.6%
L. 1+3	96.7%	95.9%	94.2%	99.5%	99.5%	99.6%	95.6%	99.8%	98.5%	99.6%	97.8%
L. 1+4	97.1%	93.9%	94.2%	97.9%	99.0%	100.0%	97.4%	99.8%	98.5%	99.6%	97.5%
L. 2+3	92.7%	91.8%	97.4%	97.4%	99.5%	98.2%	93.4%	98.8%	98.2%	99.3%	95.7%
L. 2+4	90.9%	92.3%	90.9%	95.9%	98.5%	97.8%	94.1%	97.8%	97.8%	98.9%	95.3%
L. 3+4	86.5%	93.9%	92.3%	92.3%	96.9%	97.8%	89.7%	96.9%	94.5%	97.1%	94.1%
L. 1+2+3	96.7%	97.6%	96.4%	99.0%	99.5%	99.6%	97.1%	99.8%	98.9%	100.0%	98.5%
L. 1+2+4	96.7%	97.4%	95.3%	97.9%	99.5%	99.6%	97.4%	99.7%	99.3%	100.0%	98.3%
L. 1+3+4	96.7%	98.0%	96.0%	98.5%	99.5%	99.6%	96.7%	99.8%	98.5%	99.6%	98.4%
L. 2+3+4	94.2%	93.4%	98.5%	99.5%	99.5%	98.5%	93.0%	98.3%	97.8%	99.3%	96.2%
ALL	99.3%	98.3%	98.5%	99.0%	99.5%	99.6%	98.5%	99.8%	99.3%	100.0%	98.9%

The classification results of all the methods are depicted in Figure 12. It can easily be found that the classification accuracy of the proposed method was much higher than most of the traditional handcrafted features. Although the accuracy of the proposed model was a bit lower than the ASCM feature that achieved state-of-the-art results with the scattering center model, the result obtained by the proposed model was still comparable and can adaptively extract features without manual intervention. Compared with most of the deep representation learning methods (e.g., the ED-AE, the Triple-DAE, the MGAN-CNN, and the ESENet), the proposed model also yielded much better performance. In comparison with the CNN with SVM, the A-Convnet and the ASC-CNN that achieved state-of-the-art results, the proposed MSCAE was also competitive. The results achieved by the CNN with SVM and the A-Convnet without DA preprocess demonstrated that their high classification rates mainly relied on the DA operations. Although their DA processes did improve the performance, they induced certain problems including bringing in man-made uncertainty and unstable performance, amplifying the sampling biases in the original dataset, and high computational complexity. The results obtained by the ASC-CNN devised a complex decision fusion strategy to improve the accuracy obtained by the ASC and the CNN separately. Although the performance of the proposed model was better than the proposed model, it requires a complicated process to extract the ASC features and higher computational complexity.

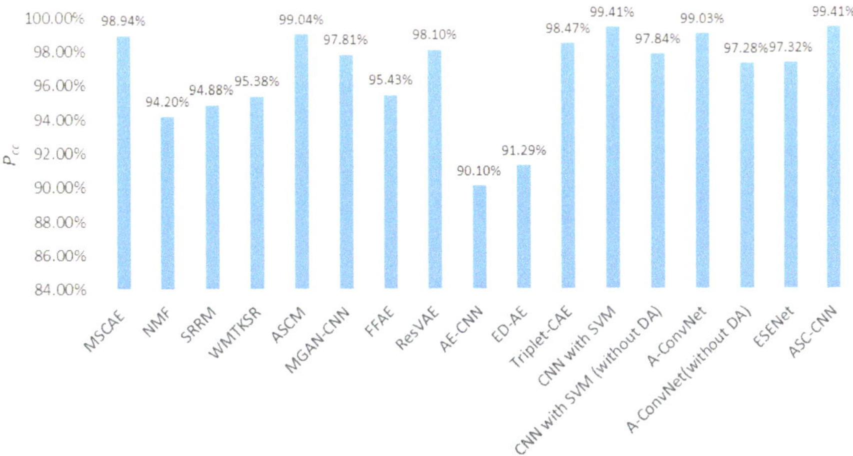

Figure 12. Performance comparison on the ten-target dataset with different handcrafted feature extraction methods and deep representation models.

3.6. Classification Experiment with Noise Corruption

An important characteristic of SAR data is that serious noise can often be observed in the images, which is a major factor causing performance deterioration in SAR ATR. Accordingly, to demonstrate the robustness of the proposed model, the SAR images corrupted by different levels of SNRs were simulated to evaluate the model's robustness to noise. The original MSTAR images that had an SNR over 30 dB were considered as noise-free sources. To obtain the noise-contaminated images, the original MSTAR patches were first transformed into the frequency-aspect domain with the 2D inverse discrete Fourier transform (IDFT), and different levels of additive complex Gaussian noises were added to the transformed images with the SNR defined in Equation (4) in accordance with Reference [5].

$$NR(dB) = 10log_{10}\frac{\sum_{u=0}^{U-1}\sum_{v=0}^{V-1}\left|f(u,v)\right|^2}{HW\sigma^2} \tag{4}$$

where $f(h,w)$ denotes the complex RCS computed by the EM code; σ^2 is the variance of the complex noise. By transforming the noisy RCS into the image domain using the same imaging process, the noise-contaminated images can be generated for experimental evaluation. Figure 13 presents some contaminated images with different SNRs.

Some input images and the corresponding reconstruction results at 10 dB and −10 dB SNR are presented in Figure 14 for comparison. Although many inputs at −10 dB SNR were seriously contaminated such that the targets in the patches can merely be observed, the output images of the trained model successfully reconstructed the major parts of the targets, demonstrating the excellent noise suppression capability of the proposed model. The average classification results and the corresponding standard deviations with noise-contaminated data under different SNRs are shown in Figure 15. The average experimental results with other methods are also presented in the figure including the Triplet-CAE, the CNN with SVM, and the A-Convnet. With the decreasing SNR of the input images, the classification accuracy of all the models suffers different degrees of deterioration and the highest was obtained by the proposed model at nearly every SNR level. When the SNR was higher than 0 dB such that the geometric and scattering characteristics were not seriously interrupted by the noise, each model reported a classification rate higher than 85%, and the proposed model achieved the highest accuracy. Even when the noise level was −10 dB such that most of the targets were concealed in the noise, as presented in Figures 13 and 14, the classification rate of the MSCAE still yielded a better performance than the other reference models in Figure 15, demonstrating the robustness of the proposed model under serious noise interruption.

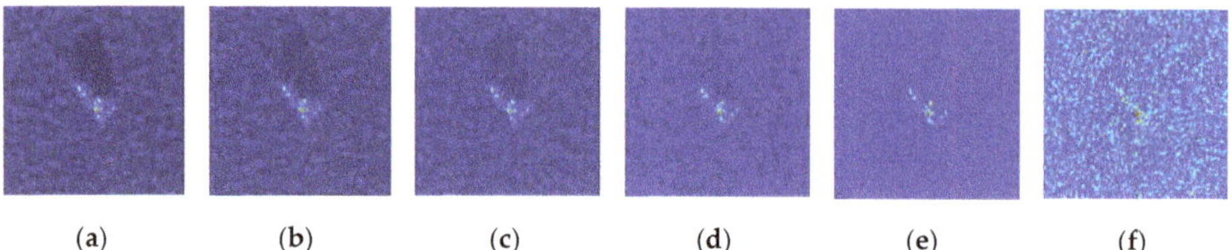

Figure 13. The noise interrupted images with different signal-to-noise ratios (SNRs). (**a**) The original image, (**b**) the SNR at 10 dB, (**c**) the SNR at 5 dB, (**d**) the SNR at 0 dB, (**e**) the SNR at −5 dB, and (**f**) the SNR at −10 dB.

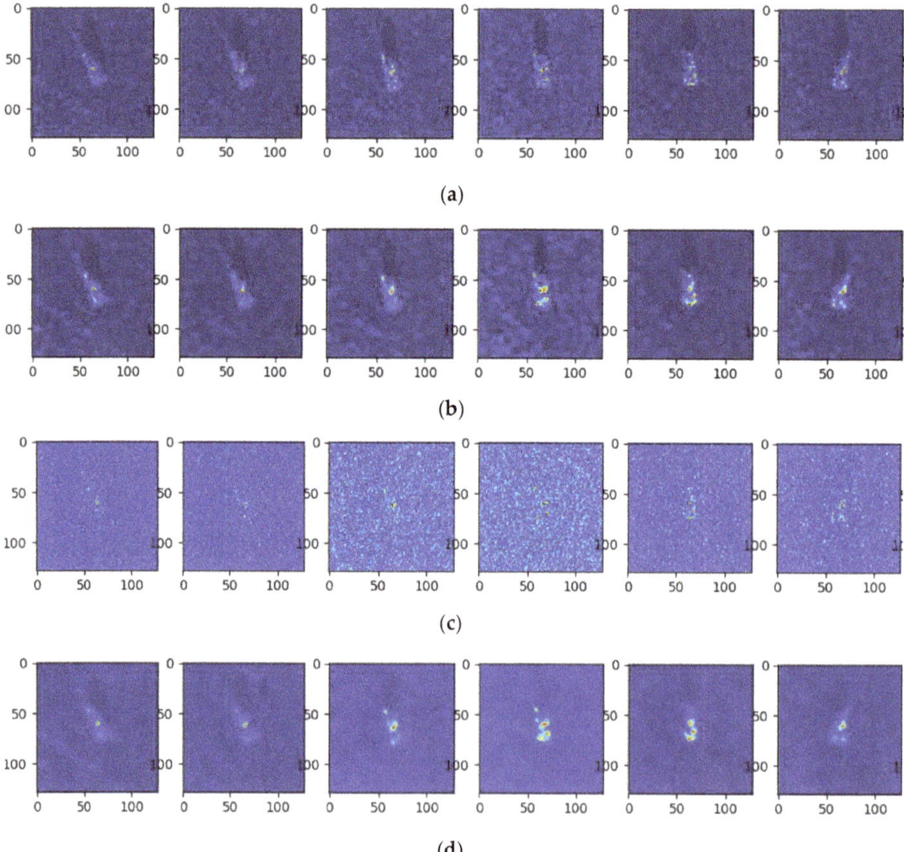

Figure 14. The input images and reconstruction results of the trained model at the noise levels of 10 dB SNR and −10 dB SNR. (**a**) Input images at a noise level of 10 dB SNR, (**b**) the reconstructed results of (**a**), (**c**) the input images at a noise level of -10dB SNR, (**d**) the reconstructed results of (**c**).

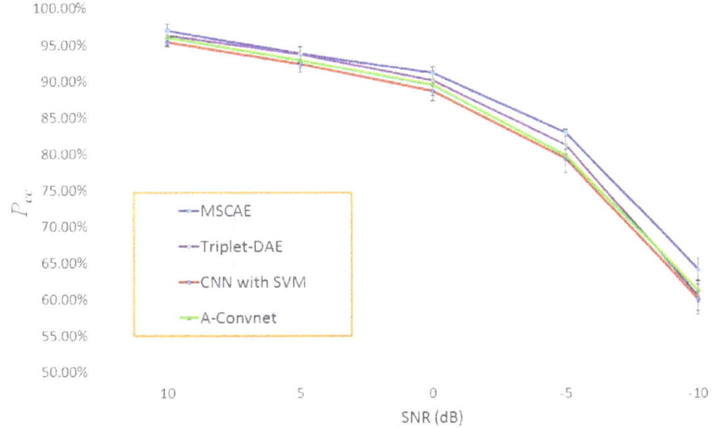

Figure 15. Classification results at different SNR levels.

3.7. Classification Experiment with Resolution Variance

The proposed model was subsequently evaluated concerning resolution variance. Theoretically, the range resolution and azimuth resolution of the SAR imagery was determined by the bandwidth of the transmitted wave and the synthetic aperture angle. However, due to the instability of the radars, the actual resolution of the measured SAR images would fluctuate around the theoretical values. Meanwhile, it was infeasible to train and maintain models at every possible resolution. Consequently, the robustness of resolution variation is also an important factor for model performance evaluation. Because the resolution of all target patches in the MSTAR dataset was 0.3 m × 0.3 m, the target patches with varied resolution should be simulated from the original images in the dataset. The spatial SAR images were converted into the frequency-aspect domain by the 2D-IDFT, and the sub-band was extracted. The sub-band data were subsequently resampled by zero-padding in the frequency domain and turned back to the spatial domain.

In the evaluation experiment, the resolution of the simulated data varied from 0.3 m × 0.3 m to 0.7 m × 0.7 m, and some images at different resolutions are presented in Figure 16. Similar to the configuration of the noise interruption experiment, the classification results of the proposed model are compared with the three reference models including the Triplet-DAE, the CNN with SVM, and the A-Convnet. At each resolution level, the experiment of each model was executed ten times to alleviate the influence of randomness caused by the model initialization and optimization. The average experimental results of each model are plotted in Figure 17. As shown in the figure, limited resolution deterioration did not seriously affect the performance of all the models. Even when the resolution was 0.6 m × 0.6 m, their average accuracy was still higher than 90%. However, the proposed model still gained the highest classification rate in comparison with the reference models at almost all the resolutions, illustrating its robustness under the extended operation condition of resolution variance.

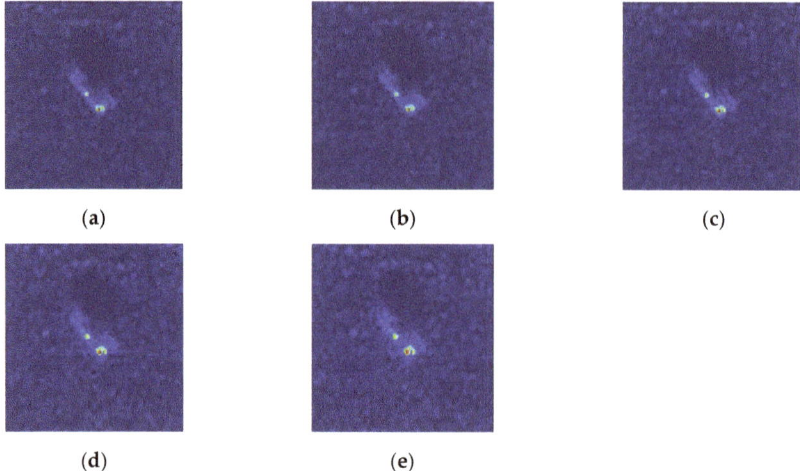

Figure 16. MSTAR data at different resolution. (**a**) 0.3 m × 0.3 m, (**b**) 0.4 m × 0.4 m, (**c**) 0.5 m × 0.5 m, (**d**) 0.6 m × 0.6 m, (**e**) 0.7 m × 0.7 m.

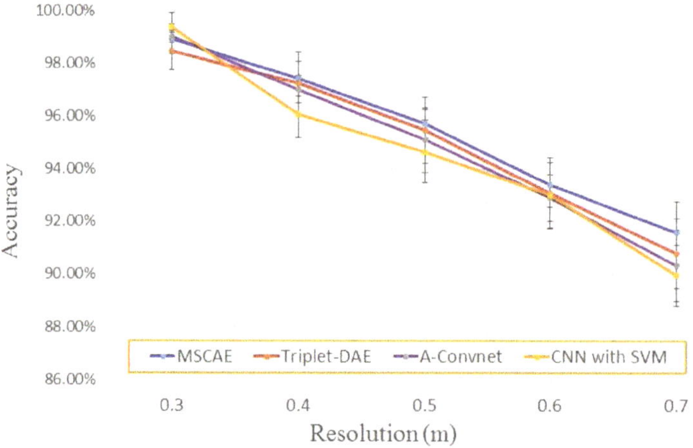

Figure 17. Classification results at different resolutions.

4. Conclusions and Future Work

In this paper, an unsupervised representation learning model was proposed, providing an effective way to learn the multi-scale representation of targets in SAR images via its U-shaped architecture, the CSeConv and the PPM blocks, and the modified loss function based on the SSIM and the restriction of speckle suppression. The major contributions of our work include:

(1) A proposed unsupervised multi-scale representation learning framework for feature extraction in SAR ATR. The utilization of the U-shaped multi-scale architecture and the PPM blocks simultaneously obtained abstract features and local detailed characteristics of targets, boosting the representational power of the proposed model;

(2) An objective function composed of a modified reconstruction loss and a speckle suppression restriction. The reconstruction loss based on SSIM and ILSF forces the MSCAE to learn adaptive

speckle suppression capability, while the restriction guarantees that the speckle filtering procedure was implemented in the feature learning step;

(3) The CSeConv and the CSeDeConv decreased the trainable parameters and calculation consumption, avoiding overfitting caused by insufficient samples. Moreover, they introduced more nonlinearity and slightly improved the performance of the MSCAE.

The MSTAR dataset was utilized to evaluate the performance of the proposed model. The proposed method was tested under both standard operating conditions and several extended operating conditions with both the three-target dataset and the ten-target dataset including the version variants, the noise corruption, and the resolution variance. Evaluation experiments demonstrated that the proposed method outperformed most of the conventional and deep learning algorithms and achieved comparable accuracy to the state-of-the-art results without any supervised information.

Author Contributions: S.T., W.G. and Y.L. conceived the methodology and conducted the entire experiments. S.T. and Y.L. wrote the manuscript. C.W. supervised the experiments and helped discuss the proposed method. H.Z. contributed to the organization of the paper and also the experimental analysis. All authors have read and agreed to the published version of the manuscript.

Funding: This research was funded in part by the Key Program of National Natural Science Foundations of China (Grant No. 41930110), the National Natural Science Foundations of China (Grant No. 41501356) and the Natural Science Foundation of Jiangsu Province under Grants No. BK20150774.

Acknowledgments: The authors thank the US Air Force Research Lab for providing the public MSTAR data. In addition, we are grateful to anonymous referees for their instructive comments.

Conflicts of Interest: The authors declare no conflict of interest.

References

1. Zhu, J.-W.; Qiu, X.; Pan, Z.; Zhang, Y.-T.; Lei, B. An Improved Shape Contexts Based Ship Classification in SAR Images. *Remote. Sens.* **2017**, *9*, 145. [CrossRef]
2. Mishra, A.K.; Motaung, T. Application of linear and nonlinear PCA to SAR ATR. In Proceedings of the 2015 25th International Conference Radioelektronika (RADIOELEKTRONIKA), Pardubice, Czech Republic, 21–22 April 2015; pp. 349–354.
3. Yin, K.; Jin, L.; Zhang, C.; Guo, Y. A method for automatic target recognition using shadow contour of SAR image. *IETE Tech. Rev.* **2013**, *30*, 313. [CrossRef]
4. Li, T.; Du, L. Target Discrimination for SAR ATR Based on Scattering Center Feature and K-center One-Class Classification. *IEEE Sensors J.* **2018**, *18*, 2453–2461. [CrossRef]
5. Ding, B.; Wen, G. Target Reconstruction Based on 3-D Scattering Center Model for Robust SAR ATR. *IEEE Trans. Geosci. Remote. Sens.* **2018**, *56*, 3772–3785. [CrossRef]
6. Li, Y.-B.; Zhou, C.; Wang, N. A survey on feature extraction of SAR Images. In Proceedings of the International Conference on Computer Application and System Modeling (ICCASM), Taiyuan, China, 22–24 October 2010.
7. Ding, B.; Wen, G.; Huang, X.; Ma, C.; Yang, X. Target Recognition in Synthetic Aperture Radar Images via Matching of Attributed Scattering Centers. *IEEE J. Sel. Top. Appl. Earth Obs. Remote. Sens.* **2017**, *10*, 3334–3347. [CrossRef]
8. Yoshua, B. Learning Deep Architectures for AI. *Found. Trends Mach. Learn.* **2009**, *2*, 1–127.
9. Dong, G.; Liao, G.; Liu, H.; Kuang, G. A Review of the Autoencoder and Its Variants: A Comparative Perspective from Target Recognition in Synthetic-Aperture Radar Images. *IEEE Geosci. Remote. Sens. Mag.* **2018**, *6*, 44–68. [CrossRef]
10. Li, H.; Gong, M.; Wang, C.; Miao, Q. Self-paced stacked denoising autoencoders based on differential evolution for change detection. *Appl. Soft Comput.* **2018**, *71*, 698–714. [CrossRef]
11. Gao, F.; Yang, Y.; Wang, J.; Sun, J.; Yang, E.; Zhou, H. A Deep Convolutional Generative Adversarial Networks (DCGANs)-Based Semi-Supervised Method for Object Recognition in Synthetic Aperture Radar (SAR) Images. *Remote. Sens.* **2018**, *10*, 846. [CrossRef]
12. Gao, F.; Liu, Q.; Sun, J.; Hussain, A.; Zhou, H. Integrated GANs: Semi-Supervised SAR Target Recognition. *IEEE Access* **2019**, *7*, 113999–114013. [CrossRef]

13. Jia, C.N.; Yue, L.X. SAR automatic target recognition based on a visual cortical system. In Proceedings of the 2013 6th International Congress on Image and Signal Processing (CISP), Hangzhou, China, 16–18 December 2013; pp. 778–782.
14. Geng, J.; Fan, J.; Wang, H.; Ma, X.; Li, B.; Chen, F. High-Resolution SAR Image Classification via Deep Convolutional Autoencoders. *IEEE Geosci. Remote. Sens. Lett.* **2015**, *12*, 2351–2355. [CrossRef]
15. Kang, M.; Ji, K.; Leng, X.; Xing, X.; Zou, H. Synthetic Aperture Radar Target Recognition with Feature Fusion Based on a Stacked Autoencoder. *Sensors* **2017**, *17*, 192. [CrossRef] [PubMed]
16. Gleich, D.; Planinsic, P. SAR patch categorization using dual tree orientec wavelet transform and stacked autoencoder. In Proceedings of the 2017 International Conference on Systems, Signals and Image Processing (IWSSIP), Poznan, Poland, 22–24 May 2017; pp. 1–4.
17. Zhang, L.; Ma, W.; Zhang, D. Stacked Sparse Autoencoder in PolSAR Data Classification Using Local Spatial Information. *IEEE Geosci. Remote. Sens. Lett.* **2016**, *13*, 1359–1363. [CrossRef]
18. Zhang, L.; Jiao, L.; Ma, W.; Duan, Y.; Zhang, D. PolSAR image classification based on multi-scale stacked sparse autoencoder. *Neurocomputing* **2019**, *351*, 167–179. [CrossRef]
19. Hou, B.; Kou, H.; Jiao, L. Classification of Polarimetric SAR Images Using Multilayer Autoencoders and Superpixels. *IEEE J. Sel. Top. Appl. Earth Obs. Remote. Sens.* **2016**, *9*, 3072–3081. [CrossRef]
20. Chen, Y.; Jiao, L.; Li, Y.; Zhao, J. Multilayer Projective Dictionary Pair Learning and Sparse Autoencoder for PolSAR Image Classification. *IEEE Trans. Geosci. Remote. Sens.* **2017**, *55*, 6683–6694. [CrossRef]
21. Lv, N.; Chen, C.; Qiu, T.; Sangaiah, A.K. Deep Learning and Superpixel Feature Extraction Based on Contractive Autoencoder for Change Detection in SAR Images. *IEEE Trans. Ind. Informatics* **2018**, *14*, 5530–5538. [CrossRef]
22. Xu, Y.; Zhang, G.; Wang, K.; Leung, H. SAR Target Recognition Based On Variational Autoencoder. In Proceedings of the 2019 IEEE MTT-S International Microwave Biomedical Conference (IMBioC), Nanjing, China, 6–8 May 2019; Volume 1, pp. 1–4.
23. Song, Q.; Xu, F.; Jin, Y.-Q. SAR Image Representation Learning With Adversarial Autoencoder Networks. In Proceedings of the IGARSS 2019—2019 IEEE International Geoscience and Remote Sensing Symposium, Yokohama, Japan, 28 July–2 August 2019; pp. 9498–9501.
24. Kim, H.; Hirose, A. Unsupervised Fine Land Classification Using Quaternion Autoencoder-Based Polarization Feature Extraction and Self-Organizing Mapping. *IEEE Trans. Geosci. Remote. Sens.* **2017**, *56*, 1839–1851. [CrossRef]
25. Geng, J.; Wang, H.; Fan, J.; Ma, X. Classification of fusing SAR and multispectral image via deep bimodal autoencoders. In Proceedings of the 2017 IEEE International Geoscience and Remote Sensing Symposium (IGARSS), Fort Worth, TX, USA, 23–28 July 2017; pp. 823–826.
26. Huang, Z.; Pan, Z.; Lei, B. Transfer Learning with Deep Convolutional Neural Network for SAR Target Classification with Limited Labeled Data. *Remote. Sens.* **2017**, *9*, 907. [CrossRef]
27. Rostami, M.; Kolouri, S.; Eaton, E.; Kim, K. Deep Transfer Learning for Few-Shot SAR Image Classification. *Remote. Sens.* **2019**, *11*, 1374. [CrossRef]
28. Shao, Z.; Zhang, L.; Wang, L. Stacked Sparse Autoencoder Modeling Using the Synergy of Airborne LiDAR and Satellite Optical and SAR Data to Map Forest Above-Ground Biomass. *IEEE J. Sel. Top. Appl. Earth Obs. Remote. Sens.* **2017**, *10*, 5569–5582. [CrossRef]
29. De, S.; Bruzzone, L.; Bhattacharya, A.; Bovolo, F.; Chaudhuri, S. A Novel Technique Based on Deep Learning and a Synthetic Target Database for Classification of Urban Areas in PolSAR Data. *IEEE J. Sel. Top. Appl. Earth Obs. Remote. Sens.* **2018**, *11*, 154–170. [CrossRef]
30. Deng, S.; Du, L.; Li, C.; Ding, J.; Liu, H. SAR Automatic Target Recognition Based on Euclidean Distance Restricted Autoencoder. *IEEE J. Sel. Top. Appl. Earth Obs. Remote. Sens.* **2017**, *10*, 3323–3333. [CrossRef]
31. Tian, S.; Wang, C.; Zhang, H.; Bhanu, B. SAR object classification using the DAE with a modified triplet restriction. *IET Radar Sonar Navig.* **2019**, *13*, 1081–1091. [CrossRef]
32. Xie, W.; Jiao, L.; Hou, B.; Ma, W.; Zhao, J.; Zhang, S.; Liu, F. POLSAR Image Classification via Wishart-AE Model or Wishart-CAE Model. *IEEE J. Sel. Top. Appl. Earth Obs. Remote. Sens.* **2017**, *10*, 3604–3615. [CrossRef]
33. Wang, J.; Hou, B.; Jiao, L.; Wang, S. POL-SAR Image Classification Based on Modified Stacked Autoencoder Network and Data Distribution. *IEEE Trans. Geosci. Remote. Sens.* **2020**, *58*, 1678–1695. [CrossRef]
34. Liu, G.; Li, L.; Jiao, L.; Dong, Y.; Li, X. Stacked Fisher autoencoder for SAR change detection. *Pattern Recognit.* **2019**, *96*, 106971. [CrossRef]

35. Wang, L.; Bai, X.; Zhou, F. SAR ATR of Ground Vehicles Based on ESENet. *Remote. Sens.* **2019**, *11*, 1316. [CrossRef]
36. Shao, J.; Qu, C.; Li, J.; Peng, S. A Lightweight Convolutional Neural Network Based on Visual Attention for SAR Image Target Classification. *Sensors* **2018**, *18*, 3039. [CrossRef]
37. Wagner, S. SAR ATR by a combination of convolutional neural network and support vector machines. *IEEE Trans. Aerosp. Electron. Syst.* **2016**, *52*, 2861–2872. [CrossRef]
38. Chen, S.; Wang, H.; Xu, F.; Jin, Y.-Q. Target Classification Using the Deep Convolutional Networks for SAR Images. *IEEE Trans. Geosci. Remote. Sens.* **2016**, *54*, 4806–4817. [CrossRef]
39. Jiang, C.; Zhou, Y. Hierarchical Fusion of Convolutional Neural Networks and Attributed Scattering Centers with Application to Robust SAR ATR. *Remote. Sens.* **2018**, *10*, 819. [CrossRef]
40. Lee, J.-S.; Wen, J.-H.; Ainsworth, T.L.; Chen, K.-S.; Chen, A. Improved Sigma Filter for Speckle Filtering of SAR Imagery. *IEEE Trans. Geosci. Remote. Sens.* **2008**, *47*, 202–213.
41. Wissinger, J.; Ristroph, R.; Diemunsch, J.R.; Severson, W.E.; Fruedenthal, E. MSTAR's extensible search engine and model-based inferencing toolkit. *Proc. SPIE Int. Soc. Opt. Eng.* **1999**, *3721*, 554–570.
42. Ross, T.D.; Worrell, S.W.; Velten, V.J.; Mossing, J.C.; Bryant, M.L. Standard SAR ATR evaluation experiments using the MSTAR public release data set. In Proceedings of the Algorithms for Synthetic Aperture Radar Imagery V, Orlando, FL, USA, 14–17 April 1998.
43. Dumoulin, V.; Visin, F. A Guide to Convolution Arithmetic for Deep Learning. *arXiv* **2016**, arXiv:1603.07285.
44. Chollet, F. Xception: Deep Learning with Depthwise Separable Convolutions. In Proceedings of the 2017 IEEE Conference on Computer Vision and Pattern Recognition (CVPR), Honolulu, HI, USA, 21–26 July 2017; pp. 1800–1807.
45. Simonyan, K.; Zisserman, A. Very Deep Convolutional Networks for Large-Scale Image Recognition. In Proceedings of the International Conference on Learning Representations (ICLR 15), San Diego, CA, USA, 7–9 May 2015.
46. He, K.; Zhang, X.; Ren, S.; Sun, J. Delving Deep into Rectifiers: Surpassing Human-Level Performance on ImageNet Classification. In Proceedings of the 2015 IEEE International Conference on Computer Vision (ICCV), Santiago, Chile, 7–13 December 2015; pp. 1026–1034.
47. Boureau, Y.L.; Ponce, J.; Lecun, Y. A Theoretical Analysis of Feature Pooling in Visual Recognition. In Proceedings of the International Conference on Machine Learning (ICML-10), Haifa, Israel, 21–24 June 2010.
48. He, K.; Zhang, X.; Ren, S.; Sun, J. Spatial Pyramid Pooling in Deep Convolutional Networks for Visual Recognition. *IEEE Trans. Pattern Anal. Mach. Intell.* **2015**, *37*, 1904–1916. [CrossRef]
49. Wang, Z.; Bovik, A.C.; Sheikh, H.R.; Simoncelli, E.P. Image Quality Assessment: From Error Visibility to Structural Similarity. *IEEE Trans. Image. Process* **2004**, *13*, 600–612. [CrossRef]
50. Tian, S.; Yin, K.; Wang, C.; Zhang, H. An SAR ATR Method Based on Scattering Centre Feature and Bipartite Graph Matching. *IETE Tech. Rev.* **2015**, *32*, 1–12. [CrossRef]
51. Kingma, D.; Ba, J. Adam: A Method for Stochastic Optimization. In Proceedings of the International Conference for Learning Representations, San Diego, CA, USA, 7–9 May 2014.
52. Wang, Y.; Han, P.; Lu, X.; Wu, R.; Huang, J. The Performance Comparison of Adaboost and SVM Applied to SAR ATR. In Proceedings of the 2006 CIE International Conference on Radar, Shanghai, China, 16–19 October 2006; pp. 1–4.
53. Zhang, H.; Nasrabadi, N.M.; Zhang, Y.; Huang, T. Multi-View Automatic Target Recognition using Joint Sparse Representation. *IEEE Trans. Aerosp. Electron. Syst.* **2012**, *48*, 2481–2497. [CrossRef]
54. Dungan, K.E. Feature-Based Vehicle Classification in Wide-Angle Synthetic Aperture Radar. Ph.D. Thesis, The Ohio State University, Columbus, OH, USA, 2010.
55. Cui, Z.; Feng, J.; Cao, Z.; Ren, H.; Yang, J. Target recognition in synthetic aperture radar images via non-negative matrix factorisation. *IET Radar Sonar Navig.* **2015**, *9*, 1376–1385. [CrossRef]
56. Cui, Z.; Cao, Z.; Yang, J.; Ren, H. Hierarchical Recognition System for Target Recognition from Sparse Representations. *Math. Probl. Eng.* **2015**, *2015*, 1–6. [CrossRef]
57. Ding, J.; Chen, B.; Liu, H.; Huang, M. Convolutional Neural Network With Data Augmentation for SAR Target Recognition. *IEEE Geosci. Remote. Sens. Lett.* **2016**, *13*, 1–5. [CrossRef]
58. Chen, S.; Wang, H. SAR target recognition based on deep learning. In Proceedings of the 2014 International Conference on Data Science and Advanced Analytics (DSAA), Shanghai, China, 30 October–1 November 2014; pp. 541–547.

59. Dong, G.; Kuang, G. Target Recognition in SAR Images via Classification on Riemannian Manifolds. *IEEE Geosci. Remote. Sens. Lett.* **2014**, *12*, 199–203. [CrossRef]
60. Ning, C.; Liu, W.; Zhang, G.; Wang, X. Synthetic Aperture Radar Target Recognition Using Weighted Multi-Task Kernel Sparse Representation. *IEEE Access* **2019**, *7*, 181202–181212. [CrossRef]
61. Zheng, C.; Jiang, X.; Liu, X. Semi-Supervised SAR ATR via Multi-Discriminator Generative Adversarial Network. *IEEE Sens. J.* **2019**, *19*, 7525–7533. [CrossRef]

Article

A MIMO-SAR Tomography Algorithm Based on Fully-Polarimetric Data

Lingyu Kong and Xiaojian Xu *

School of Electronics and Information Engineering, Beihang University, Beijing 100191, China; konglingyu@buaa.edu.cn

* Correspondence: xiaojianxu@buaa.edu.cn; Tel.: +86-010-8231-6065

Received: 4 September 2019; Accepted: 4 November 2019; Published: 6 November 2019

Abstract: A fully-polarimetric unitary multiple signal classification (UMUSIC) tomography algorithm is proposed, which can be used for acquiring high-resolution three-dimensional (3D) imagery, in a polarimetric multiple-input multiple-output synthetic aperture radar (MIMO-SAR) with a small number of baselines. In terms of the elevation resolution, UMUSIC provides an improvement over standard MUSIC by utilizing the conjugate of the complex sample data and converting the complex covariance matrix into a real matrix. The combination of UMUSIC and fully-polarimetric data permits a further reduction of the noise of the sample covariance matrix, which is obtained through pixel averaging of multiple two-dimensional (2D) images. Considering the consistency of four polarizations, this algorithm not only makes scattering centers have the same estimated height in four polarizations, but it also improves the estimation accuracy. Simulation results show that this algorithm outperforms the popular distributed compressed sensing (DCS). Image processing of measured data of an aircraft model using a multiple-input multiple-output synthetic aperture radar (MIMO-SAR) with six baselines is presented to validate the proposed algorithm.

Keywords: polarimetric; SAR tomography; MIMO radar

---

1. Introduction

Multiple-input multiple-output synthetic aperture radar (MIMO-SAR) is an enabling technique capable of imaging a target [1–7], which is different from a rail synthetic aperture radar (SAR) and a turntable inverse synthetic aperture radar (ISAR). Two-dimensional (2D) virtual apertures can be synthesized through different combinations of transceiver antenna elements. A large number of virtual apertures in the cross-range and elevation directions are beneficial to obtain high-resolution three-dimensional (3D) radar images [4–6], but they also result in a high cost and large size of radar systems due to the increase of the number of antenna elements. When the measured target has several scattering centers in an elevation direction, such as airplanes, an affordable array strategy can be adopted: the priority is given to ensuring adequate cross-range virtual apertures for high-resolution two-dimensional (2D) radar images [7], and then a small number of elevation virtual apertures ensure high-resolution 3D radar images by SAR tomography [8–18].

The reconstruction quality of SAR tomography depends on the product of the number of baselines and the signal-to-noise ratio (SNR) [8]. When the number of baselines is limited, the SNR of radar images can be equivalently improved by filtering [8,9], auxiliary information [10–12], and polarization [13–17]. The SNR can be improved by integrating nonlocal filtering into the compressed sensing (CS) algorithm and a reasonable reconstruction of buildings from only seven baselines is feasible [8]. In addition, [9] investigates the possibility related to the use of a multi-looking approach for fine resolution analysis of ground structures that combines SAR tomography. Filtering consisting of averaging pixels is bound to reduce the range-azimuth resolution and therefore is not suitable for high-resolution radar images of artificial targets. The auxiliary information added to the standard Capon and multiple signal classification (MUSIC) algorithms can be exploited to reduce the ambiguity and resolve the superimposition of the scatterers in the case of a limited number of radar images [10]. Atmospheric phases for SAR tomography in mountainous regions are regressed against the spatial coordinates in map geometry at persistent scatterers locations [11]. The high-resolution 3D positions of a large amount of natural scatterers are obtained by a geodetic SAR tomography framework that fuses SAR tomography and SAR image geodesy compensating SAR measurement error [12]. The auxiliary information can effectively improve the image quality, but it needs to be obtained by other technologies which increases the complexity of the algorithm and the cost of the imaging. A distributed compressed sensing (DCS) algorithm based on fully-polarimetric data is proposed in [13–16] to improve the accuracy of the estimation. However, the CS algorithm suffers from a high computational expense and is hard to extend to fast practice [18]. In [17], a comparison among tomograms obtained in different polarizations is made to analyze how polarimetry can enhance target signatures.

To address these problems, the combination of spectral analysis and full polarization is an attractive way to improve resolution and processing speed for a small number of baselines. This paper explores a fully-polarimetric unitary multiple signal classification (UMUSIC) technique for polarimetric MIMO-SAR tomography [7]. The remainder of the paper is organized as follows. Section 2 describes a signal model based on fully-polarimetric data. A fast and high-resolution UMUSIC algorithm is developed in Section 3. In Section 4, two algorithms are compared through simulation of different point scatterers. Finally, Section 5 contains measured tomography results of an aircraft model to validate the proposed algorithm.

2. Polarization Signal Model

SAR tomography allows us to obtain 3D imagery to describe the electromagnetic property of illuminated objects. The geometry of MIMO-SAR tomography is shown in Figure 1, where x,y,z denote coordinates originating from the center of the imagery scene, R_n represents the distance from the target to the nth baseline, and R_0 is the projection distance from the radar to the center of the imagery scene on the y-axis. The orange triangles and blue circles denote receivers and transmitters, respectively. Each baseline represents a linear array, where transmitters are at both ends of the array and receivers are in the middle of the array. The 2D image for the nth baseline is represented by the following form [19].

$$g_n(x,y,z_n) = \int_{-h/2}^{h/2} \sqrt{\widehat{\sigma}(x,y,z)} e^{-j\frac{4\pi}{\lambda}R_n} dz \tag{1}$$

where $\sqrt{\widehat{\sigma}(x,y,z)}$ denotes the target scattering function that needs to be solved, h is the height of the imagery scene, λ represents the wavelength and z_n refers to the z-coordinates of the nth baseline. Under the Born weak-scattering approximation, R_n representing the distance from the point target at (x, y, z) to the nth baseline, is approximated as:

$$R_n = \sqrt{(y+R_0)^2 + (z-z_n)^2} \approx y + R_0 + \frac{z_n^2 + z^2 - 2z_n z}{2R_0} \tag{2}$$

The first three terms in (2) are irrelevant to z, the fourth term is the residual phase term that can be merged into $\sqrt{\widehat{\sigma}(x,y,z)}$, and the fifth term is the phase term for imaging in the elevation direction. We can choose one of N baselines as a reference baseline as:

$$g_{ref}(x,y,z_{ref}) = \mathrm{e}^{-j\frac{4\pi}{\lambda}(y+R_0+\frac{z_n^2}{2R_0})} \tag{3}$$

After phase compensation, the N 2D images $g'_n(x,y,z_n)$ and $\widehat{\sigma}'(x,y,z)$ are in a Fourier transform relation.

$$g'_n(x,y,z_n) = g_n(x,y,z_n) g^*_{ref}(x,y,z_{ref}) = \int_{-h/2}^{h/2} \widehat{\sigma}'(x,y,z) e^{-j2\pi w_n z} dz \quad (4)$$

with

$$w_n = \frac{2z_n}{\lambda R_0} \quad (5)$$

$$\widehat{\sigma}'(x,y,z) = \sqrt{\widehat{\sigma}(x,y,z)} e^{-j\frac{2\pi z^2}{\lambda R_0}} \quad (6)$$

where $\widehat{\sigma}'(x,y,z)$ integrated with the residual phase term still has the same amplitude as $\sqrt{\widehat{\sigma}(x,y,z)}$, which has no effect on the 3D imagery. $g'_n(x,y,z_n)$ can be discretized as the multiplication of two matrices.

$$g'_n(x,y,z_n) = \sum_{l=1}^{L} \widehat{\sigma}'(x,y,z_l) \mathrm{e}^{-j2\pi w_n z_l} = \mathbf{a}_n \mathbf{s} \quad (7)$$

with

$$\mathbf{a}_n = \left[\mathrm{e}^{-j2\pi w_n z_1}, \mathrm{e}^{-j2\pi w_n z_2}, \ldots, \mathrm{e}^{-j2\pi w_n z_L}\right] \quad (8)$$

$$\mathbf{s} = \left[\widehat{\sigma}'(x,y,z_1), \widehat{\sigma}'(x,y,z_2), \ldots, \widehat{\sigma}'(x,y,z_L)\right]^T \quad (9)$$

where L represents the number of scattering centers in the elevation direction. Before imaging, the imagery scene in the elevation direction needs to be divided into many discrete points to represent the range of L. As a consequence, the matrix $\mathbf{a}_n$ is very large, which is the reason why the imaging algorithm takes a long time to locate scattering centers and determine L in the simulation and measurement. In combination with N 2D images, the polarimetric tomography model is given by:

$$\mathbf{g} = \begin{bmatrix} g'_1(x,y,z_1) \\ g'_2(x,y,z_2) \\ \vdots \\ g'_N(x,y,z_N) \end{bmatrix} = \begin{bmatrix} \mathrm{e}^{-j2\pi w_1 z_1} & \mathrm{e}^{-j2\pi w_1 z_2} & \cdots & \mathrm{e}^{-j2\pi w_1 z_L} \\ \mathrm{e}^{-j2\pi w_2 z_1} & \mathrm{e}^{-j2\pi w_2 z_2} & \cdots & \mathrm{e}^{-j2\pi w_2 z_L} \\ \vdots & \vdots & \ddots & \vdots \\ \mathrm{e}^{-j2\pi w_N z_1} & \mathrm{e}^{-j2\pi w_N z_2} & \cdots & \mathrm{e}^{-j2\pi w_N z_L} \end{bmatrix} \mathbf{s} = \mathbf{A}\mathbf{s} \quad (10)$$

(10) can be further developed by merging with the fully-polarimetric data.

$$\mathbf{G} = \begin{bmatrix} \mathbf{g}_{HH} & \mathbf{g}_{HV} & \mathbf{g}_{VH} & \mathbf{g}_{VV} \end{bmatrix} = \left[\mathbf{a}_1, \mathbf{a}_2, \ldots, \mathbf{a}_N\right]^T \begin{bmatrix} \mathbf{s}_{HH} & \mathbf{s}_{HV} & \mathbf{s}_{VH} & \mathbf{s}_{VV} \end{bmatrix} = \mathbf{A}\mathbf{S} \quad (11)$$

where $\mathbf{G}$ denotes the $N \times 4$ 2D imagery matrix, $\mathbf{S}$ refers to the $L \times 4$ 3D imagery matrix, and $\mathbf{A}$ is the $N \times L$ transformation matrix.

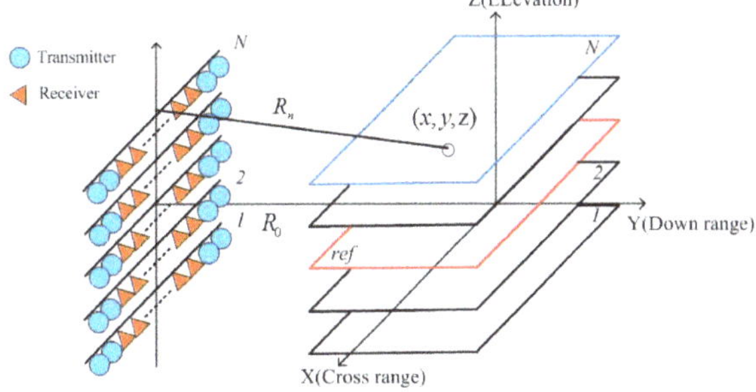

Figure 1. Geometry of multiple-input multiple-output synthetic aperture radar (MIMO-SAR) tomography.

3. Tomography Algorithm

Multiple signal classification (MUSIC) is a spectral analysis algorithm based on the eigen decomposition of the sample covariance matrix. It is necessary to reduce the matrix noise by some techniques, including snapshot in direction-of-arrival and multi-looking in SAR tomography that also leads to a decrease in range-azimuth resolution [13]. In order to dispel the influence, we employ fully-polarimetric data and their conjugation to obtain the matrix.

The polarization tomography model (11) including Gaussian white noise matrix $\mathbf{W}$ is rewritten as:

$$\mathbf{G} = \mathbf{AS} + \mathbf{W} \tag{12}$$

The covariance matrix $\mathbf{R}$ is given by:

$$\begin{aligned} \mathbf{R} &= E\{\mathbf{GG}^H\} = E\{[\mathbf{AS}+\mathbf{W}][\mathbf{S}^H\mathbf{A}^H+\mathbf{W}^H]\} \\ &= \mathbf{A}E\{\mathbf{SS}^H\}\mathbf{A}^H + E\{\mathbf{WW}^H\} = \mathbf{APA}^H + \sigma^2\mathbf{I} \end{aligned} \tag{13}$$

where $\sigma^2\mathbf{I} = diag\{|\sigma_1|^2 \cdots |\sigma_N|^2\}$. The Hermite matrix $\mathbf{APA}^H$ composed of the positive definite diagonal matrix $\mathbf{P}$ and the column full rank matrix $\mathbf{A}$ can be eigen decomposed to obtain the noise subspace.

Multi-sample data is critical for the MUSIC algorithm to ensure a high-quality covariance matrix. In SAR tomography, the average of pixels with the same scattering characteristics is called multi-looking. However, the number of these pixels is limited, and the average processing also reduces the range-azimuth resolution. The combination of observed data and their conjugation can be equivalent to doubling the number of these pixels, which not only improves the estimation accuracy but also solves the problem of coherent signal estimation. To force the Hermite property of $\mathbf{APA}^H$, the average of the covariance matrices is computed from forward and backward data samples [20].

$$\mathbf{R}_M = \frac{1}{2}(\mathbf{R} + \mathbf{J}_N\mathbf{R}^*\mathbf{J}_N) = \mathbf{A}\tilde{\mathbf{P}}\mathbf{A}^H + \sigma^2\mathbf{I} \tag{14}$$

with

$$\widetilde{\mathbf{P}} = \frac{1}{2}(\mathbf{P}+\mathbf{DP}^*\mathbf{D}^H) \tag{15}$$

$$\mathbf{D} = diag\{e^{j2\pi(w_N+w_1)z_1} \ldots e^{j2\pi(w_N+w_1)z_L}\} \tag{16}$$

where superscript * represents the conjugation, $\mathbf{J}_N$ denotes the $N \times N$ exchange matrix with ones on its antidiagonal and zeros elsewhere. In order to reduce the computational complexity, $\mathbf{R}_M$ can be transformed into a real covariance matrix by unitary transformation.

$$\mathbf{R}_U = \mathbf{Q}_N^H\mathbf{R}_M\mathbf{Q}_N = \mathbf{Q}_N^H\mathbf{A}\tilde{\mathbf{P}}\mathbf{A}^H\mathbf{Q}_N + \sigma^2\mathbf{I} \tag{17}$$

where $\mathbf{Q}_N$ is any $N \times N$ unitary matrix to satisfy column conjugate symmetry. A simple form can be chosen as [21]

$$\mathbf{Q}_{N:\text{even}} = \frac{1}{\sqrt{2}}\begin{bmatrix} \mathbf{I}_{N/2} & \mathrm{j}\mathbf{I}_{N/2} \\ \mathbf{J}_{N/2} & -\mathrm{j}\mathbf{J}_{N/2} \end{bmatrix} \tag{18}$$

$$\mathbf{Q}_{N:odd} = \frac{1}{\sqrt{2}}\begin{bmatrix} \mathbf{I}_{(N-1)/2} & 0 & \mathrm{j}\mathbf{I}_{(N-1)/2} \\ 0 & \sqrt{2} & 0 \\ \mathbf{J}_{(N-1)/2} & 0 & -\mathrm{j}\mathbf{J}_{(N-1)/2} \end{bmatrix} \tag{19}$$

As mentioned in (14) and (17), the real covariance matrix $\mathbf{R}_U$ can be rewritten as:

$$\begin{aligned} \mathbf{R}_U &= \tfrac{1}{2}(\mathbf{Q}_N^H\mathbf{R}\mathbf{Q}_N + \mathbf{Q}_N^H\mathbf{J}_N\mathbf{R}^*\mathbf{J}_N\mathbf{Q}_N) \\ &= \tfrac{1}{2}(\mathbf{Q}_N^H\mathbf{R}\mathbf{Q}_N + (\mathbf{Q}_N^H)^*\mathbf{R}^*\mathbf{J}_N\mathbf{Q}_N^*) = \mathrm{Re}\{\mathbf{Q}_N^H\mathbf{R}\mathbf{Q}_N\} \end{aligned} \tag{20}$$

The real matrix is eigen decomposed as:

$$\mathrm{Re}\{\mathbf{Q}_N^H\mathbf{R}\mathbf{Q}_N\} = \sum_{i=1}^{N}\lambda_i u_i u_i^H + \sigma^2\sum_{i=1}^{N}u_i u_i^H = \sum_{i=1}^{L}\lambda_i u_i u_i^H + \sigma^2\sum_{i=1}^{N}u_i u_i^H \tag{21}$$

where $\lambda_1, \ldots, \lambda_N$ are eigenvalues and $u_1, \ldots, u_N$ represent corresponding orthogonal normalized eigenvectors. Among N eigenvectors of the $\mathbf{R}_U$, L eigenvalues are related to the signal, and N-L eigenvalues are related to the noise. By using the noise subspace $\mathbf{E}_N = span\{u_{N-L}, \ldots, u_N\}$, the fully-polarimetric pseudo-spectrum is expressed as:

$$\mathbf{P}_{MUSIC}^{FP}(w) = \frac{1}{\mathbf{A}^H(w)\mathbf{E}_N\mathbf{E}_N^H\mathbf{A}(w)} \tag{22}$$

We can find out the peaks of the spectrums to locate different scattering centers in the elevation dimension and estimate the scattering intensity by the least square method (LSM) in four polarizations.

$$\mathbf{S} = \left(\mathbf{A}^H\mathbf{A}\right)^{-1}\mathbf{A}^H\mathbf{G} \tag{23}$$

where $\mathbf{A}$ is updated according to positions of the scattering centers.

4. Simulation

We adopt the fully-polarimetric DCS as the comparison item. Considering crosstalk, noise, and dispersion, point scatterers with typical polarimetric scattering matrices (PSMs) are simulated to generate return signals. The simulation scene is shown in Figure 2, where scatterers with different heights are located in the coordinate origin. The working frequency is 8 GHz–12 GHz. The down-range, cross-range and elevation Rayleigh limits of the MIMO radar are 0.037 m, 0.047 m, and 0.188 m, respectively. As shown in Figure 3, we simulate three cases to compare the two algorithms.

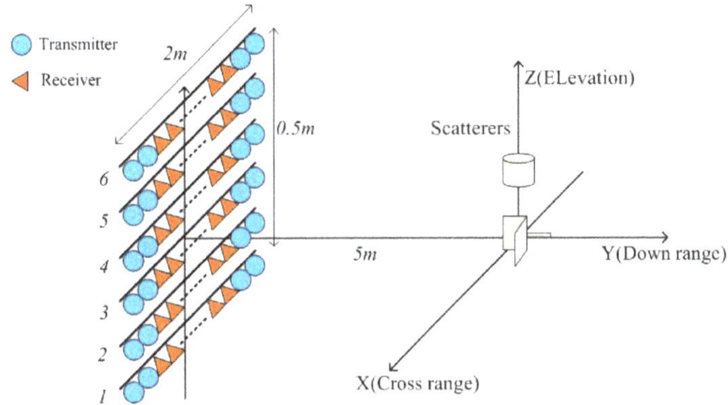

Figure 2. Simulation scene of the MIMO-SAR tomography.

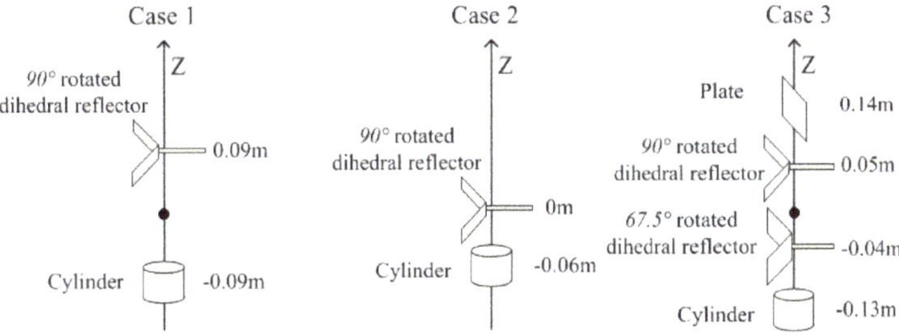

Figure 3. Positions of point scatterers in three cases.

4.1. Case 1: Two Point Scatterers with a Spacing of 0.18 m

A cylinder and a 90° rotated dihedral reflector are located at −0.09 m and 0.09 m in the elevation dimension, respectively. When the two scatterers spacing is 0.18 m (close to the Rayleigh limit), the simulation results of the two algorithms are seen in Figure 4, where lines represent pseudo-spectrums and points denote estimated results including height and scattering intensity of scatterers.

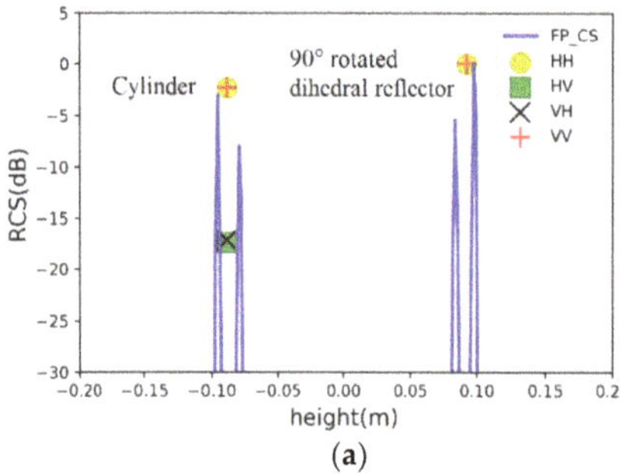

(a)

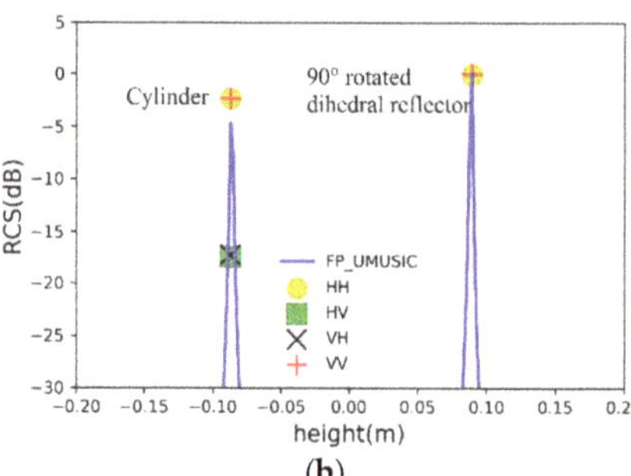

(b)

Figure 4. Estimated results of two point scatterers with a spacing of 0.18 m by (**a**) fully-polarimetric distributed compressed sensing (DCS) and (**b**) fully-polarimetric unitary multiple signal classification (UMUSIC).

Two fully-polarimetric algorithms make scatterers have the same estimated height in four polarizations. The specific estimated results are listed in Table 1. The PSMs of the two scatterers are estimated, where the scattering intensity of the cylinder return signal at −0.09 m is inconsistent with the truth because of polarimetric distortion. In Figure 4a, it is noteworthy that the pseudo-spectrum of CS is leaked to form false scattering points. There are two main reasons for signal leakage [13]: on the one hand, if the regularization parameter is too small in the optimization model, it can lead to over-fitting of data; on the other hand, the observed data does not satisfy the sparsity in the unit orthogonal basis. Therefore, it is necessary to use a sliding window to suppress signal leakage.

Table 1. Estimated results of two point scatterers with a spacing of 0.18 m.

			Cylinder (−0.09 m)	**90° Rotated Dihedral Reflector (0.09 m)**
Fully-polarimetric DCS	Height		−0.091 m	0.091 m
	Scattering intensity	HH	−1.8 dB	0 dB
		HV	−18.5 dB	−39.6 dB
		VH	−18.5 dB	−36.5 dB
		VV	−1.8 dB	0 dB
Fully-polarimetric UMUSIC	Height		−0.090 m	0.090 m
	Scattering intensity	HH	−1.8 dB	0 dB
		HV	−18.5 dB	−39.1 dB
		VH	−18.5 dB	−36.2 dB
		VV	−1.8 dB	0 dB

4.2. Case 2: Two Point Scatterers with a Spacing of 0.06 m

When the spacing is reduced to 0.06 m (one-third of elevation Rayleigh limit), the estimated results of the two algorithms are displayed in Figure 5 and Table 2. Two fully-polarimetric algorithms still have high-resolution. Furthermore, polarimetric distortion of the cylinder return signals becomes more severe as the spacing decreases. Consequently, polarimetric calibration is necessary for a fully-polarimetric radar system.

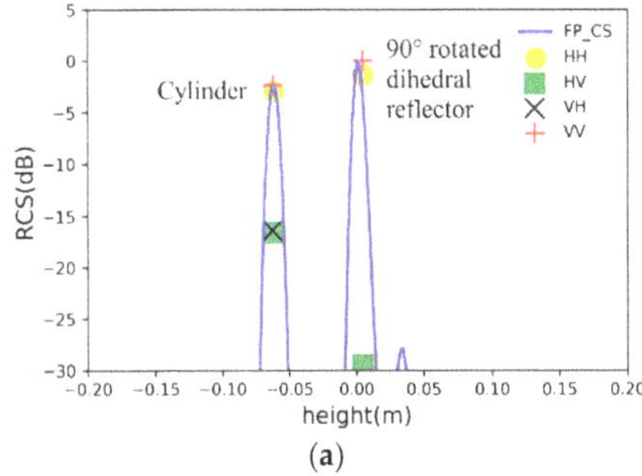

(a)

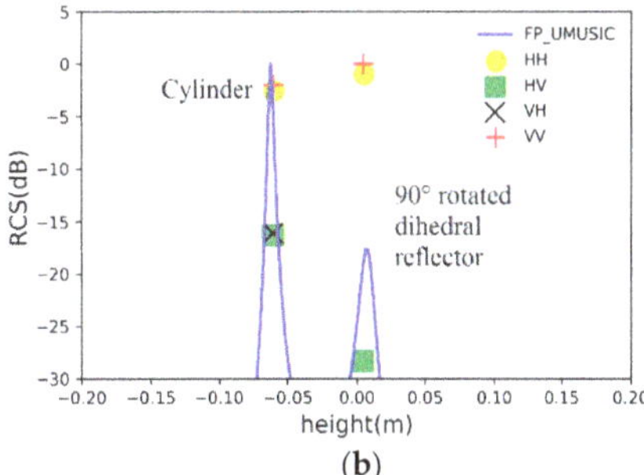

(b)

Figure 5. Estimated results of two point scatterers with a spacing of 0.06 m by (**a**) fully-polarimetric DCS and (**b**) fully-polarimetric UMUSIC.

Table 2. Estimated results of two point scatterers with a spacing of 0.06 m.

			Cylinder (−0.06 m)	90° Rotated Dihedral Reflector (0 m)
Fully-polarimetric DCS	Height		−0.062 m	0.004 m
	Scattering intensity	HH	−3.0 dB	−1.6 dB
		HV	−17.6 dB	−29.5 dB
		VH	−17.6 dB	−37.5 dB
		VV	−3.2 dB	0 dB
Fully-polarimetric UMUSIC	Height		−0.060 m	0.004 m
	Scattering intensity	HH	−2.9 dB	−0.7 dB
		HV	−17.6 dB	−28.7 dB
		VH	−17.6 dB	−36.4 dB
		VV	−3.1 dB	0 dB

4.3. Case 3: Four Point Scatterers with a Spacing of 0.09 m

The four point scatterers are a cylinder, a 67.5° rotated dihedral reflector, a 90° rotated dihedral reflector, and a plate and their PSMs are listed in Table 3. According to the MIMO configuration, the bistatic angles of all transceiver channels are less than 10°. To simplify the simulation, we assume that the PSMs listed in Table 3 are applicable to all transceiver channels. It can be seen from Figure 6 that the pseudo-spectrums of two fully-polarimetric algorithms are not affected when the number of scatterers increase. We summarize the estimation results in Table 4, which demonstrates the estimation accuracy of fully-polarimetric UMUSIC is higher than that of the fully-polarimetric DCS. The CS, which is essentially an optimization problem, needs to be solved iteratively, therefore, its processing speed is bound to be limited by the number of iterations. The simulation results show that for a pixel, the processing speed of the fully-polarimetric UMUSIC is more than five times faster than that of the fully-polarimetric DCS in the same computing condition.

Table 3. Polarimetric scattering matrix (PSM) of four point scatterers

Parameters	Cylinder	67.5° Rotated Dihedral Reflector	90° Rotated Dihedral Reflector	Plate
HH	−1	$\sqrt{2}/2$	1	−1
HV	0	$\sqrt{2}/2$	0	0
VH	0	$\sqrt{2}/2$	0	0
VV	−1	$-\sqrt{2}/2$	−1	−1

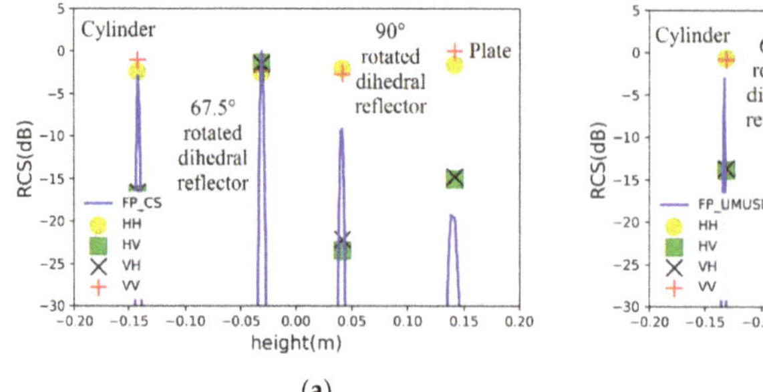

(a)

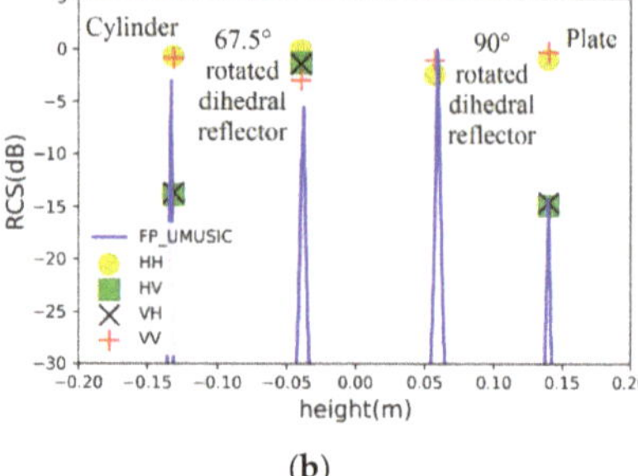

(b)

Figure 6. Estimated heights of four point scatterers with a spacing of 0.09 m by (**a**) fully-polarimetric DCS and (**b**) fully-polarimetric UMUSIC.

Table 4. Estimated results of two point scatterers with a spacing of 0.09 m.

			Cylinder (−0.13 m)	67.5° Rotated Dihedral Reflector (−0.04 m)	90° Rotated Dihedral Reflector (0.05 m)	Plate (0.14 m)
Fully-polarimetric DCS	Height		−0.141 m	−0.030 m	0.041 m	0.141 m
	Scattering intensity	HH	−2.4 dB	−2.5 dB	−2.4 dB	−2.2 dB
		HV	−16.2 dB	−1.6 dB	−24.5 dB	−15.1 dB
		VH	−16.2 dB	−1.6 dB	−24.1 dB	−15.1 dB
		VV	−1.2 dB	−2.4 dB	−2.5 dB	0 dB
Fully-polarimetric UMUSIC	Height		−0.131 m	−0.039 m	0.057 m	0.140 m
	Scattering intensity	HH	−0.3 dB	0 dB	−1.7 dB	−0.5 dB
		HV	−14.6 dB	−2.4 dB	−36.8 dB	−15.1 dB
		VH	−14.6 dB	−2.4 dB	−37.2 dB	−15.1 dB
		VV	−0.3 dB	−4.0 dB	−3.0 dB	0 dB

5. Experiment

An experimental polarimetric MIMO array has been upgraded based on the radar system in [7], and baselines with different heights are controlled by an elevator. It can be seen from Figure 7 that the polarimetric MIMO array consists of 20 receive elements and 6 transmit elements, where the combinations among them synthesize 80 transceiver channels. The measured target is an aircraft model with an elevation angle of 16 degrees on a foam support, as shown in Figure 8. M1, M2, and M3 represent three missile models mounted on the wing, respectively. To avoid complex scattering properties of cavity structures, the inlet of the aircraft model is sealed with copper foils. The measurement parameters are the same as the simulation parameters in Section 4.

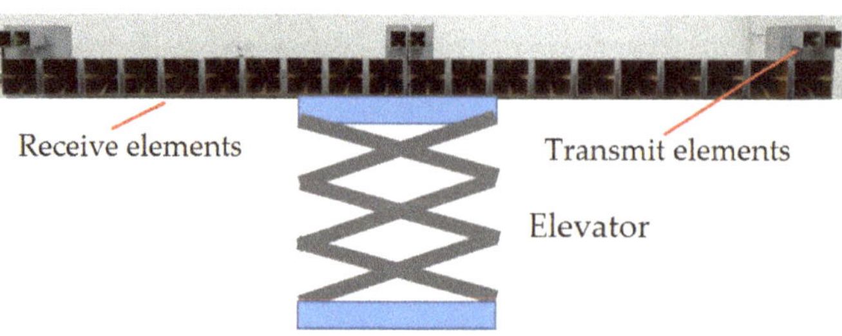

Figure 7. An experimental polarimetric MIMO array.

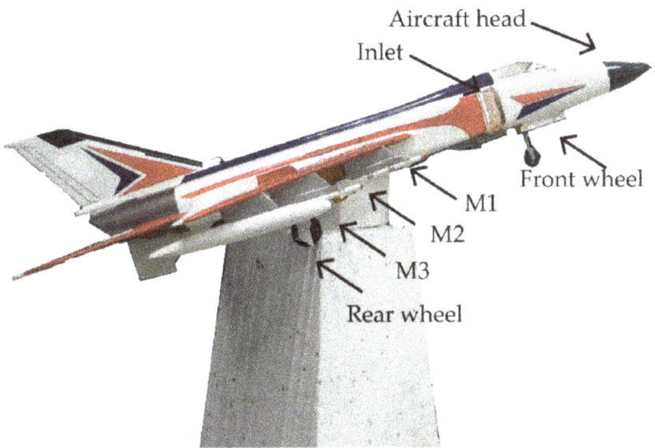

Figure 8. Aircraft model.

It can be seen from Figure 9 that the scattering mechanisms of the aircraft model are different in four polarizations. In the HH image, there are three strong scattering centers including two parts that are not distinguished (see Figure 9a). Compared with the HH image, more components can be distinguished from the VV image. The scattering intensity of the two cross-polarization images is low. Figures 10–13 illustrate the 3D point cloud maps obtained from 24 2D images. The top views are similar to the 2D image, which proves that the 3D scattering intensity can be estimated by LSM. It can be seen from the bottom and side views that scattering centers with different heights are basically consistent with the aircraft model. In addition, the scattering intensity in front of the fuselage is higher than that of the fuselage tail due to the shielding of the supporting foam. By comparing the 3D point cloud maps in different polarizations, we can analyze its scattering mechanism.

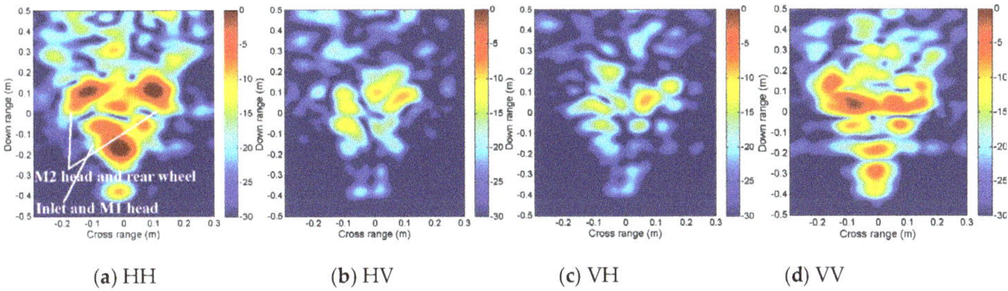

Figure 9. Four 2D images obtained from a single baseline: (**a**) HH, (**b**) HV, (**c**) VH, and (**d**) VV.

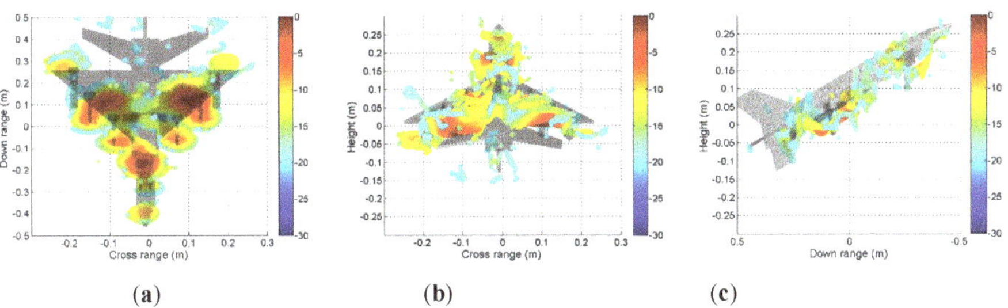

Figure 10. Tomography results in HH polarization. Three views of the airplane model are shown: (**a**) top view; (**b**) bottom view; (**c**) side view.

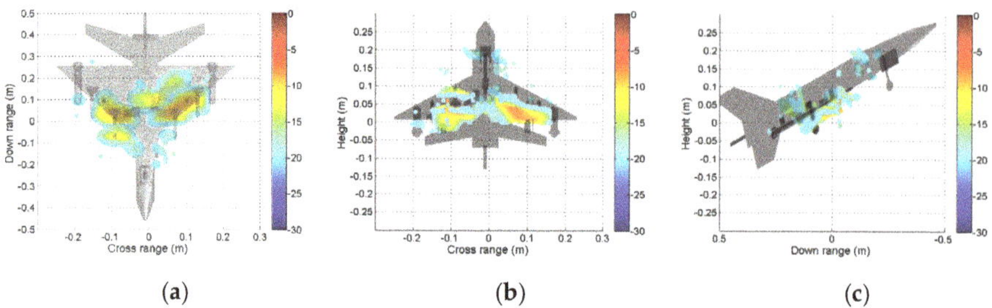

(a) (b) (c)

Figure 11. Tomography result in HV polarization. Three views of the airplane model are shown: (**a**) top view; (**b**) bottom view; (**c**) side view.

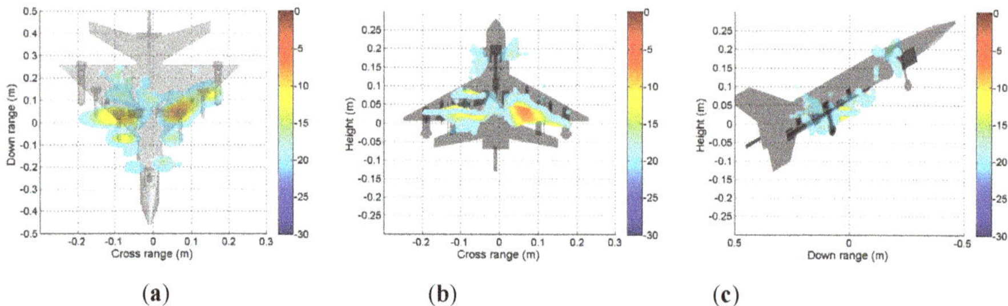

(a) (b) (c)

Figure 12. Tomography result in VH polarization. Three views of the airplane model are shown: (**a**) top view; (**b**) bottom view; (**c**) side view.

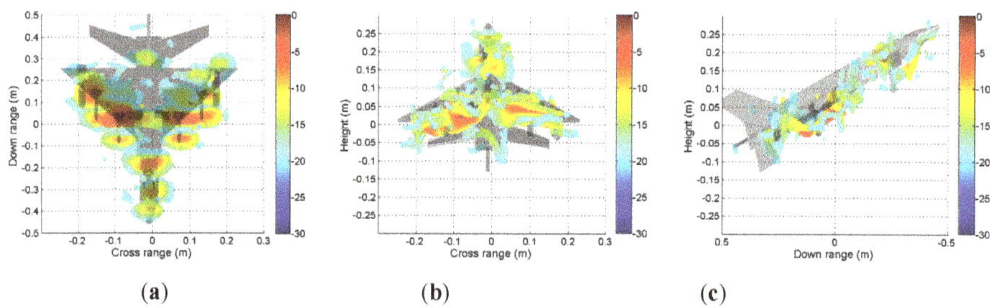

(a) (b) (c)

Figure 13. Tomography result in VV polarization. Three views of the airplane model are shown: (**a**) top view; (**b**) bottom view; (**c**) side view.

Figure 14 illustrates tomographic image slices along the down range for HH (Figure 10), HV (Figure 11), VH (Figure 12), and VV (Figure 13). It can be seen from the figures that scattering of the model shows obvious variety with heights. We summarize components of the model in Table 5, where M1 tail and M2 head cannot be distinguished because they have the same height, so do M2 tail and rear wheel.

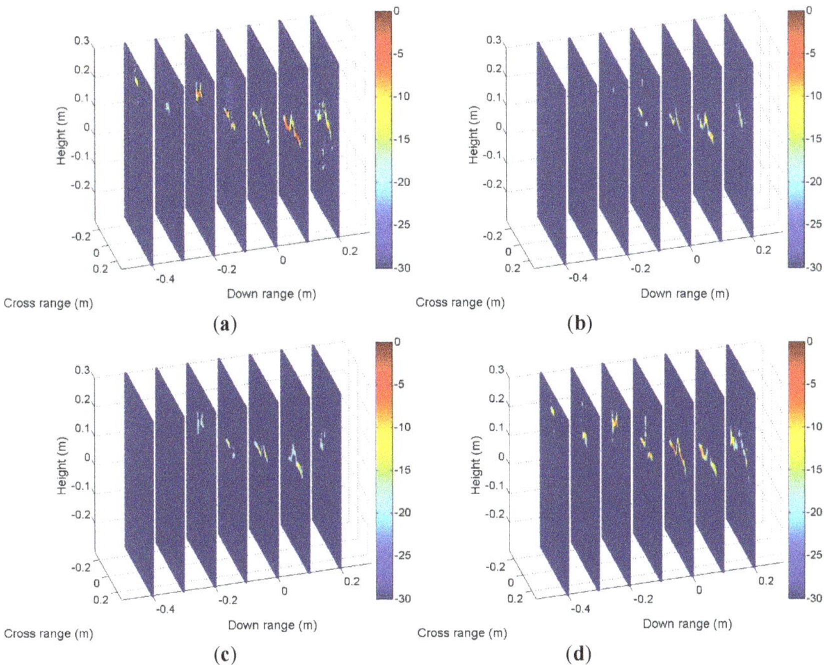

Figure 14. Tomographic image slices along the downrange: (**a**) HH, (**b**) HV, (**c**) VH, and (**d**) VV.

Table 5. Components of the aircraft in tomographic image slices.

Down Range	**−0.4 m**	**−0.3 m**	**−0.2 m**	**−0.1 m**	**0 m**	**0.1 m**	**0.2 m**
Components	Aircraft head	Front wheel	Inlet	M1 head	M1 tail+ M2 head	M2 tail+ Rear wheel	M3 tail
HH	x	x	x	x	x	x	x
HV				x	x	x	
VH			x	x	x	x	
VV	x	x	x	x	x	x	x

6. Conclusions

This paper proposes a fully-polarimetric UMUSIC tomography algorithm to acquire high-resolution 3D radar imagery for a MIMO-SAR with a small number of baselines. In order to mitigate the effect of multi-looking on the range-azimuth resolution, we employ fully-polarimetric data and their conjugation to obtain the sample covariance matrix. Two algorithms including the fully-polarimetric DCS and the fully-polarimetric UMUSIC, are compared through numeric simulation of different point scatterers. Simulation results demonstrate that the fully-polarimetric UMUSIC outperforms the popular fully-polarimetric DCS in processing speed and estimation accuracy. Measurements for an aircraft model are conducted using an X-band experimental polarimetric MIMO-SAR which was upgraded from a previous system [7]. The resulting 3D images using six baselines demonstrate the usefulness of the algorithm for 3D imagery of complex radar targets.

Author Contributions: Methodology, L.K.; validation, L.K. and X.X.; writing—original draft preparation, L.K.; writing—review and editing, L.K. and X.X.

Funding: This research was funded by National Natural Science Foundation of China: 61371005.

Conflicts of Interest: The authors declare no conflict of interest.

References

1. Li, J.; Stoica, P. MIMO radar with collocated antennas. *IEEE Signal Process. Mag.* **2007**, *24*, 106–114. [CrossRef]
2. Haimovich, A.M.; Blum, R.S.; Cimini, L.J. MIMO radar with widely separated antennas. *IEEE Signal Process. Mag.* **2008**, *25*, 116–129. [CrossRef]
3. Ma, Y.; Miao, C.; Zhao, Y.Y.; Wu, W. An MIMO Radar System Based on the Sparse-Array and Its Frequency Migration Calibration Method. *Sensors* **2019**, *19*, 3580. [CrossRef] [PubMed]
4. Narayanan, R.M.; Gebhardt, E.T.; Broderick, S.P. Through-Wall Single and Multiple Target Imaging Using MIMO Radar. *Electronics* **2017**, *6*, 70. [CrossRef]
5. Zhao, D.Z.; Jin, T.; Dai, Y.P.; Song, Y.P.; Su, X.C. A Three-Dimensional Enhanced Imaging Method on Human Body for Ultra-Wideband Multiple-Input Multiple-Output Radar. *Electronics* **2018**, *7*, 101. [CrossRef]
6. Zhou, J.; Zhu, R.; Jiang, G.; Zhao, L.; Cheng, B. A Precise Wavenumber Domain Algorithm for Near Range Microwave Imaging by Cross MIMO Array. *IEEE Trans. Microw. Theory Technol.* **2019**, *67*, 1316–1326. [CrossRef]
7. Liu, Y.Z.; Xu, X.J.; Xu, G.Y. MIMO radar calibration and imagery for near-field target scattering diagnosis. *IEEE Trans. Aerosp. Electron. Syst.* **2018**, *54*, 442–452. [CrossRef]
8. Shi, Y.; Zhu, X.; Bamler, R. Nonlocal Compressive Sensing-Based SAR Tomography. *IEEE Trans. Geosci. Remote Sens.* **2019**, *57*, 3015–3024. [CrossRef]
9. Fornaro, G.; Pauciullo, A.; Reale, D.; Verde, S. Multilook SAR Tomography for 3-D Reconstruction and Monitoring of Single Structures Applied to COSMO-SKYMED Data. *IEEE J. Sel. Top. Appl. Earth Obs. Remote Sens.* **2014**, *7*, 2776–2785. [CrossRef]
10. Aguilera, E.; Nannini, M.; Reigber, A. Multisignal Compressed Sensing for Polarimetric SAR Tomography. *IEEE Geosci. Remote Sens. Lett.* **2012**, *9*, 871–875. [CrossRef]
11. Muhammad, A.S.; Tazio, S.; Irena, H.; Othmar, F. A Case Study on the Correction of Atmospheric Phases for SAR Tomography in Mountainous Regions. *IEEE Trans. Geosci. Remote Sens.* **2019**, *57*, 18–35.
12. Zhu, X.X.; Montazeri, S.; Gisinger, C.; Hanssen, R.F.; Bamler, R. Geodetic SAR Tomography. *IEEE Trans. Geosci. Remote Sens.* **2016**, *54*, 18–35. [CrossRef]
13. Xing, S.Q.; Li, Y.Z.; Dai, D.H.; Wang, X.S. Three-Dimensional Reconstruction of Man-MadeObjects Using Polarimetric Tomographic SAR. *IEEE Trans. Geosci. Remote Sens.* **2013**, *51*, 3694–3705. [CrossRef]
14. Liang, L.; Guo, H.D.; Li, X.W. Three-Dimensional Structural Parameter Inversion of Buildings by Distributed Compressive Sensing-Based Polarimetric SAR Tomography Using a Small Number of Baselines. *IEEE J. Sel. Top. Appl. Earth Obs. Remote Sens.* **2014**, *7*, 4218–4230. [CrossRef]
15. Aghababaee, H.; Ferraioli, G.; Schirinzi, G.; Pascazio, V. Regularization of SAR Tomography for 3-D Height Reconstruction in Urban Areas. *IEEE J. Sel. Top. Appl. Earth Obs. Remote Sens.* **2019**, *12*, 648–659. [CrossRef]
16. Li, X.W.; Liang, L.; Guo, H.; Huang, Y. Compressive Sensing for Multibaseline Polarimetric SAR Tomography of Forested Areas. *IEEE Trans. Geosci. Remote Sens.* **2016**, *54*, 153–166. [CrossRef]
17. Nannini, M.; Scheiber, R.; Horn, R.; Moreira, A. First 3-D Reconstructions of Targets Hidden Beneath Foliage by Means of Polarimetric SAR Tomography. *IEEE Geosci. Remote Sens. Lett.* **2012**, *9*, 60–64. [CrossRef]
18. Yang, J.G.; Jin, T.; Xiao, C.; Huang, X.T. Compressed Sensing Radar Imaging: Fundamentals, Challenges, and Advances. *Sensors* **2019**, *19*, 3100. [CrossRef]
19. Fornaro, G.; Lombardini, F.; Serafino, F. Three-Dimensional Multipass SAR Focusing Experiments with Long-Term Spaceborne Data. *IEEE Trans. Geosci. Remote Sens.* **2005**, *43*, 702–714. [CrossRef]
20. Pesavento, M.; Gershman, A.B.; Haardt, M. Unitary Root-MUSIC with a Real-Valued Eigendecomposition: A Theoretical and Experimental Performance Study. *IEEE Trans. Signal Process.* **2000**, *48*, 1306–1314. [CrossRef]
21. Linebarger, D.A.; DeGroat, R.D.; Dowling, E.M. Efficient direction-finding methods employing forward–backward averaging. *IEEE Trans. Signal Process.* **1994**, *42*, 2136–2145. [CrossRef]

Article

Target Localization Using Double-Sided Bistatic Range Measurements in Distributed MIMO Radar Systems

Hyuksoo Shin and Wonzoo Chung *

Division of Computer and Communications Engineering, Korea University, Seoul 02841, Korea; shs727@korea.ac.kr
* Correspondence: wchung@korea.ac.kr

Received: 22 April 2019; Accepted: 30 May 2019; Published: 2 June 2019

Abstract: We develop a novel approach improving existing target localization algorithms for distributed multiple-input multiple-output (MIMO) radars based on bistatic range measurements (BRMs). In the proposed algorithms, we estimate the target position with auxiliary parameters consisting of both the target–transmitter distances and the target–receiver distances (hence, "double-sided") in contrast to the existing BRM methods. Furthermore, we apply the double-sided approach to multistage BRM methods. Performance improvements were demonstrated via simulations and a limited theoretical analysis was attempted for the ideal two-dimensional case.

Keywords: distributed MIMO radar; target localization; double-sided bistatic range (BR)

1. Introduction

In distributed multiple-input multiple-output (MIMO) radar systems, target localization based on the time delays between transmitters and receivers is an attractive research topic due to its high accuracy and simplicity [1–3]. As target-mediated time delays are nonlinear, estimation of target location via direct analysis of these delays is difficult. Hence, several approaches seeking to linearize the relationship between the target and the time delays have been proposed [4–15]. Of these, algorithms based on bistatic range measurements (BRMs), which are the sum of target–transmitter and target–receiver distances, are introduced in [6–15].

A single stage algorithm based on BRM, introduced first in [6,7], estimates the target position with the help of auxiliary parameters (distances between the target and transmitters or distances between the target and receivers). Multistage algorithms, such as those in [8–15], further refine the target position by re-using the estimates of the first-stage BRM method and exploiting their relationships, and asymptotically attain the Cramer–Rao lower bound (CRLB) [12] assuming accurate estimates of the first stage. A recent study [15] shows that the choice of auxiliary parameters (target–transmitter side or target–receiver side) in BRM methods affects the target estimation accuracy. Therefore, a systematic approach that utilizes all available auxiliary parameters optimally is desirable.

In this paper, we propose a novel approach that utilizes both target–transmitter distances and target–receiver distances as the auxiliary parameters, to improve the mean square error (MSE) performance. Furthermore, the proposed approach can be applied to the second-stage of the multistage BRM algorithms, such as in those of [8–15]. The existing multistage algorithms can be divided into two types depending on the way of linearizing the nonlinear relations between target position and auxiliary parameters estimated in the first stage: the algorithms in [8–12] linearize nonlinear relationships by squaring them and the algorithms in [13–15] use first-order Taylor expansion to this end. We present two types of double-sided two-stage BRM algorithms by applying our approach to the most recent multistage BRM algorithms, i.e., two-stage methods using squared Taylor approximated relationships. The improved MSE performances of the proposed algorithms were demonstrated by simulations and limited theoretical analysis was attempted for an ideal two-dimensional case.

The remainder of this paper is organized as follows. We briefly review the BRM method with a distributed MIMO radar system model in Section 2. In Section 3, we develop double-sided, single- and two-stage BRM algorithms. A theoretical analysis for ideal two-dimensional target/antenna positions presented in Section 4 shows the improved MSE performance afforded by the double-sided BRM algorithm. The simulations of practical three-dimensional target/antenna positions presented in Section 5 confirm that our algorithms improve MSE performance. Our conclusions are presented in Section 6.

Table 1 lists the notations used in this paper.

Table 1. List of notations.

Notations	Definition
$\mathbf{0}_{i\times j}$	$i \times j$ matrices, all elements of which are zero
$\mathbf{1}_{i\times j}$	$i \times j$ matrices, all elements of which are unity
$\mathbf{I}_i$	$i \times i$ identity matrix
$\text{diag}(\cdot)$	Diagonal matrix generated from an input vector
$\text{blkdiag}(\cdot)$	Block diagonal matrix generated from input vectors (or matrices)
$\otimes$	Kronecker product
$\odot$	Element-wise product
$\text{sgn}(\cdot)$	sign function
$\sqrt{\cdot}$	element-wise square root of the input vector

2. System Model for BRM Based Target Localization and Problem Formulation

We consider a three-dimensional, widely separated MIMO radar system consisting of a single target located at an unknown position $\mathbf{x}_o = [x_o, y_o, z_o]^T$ with M transmitting antennae (Tx) and N receiving antennae (Rx) located at known positions $\mathbf{x}_t(m) = [x_t(m), y_t(m), z_t(m)]^T, m = 1, \cdots, M$ and $\mathbf{x}_r(n) = [x_r(n), y_r(n), z_r(n)]^T$, $n = 1, \cdots, N$, respectively, and, we denote the positions of antennae as $\mathbf{X}_t = [\mathbf{x}_t(1), \cdots, \mathbf{x}_t(M)]$ and $\mathbf{X}_r = [\mathbf{x}_r(1), \cdots, \mathbf{x}_r(N)]$, together.

The bistatic range (BR) between the mth Tx and the nth Rx, denoted by r_{mn}, is defined as the sum of the distance from the mth Tx to the target, denoted by $d_t(m) = \|\mathbf{x}_o - \mathbf{x}_t(m)\|$, and the distance from the target to the nth Rx, denoted by $d_r(n) = \|\mathbf{x}_o - \mathbf{x}_r(n)\|$ ([16]):

$$r_{mn} = d_t(m) + d_r(n). \tag{1}$$

Each BR is measured by converting the estimated time delay between a Tx and an Rx to a distance. Any BR measurement (BRM) between the mth Tx and the nth Rx, denoted by $\hat{r}_{mn}$, is often corrupted by measurement error, denoted by ω_{mn} and modeled as an i.i.d., zero-mean white Gaussian noise with variance σ_ω^2 ([4]):

$$\hat{r}_{mn} = r_{mn} + \omega_{mn}. \tag{2}$$

The goal of BRM based target localization is to estimate the target location $\mathbf{x}_o$ from the BRMs $\{\hat{r}_{mn}\}_{m=1,\cdots,M,n=1,\cdots,N}$.

The BRM method in [6,7] jointly estimates the target location, $\mathbf{x}_o$, and the distances from Txs to the target, denoted by $\mathbf{d}_t = [d_t(1), \cdots, d_t(M)]^T$, from the BRMs, using the following linear model in the presence of noise:

$$\mathbf{b}_t = [\mathbf{1}_{M\times 1} \otimes \mathbf{X}_r^T - \mathbf{X}_t^T \otimes \mathbf{1}_{N\times 1}, -\mathbf{R}_t][\mathbf{x}_o^T, \mathbf{d}_t^T]^T + \varepsilon_t, \tag{3}$$

where

$$\mathbf{b}_t = \frac{1}{2}\begin{bmatrix} \|\mathbf{x}_r(1)\|^2 - \hat{r}_{11}^2 - \|\mathbf{x}_t(1)\|^2 \\ \vdots \\ \|\mathbf{x}_r(N)\|^2 - \hat{r}_{MN}^2 - \|\mathbf{x}_t(M)\|^2 \end{bmatrix} \tag{4}$$

$$\mathbf{R}_t = \mathrm{blkdiag}(\mathbf{r}_1, \cdots, \mathbf{r}_M), \tag{5}$$

where $\mathbf{r}_m = [\hat{r}_{m1}, \cdots, \hat{r}_{mN}]^T$ and ε_t is a vector reflecting BR measurement error ([7]).

Alternatively, the BRM equation can be constructed using the distances from the target to the Rxs, denoted by $\mathbf{d}_r = [d_r(1), \cdots, d_r(N)]^T$, instead of the $\mathbf{d}_t$ values:

$$\mathbf{b}_r = [\mathbf{X}_t^T \otimes \mathbf{1}_{N\times 1} - \mathbf{1}_{M\times 1} \otimes \mathbf{X}_r^T, -\mathbf{R}_r][\mathbf{x}_o^T, \mathbf{d}_r^T]^T + \varepsilon_r, \tag{6}$$

where

$$\mathbf{b}_r = \frac{1}{2}\begin{bmatrix} \|\mathbf{x}_t(1)\|^2 - \hat{r}_{11}^2 - \|\mathbf{x}_r(1)\|^2 \\ \vdots \\ \|\mathbf{x}_t(M)\|^2 - \hat{r}_{MN}^2 - \|\mathbf{x}_r(N)\|^2 \end{bmatrix} \tag{7}$$

$$\mathbf{R}_r = [\mathrm{diag}(\mathbf{r}_1), \cdots, \mathrm{diag}(\mathbf{r}_M)]^T, \tag{8}$$

where ε_r is a vector reflecting BR measurement error [7]. Note that the estimated auxiliary parameters $\hat{\mathbf{d}}_t$ or $\hat{\mathbf{d}}_r$ contain the target information $\mathbf{x}_o$. Multistage algorithms further refine the target position by exploiting this information.

The two-stage BRM method using the squared relationships ([12]) estimates the squared target position, $\mathbf{x}_o \odot \mathbf{x}_o$, using $\hat{\mathbf{x}}_o$ and $\hat{\mathbf{d}}_t$ yielded by the first-stage BRM method based on the following linear model (which reflects the relationship between $[\mathbf{x}_o^T, \mathbf{d}_t^T]^T$ and $\mathbf{x}_o \odot \mathbf{x}_o$):

$$\begin{bmatrix} \hat{\mathbf{x}}_o \odot \hat{\mathbf{x}}_o \\ \hat{\mathbf{d}}_t \odot \hat{\mathbf{d}}_t + 2\mathbf{X}_t^T\hat{\mathbf{x}}_o - (\mathbf{X}_t^T \odot \mathbf{X}_t^T)\mathbf{1}_{3\times 1} \end{bmatrix} = \begin{bmatrix} \mathbf{I}_3 \\ \mathbf{1}_{M\times 3} \end{bmatrix} (\mathbf{x}_o \odot \mathbf{x}_o) + \varepsilon_{S,t}, \tag{9}$$

where $\varepsilon_{S,t}$ is the error vector due to the first-stage estimation error ([12]).

Alternatively, we obtain the following linear model reflecting the relationship between $[\mathbf{x}_o^T, \mathbf{d}_r^T]^T$ and $\mathbf{x}_o \odot \mathbf{x}_o$:

$$\begin{bmatrix} \hat{\mathbf{x}}_o \odot \hat{\mathbf{x}}_o \\ \hat{\mathbf{d}}_r \odot \hat{\mathbf{d}}_r + 2\mathbf{X}_r^T\hat{\mathbf{x}}_o - (\mathbf{X}_r^T \odot \mathbf{X}_r^T)\mathbf{1}_{3\times 1} \end{bmatrix} = \begin{bmatrix} \mathbf{I}_3 \\ \mathbf{1}_{N\times 3} \end{bmatrix} (\mathbf{x}_o \odot \mathbf{x}_o) + \varepsilon_{S,r}, \tag{10}$$

where $\varepsilon_{S,r}$ is the error vector due to the first-stage estimation error [12].

Let $\widehat{\mathbf{x}_o \odot \mathbf{x}_o}$ denote the $\mathbf{x}_o \odot \mathbf{x}_o$ estimated by the linear model of (9) (or (10)); then, the refined target location, denoted by $\hat{\mathbf{x}}_{o,S}$, is:

$$\hat{\mathbf{x}}_{o,S} = \text{sgn}(\hat{\mathbf{x}}_o) \odot \sqrt{\widehat{\mathbf{x}_o \odot \mathbf{x}_o}}. \tag{11}$$

The two-stage BRM method using Taylor approximated relationships [15] considers the first-order Taylor expansion of $d_t(m)$ at $\hat{\mathbf{x}}_o$ to be

$$\begin{aligned} d_t(m) = \hat{d}_t(m) - \triangle d_t(m) &= \|\hat{\mathbf{x}}_o - \triangle \mathbf{x}_o - \mathbf{x}_t(m)\| \\ &\simeq \|\hat{\mathbf{x}}_o - \mathbf{x}_t(m)\| - \frac{\hat{\mathbf{x}}_o^T - \mathbf{x}_t^T(m)}{\|\hat{\mathbf{x}}_o - \mathbf{x}_t(m)\|} \triangle \mathbf{x}_o \end{aligned} \quad \text{for} \quad m = 1, \cdots, M, \tag{12}$$

where $\triangle \mathbf{x}_o, \triangle d_t(1), \cdots, \triangle d_t(M)$ are the estimation errors at the $\hat{\mathbf{x}}_o$. The linear model reflecting the relationships of (12) is

$$\begin{bmatrix} \mathbf{0}_{3\times 1} \\ \hat{d}_t(1) - \|\hat{\mathbf{x}}_o - \mathbf{x}_t(1)\| \\ \vdots \\ \hat{d}_t(M) - \|\hat{\mathbf{x}}_o - \mathbf{x}_t(M)\| \end{bmatrix} = \begin{bmatrix} -\mathbf{I}_3 \\ (\hat{\mathbf{x}}_o^T - \mathbf{x}_t^T(1))/\|\hat{\mathbf{x}}_o - \mathbf{x}_t(1)\| \\ \vdots \\ (\hat{\mathbf{x}}_o^T - \mathbf{x}_t^T(M))/\|\hat{\mathbf{x}}_o - \mathbf{x}_t(M)\| \end{bmatrix} \triangle \mathbf{x}_o + \begin{bmatrix} \triangle \mathbf{x}_o \\ \triangle d_t(1) \\ \vdots \\ \triangle d_t(M) \end{bmatrix}. \tag{13}$$

Alternatively, we obtain the following linear model using $\mathbf{d}_r$ instead of $\mathbf{d}_t$:

$$\begin{bmatrix} \mathbf{0}_{3\times 1} \\ \hat{d}_r(1) - \|\hat{\mathbf{x}}_o - \mathbf{x}_r(1)\| \\ \vdots \\ \hat{d}_r(N) - \|\hat{\mathbf{x}}_o - \mathbf{x}_r(N)\| \end{bmatrix} = \begin{bmatrix} -\mathbf{I}_3 \\ (\hat{\mathbf{x}}_o^T - \mathbf{x}_r^T(1))/\|\hat{\mathbf{x}}_o - \mathbf{x}_r(1)\| \\ \vdots \\ (\hat{\mathbf{x}}_o^T - \mathbf{x}_r^T(N))/\|\hat{\mathbf{x}}_o - \mathbf{x}_r(N)\| \end{bmatrix} \triangle \mathbf{x}_o + \begin{bmatrix} \triangle \mathbf{x}_o \\ \triangle d_r(1) \\ \vdots \\ \triangle d_r(N) \end{bmatrix}. \tag{14}$$

We make an intermediate estimation of the error $\triangle \mathbf{x}_o$ of the first stage to refine the target position. Let $\widehat{\triangle \mathbf{x}_o}$ denote the $\triangle \mathbf{x}_o$ estimated by the linear model of (13) (or (14)); then, the refined target position, denoted by $\hat{\mathbf{x}}_{o,A}$, is:

$$\hat{\mathbf{x}}_{o,A} = \hat{\mathbf{x}}_o - \widehat{\triangle \mathbf{x}_o}. \tag{15}$$

In the existing single-sided BRM methods, $\mathbf{x}_o$ and $\mathbf{d}_t$, or $\mathbf{x}_o$ and $\mathbf{d}_r$, are used exclusively. As the BRMs are the sum of the $\mathbf{d}_t$ and $\mathbf{d}_r$ values, target estimation accuracy can be improved by simultaneously estimating $\mathbf{x}_o$, $\mathbf{d}_t$ and $\mathbf{d}_r$ in the first stage, and by fully utilizing these values in the second stage. Thus, the goal of our paper is to develop target estimation schemes that use both the Tx- and Rx-sided linear models simultaneously.

3. The Double-Sided BRM Approach

3.1. The Double-Sided Single-Stage BRM Algorithm

The target estimation performance of the BRM algorithm depends on the choice of auxiliary parameters (the transmitter-side parameters $\mathbf{d}_t$ or the receiver-side parameters $\mathbf{d}_r$), as shown in [15]. Such dependency implies that the linear models in (3) and (6) cannot fully exploit the target information in BRM observations. Thus, by merging the two linear models in (3) and (6) into a single linear model and, consequently, simultaneously estimating the target, $\mathbf{d}_t$ and $\mathbf{d}_r$ values, we fully utilize all BR information for the target estimation.

To simultaneously estimate $\mathbf{x}_o$, $\mathbf{d}_t$ and $\mathbf{d}_r$, we rewrite the two linear models of (3) and (6) as equivalent linear models with respect to $[\mathbf{x}_o^T, \mathbf{d}_t^T, \mathbf{d}_r^T]^T$, by inserting $\mathbf{0}_{MN\times N}\mathbf{d}_r$ and $\mathbf{0}_{MN\times M}\mathbf{d}_t$:

$$\mathbf{b}_t = [\mathbf{1}_{M\times 1} \otimes \mathbf{X}_r^T - \mathbf{X}_t^T \otimes \mathbf{1}_{N\times 1}, -\mathbf{R}_t, \mathbf{0}_{MN\times N}][\mathbf{x}_o^T, \mathbf{d}_t^T, \mathbf{d}_r^T]^T + \varepsilon_t, \tag{16}$$

$$\mathbf{b}_r = [\mathbf{X}_t^T \otimes \mathbf{1}_{N\times 1} - \mathbf{1}_{M\times 1} \otimes \mathbf{X}_r^T, \mathbf{0}_{MN\times M}, -\mathbf{R}_r][\mathbf{x}_o^T, \mathbf{d}_t^T, \mathbf{d}_r^T]^T + \varepsilon_r. \tag{17}$$

Using the above linear equations, we construct a single linear model with respect to $[\mathbf{x}_o^T, \mathbf{d}_t^T, \mathbf{d}_r^T]^T$ as follows:

$$\mathbf{b} = \mathbf{H}[\mathbf{x}_o^T, \mathbf{d}_t^T, \mathbf{d}_r^T]^T + \varepsilon, \tag{18}$$

where $\mathbf{b} = [\mathbf{b}_t^T, \mathbf{b}_r^T]^T$, $\varepsilon = [\varepsilon_t^T, \varepsilon_r^T]^T$, and

$$\mathbf{H} = \begin{bmatrix} \mathbf{1}_{M\times 1} \otimes \mathbf{X}_r^T - \mathbf{X}_t^T \otimes \mathbf{1}_{N\times 1} & -\mathbf{R}_t & \mathbf{0}_{MN\times N} \\ \mathbf{X}_t^T \otimes \mathbf{1}_{N\times 1} - \mathbf{1}_{M\times 1} \otimes \mathbf{X}_r^T & \mathbf{0}_{MN\times M} & -\mathbf{R}_r \end{bmatrix}. \tag{19}$$

The weighted least squares (WLS) solution of (18), denoted by $[\hat{\mathbf{x}}_o^T, \hat{\mathbf{d}}_t^T, \hat{\mathbf{d}}_r^T]^T$, is:

$$[\hat{\mathbf{x}}_o^T, \hat{\mathbf{d}}_t^T, \hat{\mathbf{d}}_r^T]^T = (\mathbf{H}^T\mathbf{W}\mathbf{H})^{-1}\mathbf{H}^T\mathbf{W}\mathbf{b}, \tag{20}$$

where the diagonal weighting matrix $\mathbf{W}$ is:

$$\mathbf{W} = \mathrm{diag}\left(\sigma_\omega^2 \begin{bmatrix} (\mathbf{d}_r \odot \mathbf{d}_r) \otimes \mathbf{1}_{M\times 1} \\ \mathbf{1}_{N\times 1} \otimes (\mathbf{d}_t \odot \mathbf{d}_t) \end{bmatrix}\right)^{-1}. \tag{21}$$

In practice, we apply the approximated $\mathbf{W}$ using estimated $\mathbf{d}_t$ and $\mathbf{d}_r$ via a least square (LS) approach (substituting an identity matrix for W in (20)) as in previous methods [6–15]. Note that, instead of error covariance matrix, $\mathrm{Cov}[\varepsilon]$, we use the diagonal terms of $\mathrm{Cov}[\varepsilon]$ for W, since $\mathrm{Cov}[\varepsilon]$ is not invertible here.

The analysis of Section 4 shows that our double-sided BRM method enhances the MSE of target location estimated by the existing BRM method by a factor of two, given ideal two-dimensional target/antenna positions. The numerical simulations presented in Section 5 show that our method affords a better MSE performance than the existing BRM method when dealing with practical target/antenna positions.

3.2. The Double-Sided Two-Stage BRM Algorithms

In this subsection, we develop two double-sided two-stage BRM algorithms by modifying the above single-sided two-stage BRM algorithms using the squared relationships [12] and the Taylor approximation [15] to fully utilize the parameters ($\hat{\mathbf{x}}_o$, $\hat{\mathbf{d}}_t$, and $\hat{\mathbf{d}}_r$) estimated by the first stage double-sided BRM algorithm.

3.2.1. Proposed Double-Sided Two-Stage BRM Algorithm Using the Squared Relationships

As for the single-stage algorithm, we construct an extended linear model reflecting the relationships between $\mathbf{d}_t$, $\mathbf{d}_r$ and $\mathbf{x}_o \odot \mathbf{x}_o$ by merging the two single-sided linear models of (9) and (10) as the following:

$$\begin{bmatrix} \hat{\mathbf{x}}_o \odot \hat{\mathbf{x}}_o \\ \hat{\mathbf{d}}_t \odot \hat{\mathbf{d}}_t + 2\mathbf{X}_t^T\hat{\mathbf{x}}_o - (\mathbf{X}_t^T \odot \mathbf{X}_t^T)\mathbf{1}_{3\times1} \\ \hat{\mathbf{d}}_r \odot \hat{\mathbf{d}}_r + 2\mathbf{X}_r^T\hat{\mathbf{x}}_o - (\mathbf{X}_r^T \odot \mathbf{X}_r^T)\mathbf{1}_{3\times1} \end{bmatrix} = \begin{bmatrix} \mathbf{I}_3 \\ \mathbf{1}_{M\times3} \\ \mathbf{1}_{N\times3} \end{bmatrix} (\mathbf{x}_o \odot \mathbf{x}_o) + \boldsymbol{\varepsilon}_p, \tag{22}$$

where $\boldsymbol{\varepsilon}_p$ is error vector due to the estimation error. The method of (22) provides an estimate of the squared target location, $\mathbf{x}_o \odot \mathbf{x}_o$, using all $[\hat{\mathbf{x}}_o^T, \hat{\mathbf{d}}_t^T, \hat{\mathbf{d}}_r^T]^T$ given by the first-stage double-sided BRM algorithm. Denote $[\mathbf{I}_3, \mathbf{1}_{M\times3}^T, \mathbf{1}_{N\times3}^T]^T$ as $\mathbf{H}_p$; then, the WLS solution of (22), denoted by $\widehat{\mathbf{x}_o \odot \mathbf{x}_o}$, is:

$$\widehat{\mathbf{x}_o \odot \mathbf{x}_o} = (\mathbf{H}_p^T\mathbf{W}_p\mathbf{H}_p)^{-1}\mathbf{H}_p^T\mathbf{W}_p \begin{bmatrix} \hat{\mathbf{x}}_o \odot \hat{\mathbf{x}}_o \\ \hat{\mathbf{d}}_t \odot \hat{\mathbf{d}}_t + 2\mathbf{X}_t^T\hat{\mathbf{x}}_o - (\mathbf{X}_t^T \odot \mathbf{X}_t^T)\mathbf{1}_{3\times1} \\ \hat{\mathbf{d}}_r \odot \hat{\mathbf{d}}_r + 2\mathbf{X}_r^T\hat{\mathbf{x}}_o - (\mathbf{X}_r^T \odot \mathbf{X}_r^T)\mathbf{1}_{3\times1} \end{bmatrix}. \tag{23}$$

The weighting matrix W_p is:

$$\mathbf{W}_p = (\mathbf{T}(\mathbf{H}^T\mathbf{W}\mathbf{H})^{-1}\mathbf{T}^T)^{-1}, \tag{24}$$

where

$$\mathbf{T} = 2\begin{bmatrix} \mathrm{diag}(\mathbf{x}_o) & \mathbf{0}_{3\times(M+N)} \\ \mathbf{A}^T & \mathrm{diag}([\mathbf{d}_r^T, \mathbf{d}_r^T]^T) \end{bmatrix} \tag{25}$$

$$\mathbf{A} = [\mathbf{X}_t, \mathbf{X}_r]. \tag{26}$$

The final target position estimate, denoted by $\hat{\mathbf{x}}_{o,DS}$, is:

$$\hat{\mathbf{x}}_{o,DS} = \mathrm{sgn}(\hat{\mathbf{x}}_o) \odot \sqrt{\widehat{\mathbf{x}_o \odot \mathbf{x}_o}}. \tag{27}$$

3.2.2. Proposed Double-Sided Two-Stage BRM Algorithm Using the Taylor Approximated Relationships

To utilize all $[\hat{\mathbf{x}}_o^T, \hat{\mathbf{d}}_t^T, \hat{\mathbf{d}}_r^T]^T$ values given by the first-stage double-sided BRM algorithm, we construct the following extended linear model which reflects the Taylor approximated relationships between $\hat{\mathbf{d}}_t$, $\hat{\mathbf{d}}_r$ and $\hat{\mathbf{x}}_o$ by merging the linear models in (13) and (14):

$$\begin{bmatrix} \mathbf{0}_{3\times1} \\ \hat{d}_t(1) - \|\hat{\mathbf{x}}_o - \mathbf{x}_t(1)\| \\ \vdots \\ \hat{d}_t(M) - \|\hat{\mathbf{x}}_o - \mathbf{x}_t(M)\| \\ \hat{d}_r(1) - \|\hat{\mathbf{x}}_o - \mathbf{x}_r(1)\| \\ \vdots \\ \hat{d}_r(N) - \|\hat{\mathbf{x}}_o - \mathbf{x}_r(N)\| \end{bmatrix} = \begin{bmatrix} -\mathbf{I}_3 \\ (\hat{\mathbf{x}}_o^T - \mathbf{x}_t^T(1))/\|\hat{\mathbf{x}}_o - \mathbf{x}_t(1)\| \\ \vdots \\ (\hat{\mathbf{x}}_o^T - \mathbf{x}_t^T(M))/\|\hat{\mathbf{x}}_o - \mathbf{x}_t(M)\| \\ (\hat{\mathbf{x}}_o^T - \mathbf{x}_r^T(1))/\|\hat{\mathbf{x}}_o - \mathbf{x}_r(1)\| \\ \vdots \\ (\hat{\mathbf{x}}_o^T - \mathbf{x}_r^T(N))/\|\hat{\mathbf{x}}_o - \mathbf{x}_r(N)\| \end{bmatrix} \triangle\mathbf{x}_o + \begin{bmatrix} \triangle\mathbf{x}_o \\ \triangle d_t(1) \\ \vdots \\ \triangle d_t(M) \\ \triangle d_r(1) \\ \vdots \\ \triangle d_r(N) \end{bmatrix}, \tag{28}$$

where $\triangle \mathbf{x}_o, \triangle d_t(1), \cdots, \triangle d_t(M), \triangle d_r(1), \cdots, \triangle d_r(N)$ are the estimation errors at $\hat{\mathbf{x}}_o$. The method of (28) provides an estimate of $\triangle \mathbf{x}_o$. Let us denote

$$\mathbf{H}_p = \begin{bmatrix} -\mathbf{I}_3 \\ (\hat{\mathbf{x}}_o^T - \mathbf{x}_t^T(1))/\|\hat{\mathbf{x}}_o - \mathbf{x}_t(1)\| \\ \vdots \\ (\hat{\mathbf{x}}_o^T - \mathbf{x}_t^T(M))/\|\hat{\mathbf{x}}_o - \mathbf{x}_t(M)\| \\ (\hat{\mathbf{x}}_o^T - \mathbf{x}_r^T(1))/\|\hat{\mathbf{x}}_o - \mathbf{x}_r(1)\| \\ \vdots \\ (\hat{\mathbf{x}}_o^T - \mathbf{x}_r^T(N))/\|\hat{\mathbf{x}}_o - \mathbf{x}_r(N)\| \end{bmatrix}; \tag{29}$$

then, the WLS solution of (28), denoted by $\widehat{\triangle \mathbf{x}}_o$, is:

$$\widehat{\triangle \mathbf{X}}_o = (\mathbf{H}_p^T \mathbf{W}_p \mathbf{H}_p)^{-1} \mathbf{H}_p^T \mathbf{W}_p \begin{bmatrix} \mathbf{0}_{3\times 1} \\ \hat{d}_t(1) - \|\hat{\mathbf{x}}_o - \mathbf{x}_t(1)\| \\ \vdots \\ \hat{d}_t(M) - \|\hat{\mathbf{x}}_o - \mathbf{x}_t(M)\| \\ \hat{d}_r(1) - \|\hat{\mathbf{x}}_o - \mathbf{x}_r(1)\| \\ \vdots \\ \hat{d}_r(N) - \|\hat{\mathbf{x}}_o - \mathbf{x}_r(N)\| \end{bmatrix} \tag{30}$$

where the weighting matrix $\mathbf{W}_p$ is

$$\mathbf{W}_p = (\mathbf{H}^T \mathbf{W} \mathbf{H})^{-1}. \tag{31}$$

The final target position estimate, denoted by $\hat{\mathbf{x}}_{o,DA}$, is:

$$\hat{\mathbf{x}}_{o,DA} = \hat{\mathbf{x}}_o - \widehat{\triangle \mathbf{x}}_o. \tag{32}$$

Unfortunately, theoretical performance analysis of (27) and (32) are virtually impossible given their complexity. However, the simulation results presented in Section 5 support the suggestion that our double-sided BRM method improves existing algorithms.

Table 2 compares the overall complexity of the double-sided algorithms to that of single-sided algorithms in terms of the number of multiplications.

Table 2. Complexity table of the target localization algorithms.

Methods	Number of Multiplications
Single-sided BRM algorithm [7]	$(M+3)^3 + (2MN+2M+7)MN(M+3)$
Single-sided two-stage BRM algorithms ([12,15])	$(M+3)^3 + (2MN+2M+7)MN(M+3)$ $+3^3 + 3(M+3)(2M+13)$
Double-sided BRM algorithm	$(M+N+3)^3 + 2(4MN+2M+2N+7)MN(M+N+3)$
Double-sided two-stage BRM algorithms	$(M+N+3)^3 + 2(4MN+2M+2N+7)MN(M+N+3)$ $+3^3 + 3(M+N+3)(2M+2N+13)$

The extra complexity of the double-sided algorithms is attributable principally to the larger matrix used for WLS computation. The increased computation cost scales polynomially, but is acceptable given the performance gain demonstrated by the simulations presented in Section 5.

4. Performance Analysis of Double-Sided BRM Method for Ideal Target/Antennae Positions

Here, we derive target estimation MSEs of our double-sided BRM method and the BRM method of Noroozi [7] when the two-dimensional target/antenna positions are ideal. Derivation of general, theoretical MSEs of target estimations is extremely complicated; the existing study in [7] assumes that the target/antenna distributions in the x-y plane are ideal. Accepting this, let the target be at (without loss of generality) $\mathbf{x}_o = [0,0]^T$, and let the antennae be located uniformly around the target:

$$\begin{aligned} \mathbf{x}_t(m) &= d\left[\cos\left(\theta_0 + \frac{2\pi m}{M}\right), \sin\left(\theta_0 + \frac{2\pi m}{M}\right)\right]^T, \\ \mathbf{x}_r(n) &= d\left[\cos\left(\phi_0 + \frac{2\pi n}{N}\right), \sin\left(\phi_0 + \frac{2\pi n}{N}\right)\right]^T, \end{aligned} \quad (33)$$

where d is the common distance between the target and the various antennae, and θ_0 and ϕ_0 are distinct angles.

Assuming small BR errors, the error covariance matrix of the WLS estimator can be derived from [17,18]:

$$Cov[[\hat{\mathbf{x}}_o^T, \hat{\mathbf{d}}_t^T, \hat{\mathbf{d}}_r^T]^T - [\mathbf{x}_o^T, \mathbf{d}_t^T, \mathbf{d}_r^T]^T] = (\mathbf{H}_o^T\mathbf{W}\mathbf{H}_o)^{-1}\mathbf{H}_o^T\mathbf{W}Cov[\boldsymbol{\varepsilon}]\mathbf{W}\mathbf{H}_o(\mathbf{H}_o^T\mathbf{W}\mathbf{H}_o)^{-1}, \quad (34)$$

where $\mathbf{H}_o$ is the noise-free version of $\mathbf{H}$ (derived by substituting r_{mn} for $\hat{r}_{mn}$ in (19)). Accepting the above assumption, $\mathbf{d}_t$ and $\mathbf{d}_r$ simplify to $d\mathbf{1}_{M\times 1}$ $d\mathbf{1}_{N\times 1}$, respectively, hence, the weighting matrix $\mathbf{W}$ of (21) and the covariance matrix of $\boldsymbol{\varepsilon} = [\boldsymbol{\varepsilon}_t^T, \boldsymbol{\varepsilon}_r^T]^T$, $Cov[\boldsymbol{\varepsilon}]$, simplify to:

$$\mathbf{W} = 1/(d^2\sigma_\omega^2)\mathbf{I}_{2MN} \quad (35)$$

$$Cov[\boldsymbol{\varepsilon}] = d^2\sigma_\omega^2 \begin{bmatrix} \mathbf{I}_{MN} & \mathbf{I}_{MN} \\ \mathbf{I}_{MN} & \mathbf{I}_{MN} \end{bmatrix}. \quad (36)$$

As the antennas are uniformly located on a circle of radius d, the assumption further yields the following properties (the results for Rxs are the same):

$$\textstyle\sum_{m=1}^{M} x_t(m) = \sum_{m=1}^{M} y_t(m) = 0 \quad (37)$$

$$\textstyle\sum_{m=1}^{M} x_t^2(m) = \sum_{m=1}^{M} y_t^2(m) = Md^2/2. \quad (38)$$

Using (37) and (38), each term of (34), $(\mathbf{H}_o^T\mathbf{W}\mathbf{H}_o)^{-1}$ and $\mathbf{H}_o^T\mathbf{W}Cov[\boldsymbol{\varepsilon}]\mathbf{W}\mathbf{H}_o$, can be simplified as follows:

$$(\mathbf{H}_o^T\mathbf{W}\mathbf{H}_o)^{-1} = \frac{\sigma_\omega^2}{MN}\begin{bmatrix} \mathbf{I}_2 & \frac{1}{2d}\mathbf{A} \\ \frac{1}{2d}\mathbf{A}^T & \mathbf{B} \end{bmatrix} \quad (39)$$

$$\mathbf{H}_o^T\mathbf{W}Cov[\boldsymbol{\varepsilon}]\mathbf{W}\mathbf{H}_o = \frac{1}{\sigma_\omega^2}\begin{bmatrix} \mathbf{0}_{2\times 2} & \mathbf{0}_{2\times(M+N)} \\ \mathbf{0}_{(M+N)\times 2} & \mathbf{D} \end{bmatrix} \quad (40)$$

where **A** is that of (26), and

$$\mathbf{D} = \begin{bmatrix} 4N\mathbf{I}_M & 4\mathbf{1}_{M\times N} \\ 4\mathbf{1}_{N\times M} & 4M\mathbf{I}_N \end{bmatrix}, \tag{41}$$

hence,

$$\begin{aligned} &Cov[[\hat{\mathbf{x}}_o^T, \hat{\mathbf{d}}_t^T, \hat{\mathbf{d}}_r^T]^T - [\mathbf{x}_o^T, \mathbf{d}_t^T, \mathbf{d}_r^T]^T] \\ &\qquad = \sigma_\omega^2 \begin{bmatrix} \frac{1}{4d^2}\mathbf{ADA}^T & \frac{1}{2d}\mathbf{ADB} \\ \frac{1}{2d}\mathbf{BDA} & \mathbf{BDB} \end{bmatrix}. \end{aligned} \tag{42}$$

As the MSEs of the x and y components are the $(1,1)$ and $(2,2)$ elements of $Cov[[\hat{\mathbf{x}}_o^T, \hat{\mathbf{d}}_t^T, \hat{\mathbf{d}}_r^T]^T - [\mathbf{x}_o^T, \mathbf{d}_t^T, \mathbf{d}_r^T]^T]$, we are interested only in $(1/4d^2)\mathbf{ADA}^T$. Using (38) once more, $(1/4d^2)\mathbf{ADA}^T$ is:

$$\frac{1}{4d^2}\mathbf{ADA}^T = \frac{1}{MN}\mathbf{I}_2. \tag{43}$$

Thus, we finally obtain

$$E\left\{(\hat{x}_o - x_o)^2\right\} = E\left\{(\hat{y}_o - y_o)^2\right\} = \frac{\sigma_\omega^2}{MN}. \tag{44}$$

Meanwhile, under the same assumption, the MSEs of the existing BRM method in [7] are:

$$E\left\{(\hat{x}_{BRM} - x_o)^2\right\} = E\left\{(\hat{y}_{BRM} - y_o)^2\right\} = \frac{2\sigma_\omega^2}{MN}. \tag{45}$$

A comparison of (44) and (45) shows that our method improves the MSE performance of the BRM method by a factor of two, given the assumed two-dimensional target/antenna positioning. As presented in the following section, simulations highlighted the improvements afforded by our algorithms when practical target/antenna settings were evaluated.

5. Numerical Simulation for Practical Target/Antennae Positions

Figure 1 presents the MSE performances of the proposed algorithms for the antenna positions specified in Table 3 and a target located at $\mathbf{x}_o = [0m, 0m, 0m]^T$. The results in Figure 1a show that our double-sided BRM method consistently affords better MSE performance than the single-sided BRM method of Noroozi [7], and the results in Figure 1b,c show that the double-sided two-stage BRM algorithms afford better MSE performance than the single-sided two-stage BRM methods of Amiri [12] and Wang [15].

Figure 2 presents the MSEs of target estimations when the target moves along the x-axis with the y and z target positions fixed at $y_o = 400$ m and $z_o = 100$ m, and antennas positioned as specified in Table 4. Here, the noise variance, σ_ω, was considered to be 5 m^2. The simulations shown in Figure 2 revealed that our algorithms afforded better MSE performance than existing algorithms for all target positions tested.

Table 3. Transmitters and receiver Positions (m).

k	$x_t(k)$	$y_t(k)$	$z_t(k)$	$x_r(k)$	$y_r(k)$	$z_r(k)$
1	250	300	180	−250	−300	−180
2	300	350	120	−300	−350	−120
3	300	250	160	−300	−250	−160
4	200	320	150	−200	−320	−150
5	250	200	150	−250	−200	−150
6	200	200	200	-	-	-
7	300	300	300	-	-	-

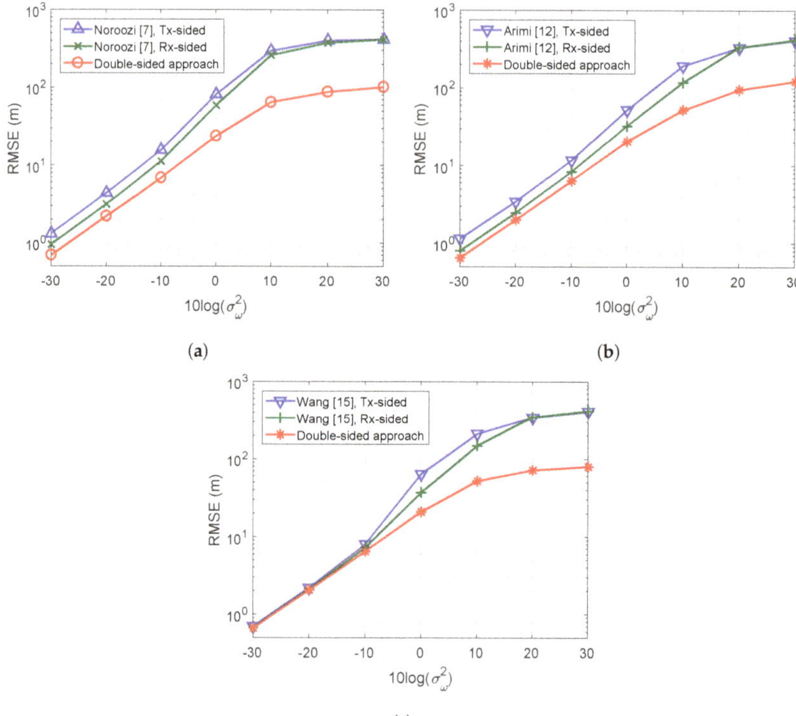

Figure 1. Target estimation MSE of the double-sided and single-sided algorithms with respect to noise variance: (**a**) single-stage; (**b**) two-stage using squared relations; and (**c**) two-stage using approximated relations.

Table 4. Transmitters and Receiver Positions (m).

k	$x_t(k)$	$y_t(k)$	$z_t(k)$	$x_r(k)$	$y_r(k)$	$z_r(k)$
1	0	0	15	−450	−450	20
2	−300	−200	15	−450	450	30
3	−300	200	10	450	−450	40
4	−200	−300	20	450	450	10
5	−200	300	10	0	600	20
6	200	−300	10	600	0	10
7	200	300	8	−600	0	15
8	300	−200	12	0	−600	10
9	300	200	16	-	-	-

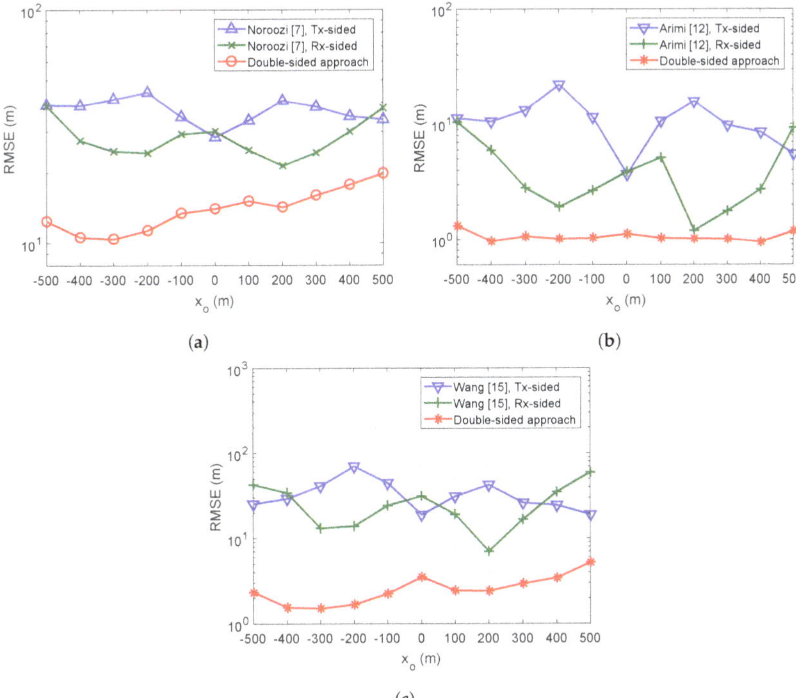

Figure 2. Target estimation MSE of the double-sided and single-sided algorithms with respect to the target position: (**a**) single-stage; (**b**) two-stage using squared relations; and (**c**) two-stage using approximated relations.

6. Conclusions

Here, we develop a novel target localization approach improving the target estimation accuracy of existing BRM based algorithms for distributed MIMO radars. The proposed double-sided BRM method estimates target, target–transmitter, and target–receiver distances simultaneously. We also took a double-sided approach to two-stage BRM methods. The improvements afforded by the proposed algorithms were confirmed theoretically for an ideal scenario, and via numerical simulations for practical scenarios.

Author Contributions: All authors contributed equally to this work. The final manuscript has been read and approved by all authors for submission.

Funding: This work was supported by the National Research Foundation of Korea (NRF) grant funded by the Korea government (MEST) (grant No. 2018R1A2B6001456).

Conflicts of Interest: The authors declare that they have no competing interests.

References

1. Haimovich, A.M.; Blum, R.S.; Cimini, L.J. MIMO Radar with Widely Separated Antennas. *IEEE Signal Process. Mag.* **2008**, *25*, 116–129. [CrossRef]
2. Bar-Shalom, O.; Weiss A.J. Direct positioning of stationary targets using MIMO radar. *Elsevier J. Signal Process.* **2011**, *91*, 2345–2358. [CrossRef]
3. Gogineni, S.; Nehorai, A. Target Estimation Using Sparse Modeling for Distributed MIMO Radar. *IEEE Trans. Signal Process.* **2011**, *59*, 5315–5325. [CrossRef]
4. Godrich, H.; Haimovich, A.M.; Blum, R.S. Target localization accuracy gain in MIMO radar-based system. *IEEE Trans. Inf. Theory* **2010**, *56*, 2783–2803. [CrossRef]
5. Dianat, M.; Taban, M.R.; Dianat, J.; Sedighi, V. Target localization using least squares estimation for MIMO radars with widely separated antennas. *IEEE Trans. Aerosp. Electron. Syst.* **2013**, *49*, 2730–2741. [CrossRef]
6. Noroozi, A.; Sebt, M.A. Target localization from bistatic range measurements in multi-transmitter multi-receiver passive radar. *IEEE Signal Process. Lett.* **2015**, *22*, 2445–2449. [CrossRef]
7. Noroozi, A.; Sebt, M.A. Weighted least squares target location estimation in multi-transmitter multi-receiver passive radar using bistatic range measurements. *IET Radar Sonar Navig.* **2016**, *6*, 1088–1097. [CrossRef]
8. Du, Y.; Wei, P. An explicit solution for target localization in Non-coherent distributed MIMO radar systems. *IEEE Signal Process. Lett.* **2014**, *21*, 1093–1097.
9. Einemo, M.; So, H.C. Weighted least squares algorithm for target localization in distributed MIMO radar. *Signal Process.* **2015**, *115*, 144–150. [CrossRef]
10. Park, C.H.; Chang, J.H. Closed-form localization for distributed MIMO radar systems using time delay measurements. *IEEE Trans. Wirel. Commun.* **2016**, *15*, 1480–1490. [CrossRef]
11. Pishevar, S.; Mohamed-Pour, K.; Noroozi, A. A closed-form two-step target localization in MIMO radar systems using BR measurements. In Proceedings of the 2017 25th Iranian Conference on Electrical Engineering (ICEE), Tehran, Iran, 2–4 May 2017; pp. 2088–2093.
12. Amiri, R.; Behnia, F.; Zamani, H. Asymptotically efficient target localization from bistatic range measurements in distributed MIMO radars. *IEEE Signal Process. Lett.* **2017**, *24*, 299–303. [CrossRef]
13. Liu, Y.; Guo, F.; Yang, L. An improved algebraic solution for TDOA localization with sensor position errors. *IEEE Commun. Lett.* **2015**, *19*, 2218–2221. [CrossRef]
14. Amiri, R.; Behnia, F. An efficient weighted least squares estimator for elliptic localization in distributed MIMO radars. *IEEE Signal Process. Lett.* **2017**, *24*, 902–906. [CrossRef]
15. Wang, J.; Qin, Z.; Wei, S.; Sun, Z.; Xiang, H. Effects of nuisance variables selection on target localisation accuracy in multistatic passive radar. *Electron. Lett.* **2018**, *54*, 1139–1141. [CrossRef]
16. Malanowski, M.; Kulpa, K. Two Methods for Target Localization in Multistatic Passive Radar. *IEEE Trans. Aerosp. Electron. Syst.* **2012**, *48*, 572–580. [CrossRef]
17. Barkat, M. *Signal Detection and Estimation*; Artech House: Norwood, NJ, USA, 2005.
18. Chan, Y.T.; Ho, K.C. A simple and efficient estimator for hyperbolic location. *IEEE Trans. Signal Process.* **1994**, *42*, 1905–1915. [CrossRef]

Article

Research of a Radar Imaging Algorithm Based on High Pulse Repetition Random Frequency Hopping Synthetic Wideband Waveform

Songhua He * and Xiaotian Wu *

School of Information Science and Engineering, Hunan University, Changsha 410082, China
* Correspondence: hesonghua@hnu.edu.cn (S.H.); xiaotian_w@hnu.edu.cn (X.W.)

Received: 17 November 2019; Accepted: 6 December 2019; Published: 9 December 2019

Abstract: Aiming at the imaging algorithm of high-pulse-repetition random-frequency-hopping synthetic wideband radar on a supersonic/hypersonic aircraft platform, this study established an echo simulation model of target and clutter, analyzed the special range-Doppler coupling effect and its influence on imaging, and proposes a method of imaging with pipeline-parallel processing based on generalized 2D matched-filtering and Doppler pre-processing. In the method, Doppler-beam-sharpening was advanced to be performed with the pulse compression process in each frame, and the special range-Doppler coupling effect caused by high dynamic motion of platform and random frequency hopping in bandwidth synthesis was well suppressed; several modes of random frequency hopping were designed and the pipeline-parallel image processing algorithm was optimized for each mode. Theoretical analysis and simulation results show that the proposed imaging method can effectively avoid the divergence of 2D range-Doppler images in the range direction, and can meet the requirements of real-time imaging.

Keywords: high pulse repetition frequency (HPRF); random frequency hopping (RFH); radar imaging; hypersonic aircraft

---

1. Introduction

In order to improve the performances of target-detection, low-probability-of-intercept, and anti-jamming, the high pulse repetition frequency (HPRF) synthetic wideband waveform has been used for radar imaging in supersonic/hypersonic aircraft guidance [1–5]. HPRF can decrease the velocity ambiguity, reduce the folding effect of clutter in Doppler direction, and then improve the signal-to-clutter ratio (SCR) and target detection ability. In addition, combined with random frequency hopping (RFH), HPRF can increase the number of accumulated pulses per unit time and improve the signal accumulation gain ratio of detector (coherent detection) to interceptor (non-coherent detection), which can decrease the peak power of the transmitting signal and improve the low-interception ability of radar. Furthermore, wideband HPRF RFH can improve the anti-jamming ability of guidance radar. RFH makes the reconnaissance jammer unable to predict the frequency-hopping pattern adopted in each frame; and makes it difficult to implement the answering-deception-jamming and the narrowband-blocking-jamming at each frequency point. HPRF and RFH enable the radar receiver to select echoes of a target from a relatively narrow range region according to the number of periods of return delay, and can suppress the deceptive-repeater-jamming, which is not in the same range region as the target, especially from the side-lobe direction.

In the case of RFH, the frequency domain sampling of radar signal in each frame is random and non-uniform. Because the basic frequency changes randomly between frames, the equivalent time-domain sampling of the inter-frame Doppler processing is also non-uniform and random. It is difficult to adopt the fast imaging algorithm based on Inverse Discrete Fourier Transform (IDFT) whether in pulse-compression processing at range direction or beam-sharpening processing in the Doppler direction. In addition, in the case of supersonic/hypersonic application, due to the large broadening in clutter Doppler spectrum, the special range-Doppler coupling effect and the large motion-compensation residue in the echoes, it is difficult to use the conventional method of fractal-dimension processing in the two directions of range and Doppler. Therefore, it is important to develop new imaging methods and fast algorithms for RFH radar. The existing non-uniform DFT (NU-DFT) fast algorithms, such as Vandermonde determinant method [6,7], regular Fourier matrix method [8,9], and min-max interpolation method [10,11], have specific constraints on the structural characteristics of non-uniform sampling signals, and their versatility is relatively poor. Compressed sensing technology [12] has also been widely used in the field of RFH synthetic wideband imaging [13–19]. According the theory of compressed sensing, it requires that the target and clutter background meet the basic conditions of sparsity. In the cases of wideband high-range-resolution (HRR) and high signal-to-noise ratio (SNR) or high SCR, compared with the number of range resolution units, the number of the observed targets or the number of scattering centers of the targets is limited. Therefore, it can provide a good guarantee for the sparsity in the high-resolution range profiles. The advantage of the compressed sensing method is that it can reconstruct the range profile through a small number of observations, which means in RFH radar, that it can effectively reduce the number of transmitting pulses without decreasing the resolution of the range profile. However, the compressed sensing algorithm needs a large number of matrix inversion operations, which makes it difficult to meet the real-time requirement, especially in hypersonic-platform-borne (HPB) radar imaging application of limited computing resources and extremely short platform-target intersection time. In addition, in the case of strong clutter and low SCR, the sparsity required by compressed sensing is also difficult to meet.

In view of the above problems, we established the scattering echo model of target and clutter for an HPRF RFH radar system and supersonic/hypersonic aircraft platform, and analyzed the special range-Doppler coupling effect and its influence on imaging. We also proposed the 2D range-Doppler imaging method and the 1D HRR imaging method based on Doppler pre-processing and 2D generalized matched-filtering (GMF) processing. Additionally, we designed several RFH modes, and proposed the corresponding pipeline-parallel processing, fast, real-time imaging algorithm for a different RFH mode. Theoretical analysis and simulation experiments showed that the proposed imaging method could effectively suppress the special range Doppler-coupling effect, achieve good imaging performances, and easily meet the real-time imaging requirements of supersonic/hypersonic aircraft-borne application.

2. Echo Modeling of RFH Synthetic Wideband Radar

Set the imaging processing time period of RFH radar as $M*N*T$. M is the number of sub-frames, which is the number of accumulated frames required for Doppler processing, the size of which defines the speed resolution; N is the number of frequency hopping points per frame, the synthetic bandwidth of which defines the range resolution; T is the pulse repetition period. In the case of HPRF, the echo delay τ_H of targets and sea/land clutters from the main-lobe direction is much greater than T, $n_sT < \tau_H < (n_s + 1)T$, where n_s is integer. Here we assume that each receiving complex frame lags behind the transmitting complex frame for n_s periods and the period in the receiving complex frame is numbered $(n|m)$, where m is the number of the frame ($m = 0, 1, \ldots, M - 1$) and n is the number of pulse period in each frame ($n = 0, 1, \ldots, N - 1$). In each pulse period, the receiving signal is sampled and the sampling interval is equal to the pulse-width τ; then, the total number of sampling points in each period is $K = INT[T/\tau]$ ($INT[.]$ represents rounding down). Taking the starting time of each period as the reference, the corresponding sampling time is $\tau, 2\tau, \ldots, K\tau$, which is numbered $k = 0, 1, \ldots, K - 1$ respectively, and k is the number of sampling unit. By using the n_s-period-delayed frequency hopping pattern to construct the local reference signal, the echo signal is coherently received, and the echo signal with a range of cn_sT-$c(n_s + 1)T$ (c is the speed of light) is selected by the IF filter of receiver.

For any scattering point in the selected range, according to the radar principle, the received/sampled signal can be expressed as follows:

$$x(n|m,k) = Aexp\{j2\pi[(2R/c)\Delta f_d i_{mn} - (2f_m VNT/c)m - (2f_m VT/c)n - (2\Delta f_d VT/c)ni_{mn} - (2\Delta f_d VNT/c)mi_{mn} - (2\Delta f_d V\tau/c)ki_{mn} - (2\Delta f_d V/c)\tau i_{mn} - (2f_m V\tau/c)k + 2f_m(R - V\tau)/c]\}. \quad (1)$$

Here, the pure RFH mode (fast hopping intra frame/wide hopping inter frame; other modes are special cases of this mode) is investigated, where f_m is the basic frequency of the transmitting signal at the m-th frame; Δf_d is the minimum frequency jump interval determined by the minimum quantization level of direct digital synthesizer (DDS); i_{mn} is an integer which is randomly selected according to a certain frequency hopping pattern in the range of integer set [0, 1, 2, ... , $I - 1$], where $I = \Delta F/\Delta f_d$ (the same value cannot be repeated in the same frame); ΔF is the synthetic bandwidth; $f_m + \Delta f_d i_{mn}$ is the carrier frequency of the transmitting signal in the n-th period of the m-th frame; R', R, and V are, respectively, the actual range, ambiguous range, and radial velocity of the point target at the starting time of the first pulse period of each receiving complex frame, $R' = R + cn_s T/2$ and $0 \le R < cT/2$. The velocity is defined as positive for movement facing the radar and is assumed to remain unchanged within MNT (the time of a complex frame, usually several milliseconds). Here we ignore the influence of the slight change of V within a short time of milliseconds.

As shown in Figure 1, in the n-th pulse period of the m-th frame in each complex frame, the time delay of the receiving echo is $2(R - VmNT - VnT - Vn_sT)/c$ relative to the starting time $mNT + nT$ of this period. Because the k-th sampling unit in each period can only acquire data of echo which has delay in the range of $k\tau - (k+1)\tau$, $k\tau < 2(R - VmNT - VnT - Vn_sT)/c < (k+1)\tau$, and the echo is sampled by the k-th sampling unit.

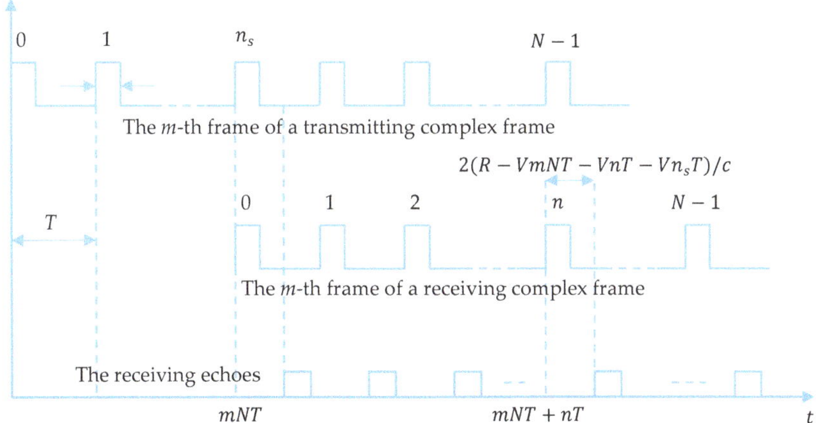

Figure 1. Sequence chart of each frame.

It is apparent that if the scatter has a facing range-walk of δR in a complex frame and $0 < R - ck\tau < \delta R$, then the echo is sampled first by the k-th unit in some frames, and then by the $(k-1)$-th unit in the remaining frames, which is called cross-sampling-unit movement. For supersonic/hypersonic applications, considering the large-scale range-walk of cross-sampling unit in a complex frame, the following requirement must be met for each scattering point:

$$kc\tau/2 < R - VmNT - VnT - Vn_sT < (k+1)c\tau/2.$$

For $(n|m, k)$ combinations that do not satisfy the above equation, $x(n|m, k) = 0$.

The advantage of the simulation model shown in Equation (1) is that it can fully describe the actual cross-sampling-unit movement in supersonic/hypersonic applications. In addition, it is suitable for panoramic simulation of echoes from the area that is illuminated by the main-lobe of radar beam. The panoramic clutter area can be divided into many grids, and echo from each grid can be simulated as point scattering by using Equation (1). The target can be simulated as multiple scattering centers, and echo from each scattering center can be simulated by using Equation (1). The target may be close to the junction of two sampling unit, and echoes from the target may appear successively at two adjacent sampling units (which is the so-called cross-sampling-unit range-walk). The clutter echoes appear at most sampling units (in the case of HPRF, $cT/2$ is larger than but very close to the radial length of the illuminated area of main-lobe). According to Equation (1), the return data containing both clutter-background and target can be simulated as follows:

$$x_S(n|m,k) = x_T(n|m,k) + x_C(n|m,k),$$

where $x_T(n|m, k)$ is the target echo, which can be expressed as the sum of the point scattering echoes of multiple scattering centers; $x_C(n|m, k)$ is the clutter echo, which can be expressed as the sum of the point scattering echoes of each clutter grid in the main-lobe illuminated area. The amplitude of each clutter scattering point is randomly selected according to the Rayleigh distribution, and the parameter σ^2 of the Rayleigh distribution is controlled according to the required SCR.

According to the moving speed of the platform, angle between the moving direction of the platform, and the illumination direction of the beam, the estimated value V_C of the radial speed of the center of the main-lobe clutter can be obtained. Clutter-center velocity compensation is applied to data acquired in each complex frame as follows:

$$\begin{gathered} y_S(n|m,k) = x_S(n|m,k) \times exp\{j2\pi[(2f_m V_C NT/c)m + (2f_m V_C T/c)n + \\ (2\Delta f_d V_C T/c)ni_{mn} + (2\Delta f_d V_C NT/c)mi_{mn} + (2\Delta f_d V_C \tau/c)ki_{mn} + (2\Delta f_d V_C/c)\tau i_{mn} + \\ (2f_m V_C \tau/c)k + 2f_m V_C \tau/c]\}. \end{gathered} \tag{2}$$

Considering the Doppler broadening effect and the velocity estimation error of the moving platform, let $v = V - V_C$ be the velocity surplus of the scattered relative to the clutter center. After clutter-center velocity compensation, the sampled signal of each scatter can be expressed as follows:

$$\begin{gathered} y(n|m,k) = Aexp\{j2\pi[(2R/c)\Delta f_d i_{mn} - (2f_m vNT/c)m - (2f_m vT/c)n - \\ (2\Delta f_d vT/c)ni_{mn} - (2\Delta f_d vNT/c)mi_{mn} + 2f_m R/c]\}exp\{j\varphi(f_0, R, v)\}, \end{gathered} \tag{3}$$

where $\varphi(f_0, R, v)$ is a constant term independent of $(n|m, k)$.

Because the velocity surplus of clutter or target is far less than the platform velocity, some phase terms in the compensated signal can be ignored, which can simplify the subsequent imaging process. The ignored phase terms are $j2\pi[-(2\Delta f_d v\tau/c)ki_{mn}]$, $j2\pi[-(2\Delta f_d v/c)\tau i_{mn}]$, $j2\pi(-2f_m v\tau/c)$, and $j2\pi(-2f_m v\tau/c)k$, the variation range of which is not more than $\pi/4$ in a complex frame.

3. High Quality Real-Time Imaging of HPB HPRF RFH Radar

3.1. Special Range-Doppler Coupling Effect and Its Suppression

In the case of conventional stepped-frequency (SF) synthetic wideband radar system, $i_{mn}\Delta f_d = n\Delta f$ and $f_m = f_0$; in Equation (3), Δf is the frequency interval between adjacent pulses and $N\Delta f$ is the synthetic bandwidth. Each frame has the same basic frequency f_0 and the same stepped-frequency hopping. According to Equation (3), in any frame-m, the change of signal phase between pulses mainly depends on the phase term $2\pi(2R/c)\Delta fn$, which is only related to range-R. Therefore, FFT processing or pulse compression processing in each frame can be used to obtain the distribution of scatters in the range direction; i.e., target range profile [20]. The range resolution determined by DFT is $c/(2N\Delta f)$. The second-order phase term $2\pi(2\Delta fvT/c)n^2$ in Equation (3) may cause energy diffusion of the scattering center in range profile, but the diffusion can be ignored because the synthetic bandwidth $N\Delta f$ is far smaller than the carrier frequency f_0 and the phase variation of the second-order phase term is very small. The other velocity-related phase terms $2\pi(2f_0vT/c)n$ and $2\pi(2\Delta fvNT/c)mn$ are linear with n, and their influence on pulse compression is that the position of the scatter on the FFT spectrum is shifted by an offset of $f_0v(1+mN)T/\Delta f$, which is called range-Doppler coupling effect. The range-Doppler coupling effect in an SF radar system can cause error in range measurement, but cannot cause significant diffusion of energy or degradation of imaging quality. After pulse compression and envelope alignment of a range profile in each frame, the inter-frame phase change of each range unit mainly depends on the phase term $2\pi(2f_0vNT/c)m$. Therefore, FFT processing or Doppler processing in each range unit can be used to obtain the distribution of scatters in velocity or Doppler direction. The distribution of the scatters on the 2D range-Doppler plane can be obtained by synthesizing the distributions of all the range units. As described above, in the conventional SF system, the imaging processing method of pulse compression in each frame at first, and then Doppler processing in each range resolution unit, are generally adopted.

It can be seen from Equation (3) that the phase of the signal is complexly related to the range and speed of the scatter due to HPRF and RFH. In each frame-m, the range-related phase term $2\pi(2R/c)\Delta f_d i_{mn}$, changes randomly and nonlinearly between pulses because i_{mn} changes randomly. In order to use traditional FFT for pulse compression in each frame, it is necessary to rearrange the data in order of frequency from small to large, and then interpolate the non-uniform frequency-sampled data into uniform frequency-sampled data. The randomly-changed phase term $2\pi(2R/c)\Delta f_d i_{mn}$ is transformed to linearly-changed phase term $2\pi(2R/c)\Delta fn$ after rearrangement and interpolation. However, data rearrangement randomizes the original linear range-Doppler coupling phase term $2\pi(2f_m vT/c)n$. For supersonic/hypersonic applications, even if the clutter-center velocity compensation is made by using Equation (2), the phase change of the coupling phase term caused by the velocity residual v is still large for the scatters that are not at the direction of beam-center, and it can be close to or even more than 2π in one frame. The random change of phase is equivalent to adding multiplicative noise to the signal, and it seriously reduces the coherence of the rearranged data, which leads to serious energy-divergence of scatters and degradation of imaging quality. For inter-frame Doppler processing, because i_{mn} changes randomly, the velocity-related phase term $(2f_m vNT/c)m$ changes randomly and nonlinearly between frames, which makes the Doppler processing complicated. The random and nonlinear range-Doppler coupling effect not only exists within the frame but occurs between frames, so it is difficult to carry out fractal-dimension processing in range and Doppler directions respectively.

In this paper, the above phenomenon is called the special range-Doppler coupling effect of RFH synthetic wideband radar in a highly dynamic application. Because of the above special effect, the conventional imaging processing method of pulse compression in each frame at first and then Doppler processing in each range resolution unit cannot be adopted in HPRF RFH radar. In order to suppress the special range-Doppler coupling effect, Doppler processing must be advanced to each frame and be synchronous with the pulse compression processing, which is called Doppler pre-processing in this paper.

The imaging algorithm of Doppler pre-processing is based on the 2D GMF algorithm, which can be executed by means of pipeline-parallel processing, and the computation can be dispersed to each frame in combination with the data acquisition process. The algorithm can be optimized in real-time ability according to different RFH modes.

As an example, the basic principle of suppressing the above special coupling effect through Doppler pre-processing is illustrated by the following intra-frame pseudo RFH mode where $i_{mn}\Delta f_d = i_n\Delta f$ and $f_m = f_0$ (the basic frequency of pulse signal remains unchanged between frames; for different n, i_n randomly takes different values in $[0,1,2,\ldots,N-1]$ without repetition); then,

$$\begin{aligned}y(n|m,k) = Aexp\{j2\pi[(2R/c)\Delta fi_n - (2f_0vNT/c)m - (2f_0vT/c)n- \\ (2\Delta fvT/c)ni_n - (2\Delta fvNT/c)mi_n + 2f_0R/c]\}exp\{j\varphi(f_0,R,v)\}.\end{aligned}$$

Obviously, for any fixed period number n, the phase terms $2\pi(2f_0vNT/c)m$ and $(2\Delta fvNT/c)mi_n$ vary non-randomly and linearly with the frame number m. Therefore, the M sampled data $\{y_S(n|m,k)|m = 0,1,\ldots,M-1\}$ of the same period number n and the same sampling unit number k can be processed first by using FFT (Doppler pre-processing). The velocity resolution converted from the DFT spectral resolution is $\Delta v = c/(2f_0NMT)$, and the distribution of the scatters in the velocity direction or the Doppler direction can be obtained. In the Doppler pre-processed data $\{Y_S(n|l_v,k)|l_v = 0,1,\ldots,M-1\}$, the phase terms of the signal on the l_v-th speed channel changes with n are mainly $2\pi(2R/c)\Delta fi_n$ and $2\pi(2f_0vT/c)n$. Due to the accumulation or filtering effect of DFT, the change range of v in this channel is $[l_v\Delta v - \Delta v/2, l_v\Delta v + \Delta v/2]$. If the signal on the l_v-th speed channel is phase-compensated by the phase factor $exp\{j2\pi(2f_0l_v\Delta vT/c)n\}$ during or after Doppler processing, the phase term $2\pi(2f_0vT/c)n$ becomes $2\pi(2f_0v'T/c)n$, where $-\Delta v/2 < v' < \Delta v/2$. As long as the accumulation time MNT is long enough or the resolution of velocity is high enough, Δv is small enough, and the change range of the term $2\pi(2f_0v'T/c)n$ can be far less than $\pi/4$. Rearrange the data $\{Y_S(n|l_v,k)|n = 0,1,\ldots,N-1\}$ on each speed channel l_v in the order of i_n from small to large; then, the phase term $2\pi(2R/c)\Delta fi_n$ becomes $2\pi(2R/c)\Delta fn$, while $2\pi(2f_0v'T/c)n$ becomes a random phase term of small value, which can be ignored.

By FFT processing of the rearranged data on each speed channel, the distribution of the scatters along range direction of each speed channel can be obtained. By synthesizing all the speed channels, the distribution of the scatters on the 2D range-Doppler plane can be obtained. It has almost the same imaging effect as the conventional imaging algorithm used in SF Radar.

For other RFH modes, Doppler pre-processing can also suppress the above special range-Doppler coupling effect, and different fast 2D range-Doppler imaging algorithms can be obtained.

3.2. Image Processing Based on 2D GMF

According to the theory of matched filtering, for any transmitting signal waveform, as long as it has a certain bandwidth and a certain time width, the 2D range-Doppler image of the detected area can be obtained from the return signal through 2D matched filtering processing at the receiving end. The range resolution of the image depends on the effective bandwidth of the transmitting signal, and the speed resolution depends on the effective time-width of the signal. If the random frequencies are uniformly-distributed, the effective bandwidth is proportional to the synthetic bandwidth ΔF.

Supposing the required non-ambiguous range depth of imaging at each sampling unit is R_p, the parameters ΔF and N are designed to satisfy $R_p = cN/(2\Delta F)$. If range depth of R_p is divided into N range cells, the corresponding range width of each cell is $c/(2\Delta F)$, which is exactly the nominal range resolution corresponding to the synthetic bandwidth ΔF of RFH signal. The non-ambiguous velocity measurement range $[0, c/(2f_0MT)]$ is divided into M velocity cells, and the velocity width corresponding to each cell is $c/(2f_0MNT)$, which is exactly the velocity resolution corresponding to the accumulation time MNT of a complex frame. According to Equation (3) and the principle of 2D GMF, the processing of 2D range-Doppler segment imaging at each sampling unit k can be described as follows

$$\begin{gathered}P(l_u,l_v,k) = \sum_{m=0}^{M-1}\sum_{n=0}^{N-1} W_\Omega(m,n)y_S(n|m,k)exp\{-j2\pi[(f_m+\Delta f_d i_{mn})k\tau]\}\times \\ exp\left\{-j2\pi\left[\frac{f_m+\Delta f_d i_{mn}}{\Delta F}l_u\right]\right\}\times exp\left\{j2\pi\left[\frac{f_m+\Delta f_d i_{mn}}{f_0}\left(l_v-\frac{M}{2}\right)\frac{m}{M}\right]\right\}\times exp\left\{j2\pi\left[\frac{f_m+\Delta f_d i_{mn}}{f_0}\left(l_v-\frac{M}{2}\right)\frac{n}{MN}\right]\right\},\end{gathered} \tag{4}$$

where $\{P(l_u,l_v,k)|l_u = 0,1,\ldots,N-1; l_v = 0,1,\ldots,M-1\}$ is called the 2D segment image of the target area obtained by the sampling point k. $P(l_u,l_v,k)$ is the value (complex number) at the pixel point (l_u,l_v), where l_u is the number of pixel points in the range direction; l_v is the number of pixel points in the speed direction. The total number of pixels in the segment image is MN.

The function of phase term $exp\{-j2\pi[(f_m+\Delta f_d i_{mn})k\tau]\}$ is to calibrate the segment image, so that the starting position $l_u = 0$ and the ending position $l_u = N-1$ of the segment image in the range direction correspond to the starting position $ck\tau/2$ and the ending position $ck\tau/2 + (N-1)R_p/N$.

The purpose of calibration is to ensure that the segment image acquired at different sampling units is not ambiguous. $exp\{-j2\pi[(f_m + \Delta f_d i_{mn})l_u/\Delta F]\}$ is the frequency-domain, non-uniform-sampling Fourier transform factor in the direction of range. The range-domain sampling after transformation is uniform, and the sampling interval is $c/(2\Delta f)$, but the frequency-domain sampling interval defined by $i_{mn}\Delta f_d$ before transformation is non-uniform and random. $exp\{j2\pi(f_m + \Delta f_d i_{mn})(l_v - M/2)m/(Mf_0)\}$ is the time-domain, non-uniform sampling Fourier transform factor in the velocity direction. The equivalent time-domain sampling before transformation is non-uniform because of variation of between frame, and the Doppler- frequency-domain sampling or velocity sampling after the transformation is uniform, with an interval of $c/(2f_0MNT)$. $exp\{j2\pi[(f_m + \Delta f_d i_{mn})(l_v - M/2)n/MNf_0]\}$ is the range-Doppler coupling compensation factor, which can realize the phase compensation of the range movement within the sampling unit and across the sampling unit. Obviously, the motion compensation and 2D Fourier transform are carried out synchronously, and different phase compensation is used in different velocity channels to improve the compensation accuracy.

Obviously, when $i_{mn}\Delta f_d = n\Delta f$, $f_m = f_0$, the RFH synthesis wideband system degenerates to the conventional SF synthesis wideband system, and the 2D GMF of Equation (4) degenerates to the conventional 2D windowed DFT operation, which can be implemented by 2D fast Fourier transform.

However, the computation complexity of 2D GMF is much higher than that of 2D FFT, so it is necessary to combine different RFH modes and use fast algorithms to realize 2D GMF to meet the real-time needs of high-speed platform-borne application.

Since the range width of echoes in each sampling unit is $c\tau/2$, if the non-ambiguous range depth of the segment image is R_p, it is required that $R_p \geq c\tau/2 + R_I$, so that the echoes of scatters which move across sampling unit can be accumulated in-phase at the same point of panoramic image, and this is of importance in supersonic/hypersonic applications. R_I is the maximum moving range of the scatters in an imaging period of MNT. The overlapping width of the segment image of adjacent sampling units in the range direction is $R_p - c\tau/2$, and the number of non-overlapping range resolution cells is $K_d = Nc\tau/(2R_p)$. Then, the panoramic image in the beam irradiation area can be obtained from the segment image of all the sampling units as follows:

$$Z(i,j) = \sum_{k=0}^{K-1} P(i - kK_d, j, k)U(i - kK_d)i = 0,1,\ldots,K_d + (K-1)(N - K_d) - 1; j = 0,1,\ldots,M-1,$$

where $U(i)$ is a rectangular function with length N, defined as:

$$U(i) = \begin{cases} 1, & 0 \leq i \leq N-1 \\ 0, & otherwise \end{cases}$$

In summary, the procedure of imaging process is shown in Figure 2.

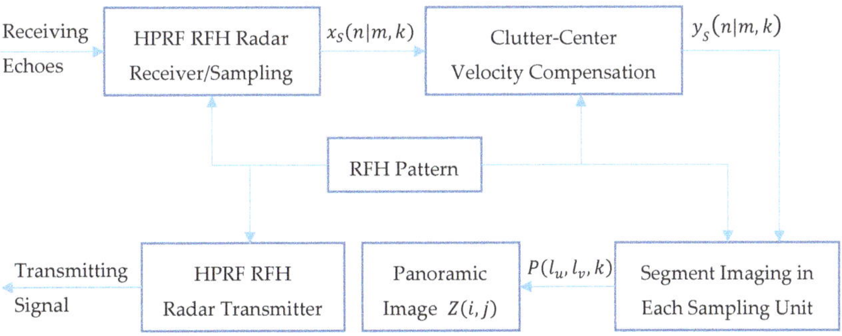

Figure 2. The procedure of imaging processing.

3.3. The Tradeoff between Randomness and Real-Time Performance

For the conventional SF radar, there is fast imaging algorithm of 2D FFT because of uniform sampling in both frequency-domain and slow-time-domain. Theoretically speaking, for the RFH radar system, the fast imaging algorithm depends on the structural characteristics of the RFH pattern. Because there are too many RFH patterns, it is impossible to design optimal imaging algorithm that has the least amount of computation for every RFH pattern. However, it is possible to design pipeline-parallel processing real-time imaging for different RFH modes combined with the data acquisition process.

In this paper, several RFH modes are defined as follows.

3.3.1. Intra-Complex-Frame Pure-RFH

In the m-th frame and the n-th pulse period of a complex frame, the frequency f_{mn} of the transmitting signal is randomly selected according to certain algorithm in the frequency band $(f_m, f_m + \Delta F)$. The basic frequency f_m can hop randomly in a wide frequency range between frames. In this mode of RFH, $f_{mn} = f_m + \Delta f_d i_{mn}$, where i_{mn} is the sequence number corresponding to the carrier frequency of the m-th frame and the n-th period. i_{mn} is randomly and un-repeatedly selected according to a certain probability density distribution in the integer set $[0, 1, 2, \ldots, I-1]$, where $I = \Delta F/\Delta f_d$. Different frame-m adopts different baseband frequency point set $\{\Delta f_d i_{mn} | n = 0, 1, \ldots, N-1\}$. This mode has the best performance of randomness, low-interception, and anti-interference.

3.3.2. INTRA-Frame Pure-RFH

In each frame of a complex frame, the same frequency point set and the same hopping-order are adopted. For any frame m_1 and m_2, $f_{m_1 n} = f_{m_2 n} = f'_n$, where f'_n is randomly selected according to certain algorithm in the frequency band $(f_0, f_0 + \Delta F)$. Because the frequency is the same between frames, the initial phase φ_{mn} of the transmitting signal must be randomly selected according to certain algorithm between 0 and π, so as to reduce the cyclic autocorrelation of the transmitting signal and maintain the low interception performance. For different complex frames, the basic frequency f_0 randomly changes in a large range as much as possible. The same frequency between frames can simplify the Doppler processing and improve the real-time performance. However, compared with the intra-complex-frame pure-RFH mode, this mode loses performance of randomness, low-interception, and anti-interference because the same frequency point set and the same hopping-order are adopted in each frame.

In this mode of RFH, $f_m = f_0$, $i_{mn} = i_n$ and $f_{mn} = f_0 + \Delta f_d i_n$, where i_n is randomly and un-repeatedly selected according to a certain probability density distribution in the range of integer set $[0, 1, 2, \ldots, I-1]$, where $I = \Delta F/\Delta f_d$.

3.3.3. Intra-Complex-Frames Pseudo-RFH

The baseband frequency points are obtained by uniform sampling in $(0, \Delta F)$, $f_{mn} = f_m + i'_{mn}\Delta F/N$, where $n = 0, 1, 2, \ldots, N-1$. The basic frequency f_m can hop randomly in a wide frequency range between frames, and i'_{mn} can be randomly selected according to certain algorithm in the range of $\{0, 1, 2, \ldots, N-1\}$. In this mode, different frames can adopt different basic frequencies and different hopping-orders but the same uniformly sampled baseband frequency point set, and the pulse compression processing in the range direction can be done by fast Fourier transform after higher order motion compensation and data rearrangement, which improves the real-time performance. Compared with the above two modes, the shortcoming of this mode is that the non-ambiguous range depth of segment image at each sampling unit decreases because of the frequency-domain uniform sampling. It is necessary to increase N and reduce the frequency hopping interval $\Delta f = \Delta F/N$ to meet the design requirements of non-ambiguous range depth. In addition, this mode loses more performance of randomness, low-interception, and anti-interference.

In this mode, $\Delta f_d i_{mn} = (\Delta F/N)i'_{mn}$ and $f_{mn} = f_m + (\Delta F/N)i'_{mn}$, where i'_{mn} is randomly and un-repeatedly selected according to a certain probability density distribution in the range of integer set $[0, 1, 2, \ldots, N-1]$.

3.3.4. Intra-Frame Pseudo-RFH

The baseband frequency points are obtained by uniform sampling in $(0, \Delta F)$, $f_{mn} = f_0 + i'_n \Delta F/N$, where $n = 0, 1, 2, \ldots, N-1$. The basic frequency does not hop between frames. i'_n is randomly and un-repeatedly selected according to certain algorithm in the range of $\{0, 1, 2, \ldots, N-1\}$. The initial phase φ_{mn} is randomly selected according to certain algorithm between 0 and π. For different complex frames, the basic frequency f_0 randomly changes in a large range as much as possible. In this mode, both the pulse compression processing in the range direction and the Doppler processing in the speed direction can be done by fast Fourier transform, which further improves the real-time performance. Decreasing in non-ambiguous range depth, and more loss in performance of randomness, low-interception, and anti-interference, are also the shortcomings of this mode.

See Appendix A for the specific generation of 2D RFH patterns for those four RFH modes

3.4. Online, Fast 2D Imaging Algorithms for Different RFH Modes

As mentioned before, in order to avoid image defocusing caused by the special range-Doppler coupling in the case of RFH, Doppler pre-processing must be carried out synchronized with the pulse compression process. However, Doppler processing is a kind of inter-frame processing. It will cause a serious delay in signal processing if Doppler processing is not done until the data of the last frame is collected. Considering that the data of the RFH synthetic wideband radar is obtained in the order of frames and periods, pipeline-parallel processing can be used to divide the Doppler pre-processing and pulse compression into each frame and each period, which can reduce the delay in signal processing.

3.4.1. Intra-Complex-Frame Pseudo-RFH Mode

In the 2D matched filtering (range-Doppler imaging) equation of Equation (4), in each frame of each sampling unit, the data are rearranged in the way of frequency point from small to large. Set n' as the frequency point number after rearrangement and the corresponding number before rearrangement is n_m. Set the rearranged data as $y'_S(n'|m,k)$. According to the definition of this RFH mode $\Delta f_d i_{mn'} = n'\Delta F/N$, so the 2D matched filtering of Equation (4) can be re-written as follows:

$$\begin{aligned} P(l_u, l_v, k) = \sum_{m=0}^{M-1}\sum_{n=0}^{N-1} & W_\Omega(m, n_m) y'_S(n'|m,k) \times exp\{-j2\pi[(f_m + n'\Delta F/N)k\tau]\} \\ & \times exp\{-j2\pi n' l_u/N\} \times exp\left\{j2\pi\left[\frac{f_m + n'\Delta F/N}{f_0}\left(l_v - \frac{M}{2}\right)\frac{m}{M}\right]\right\} \\ & \times exp\left\{-j2\pi\frac{f_m}{\Delta F}l_u\right\} \times exp\left\{j2\pi\left[\frac{f_m + n'\Delta F/N}{f_0}\left(l_v - \frac{M}{2}\right)\frac{n_m}{MN}\right]\right\} \end{aligned} \quad (5)$$

Imaging Algorithm 1: FFT-based pulse-compression on multiple velocity channels

- Step 1. Initializing, set $\left\{P^{(-1)}(l_u, l_v, k) = 0 \middle| l_u = 0, 1, \ldots, N-1; l_v = 0, 1, \ldots, M-1\right\}$
- Step 2. For $m = 0, 1, \ldots, M-1$, perform the following iteration:

(A) Obtain the data of the k-th sampling unit of the m-th frame: $\{y_S(n|m,k)|n = 0, 1, \ldots, N-1\}$.

(B) Data rearrangement: $\{y_S(n|m,k)|n = 0, 1, \ldots, N-1\} \to \left\{y'_S(n'|m,k)\middle|n' = 0, 1, \ldots, N-1\right\}$.

(C) Windowing, range calibration, multi-velocity-channel motion compensation, and Doppler pre-processing:

$$y''_S(n'|m,k,l_v) = W_\Omega(m,n_m)y'_S(n'|m,k)\psi(m,n',k,l_v), \quad (6)$$

where

$$\psi(m,n',k,l_v) = exp\left\{-j2\pi[(f_m + n'\Delta F/N)k\tau] + j2\pi(f_m + n'\Delta F/N)\left(l_v - \tfrac{M}{2}\right)m/(f_0 M) + j2\pi(f_m + n'\Delta F/N)\left(l_v - \tfrac{M}{2}\right)n_m/(f_0 MN)\right\}. \quad (7)$$

(D) Multi-velocity-channel fast pulse-compression processing.

It can be obtained according to Equations (5) and (6) that

$$P(l_u,l_v,k) = \sum_{m=0}^{M-1} exp\left\{-j2\pi\frac{f_m}{\Delta F}l_u\right\}\sum_{n'=0}^{N-1} y''_S(n'|m,k,l_v) \times exp\{-j2\pi n'l_u/N\}.$$

The operation $\sum_{n'=0}^{N-1} y''_S(n'|m,k,l_v) \times exp\{-j2\pi n'l_u/N\}$ is a uniform sampling DFT, which can be realized by *FFT*:

$$\left\{Y''_S(l_u,l_v|m,k) : l_u = 0,1,\ldots,N-1\right\} = FFT\left\{y''_S(n'|m,k,l_v) : n' = 0,1,\ldots,N-1\right\}. \quad (8)$$

(E) Current-frame Doppler-accumulation processing:

$$P^{(m)}(l_u,l_v,k) = P^{(m-1)}(l_u,l_v,k) + exp\left\{-j2\pi\tfrac{f_m}{\Delta F}l_u\right\} \times Y''_S(l_u,l_v|m,k) \quad (l_u = 0,1,\ldots,N-1; l_v = 0,1,\ldots,M-1). \quad (9)$$

Obviously, $P^{(M-1)}(l_u,l_v,k) = P(l_u,l_v,k)$.

In this algorithm, gradually, a clear image is obtained through iteration. Every additional frame of data increases the sharpness of the image. Because the rearranged frequency points are uniformly sampled, the pulse compression processing on each velocity-channel can be realized by FFT, which improves the real-time performance of the imaging algorithm.

3.4.2. Intra-Frame Pseudo-RFH Mode

This RFH mode is equivalent to making all $f_m = f_0$ in the intra-complex-frame pseudo-RFH mode. After data rearrangement, the n_m corresponding to n' is the same, which is labeled as n and independent of m. In this mode, there is neither coupling phase term of l_u and m, nor coupling phase term l_v and n. Compared to imaging algorithm 1, the computational complexity can be further reduced by using fractal-dimension processing.

Imaging Algorithm 2. Fractal-dimension processing with Doppler pre-processing

- Step 1. Data rearrangement. Rearrange the data $y_S(n|m,k)$ of each frame in the order of frequency points from small to large. The rearranged data are $y'_S(n'|m,k)$, in which the number before rearrangement corresponding to n' is n.
- Step 2. Windowing and velocity calibration ($l_v = M/2$ corresponds to zero-velocity after compensation)

$$y''_S(n'|m,k) = W_\Omega(m,n_m)y'_S(n'|m,k) \times exp\left\{-j2\pi\left[\frac{f_0 + n'\Delta F/N}{f_0}\frac{m}{2}\right]\right\}(m = 0,1,\ldots,M-1). \quad (10)$$

- Step 3. For the data with the same sampling unit number k, pulse period number n', and different frame number m, carry out the non-integer sampling (l_v is an integer, but $(f_0 + n'\Delta F/N)l_v/f_0$ is not an integer) IDFT processing:

$$Y''_S(l_v|n',k) = \sum_{m=0}^{M-1} y''_S(n'|m,k) \times exp\left\{j2\pi\left[\frac{f_0 + n'\Delta F/N}{f_0} l_v \frac{m}{M}\right]\right\}(l_v = 0,1,\ldots,M-1). \tag{11}$$

- Step 4. Range calibration and multi-velocity-channel motion compensation.

$$\begin{aligned} Y'_S(l_v|n',k) &= Y''_S(l_v|n',k) \\ \times exp\Big\{-j2\pi[(f_0 + n'\Delta F/N)k\tau] &+ j2\pi\Big[\tfrac{f_0+n'\Delta F/N}{f_0}\big(l_v - \tfrac{M}{2}\big)\tfrac{n}{MN}\Big]\Big\}. \end{aligned} \tag{12}$$

- Step 5. Range-dimension pulse-compression processing on each velocity channel:

$$P(l_u, l_v, k) = \sum_{n'=0}^{N-1} Y'_S(l_v|n',k) \times exp\{-j2\pi n' l_u/N\}(l_u = 0,1,\ldots,N-1). \tag{13}$$

Obviously, the DFT processing of Equation (13) can be realized by FFT.

3.4.3. Intra-Frame Pure-RFH Mode

In this mode, $f_m = f_0$, $i_{mn} = i_n$, and the common phase factor $exp\{-j2\pi f_m l_u/\Delta F\}$ can be ignored, so Equation (4) can be written as follows:

$$\begin{aligned} P(l_u, l_v, k) = \textstyle\sum_{m=0}^{M-1}\sum_{n=0}^{N-1} W_\Omega(m,n) y_S(n|m,k) \times exp\{-j2\pi[(f_0 + \Delta f_d i_n)k\tau]\}\times \\ exp\Big\{-j2\pi\tfrac{\Delta f_d i_n}{\Delta F} l_u\Big\} \times exp\Big\{j2\pi\Big[\tfrac{f_0+\Delta f_d i_n}{f_0}\big(l_v - \tfrac{M}{2}\big)\tfrac{m}{M}\Big]\Big\} \times exp\Big\{j2\pi\Big[\tfrac{f_0+\Delta f_d i_n}{f_0}\big(l_v - \tfrac{M}{2}\big)\tfrac{n}{MN}\Big]\Big\}. \end{aligned} \tag{14}$$

Imaging Algorithm 3. Pipeline-parallel processing 2D matching filtering algorithm

- Step 1. Initialize, set $\left\{P^{(-1)}(l_u, l_v, k) = 0\middle|l_u = 0,1,\ldots,N-1; l_v = 0,1,\ldots,M-1\right\}$.
- Step 2. For $m = 0,1,\ldots,M-1$, perform the following iteration:

(A) Obtain the data of the k-th sampling unit of the m-th frame: $\{y_S(n|m,k)|n = 0,1,\ldots,N-1\}$;

(B) Windowing, range calibration, multi-velocity-channel motion compensation, and Doppler pre-processing:

$$y''_S(n|m,k,l_v) = W_\Omega(m,n) y_S(n|m,k)\psi(m,n,k,l_v), \tag{15}$$

where

$$\begin{aligned} \psi(m,n,k,l_v) = exp\Big\{-j2\pi[(f_0 + \Delta f_d i_n)k\tau] + j2\pi(f_0 + \Delta f_d i_n)\big(l_v - \tfrac{M}{2}\big)m/(f_0 M) + \\ j2\pi(f_0 + \Delta f_d i_n)\big(l_v - \tfrac{M}{2}\big)n/(f_0 MN)\Big\}. \end{aligned} \tag{16}$$

(C) Multi-velocity-channel pulse-compression processing

It can be obtained according to Equations (14) and (15) that

$$P(l_u, l_v, k) = \sum_{m=0}^{M-1}\sum_{n=0}^{N-1} y''_S(n|m,k,l_v) \times exp\left\{-j2\pi\frac{\Delta f_d i_n}{\Delta F} l_u\right\}.$$

The operation $Y''_S(l_u, l_v|m,k) = \sum_{n=0}^{N-1} y''_S(n|m,k,l_v) \times exp\{-j2\pi\Delta f_d i_n l_u/\Delta F\}$ is non-uniform sampling DFT. Even if the data are rearranged in the order of frequency points from small to large, the rearranged data are still non-uniform samples, which are difficult to be realized by FFT before inserting into uniform samples.

$$\left\{Y''_S(l_u, l_v|m,k) : l_u = 0,1,\ldots,N-1\right\} = NUDFT\left\{y''_S(n|m,k,l_v) : n = 0,1,\ldots,N-1\right\}. \quad (17)$$

(D) Current-frame Doppler accumulation processing

$$\begin{array}{c} P^{(m)}(l_u, l_v, k) = P^{(m-1)}(l_u, l_v, k) + Y''_S(l_u, l_v|m,k) \\ (l_u = 0,1,\ldots,N-1; l_v = 0,1,\ldots,M-1). \end{array} \quad (18)$$

Imaging Algorithm 4. Multi-velocity-channel pulse-compression based on data rearrangement/interpolation/range dimension FFT

In order to ensure the accuracy of interpolation, the interpolation processing must be done in each velocity channel. Due to the strong randomness of phase $(2f_0vT/c)n$ after data rearrangement, the rearranged data needs to be processed by Doppler pre-processing, so that the change range of signal phase $(2f_0\Delta vT/c)n$ on each speed channel is far less than one, where Δv is the width of velocity resolution unit. On each velocity channel, the effect of rearranged random phase term $(2f_0\Delta vT/c)n$ is negligible. The data accumulated by Doppler pre-processing are rearranged and interpolated on each velocity channel, which makes it is easy to ensure the interpolation accuracy.

- Step 1. Windowing and velocity calibration.

$$\begin{array}{c} y''_S(n|m,k) = W_\Omega(m, n_m) y_S(n|m,k) \times exp\left\{-j2\pi\left[\frac{f_0+\Delta f_d i_n}{f_0}\frac{m}{2}\right]\right\} \\ m = 0,1,\ldots,M-1. \end{array} \quad (19)$$

- Step 2. For the data with the same sampling unit number k, pulse period number n, and different frame number m, carry out the non-integer sampling (l_v is an integer, but $(f_0 + \Delta f_d i_n)l_v/f_0$ is not an integer) IDFT processing:

$$\begin{array}{c} Y''_S(l_v|n,k) = \sum\limits_{m=0}^{M-1} y''_S(n|m,k) \times exp\left\{j2\pi\left[\frac{f_0+\Delta f_d i_n}{f_0} l_v \frac{m}{M}\right]\right\} \\ (l_v = 0,1,\ldots,M-1). \end{array} \quad (20)$$

- Step 3. Range calibration, multi-velocity-channel motion compensation:

$$Y'_S(l_v|n,k) = Y''_S(l_v|n,k) \times exp\left\{-j2\pi[(f_0 + \Delta f_d i_n)k\tau] + j2\pi\left[\frac{f_0 + \Delta f_d i_n}{f_0}\left(l_v - \frac{M}{2}\right)\frac{n}{MN}\right]\right\}. \quad (21)$$

- Step 4. Data rearrangement and spline interpolation. In each sampling unit and each velocity channel, the data $Y'_S(l_v|n,k)$ is rearranged in the order of n corresponding frequency points from small to large to get the non-uniformly stepped-frequency sampled data. Then, the pre trained spline interpolation model is used to interpolate the non-uniformly sampled data to uniformly sampled data $Y_S(l_v|n',k)$, and the corresponding frequency points are unified as f_0, $f_0 + \Delta F/N$, $f_0 + 2\Delta F/N, \ldots, f_0 + (N-1)\Delta F/N$.
- Step 5. Perform range-dimension pulse-compression processing on each velocity channel

$$\begin{array}{c} P(l_u, l_v, k) = \sum\limits_{n'=0}^{N-1} Y_S(l_v|n',k) \times exp\{-j2\pi n' l_u/N\} \\ (l_u = 0,1,\ldots,N-1). \end{array} \quad (22)$$

Obviously, the DFT processing of Equation (22) can be realized by *FFT*.

3.4.4. Intra-Complex Frame Pure-RFH Mode

Imaging Algorithm 5. Pipeline-parallel processing 2D matched filtering algorithm

- Step 1. Initialize, set $\left\{P^{(-1)}(l_u, l_v, k) = 0 \middle| l_u = 0, 1, \ldots, N-1; l_v = 0, 1, \ldots, M-1\right\}$.
- Step 2. For $m = 0, 1, \ldots, M-1$, perform the following iteration:

(A) Obtain the data of the k-th sampling unit of the m-th frame: $\{y_S(n|m,k) | n = 0, 1, \ldots, N-1\}$;

(B) Windowing, range calibration, multi-speed channel motion compensation, and pre-processing of Doppler:

$$y''_S(n|m,k,l_v) = W_\Omega(m,n) y_S(n|m,k) \psi(m,n,k,l_v), \tag{23}$$

where

$$\begin{aligned}\psi(m,n,k,l_v) = exp\Big\{-j2\pi[(f_0 + \Delta f_d i_{mn})k\tau] + j2\pi(f_0 + \Delta f_d i_{mn})\big(l_v - \tfrac{M}{2}\big)m/(f_0 M) + \\ j2\pi(f_0 + \Delta f_d i_{mn})\big(l_v - \tfrac{M}{2}\big)n/(f_0 MN)\Big\}.\end{aligned} \tag{24}$$

(C) Multi-velocity-channel pulse-compression processing.

It can be obtained according to Equations (4) and (23) that

$$P(l_u, l_v, k) = \sum_{m=0}^{M-1} exp\left\{-j2\pi \frac{f_m}{\Delta F} l_u\right\} \sum_{n=0}^{N-1} y''_S(n|m,k,l_v) \times exp\left\{-j2\pi \frac{\Delta f_d i_{mn}}{\Delta F} l_u\right\}.$$

The operation $\sum_{n=0}^{N-1} y''_S(n|m,k,l_v) \times exp\{-j2\pi \Delta f_d i_{mn} l_u / \Delta F\}$ is non-uniform sampling DFT:

$$\left\{Y''_S(l_u, l_v|m,k) : l_u = 0, 1, \ldots, N-1\right\} = NUDFT\left\{y''_S(n|m,k,l_v) : n = 0, 1, \ldots, N-1\right\}. \tag{25}$$

(D) Current-frame Doppler accumulation processing

$$\begin{aligned}P^{(m)}(l_u, l_v, k) = P^{(m-1)}(l_u, l_v, k) + exp\left\{-j2\pi \tfrac{f_m}{\Delta F} l_u\right\} \times Y''_S(l_u, l_v|m,k) \\ (l_u = 0, 1, \ldots, N-1; l_v = 0, 1, \ldots, M-1).\end{aligned} \tag{26}$$

Obviously, $P^{(M-1)}(l_u, l_v, k) = P(l_u, l_v, k)$.

3.5. Real-Time 1D Hig-Range-Resolution (HRR) Imaging Algorithm of HPRF RFH Radar

The 1D HRR imaging algorithm is mainly used in the stage of target-tracking. As mentioned before, due to the special range-Doppler coupling effect of RFH synthetic wideband imaging radar in supersonic/hypersonic applications, the HRR imaging processing algorithm is essentially different from that of conventional stepped frequency synthetic wideband imaging radar, which can obtain HRR range profile by using the data of one frame. In RFH radar, if we want to obtain the range profile of the current frame, Doppler pre-processing should be executed before the current frame in order to suppress the special-range Doppler coupling effect.

The 1D HRR imaging algorithm is the same as the above online, fast, 2D imaging algorithms. The difference is that, in the target tracking stage, the target has been detected by the range-Doppler 2D image acquired in the searching stage, and the sampling unit and velocity channel where the target is located have been measured; then, the range-Doppler imaging processing algorithm only needs to be carried out on the velocity channel of the target and its adjacent channel. The multi-velocity-channel range-Doppler 2D imaging and splicing processing only need to be carried out in the corresponding sampling unit and the adjacent sampling units. That can significantly reduce the complexity of computation. In order to meet the requirements of high data rate in the tracking phase, the multi-velocity-channel range profile must be updated by frame, not by complex-frame as in 2D imaging. It is easy to design the iterative algorithm that can obtain the range profile of the next frame from that of the current frame by adding a few computations.

4. Experimental Results and Evaluation of Fast Imaging Algorithm

The diving angle of the aircraft was $\theta_M = -30°$ (the angle between the moving direction of the radar platform and the horizontal plane), the flight speed was V_M = 1750 m/s, and the height was H = 30 km. The pulse repetition period was T = 14 us, the carrier basic frequency was f_0 = 35 GHz, the frequency hopping interval was Δf = 6.25 MHz, and the pulse width was τ = 0.08 us. The number of periods in one frame was N = 16, the corresponding synthetic wideband was $N\Delta f$ = 100 MHz, the range resolution was $\Delta R = c/(2N\Delta f)$ = 1.5 m, the number of frames in one complex-frame was M = 32, the corresponding period of complex-frame was MNT = 7.168 ms, and the velocity resolution was $\Delta v = c/(2f_0MNT)$ = 0.6 m/s.

We assumed that the target was a stationary ship on the sea, and was equivalent to seven strong scattering centers. The parameters of each scattering center are shown in Table 1.

Table 1. Parameters of scattering centers of target.

Scattering Center Number	Initial Distance/m	Pitch Angle/(°)	Azimuth /(°)	Normalized Scattering Intensity
1	33,822	−59.883	10	1
2	33,828	−59.866	10	1
3	33,834	−59.848	10	1
4	33,840	−59.831	10	1
5	33,846	−59.814	10	1
6	33,852	−59.774	10	1
7	33,858	−59.747	10	1

According to the definition of resolution-cell in the above 2D GMF method, the theoretical coordinates of those seven scattering centers in spliced 2D range image are shown in Table 2.

Table 2. Theoretical positions of each scattering center of the target in image.

Scattering Center Number	Range Cell	Velocity Cell
1	148	20
2	153	20
3	157	21
4	161	21
5	165	22
6	169	23
7	173	23

Sea clutters were simulated according to average SCR of 8 dB. The intra-complex-frame pseudo-RFH mode of $i_{mn}\Delta f_d = i'_{mn}(\Delta F/N)$ and $f_m = f_0$ was investigated first, and the 2D range-Doppler normalized image obtained by 2D GMF is shown as Figure 3.

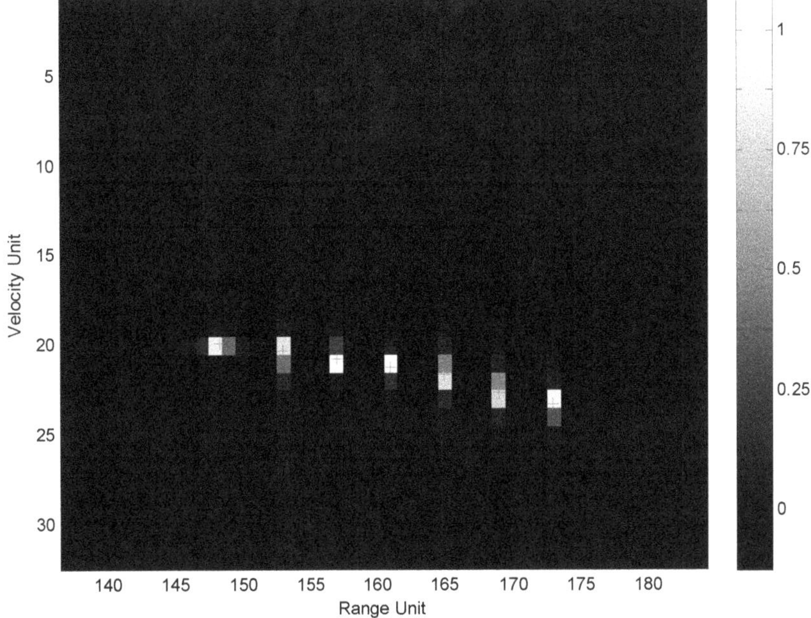

Figure 3. Image obtained by 2D GMF.

The other RFH modes were also used in the simulation experiments, and the imaging results were almost the same except the side-lobe level. The side-lobe level of pure-RFH mode was higher than that of pseudo-RFH mode. The simulation results show that the 2D GMF method can obtain high quality images in all modes of RFH. However, the computation-complexity of this method is much higher than that of 2D FFT. The 2D GMF was implemented by different pipeline-parallel algorithm according to different RFH mode in order to improve the real-time performance.

For Imaging Algorithm 1, because the rearranged data are uniformly sampled in frequency domain, the pulse-compression processing on each velocity channel can be realized by FFT, which improves the real-time performance of the imaging algorithm. The 2D GMF needs $2 \times (MN)^2$ complex multiplication operations, but the algorithm of FFT-based pulse-compression on multiple velocity channels needs only $M^2[3N + N\log(N)]$ complex multiplication operations. The total operation is reduced to $[\log(N) + 3]/N$ times of the original.

Imaging Algorithm 2 needs $MN[M + \log(N) + 3]$ complex multiplication operations. Compared with Algorithm 1, the computation is much less, but it is not convenient for pipeline processing, and the delay time of signal processing is not necessarily short.

Imaging Algorithm 3 needs $M^2N^2 + 2M^2N$ complex multiplication operations, which is more than for Imaging Algorithm 1 and Imaging Algorithm 2. However, the operations can be decomposed to each frame for execution, and the delay time of signal processing is short. It can meet the requirements of real-time imaging by multiprocessor parallel processing, and each processor is responsible for windowing, motion compensation, pulse compression, and Doppler accumulation of several velocity channels.

Regardless of the operations of low-order spline interpolation, Imaging Algorithm 4 needs $MN[M + \log(N) + 3]$ complex multiplication operations, which is equivalent to that of Algorithm 2. It is also inconvenient for pipeline processing, and the delay time of signal processing is not necessarily shorter than that of Algorithm 3.

Imaging Algorithm 5 needs $M^2N^2 + 3M^2N$ complex multiplication operations. Similar to Imaging Algorithm 3, the operations can be decomposed to each frame for execution, and the delay time of signal processing is short. It can meet the requirements of real-time imaging processing through multiprocessor parallel processing.

The imaging results of the five imaging algorithms are almost the same as that of the 2D GMF. Figure 4 shows the 2D normalized image obtained by Imaging Algorithm 1.

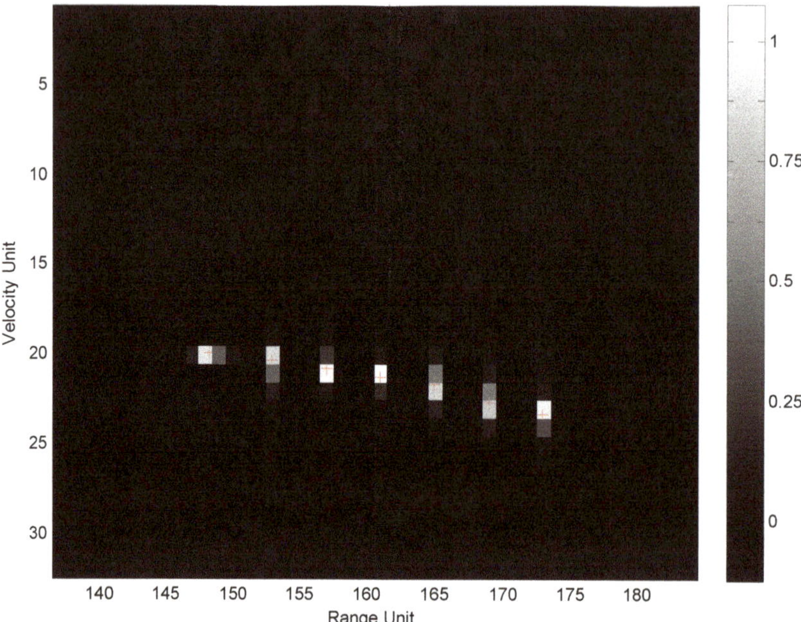

Figure 4. Image obtained by Imaging Algorithm 1.

For the intra-frame pseudo-RFH mode, as a contrast, Figure 5 shows the imaging results obtained by the traditional fractal-dimension imaging algorithm without Doppler pre-processing. In this method, the data $y_S(n|m,k)$ are re-arranged in the order of frequency points from small to large in each frame, and a new data sequence $y'_S(n|m,k)$ is obtained. Then, the range calibration is performed by multiplying the phase factor $exp\{-j2\pi[(f_0 + n\Delta F/N)k\tau]\}$, and the range profile $y'_S(l_u|m,k)$ of each frame is obtained by processing the calibrated data with N-point DFT (pulse compression). Finally, DFT (Doppler beam sharpening) processing of M-point is carried out for data of M frames in each range resolution cell l_u, and the 2D normalized image $P(l_u, l_v, k)$ is obtained.

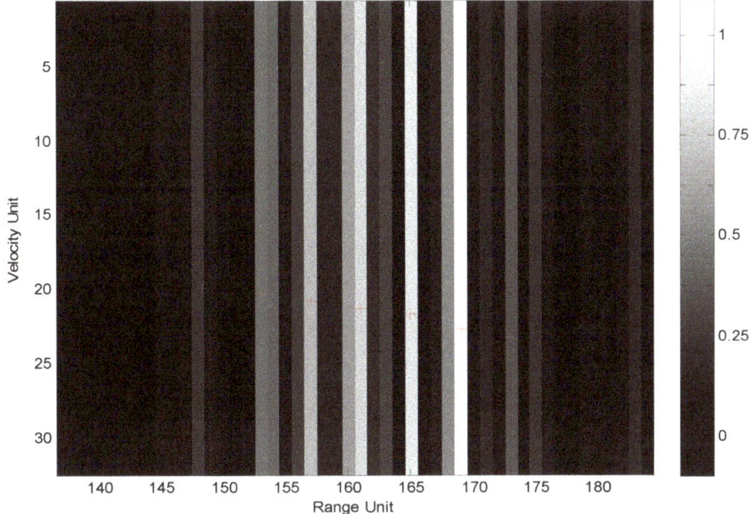

Figure 5. Image obtained by traditional algorithm without Doppler pre-processing.

Obviously, due to the special range-Doppler coupling effect caused by RFH and the lack of Doppler pre-processing in the pulse-compression process, the image is seriously divergent.

For the conventional SF mode, as a contrast, Figure 6 shows the normalized imaging results obtained by using the traditional fractal-dimension imaging algorithm.

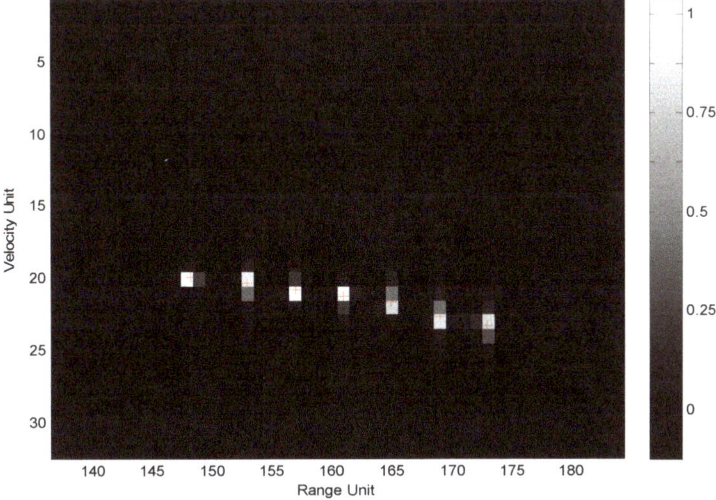

Figure 6. Image obtained by conventional stepped frequency (SF) mode.

Compared with Figures 3 and 4, it is shown that the RFH synthesis wideband system can achieve almost the same imaging effect as that of the conventional SF system, but the side-lobe level is higher in image of RFH system.

5. Discussion

In the case of RFH mode and a supersonic/hypersonic application, the conventional fractal-dimension 2D range-Doppler imaging algorithm makes it difficult to obtain high quality images because of the special range-Doppler coupling. Theoretical analysis and simulation results show that the proposed pipeline-parallel processing fast imaging algorithms based on Doppler pre-processing and 2D GMF can well suppress the above special range-Doppler coupling effect, avoid the divergence of the image in the range direction, and meet the requirements of real-time imaging. However, the side-lobe level of pure-RFH mode is higher than that of pseudo-RFH mode. Further, it is necessary to suppress the side-lobe level by optimizing the RFH pattern and the 2D window function $W_{\Omega}(m,n)$.

Author Contributions: Conceptualization, S.H. and X.W.; data curation, X.W.; formal analysis, S.H.; investigation, X.W.; methodology, S.H.; project administration, S.H.; software, X.W.; supervision, S.H.; validation, X.W.; visualization, X.W.; writing—original draft, X.W.; writing—review and editing, S.H.

Funding: This research received no external funding.

Conflicts of Interest: The authors declare no conflict of interest.

Appendix A

- Generation of 2D RFH patterns for different RFH modes

A chaos sequence is used to control the generation of the RFH pattern, which has the advantage of infinite periodicity, so it is difficult for the interceptor to decipher.

1. Intra-frame pseudo-RFH

Suppose we need to generate N stepped frequency points at $[0, \Delta F]$, where $N = \Delta F/\Delta f$ and $\Delta I = \Delta f/\Delta f_d$. Then we can use the following Bernoulli chaotic sequence to generate the RFH pattern.

$$\begin{aligned} x_n &= 2.01x_{n-1} \; mod \; 1 \\ i'_n &= INT[N \times x_n] \\ \Omega = \{i_{mn} &= i'_n \times \Delta I \big| n = 0,1,\ldots,N-1; m = 0,1,\ldots,M-1\}. \end{aligned} \quad (A1)$$

In the equation, the initial value x_{-1} can be randomly selected in the range of (0, 1).

For different complex frames, a new RFH pattern can be adopted through changing the initial value x_{-1}.

The Bernoulli chaotic pattern can also be used for the hopping of basic frequency f_0 between complex frames. If the bandwidth of the antenna is ΔF_t, and $I_{MAX} = \Delta F_t/\Delta f_d$, the basic frequency f_i of the i-th complex frames can be obtained as follows

$$\begin{aligned} z_i &= 2.01z_{i-1} \; mod \; 1 \\ j'_i &= INT[I_{MAX} \times z_i] \\ f_i &= j'_i \Delta f_d (i = 0,1,\ldots). \end{aligned} \quad (A2)$$

The initial value z_{-1} can be randomly selected in the range of (0,1).

Of course, constraints can be inserted in Equations (A1) and (A2). If $\left|i'_n - i'_{n-1}\right|$ or $\left|j'_i - j'_{i-1}\right|$ is less than a certain value, the iteration value will be discarded and the next iteration value will be selected.

2. Intra-complex-frame pseudo-RFH

In each complex frame, the frequency of the n-th pulse repetition period of the m-th frame is $f_m + i_{mn}\Delta f_d$ and the basic frequency f_m of each frame randomly changes within the allowable bandwidth of the radar antenna, and the N frequency points of each frame are randomly selected from the N uniformly stepped frequency points in $[0, \Delta F]$. Different frames use different frequency point orders. f_m and i_{mn} are generated as follows.

$$\begin{aligned} z_m &= 2.01 z_{m-1} \ mod\ 1 \\ j'_m &= INT[I_{MAX} \times z_m] \\ f_m &= j'_m \Delta f_d (i = 0, 1, \ldots, M-1) \end{aligned} \tag{A3}$$

$$\begin{aligned} x_{m,n} &= 2.01 x_{m,n-1} \ mod\ 1 \\ i'_{mn} &= INT[N \times x_{m,n}] \\ \Omega = \{f_m, i_{mn}\Delta f_d &= i'_{mn} \times \Delta I \Delta f_d | n = 0, 1, \ldots, N-1; m = 0, 1, \ldots, M-1\}, \end{aligned} \tag{A4}$$

where $x_{m,-1} = x_{m-1,M-1}$; the initial values z_{-1} and $x_{0,-1}$ can be randomly selected in the range of (0, 1). By changing the initial value, multiple groups of RFH patterns can be generated.

3. Intra-frame pure-RFH

In each complex frame, the frequency of the n-th pulse repetition period of the m-th frame is $f_0 + i_n \Delta f_d$. The basic frequency f_0 of each frame is constant, but f_0 can randomly jump within the allowable bandwidth of the radar antenna according to Equation (A2) between complex frames. Although the N frequency points of each frame are randomly selected in $[0, \Delta F]$, different frames adopt the same frequency points and order. If $I'_{MAX} = \Delta F / \Delta f_d$, the RFH pattern is generated as follows.

$$\begin{aligned} x_n &= 2.01 x_{n-1} \ mod\ 1 \\ i'_n &= INT\left[I'_{MAX} \times x_n\right] \\ \Omega = \{i_{mn}\Delta f_d &= i'_n \Delta f_d | n = 0, 1, \ldots, N-1; m = 0, 1, \ldots, M-1\}. \end{aligned} \tag{A5}$$

By changing the initial value of x_{-1}, multiple groups of RFH patterns can be generated.

4. Intra–complex-frames pure-RFH

The basic frequency f_m of each frame jumps within the frequency band allowed by the radar antenna. The N frequency points of each frame are randomly selected in the bandwidth of $[0, \Delta F]$, and different frames use different frequency points.

Algorithm: Generating two independent Bernoulli chaotic sequences $\{f_m\}$ and $\{i_{mn}\}$

$$\begin{aligned} z_m &= 2.01 z_{m-1} \ mod\ 1 \\ j'_m &= INT[I_{MAX} \times z_m] \\ f_m &= j'_m \Delta f_d (i = 0, 1, \ldots, M-1) \end{aligned} \tag{A6}$$

$$\begin{aligned} x_{m,n} &= 2.01 x_{m,n-1} \ mod\ 1 \\ i'_{mn} &= INT[I_{MAX} \times x_{m,n}] \\ \Omega = \{f_m, i_{mn}\Delta f_d &= i'_{mn} \Delta f_d | n = 0, 1, \ldots, N-1; m = 0, 1, \ldots, M-1\}, \end{aligned} \tag{A7}$$

where, the initial value z_0, $x_{0,-1}$ can be randomly selected in the range of (0, 1) and $x_{m,-1} = x_{m-1,M-1}$.

References

1. Liu, D. Research on the Correlation Technology of High Repetition Random Sequence Stepped Frequency Conversion Radar. Master's Thesis, Hunan University, Changsha, China, 2012.
2. Liao, Z. Research on Waveform Design and Imaging Method of Random Frequency Hopping Radar. Ph.D. Thesis, National University of Defense Technology, Changsha, China, 2018.
3. Zheng, L.; Zhang, S. A Novel Method of Translational Motion Compensation for Hopped-Frequency ISAR Imaging. In Proceedings of the Record of the IEEE 2000 International Radar Conference, Alexandria, VA, USA, 12 May 2000.

4. Gao, Z.; Xing, M.; Zhang, S.; Bao, Z. Experimental Results on ISAR Imaging with Stepped-Frequency Waveforms. *Electron. Lett.* **2009**, *45*, 77. [CrossRef]
5. Tian, J.; Xia, X.-G.; Cui, W.; Yang, G.; Wu, S.-L. A Coherent Integration Method via Radon-NUFrFT for Random PRI Radar. *IEEE Trans. Aerosp. Electron. Syst.* **2017**, *54*, 2101–2109. [CrossRef]
6. Liu, Q.; Nguyen, N. An Accurate Algorithm for Nonuniform Fast Fourier Transforms (NUFFT's). *IEEE Microw. Guided Wave Lett.* **2002**, *8*, 18–20. [CrossRef]
7. Li, S.; Sun, H.; Zhu, B.; Rong, L. Two-Dimensional NUFFT-Based Algorithm for Fast Near-Field Imaging. *IEEE Antennas Wirel. Propag. Lett.* **2010**, *9*, 814–817. [CrossRef]
8. Su, T.; Yu, F. A Family of Fast Hadamard-Fourier Transform Algorithms. *IEEE Signal Process. Lett.* **2012**, *19*, 583–586. [CrossRef]
9. Salishev, S.I. Regular Mixed-Radix DFT Matrix Factorization for Inplace FFT Accelerators. In Proceedings of the 2018 Systems of Signals Generating and Processing in the Field of on Board Communications, Moscow, Russia, 4–15 March 2018; pp. 1–6.
10. Fessler, J.A.; Sutton, B.P. A Min-Max Approach to the Multidimensional Nonuniform FFT: Application to Tomographic Image Reconstruction. In Proceedings of the International Conference on Image Processing, Thessaloniki, Greece, 7–10 October 2001; pp. 706–709.
11. Fessler, J.A.; Sutton, B.P. Nonuniform Fast Fourier Transforms Using Min-Max Interpolation. *IEEE Trans. Signal Process.* **2003**, *51*, 560–574. [CrossRef]
12. Candes, E.J.; Romberg, J.; Tao, T. Robust Uncertainty Principles: Exact Signal Reconstruction from Highly Incomplete Frequency Information. *IEEE Trans. Inf. Theory* **2006**, *52*, 489–509. [CrossRef]
13. Herman, M.A.; Strohmer, T. High-Resolution Radar via Compressed Sensing. *IEEE Trans. Signal Process.* **2009**, *57*, 2275–2284. [CrossRef]
14. Donoho, D.L. Compressed Sensing. *IEEE Trans. Inf. Theory* **2006**, *52*, 1289–1306. [CrossRef]
15. Liu, Z. Research on Signal Processing Method and Application of Random Modulation Radar Based on Compressed Sensing. Ph.D. Thesis, National University of Defense Technology, Changsha, China, 2013.
16. Liu, Y.; Meng, H.; Li, G.; Wang, X. Range-Velocity Estimation of Multiple Targets in Randomized Stepped-Frequency Radar. *Electron. Lett.* **2008**, *44*, 1032–1034. [CrossRef]
17. Huang, T.; Liu, Y.; Meng, H.; Wang, X. Cognitive Random Stepped Frequency Radar with Sparse Recovery. *IEEE Trans. Aerosp. Electron. Syst.* **2014**, *50*, 858–870. [CrossRef]
18. Liu, Z.; Wei, X.; Li, X. Low Sidelobe Robust Imaging in Random Frequency Hopping Wideband Radar Based on Compressed Sensing. *J. Cent. South Univ.* **2013**, *20*, 702–714. [CrossRef]
19. Yang, J.; Thompson, J.; Huang, X.; Tian, J.; Zhou, Z. Random-Frequency SAR Imaging Based on Compressed Sensing. *IEEE Trans. Geosci. Remote Sens.* **2013**, *51*, 983–994. [CrossRef]
20. Lazarov, A.; Minchev, C. ISAR Geometry, Signal Model, and Image Processing Algorithms. *IET Radar Sonar Navig.* **2017**, *11*, 1425–1434. [CrossRef]

Review

Compressed Sensing Radar Imaging: Fundamentals, Challenges, and Advances

Jungang Yang, Tian Jin, Chao Xiao * and Xiaotao Huang

College of Electronic Science and Technology, National University of Defense Technology, Changsha 410073, China

* Correspondence: xiaochao12@nudt.edu.cn

Received: 3 June 2019; Accepted: 11 July 2019; Published: 13 July 2019

Abstract: In recent years, sparsity-driven regularization and compressed sensing (CS)-based radar imaging methods have attracted significant attention. This paper provides an introduction to the fundamental concepts of this area. In addition, we will describe both sparsity-driven regularization and CS-based radar imaging methods, along with other approaches in a unified mathematical framework. This will provide readers with a systematic overview of radar imaging theories and methods from a clear mathematical viewpoint. The methods presented in this paper include the minimum variance unbiased estimation, least squares (LS) estimation, Bayesian maximum a posteriori (MAP) estimation, matched filtering, regularization, and CS reconstruction. The characteristics of these methods and their connections are also analyzed. Sparsity-driven regularization and CS based radar imaging methods represent an active research area; there are still many unsolved or open problems, such as the sampling scheme, computational complexity, sparse representation, influence of clutter, and model error compensation. We will summarize the challenges as well as recent advances related to these issues.

Keywords: radar imaging; synthetic aperture radar; compressed sensing; sparse reconstruction; regularization

1. Introduction

Radar imaging technique goes back to at least the 1950s. In the past 60 years, it has been stimulated by hardware performance, imaging theories, and signal processing technologies. Figure 1 shows the developmental history of radar imaging methods.

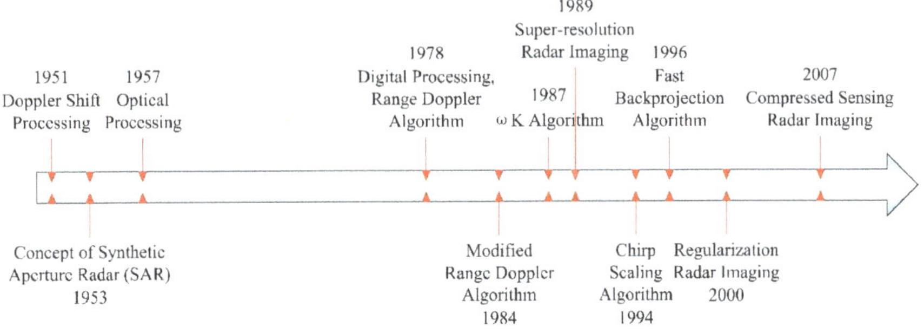

Figure 1. Developmental history of radar imaging methods.

Since the development of radar imaging techniques, the main theory that has been used has always been matched filtering [1–3]. Matched filtering is a linear process; it has the advantages of simplicity and stability. However, the drawbacks of the matched filtering method are also obvious. Since it does not exploit any prior information concerning the expected targets, its performance is limited by the signal bandwidth. It also requires a dense sampling to record the signals, according to the Shannon–Nyquist sampling theorem. Thus, the matched filtering method places significant requirements on the measured data, but only produces results with limited performance. As higher and higher imaging performance is demanded, the matched filtering method will struggle to meet the requirements.

Apart from the matched filtering framework, from a more generic mathematical viewpoint, radar imaging can be viewed as an inverse problem [4–7], whereby a spatial map of the scene is recovered using the measurements of the scattered electric field. The radar observation process is a Fredholm integral (F-I) equation of the first kind [8]. Due to observation limitations, such as limited bandwidth and limited observation angles, this inverse problem is usually ill-posed [9,10]. The classic least squares (LS) estimation method cannot solve such ill-posed inverse problems efficiently. The matched filtering method can be viewed as using an approximation to eliminate the irreversible or unstable term in the LS solution. This approximation leads to limited resolution and side-lobes in the results. Thus, matched filtering methods typically provide an image that blurs the details of the scene. Using proper models for the targets, super-resolution methods can improve the resolution of the imaging result [11,12].

Besides using approximation, the ill-posed inverse problem can be solved by another approach, i.e., adding an extra constraint to the LS formula and yielding a stable solution. This approach is called regularization [8]. In order to make the solution after regularization closer to the true value, the additional constraint should represent appropriately some prior knowledge. The regularization approach can also be explained by the Bayesian maximum a posteriori (MAP) estimation theory [6,13,14], which uses prior knowledge in a probabilistic way.

In the radar imaging scenario, imposing sparsity is one possible form of prior knowledge [15]. The advantages of the sparsity-driven regularization methods include increased image quality and robustness to limitations in data quantity. Compressed sensing (CS) refers to the use of under-sampled measurements to obtain the coefficients of a sparse expansion [16–20].

This paper summarizes the fundamentals, challenges and recent advances of sparse regularization and CS-based radar imaging methods. Using a unified mathematical model, we derive the best estimator (i.e., the minimum variance unbiased estimator), the LS estimator, the Bayesian MAP estimator, matched filtering, regularization, and CS reconstructions of the scene. The characteristics of these methods and their connections are also analyzed. Finally, we present some key challenges and recent advances in this area. These include the sampling scheme, the computational complexity, the sparse representation, the influence of clutter, and the model error compensation.

2. Mathematical Fundamentals of Radar Imaging

2.1. Radar Observation Model

In the continuous signal domain, under the Born approximation, the radar observation process can be denoted as [4]

$$s(\mathbf{r}) = \int A(\mathbf{r}, \mathbf{r}')g(\mathbf{r}')d\mathbf{r}' + n \tag{1}$$

where $s(\mathbf{r})$ denotes the observed data at the observation position of $\mathbf{r}$, $g(\mathbf{r}')$ denotes the reflectivity coefficient at $\mathbf{r}'$ in the scene, $A(\mathbf{r}, \mathbf{r}')$ denotes the system response from $\mathbf{r}'$ to $\mathbf{r}$, and n denotes noise.

Assuming the system is shift invariant, Equation (1) can be rewritten as

$$s(\mathbf{r}) = \int A(\mathbf{r} - \mathbf{r}')g(\mathbf{r}')d\mathbf{r}' + n \tag{2}$$

It can be seen that the radar observation model is a convolution process. Equation (1) is a Fredholm integral (F-I) equation of the first kind [8]. From a mathematical viewpoint, radar imaging can be viewed as the solution of the F-I equation—i.e., we want to recover $g(\mathbf{r})$ from the observed data $s(\mathbf{r})$ using the observation equation. Unfortunately, according to the theory of integral equations, solving the F-I equation is usually an ill-posed problem [8].

In practice, since digitization is commonly used, the observed data are discrete. Based on Equation (1), the discrete observation model can be written as

$$\mathbf{s} = \mathbf{A}\mathbf{g} + \mathbf{n} \tag{3}$$

where $\mathbf{s}$ is stacked from the samples of $s(\mathbf{r})$, $\mathbf{g}$ is stacked from the samples of $g(\mathbf{r}')$, $\mathbf{A}$ is formed from samples of $A(\mathbf{r}, \mathbf{r}')$, and $\mathbf{n}$ is the observation noise vector.

2.2. Best Linear Unbiased Estimate and Least Squares Estimate of the Scene

From the observation model shown in (3), radar imaging can be viewed as an estimation problem, in which the scene $\mathbf{g}$ is estimated based on the observed data $\mathbf{s}$ in a noisy environment. According to estimation theories, the minimum variance unbiased estimate is the "best" estimate in terms of estimation square error. From Equation (3), it can be seen that when the radar observation model is linear, the minimum variance unbiased estimate is the best linear unbiased estimate [13]—i.e., the expression of the best estimate of the scene is

$$\hat{\mathbf{g}} = \left(\mathbf{A}^H\mathbf{C}^{-1}\mathbf{A}\right)^{-1}\mathbf{A}^H\mathbf{C}^{-1}\mathbf{s} \tag{4}$$

where $\mathbf{C}$ is the covariance matrix of the noise term ($\mathbf{C} = E\left[\mathbf{n}\mathbf{n}^H\right]$).

In practice, a more tractable approach is LS estimation, which can be denoted as

$$\hat{\mathbf{g}} = \underset{\mathbf{g}}{\operatorname{argmin}} \, \|\mathbf{s}\text{-}\mathbf{A}\mathbf{g}\|_2^2 \tag{5}$$

Therefore, the LS estimate of the scene is

$$\hat{\mathbf{g}} = \left(\mathbf{A}^H\mathbf{A}\right)^{-1}\mathbf{A}^H\mathbf{s} \tag{6}$$

If $\mathbf{n}$ is white Gaussian noise, we have $\mathbf{C} = \sigma^2\mathbf{I}$, where $\mathbf{I}$ is the identity matrix. Under such condition, Equations (4) and (6) are the same. Therefore, the LS estimate will equal to the best estimate in white Gaussian noise [13].

If we want to use Equation (6) to calculate the best estimate of the scene, a prerequisite is that $(\mathbf{A}^H\mathbf{A})$ is invertible. However, in practice, this prerequisite is usually not satisfied, as discussed below. We assume that the size of $\mathbf{A}$ is $M \times N$, where M denotes the number of measurements and N denotes the number of unknown grid points. Then, the size of $(\mathbf{A}^H\mathbf{A})$ is $N \times N$.

One case is that $M < N$, i.e., the number of measurements is less than the unknown variables. CS is a typical example of this case. In such a case, $\operatorname{rank}(\mathbf{A}^H\mathbf{A}) = \operatorname{rank}(\mathbf{A}) \leq M < N$, i.e., $(\mathbf{A}^H\mathbf{A})$ is irreversible.

In the above case, it can be seen that due to limited number of measurements, $(\mathbf{A}^H\mathbf{A})$ is irreversible. Is it possible to make $(\mathbf{A}^H\mathbf{A})$ invertible by increasing the number of measurements (i.e., make $M > N$. As mentioned previously, due to physical limitations, such as limited bandwidth and limited observation angles, if we take more measurements, the interval between the adjacent measurements will be smaller. Thus, the coherence between the adjacent columns in $\mathbf{A}$ will increase. Consequently, $(\mathbf{A}^H\mathbf{A})^{-1}$ will probably be ill-conditioned.

In summary, the LS solution usually contains irreversible or ill-posed terms. This problem is inherent, and is derived from the property of the F-I equation of the first kind [8].

2.3. Matched Filtering Method

Examining Equation (6), it can be seen that the irreversible or ill-posed term is $(\mathbf{A}^H\mathbf{A})^{-1}$. We can multiply $(\mathbf{A}^H\mathbf{A})$ in the left side of Equation (6) to eliminate $(\mathbf{A}^H\mathbf{A})^{-1}$. In this way, we can avoid explicitly calculating the nonexistent or unstable term $(\mathbf{A}^H\mathbf{A})^{-1}$. This leads to the matched filtering method, which can be denoted as

$$\hat{\mathbf{g}}_{\mathrm{MF}} = (\mathbf{A}^H\mathbf{A})\hat{\mathbf{g}} = \mathbf{A}^H\mathbf{s} \tag{7}$$

Equation (7) can be viewed as multiplying the best estimate of the scene with $(\mathbf{A}^H\mathbf{A})$. The matrix $(\mathbf{A}^H\mathbf{A})$ is the autocorrelation of the system response, which usually has a sinc pulse shape [1,21]. The matched filtering result can be viewed as the convolution of the best estimate of the scene and the sinc function. A point target will be spread, and side-lobes will also appear in the matched filtering result [21]. This implies that the matched filtering method can only provide an image that blurs the details of the scene. The matched filtering method has a limited resolution, which depends on the autocorrelation of the system response [1].

Figure 2 shows an example of the matched filtering method. Six point targets are set in the scene. It can be seen that the matched filtering result is the convolution of the targets and the autocorrelation of the system response. As a result, an idea point target is spread into a sinc waveform. Consequently, targets will interfere with each other, and two closely spaced targets may not be resolved in the matched filtering result.

Equation (7) is the original form of the matched filtering equation. In practice, in order to reduce the computational cost and make it more convenient for implementation, some transformations and approximations are usually adopted for Equation (7). Equation (7) can represent many widely used imaging algorithms, such as backprojection algorithms, range Doppler algorithms, chirp scaling algorithms, and ωK algorithms [1].

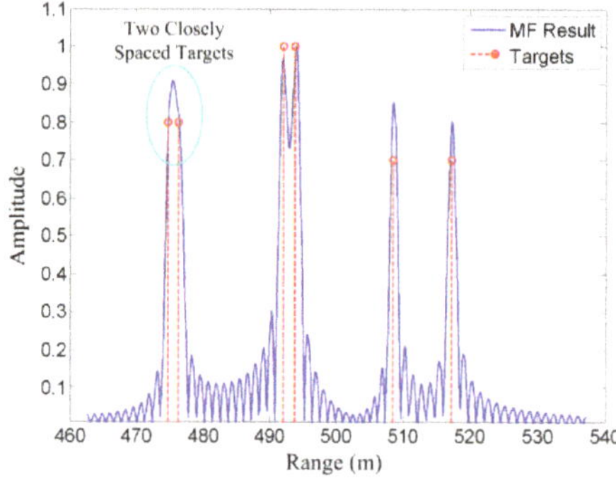

Figure 2. Matched filtering example. Two closely spaced targets cannot be resolved.

2.4. Regularization Method

Examining the LS formula (Equation (5)), it can be seen that it only relies on the observed data. In order to make the ill-posed inverse problem become well-posed, we can add an extra constraint to the LS formula [8–10]. This leads to the regularization method, which can be denoted as

$$\hat{\mathbf{g}} = \underset{\mathbf{g}}{\operatorname{argmin}} \{\|\mathbf{s}\text{-}\mathbf{A}\mathbf{g}\|_2^2 + \lambda L(\mathbf{g})\} \tag{8}$$

where λ is the regularization parameter and $L(\mathbf{g})$ is the added penalty function. In order to make the solution of Equation (8) closer to the true value, $L(\mathbf{g})$ should represent appropriate prior knowledge for the problem.

A typical choice of $L(\mathbf{g})$ is

$$L(\mathbf{g}) = ||\mathbf{g}||_p^p \tag{9}$$

where $||\cdot||_p$ denotes the ℓ_p-norm, i.e.,

$$||\mathbf{g}||_p = \begin{cases} \left(\sum_{i=1}^{N} |g_i|^p\right)^{1/p} & p > 0 \\ \text{Number of nonzero elements in } \mathbf{g} & p = 0 \end{cases} \tag{10}$$

Then, Equation (8) can be rewritten as

$$\hat{\mathbf{g}} = \underset{\mathbf{g}}{\operatorname{argmin}} \left\{ ||\mathbf{s}\text{-}\mathbf{A}\mathbf{g}||_2^2 + \lambda ||\mathbf{g}||_p^p \right\} \tag{11}$$

The choice of p can control the result of the regularization method. If we want to enforce sparsity in the result, we should choose p in the range $0 \le p \le 1$ [16,17]. For $p = 1$, Equation (11) can be compared to the Lasso solution of the CS type methods [16]. Equation (11) can be solved by gradient search algorithms, such as the Newton iteration [22].

2.5. Bayesian Maximum a Posteriori Estimation

It should be noted that in Equation (11), the added constraint term $\lambda ||\mathbf{g}||_p^p$ represents prior knowledge [17,23]. Another prior knowledge-based estimation method is Bayes theory. The main idea behind the Bayesian estimation framework is to account explicitly for the errors, and also for incomplete prior knowledge. Assuming that the noise $\mathbf{n}$ in Equation (3) is white and Gaussian, we have

$$p(\mathbf{n}) \propto \exp\left\{-\frac{1}{2\sigma^2}||\mathbf{n}||_2^2\right\} \tag{12}$$

where σ^2 is the noise variance. Then we obtain the expression of likelihood

$$p(\mathbf{s}|\mathbf{g}) \propto \exp\left\{-\frac{1}{2\sigma^2}||\mathbf{s} - \mathbf{g}||_2^2\right\} \tag{13}$$

We assume that the scene has a prior probability density function, as

$$p(\mathbf{g}) \propto \exp\left\{-\alpha ||\mathbf{g}||_p^p\right\} \tag{14}$$

If $0 \le p \le 1$, the magnitude of the scene is more likely to concentrate around zero, which implies that the scene is sparse. For a review on sparsity enforcing priors for the Bayesian estimation approach, the reader can refer to [6].

Using the prior probability density of $\mathbf{g}$ shown in (14), and according to the Bayes rule, we obtain

$$p(\mathbf{g}|\mathbf{s}) = \frac{p(\mathbf{s}|\mathbf{g})p(\mathbf{g})}{p(\mathbf{s})} \propto \frac{1}{p(\mathbf{s})}\exp\left\{-\frac{1}{2\sigma^2}||\mathbf{s} - \mathbf{g}||_2^2 - \alpha||\mathbf{g}||_p^p\right\} \tag{15}$$

Then the MAP estimate can be obtained easily as

$$\hat{\mathbf{g}} = \underset{\mathbf{g}}{\operatorname{argmax}}\, p(\mathbf{g}|\mathbf{s}) = \underset{\mathbf{g}}{\operatorname{argmin}}\, ||\mathbf{s} - \mathbf{g}||_2^2 + 2\sigma^2\alpha||\mathbf{g}||_p^p \tag{16}$$

Comparing Equations (11) and (16), it can be seen that when $\lambda = 2\sigma^2\alpha$, these two equations are equivalent, i.e., the regularization method is equivalent to Bayesian MAP estimation.

2.6. Compressed Sensing Method

For the observation model shown in Equation (3), if the scene (i.e., **g**) is sparse, according to CS theory, it can be stably reconstructed using reduced data samples. The reconstruction method can be written as [16,17]

$$\hat{\mathbf{g}} = \underset{\mathbf{g}}{\operatorname{argmin}} \, \|\mathbf{g}\|_0 \text{ s.t. } \|\mathbf{s}\text{-}\mathbf{A}\mathbf{g}\|_2^2 < \varepsilon \tag{17}$$

where s.t. means subject to and ε denotes the allowed data error in the reconstruction process.

Equation (17) is NP-hard and computationally difficult to solve [17]. Matching pursuit is an approximate method for obtaining an ℓ_0 sparse solution. In CS theory, a more tractable approach is taking the ℓ_1-norm instead of the ℓ_0-norm, which is called the ℓ_1 relaxation:

$$\hat{\mathbf{g}} = \underset{\mathbf{g}}{\operatorname{argmin}} \, \|\mathbf{g}\|_1 \text{ s.t. } \|\mathbf{s}\text{-}\mathbf{A}\mathbf{g}\|_2^2 < \varepsilon \tag{18}$$

If **g** is sparse and **A** satisfies some specific conditions, Equations (18) and (17) will have the same solution, and this solution is the exact or approximate recovery of **g** [16,17]. Equation (18) can be solved using convex programming, which is more tractable than the original ℓ_0-norm minimum problem. Unlike the matched filtering method, CS method does not have an exact or pre-defined resolution, since it is a non-linear method. Generally, the resolution capability of the CS method is much better than the matched filtering method if the targets are sparse.

Figure 3 shows an example of compressed sensing. The simulated scene is the same as the matched filtering example shown in Figure 2. Only 1/20 signal samples are used for the CS reconstruction. It can be seen that the two closely spaced targets are well resolved. This implies that the CS method can obtain better results using less data than the matched filtering method. The reason is that prior information concerning signal sparsity is utilized in the CS model.

Equation (18) is a constrained optimization problem. According to the Lagrange theory, it can be transformed into an unconstrained optimization problem, which will have the same form as Equation (11). For appropriate choices of λ and $p = 1$, Equations (11) and (18) will be equivalent [16,17]. This implies that CS is a special case of the regularization method.

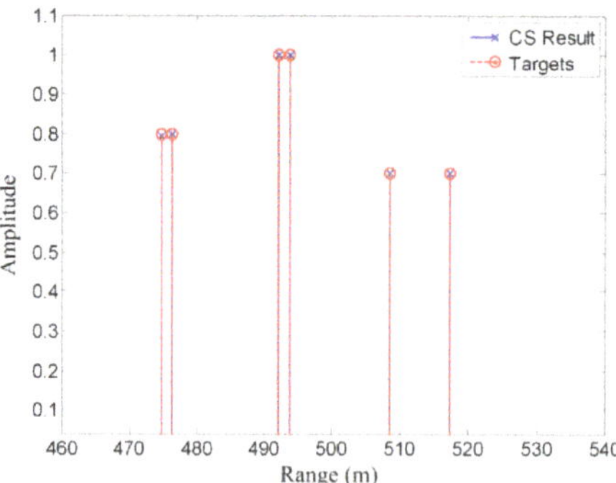

Figure 3. Compressed sensing example; closely spaced targets are well resolved.

2.7. Summary of Radar Imaging Methods

The above subsections introduced the LS estimator, matched filtering, regularization methods, Bayesian MAP estimation, and the CS method. In this subsection, we will summarize these methods and analyze their connections.

Table 1 lists the main characteristics and describes some connections between these imaging methods. The LS estimation only relies on the observed data, and cannot solve the ill-posed radar imaging problem efficiently. The matched filtering method can be viewed as using an approximation to avoid the ill-posed term in the LS solution. The regularization method, Bayesian MAP estimation, and the CS method exploit prior knowledge concerning the targets in addition to the observed data, and they are equivalent in some cases.

Table 1 also shows the equivalent geometric illustration for each method in $\mathbb{R}^2$. The observation equation can only confine the solution to a hyperplane (which becomes a line in $\mathbb{R}^2$), but cannot reliably produce a certain solution [17,23]. The other methods aim at obtaining a stable solution close to the true value, using some modifications that represent prior knowledge concerning the targets.

Figure 4 shows the block diagram and the relationship of the radar imaging methods. All of the radar imaging methods can be divided into two branches. The first branch does not use the prior information of the targets or scene, and it leads to the linear imaging methods; the most typical and widely used one in this branch is matched filtering. Another branch uses the prior information of the targets or scene. This leads to the non-linear methods. The most recently developed methods, including regularization methods, Bayesian methods, and CS methods belong to this branch.

Table 1. Characteristics and connections of radar imaging methods.

Radar Observation Model: s=Ag+n s: Observed Data, A: Measurement Matrix, g: Scene, n			
Imaging Methods	Mathematical Model	Characteristics	Equivalent Geometric Illustration
Least Squares (LS) Estimation	$\hat{\mathbf{g}} = \underset{\mathbf{g}}{\operatorname{argmin}} \lVert\mathbf{s}\text{-}\mathbf{A}\mathbf{g}\rVert_2^2$	$\hat{\mathbf{g}} = (\mathbf{A}^H\mathbf{A})^{-1}\mathbf{A}^H\mathbf{s}$, $(\mathbf{A}^H\mathbf{A})^{-1}$ is usually ill-posed or nonexistent, cannot obtain a stable solution [9,13].	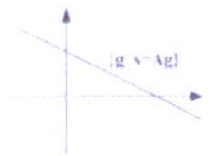
Matched Filtering	$\hat{\mathbf{g}} = \mathbf{A}^H\mathbf{s}$	Avoids the ill-posed term in the LS solution, but the resolution is limited by the system bandwidth, and side-lobes will appear in the final image [4,21].	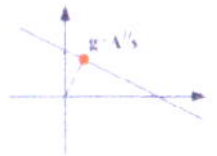
Range Doppler, Chirp Scaling, ωK, etc.	Approximations and transformations of $\mathbf{A}^H\mathbf{s}$	Approximations and transformations of the original matched filtering, in order to reduce the computational cost and make it more convenient to implement in practice [1].	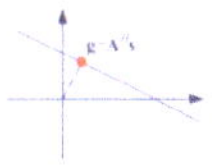
Regularization Method	$\hat{\mathbf{g}} = \underset{\mathbf{g}}{\operatorname{argmin}} \lVert\mathbf{s}\text{-}\mathbf{A}\mathbf{g}\rVert_2^2 + \lambda L(\mathbf{g})$	Add an extra constraint to the LS formula, so that the ill-posed inverse problem becomes well-posed. If the added constraint is chosen appropriately, the result will be better than that for matched filtering [8,9].	Depends on the expression of $L(\mathbf{g})$.
Sparsity-Driven Regularization	$\hat{\mathbf{g}} = \underset{\mathbf{g}}{\operatorname{argmin}} \lVert\mathbf{s}\text{-}\mathbf{A}\mathbf{g}\rVert_2^2 + \lambda\lVert\mathbf{g}\rVert_p^p$ $0 \le p \le 1$	Choose $L(\mathbf{g})$ as the ℓ_p-norm $(0 \le p \le 1)$, in order to obtain sparse reconstruction result [6,23].	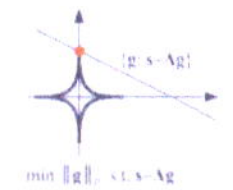
Bayesian MAP Estimation	$p(\mathbf{g}) \propto \exp\{-\alpha\lVert\mathbf{g}\rVert_p^p\}$ $\hat{\mathbf{g}} = \underset{\mathbf{g}}{\operatorname{argmin}} \lVert\mathbf{s}\text{-}\mathbf{A}\mathbf{g}\rVert_2^2 + 2\sigma^2\alpha\lVert\mathbf{g}\rVert_p^p$	For $2\sigma^2\alpha = \lambda$, the MAP estimation will be equivalent to the sparsity-driven regularization method [6,14].	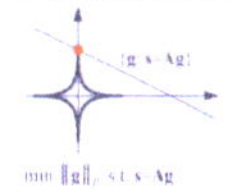
Compressed Sensing (CS) Method	$\hat{\mathbf{g}} = \underset{\mathbf{g}}{\operatorname{argmin}} \lVert\mathbf{g}\rVert_0$ s.t. $\lVert\mathbf{s}\text{-}\mathbf{A}\mathbf{g}\rVert_2^2 < \varepsilon$ or $\hat{\mathbf{g}} = \underset{\mathbf{g}}{\operatorname{argmin}} \lVert\mathbf{g}\rVert_1$ s.t. $\lVert\mathbf{s}\text{-}\mathbf{A}\mathbf{g}\rVert_2^2 < \varepsilon$	For an appropriate choice of λ, the CS method will be equivalent to the sparsity-driven regularization method [17,23].	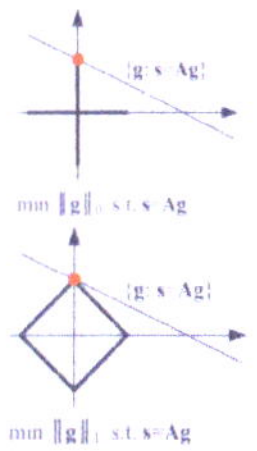

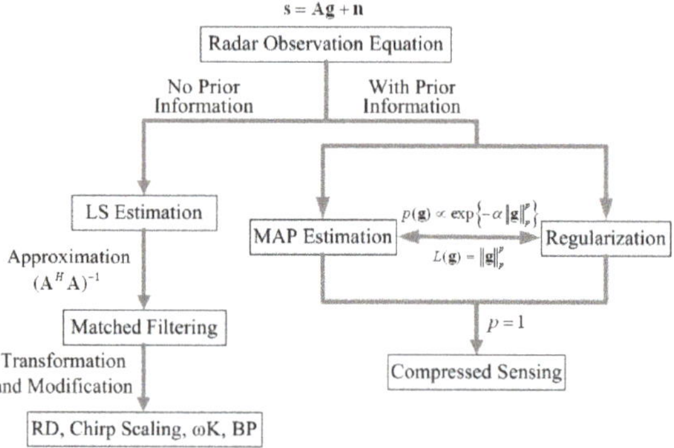

Figure 4. Block diagram and relationship of the radar imaging methods.

3. Challenges and Advances in Compressed Sensing-Based Radar Imaging

The use of regularization methods in radar imaging goes back at least to the year 2000 [21,24]. Since the CS theory was proposed in 2006, it has been explored for a wide range of radar [25–33] and radar imaging applications [4,34–38], including synthetic aperture radar (SAR) [39–42], inverse SAR (ISAR) [43–45], tomographic SAR [46–51], three-dimensional (3D) SAR [52–54], SAR ground moving target indication (SAR/GMTI) [55–61], ground penetrating radar (GPR) [62–64], and through-the-wall radar (TWR) [65–67]. In this paper, we will focus on two-dimensional (2D) imaging radar systems, i.e., SAR, GPR, and TWR.

After several years of development, although many interesting ideas have been presented in this area, there still exist a number of challenges, both in theory and practice [68]. The state of the art in this area has not yet reached the stage of practical application. We will present some challenges as well as recent advances in this part of the paper.

3.1. Sampling Scheme

CS usually involves random under-sampling [16,17]. A widely used waveform in traditional radar imaging is the linear frequency modulated (LFM) waveform. If we adopt the LFM waveform in CS-based radar imaging, a random sampling analog to digital (A/D) converter is needed, which is not easily realized in practice. This will require extra hardware components, which means that LFM waveforms are not ideally suited for CS.

Recently, many researchers have found that the stepped frequency waveform is much more suitable for CS than the LFM waveform [35,62,63,66,69]. Sparse and discrete frequencies are more convenient for hardware implementation. For a CS-based radar imaging system, a stepped frequency waveform may be the preferred choice. In practical application, a set of adjustable pseudorandom numbers can be generated to select the frequency points in the stepped frequencies. In this way, randomly generated frequencies, i.e., random and sparse measurement, can be realized, and the CS-based imaging model can be implemented.

Figures 5 and 6 show an example for CS-based stepped frequency radar imaging. The main equipment in the experimental system is a vector network analyzer (VNA). The experiment is carried out in a non-reflective microwave chamber. Five targets in the scene are shown in Figure 5. Figure 6a shows the backprojection result, using the fully sampled data (81 azimuth measurements × 2001 frequencies). Figure 6b shows the CS reconstruction result using under-sampled data (27 azimuth measurements × 128 frequencies). Considering the aspects of resolution and sidelobe levels, the CS reconstruction result is even better than the backprojection result, although it uses less sampled data. The reason is that prior information concerning signal sparsity is used in the CS model, while the backprojection method uses no prior information.

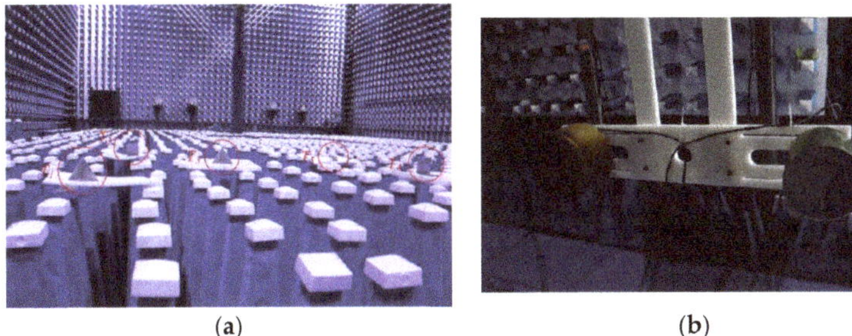

Figure 5. Experimental scene for CS-based stepped frequency radar imaging. (**a**) Five reflectors in the microwave chamber. (**b**) Transmitter and receiver antennas.

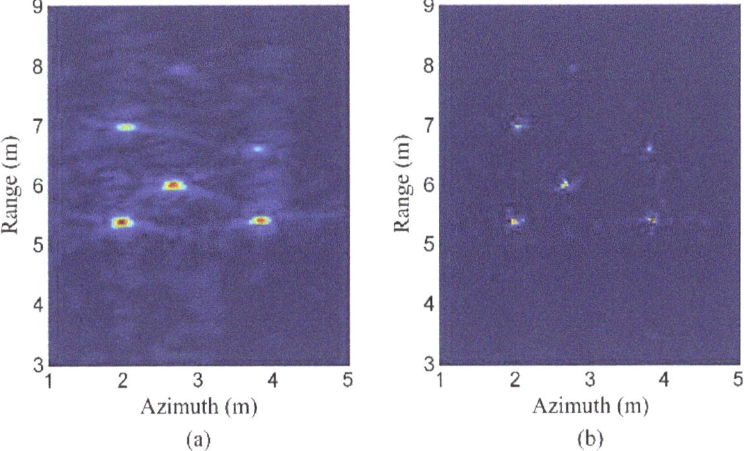

Figure 6. (**a**) Backprojection result of full data (81 azimuth measurements × 2001 frequencies). (**b**) CS result of under-sampled data (27 azimuth measurements × 128 frequencies).

3.2. Computational Complexity

In the regularization or CS model for a 2D radar imaging system, the 2D observed data and the 2D scene grid are both stacked into column vectors. This will lead to a huge size measurement matrix. For example, the original fully sampled data are 2048 × 2540 points (azimuth × range); if a 512 × 512 pixel image is reconstructed from a reduced sampling data consist of 256 × 256 points. Then the size of the matrix **A** is 65,536 × 262,144. Since regularization or CS reconstruction is a non-linear process, such a large measurement matrix will result in a huge computational burden for image reconstruction. In addition, the total memory to access the measurement matrix is 128 gigabytes (assuming float point and complex numbers are used). This is a too much memory space for normal desktop computers. Considering that data size is usually larger than the above example in practice, it is difficult for conventional methods to reconstruct a moderate-size scene by using normal computers.

A common idea for reducing computational complexity and memory occupancy is to split big data into sets of small data [70]. Based on this thought, a segmented reconstruction method for CS based SAR imaging has been proposed [71]. In this method, the whole scene is split into a set of small subscenes. Since the computational complexity is non-linear to the data size, the reconstruction time can be reduced significantly. The sensing matrices for the method proposed in [71] are much smaller than those for the conventional method. Therefore, the method also needs much less memory. Due to the short reconstruction time and lower memory requirement of the method proposed in [71], reconstructing a moderate-size scene in a short time is no longer a difficult task. The processing steps of the segmented reconstruction method are shown in Figure 7.

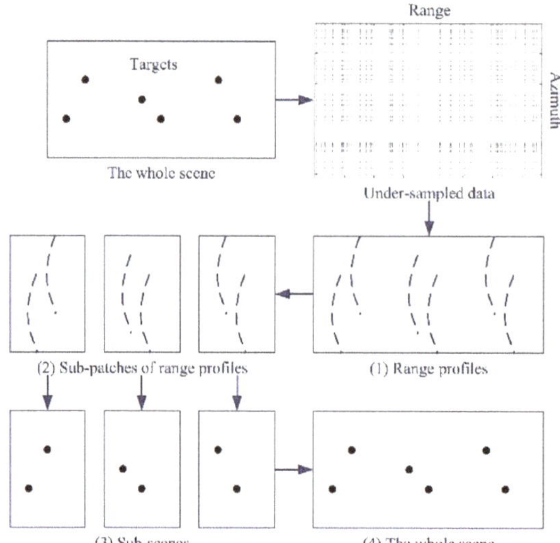

Figure 7. Processing steps of the segmented reconstruction method for CS-based synthetic aperture radar (SAR) imaging (taken from [71]).

Figures 8 and 9 show an example of the segmented reconstruction method [71]. Figure 8 shows the experimental scene of an airborne SAR system, which contains six trihedral reflectors. Figure 9a shows the conventional CS reconstruction result, where the reconstruction time is 44,032 s (12 h 14 min). The whole scene is split into five segments, and Figure 9b shows the segmented reconstruction result, where the reconstruction time is now reduced to 1498 s (25 min). It can be seen that, using the segmented reconstruction method, the reconstruction time is significantly reduced, while the reconstruction precision is nearly the same.

(**a**) Sight A of the scene. (**b**) Sight B of the scene.

Figure 8. Trihedral reflectors in the scene. Trihedral reflectors 1–4 are large, and trihedral reflectors 5 and 6 are small (taken from [71]).

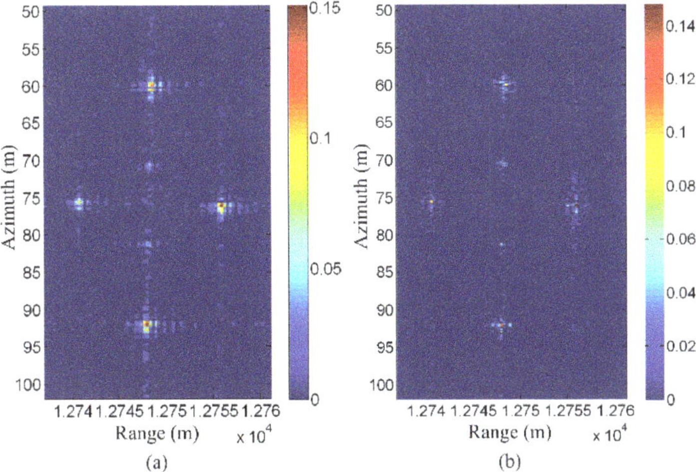

Figure 9. (**a**) Conventional CS reconstruction result (reconstruction time = 44,032 s). (**b**) Segmented reconstruction result (reconstruction time = 1498 s) (taken from [71]).

3.3. Sparsity and Sparse Representation

Sparsity of the scene is an essential requirement for sparse regularization or CS methods. For an SAR scene, an extended scene is usually not sparse in itself (not sparse in the canonical basis), except for the case of a few dominant scatterers in a low reflective background [35]. Therefore, a sparse representation is needed to use a sparsity-driven method.

CS-based optical imaging has successfully used sparse representations [72]. However, radar imaging involves complex-valued quantities; the raw data and the imaging result are both complex-valued. Since the phase of the scene are potentially random, it is very difficult to find a transform basis to sparsify a complex-valued and extended scene [73,74].

Structured dictionaries and dictionary learning ideas are proposed in [75] and [76], respectively. An alternative approach is to handle the magnitude and phase separately [41]. Although the phase of the scene is potentially random, the magnitude of the scene usually has better sparse characteristics. However, this approach has a much higher computational complexity than standard CS reconstruction. Another method investigates physical scattering behavior [4,77]. For example, a car can be represented as the superposition of responses from plate and dihedral shapes.

Figure 10 shows a simulation example for an extended and complex-valued scene. There are two extended objects in the scene, one of which has a round shape while the other has a rectangular shape. Both the two objects have random phases associated with them. It can be seen that the DCT (Discrete Cosine Transform) results of the magnitude are sparse.

Figure 11a shows the result of matched filtering. Since the random phase leads to speckle, it can be seen that although the scene has a smooth shape, the matched filtering result has obvious fluctuation. Figure 11b shows the result of conventional CS reconstruction without sparse representation. The reconstruction algorithm is SPGL1 [78]. Since the scene is not sparse in the canonical basis, the reconstruction is not accurate. Figure 11c shows the result of the method using a magnitude sparse representation [41]; it can be seen that the reconstruction result is much better than Figure 11a,b. Figure 11d shows the result of the method using the improved magnitude sparse representation method proposed in [79]. In the proposed method, besides the sparsity, the real-valued information of the magnitude and the coefficient distribution of the sparse representation are also utilized. It can be seen that both the shape and speckle are further improved.

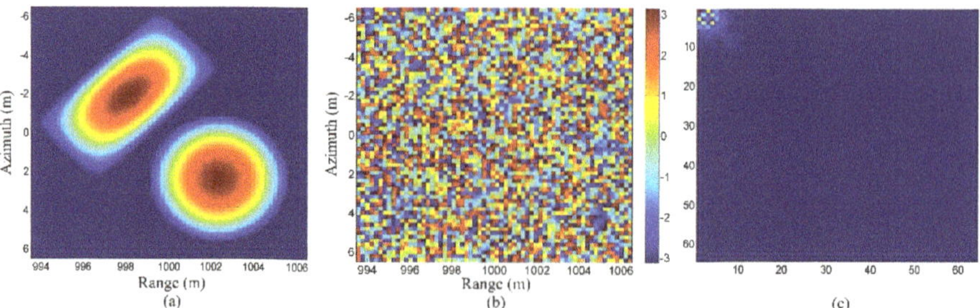

Figure 10. (**a**) Magnitude of the scene, (**b**) phase of the scene, (**c**) DCT result of the magnitude (taken from [79]).

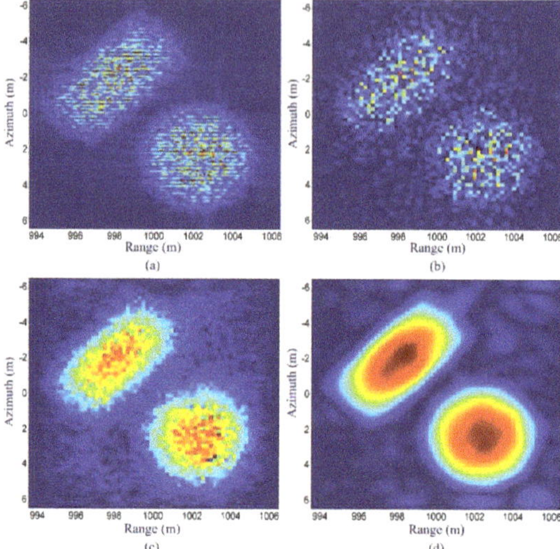

Figure 11. Simulation results: (**a**) matched filtering result, (**b**) conventional CS reconstruction result without sparse representation, (**c**) result of the method with magnitude sparse representation, and (**d**) result of the method with improved magnitude sparse representation (taken from [79]).

Figure 12 shows the real data results. The raw data is acquired by an airborne SAR system. Figure 12 contains a scene of farmland with trellises. The reflectivity from the trellises is very strong. From the real data result, it can be seen that CS with the improved magnitude sparse representation method can produce an image with less speckle and clearer edges of different regions than the previous methods.

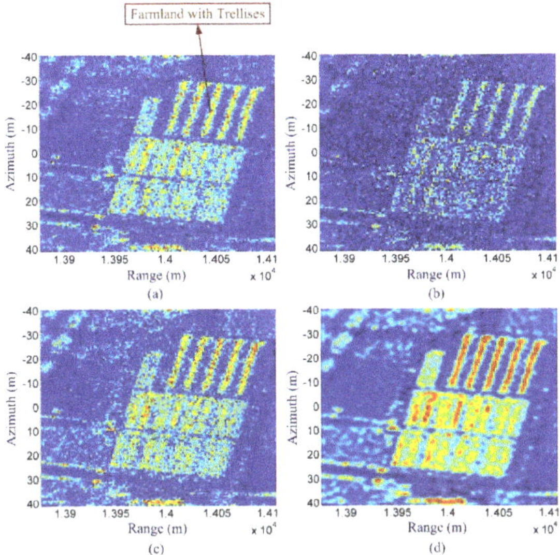

Figure 12. Real data reconstruction results (scene of farmland with trellises): (**a**) matched filtering result (full data), (**b**) conventional CS reconstruction result without sparse representation, (**c**) result of CS with magnitude sparse representation, and (**d**) result of CS with improved magnitude sparse representation (taken from [79]).

3.4. Influence of Clutter

Another practical case is when the targets of interest are sparse, but there also exists clutter in the scene. Clutter arises from reflections within the scene, so the image may no longer be sparse if significant clutter returns are present. Typical examples include GPR and TWR imaging. The interesting targets, such as landmines and humans, are usually sparse, but they are often buried in the ground surface clutter and wall clutter.

Some methods have been proposed to remove the ground surface clutter and wall clutter for downward-looking GPR and TWR [64,65]. These methods are effective in cases when the clutter is concentrated in a fixed range cell or limited to several range cells.

Another scenario is TWR/SAR imaging of moving targets. A sparsity-driven change detection method is proposed in [67]. The stationary targets and clutter are removed via change detection, and then CS reconstruction is applied to the resulting sparse scene. In [55], a SAR/GMTI method using distributed CS is proposed, which can cope with the non-sparse stationary clutter.

A more difficult case is when both the targets and clutter are stationary, and the clutter is distributed over the whole scene. Forward-looking GPR may fall into this category. Figure 13 shows a real data example for this case. In such a scenario, shrubs and rocks above the ground surface may cause strong azimuth clutter. Short range clutter is usually also strong, due to the large grazing angle and short range. Besides the strong clutter far away from the target (landmine), there is also ground surface clutter around the target. In [68], an idea is proposed to build a model in which the clutter is also taken into account as a norm in the objective function. In [80], the forward-looking clutter is suppressed in two steps. In the first step, the strong clutter outside of the reconstruction region is suppressed first. In the second step, the clutter in the reconstruction region is suppressed by selecting a proper β, which represents the ratio of the non-zeros area in the reconstructed scene. The reconstruction results are shown in Figure 14.

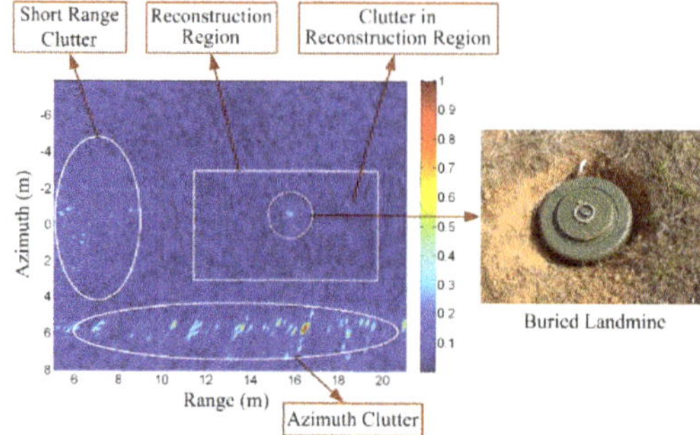

Figure 13. Real data example for the clutter problem in forward-looking GPR (backprojection result using full-sampled data). Taken from [80].

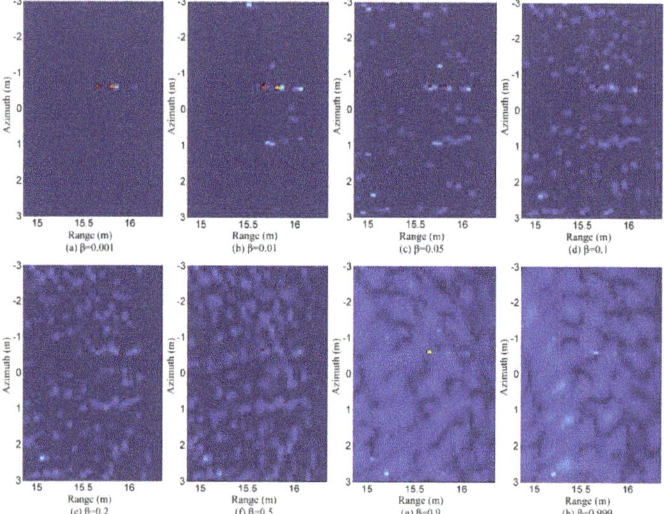

Figure 14. Reconstruction results in clutter environment with different parameters (taken from [80]).

3.5. Model Error Compensation

In the regularization or CS methods, we usually assume that the model is exact. However, in practice, the model may also contain errors. For example, imperfect knowledge of the observation position will lead to errors in the measurement matrix. This effect resembles motion errors that arise in traditional airborne SAR imaging. Figure 15 shows the geometry of the observation position errors or motion errors in SAR.

Several methods have been proposed to deal with model errors in CS-based or sparsity-driven radar imaging. A phase error correction method for sparsity-driven SAR imaging is proposed in [81]. An autofocus method for compressively sampled SAR is proposed in [82]. This method can correct phase errors in the reconstruction process. Both the methods proposed in [81,82] deal with phase errors in the observed data, or approximately treat the observation position-induced model errors as phase errors in the observed data. In [83], the platform position errors are investigated and compensated. That method considers the azimuth offset errors and also uses some approximations.

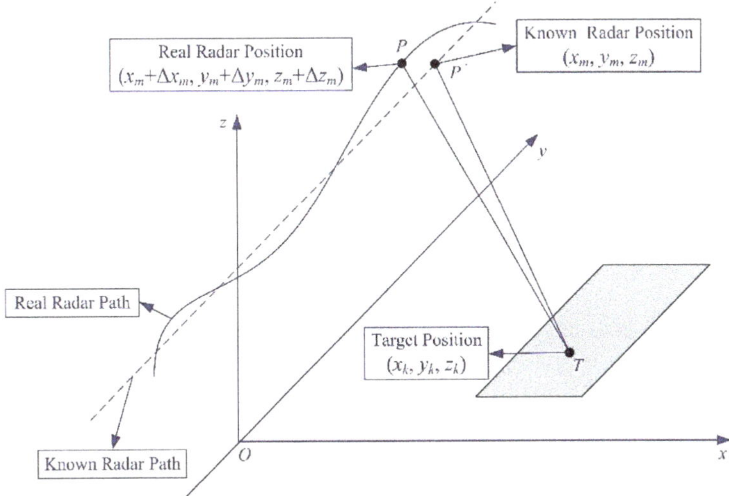

Figure 15. Geometry of the observation position errors in SAR. (Taken from [84]).

In [84], a model error compensation method is proposed. An iterative algorithm cycles through steps of target reconstruction, and observation position error estimation and compensation are used. This method can estimate the observation position error exactly, while only relying on the observed data.

Figure 16 shows a real data result using the method proposed in [84]. The data set used in this figure is the same as that used for Figure 9. In the data acquisition process, the airplane is expected to fly along a straight line. However, due to the air current's influence, the trajectory of the airplane may slightly deviate from the expected one. As a result, the observation position data inevitably contain some errors.

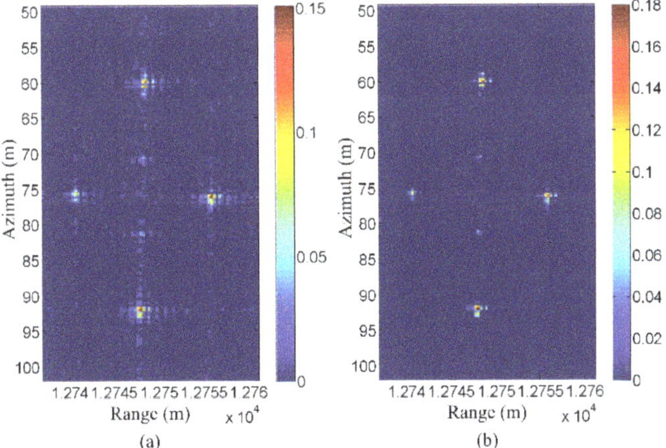

Figure 16. Observation position error compensation for airborne SAR data. (**a**) Result without observation position error compensation. (**b**) Result with observation position error compensation (taken from [84]).

Figure 16a shows the original CS reconstruction result. Since the observation position errors are not compensated, it can be seen that the targets are somewhat defocused. Figure 16b shows the corresponding CS reconstruction result with compensation for observation position error. It can be seen that the focusing quality is improved using the method proposed in [84]. The peak of the targets has an increase of about 20%, and the sidelobes are also significantly reduced.

4. Conclusions

In radar imaging area, there are many relevant techniques and methods, such as matched filtering, the range Doppler algorithm, the chirp scaling algorithm, the ωK algorithm, regularized methods, and CS methods. These techniques and methods are quite different in their forms. This paper tries to understand these techniques and methods in a unified mathematical framework.

Based on theoretical analysis, it can be seen that sparsity-driven regularization or CS-based radar imaging methods have potentially significant advantages. However, although many interesting ideas have been presented, very few of them have been verified with real data. There are still many unsolved or open problems in this area. In the issues discussed in this paper, the sampling scheme, fast reconstruction strategy, and model error problems are basically solved. However, issues concerning the sparsity or sparse representation of a complex and extended scene are still not completely solved. Strong clutter may break the sparsity of a scene, while sparse representation methods for an extended scene are currently not perfect. The state of the art in these areas has not yet reached the stage of practical application, and further investigations are needed in the future.

Funding: This work was supported in part by the National Natural Science Foundation of China under Grant 61401474, and in part by the Hunan Provincial Natural Science Foundation under Grant 2016JJ3025.

Conflicts of Interest: The authors declare no conflict of interest.

References

1. Cumming, I.; Wong, F. *Digital Processing of Synthetic Aperture Radar Data: Algorithms and Implementation*; Artech House: Boston, MA, USA, 2005.
2. Moreira, A.; Prats-Iraola, P.; Younis, M.; Krieger, G.; Hajnsek, I.; Papathanassiou, K.P. A tutorial on synthetic aperture radar. *IEEE Geosci. Remote Sens. Mag.* **2013**, *1*, 6–43. [CrossRef]
3. Mensa, D.L. *High Resolution Radar Imaging*; Artech House: Norwood, MA, USA, 1981.
4. Potter, L.C.; Ertin, E.; Parker, J.T.; Çetin, M. Sparsity and compressed sensing in radar imaging. *Proc. IEEE* **2010**, *98*, 1006–1020. [CrossRef]
5. Mohammad-Djafari, A. *Inverse Problems in Vision and 3D Tomography*; ISTE: London, UK; John Wiley and Sons: New York, NY, USA, 2010.
6. Mohammad-Djafari, A. Bayesian approach with prior models which enforce sparsity in signal and image processing. *EURASIP J. Adv. Signal Process.* **2012**, *2012*, 52. [CrossRef]
7. Mohammad-Djafari, A.; Daout, F.; Fargette, P. Fusion and inversion of SAR data to obtain a superresolution image. In Proceedings of the IEEE International Conference on Image Processing (ICIP), Cario, Egypt, 7–10 November 2009; pp. 569–572.
8. Groetsch, C.W. *The Theory of Tikhonov Regularization for Fredholm Equations of the First Kind*; Pitman Publishing Limited: London, UK, 1984.
9. Tikhonov, A.N.; Arsenin, V.Y. *Solutions of Ill-Posed Problems*; Wiley: New York, NY, USA, 1977.
10. Miller, K. Least squares methods for ill-posed problems with a prescribed bound. *SIAM J. Math. Anal.* **1970**, *1*, 52–57. [CrossRef]
11. Blacknell, D.; Quegan, S. SAR super-resolution with a stochastic point spread function. In Proceedings of the IEE Colloquium on Synthetic Aperture Radar, London, UK, 29 November 1989.
12. Pryde, G.C.; Delves, L.M.; Luttrell, S.P. Transputer based super resolution of SAR images: Two approaches to a parallel solution. In Proceedings of the IEE Colloquium on Transputers for Image Processing Applications, London, UK, 13 February 1989.
13. Kay, S.M. *Fundamentals of Statistical Signal Processing*; Prentice Hall PTR: Upper Saddle River, NJ, USA, 1993.
14. Bouman, C.; Sauer, K. A generalized Gaussian image model for edge-preserving MAP estimation. *IEEE Trans. Image Process.* **1993**, *2*, 296–310. [CrossRef] [PubMed]

15. Chang, L.T.; Gupta, I.J.; Burnside, W.D.; Chang, C.T. A data compression technique for scattered fields from complex targets. *IEEE Trans. Antennas Propag.* **1997**, *45*, 1245–1251. [CrossRef]
16. Candès, E.; Wakin, M.B. An introduction to compressive sampling. *IEEE Signal Process. Mag.* **2008**, *25*, 21–30. [CrossRef]
17. Baraniuk, R.G. Compressive sensing. *IEEE Signal Process. Mag.* **2007**, *24*, 118–121. [CrossRef]
18. Jafarpour, S.; Weiyu, X.; Hassibi, B.; Calderbank, R. Efficient and robust compressed sensing using optimized expander graphs. *IEEE Trans. Inf. Theory* **2009**, *55*, 4299–4308. [CrossRef]
19. Yuejie, C.; Scharf, L.L.; Pezeshki, A.; Calderbank, R. Sensitivity to basis mismatch in compressed sensing. *IEEE Trans. Signal Process.* **2011**, *59*, 2182–2195.
20. Raginsky, M.; Jafarpour, S.; Harmany, Z.T.; Marcia, R.F.; Willett, R.M.; Calderbank, R. Performance bounds for expander-based compressed sensing in Poisson noise. *IEEE Trans. Signal Process.* **2011**, *59*, 4139–4153. [CrossRef]
21. Logan, C.L. An Estimation-Theoretic Technique for Motion Compensated Synthetic-Aperture Array Imaging. Ph.D. Thesis, Massachusetts Institute of Technology, Cambridge, MA, USA, 2000.
22. Browne, K.E.; Burkholder, R.J.; Volakis, J.L. Fast optimization of through-wall radar images via the method of Lagrange multipliers. *IEEE Trans. Antennas Propag.* **2013**, *61*, 320–328. [CrossRef]
23. Romberg, J. Imaging via compressive sampling. *IEEE Signal Process. Mag.* **2008**, *25*, 14–20. [CrossRef]
24. Çetin, M.; Karl, W.C. Feature-enhanced synthetic aperture radar image formation based on nonquadratic regularization. *IEEE Trans. Antennas Propag.* **2001**, *10*, 623–631.
25. Kalogerias, D.; Sun, S.; Petropulu, A.P. Sparse sensing in colocated MIMO radar: A matrix completion approach. In Proceedings of the 2013 IEEE International Symposium on Signal Processing and Information Technology (ISSPIT), Athens, Greece, 12–15 December 2013; pp. 496–502.
26. Yao, Y.; Petropulu, A.P.; Poor, H.V. Measurement matrix design for compressive sensing-based MIMO radar. *IEEE Trans. Signal Process.* **2011**, *59*, 5338–5352.
27. Yao, Y.; Petropulu, A.P.; Poor, H.V. MIMO radar using compressive sampling. *IEEE J. Sel. Top. Signal Process.* **2010**, *4*, 146–163.
28. Rossi, M.; Haimovich, A.M.; Eldar, Y.C. Spatial compressive sensing for MIMO radar. *IEEE Trans. Signal Process.* **2014**, *62*, 419–430. [CrossRef]
29. Rossi, M.; Haimovich, A.M.; Eldar, Y.C. Global methods for compressive sensing in MIMO radar with distributed sensors. In Proceedings of the Forty Fifth Asilomar Conference on Signals, Systems and Computers (ASILOMAR), Pacific Grove, CA, USA, 6–9 November 2011; pp. 1506–1510.
30. Rossi, M.; Haimovich, A.M.; Eldar, Y.C. Spatial compressive sensing in MIMO radar with random arrays. In Proceedings of the 46th Annual Conference on Information Sciences and Systems (CISS), Princeton, NJ, USA, 21–23 March 2012.
31. Herman, M.; Strohmer, T. Compressed sensing radar. In Proceedings of the 2008 IEEE Radar Conference, Rome, Italy, 26–30 May 2008.
32. Herman, M.; Strohmer, T. High-resolution radar via compressed sensing. *IEEE Trans. Signal Process.* **2009**, *57*, 2275–2284. [CrossRef]
33. Herman, M.; Strohmer, T. General deviants: An analysis of perturbations in compressed sensing. *IEEE J. Sel. Top. Signal Process.* **2010**, *4*, 342–349. [CrossRef]
34. Baraniuk, R.G.; Steeghs, P. Compressive radar imaging. In Proceedings of the 2007 IEEE Radar Conference, Boston, MA, USA, 17–20 April 2007; pp. 128–133.
35. Ender, J.H.G. On compressive sensing applied to radar. *Signal Process.* **2010**, *90*, 1402–1414. [CrossRef]
36. Çetin, M.; Stojanovic, I.; Önhon, N.Ö.; Varshney, K.; Samadi, S.; Karl, W.C.; Willsky, A. Sparsity-driven synthetic aperture radar imaging: Reconstruction, autofocusing, moving targets, and compressed sensing. *IEEE Signal Process. Mag.* **2014**, *31*, 27–40. [CrossRef]
37. Zhang, B.; Hong, W.; Wu, Y. Sparse microwave imaging: Principles and applications. *Sci. China Inf. Sci.* **2012**, *55*, 1722–1754. [CrossRef]
38. Bi, H.; Zhang, B.; Zhu, X.; Jiang, C.; Hong, W. Extended chirp scaling-baseband azimuth scaling-based azimuth-range decouple L_1 regularization for TOPS SAR imaging via CAMP. *IEEE Trans. Geosci. Remote Sens.* **2017**, *55*, 3748–3763. [CrossRef]
39. Patel, V.M.; Easley, G.R.; Healy, D.M., Jr.; Chellappa, R. Compressed synthetic aperture radar. *IEEE J. Sel. Top. Signal Process.* **2010**, *4*, 244–254. [CrossRef]

40. Alonso, M.T.; López-Dekker, P.; Mallorquí, J.J. A novel strategy for radar imaging based on compressive sensing. *IEEE Trans. Geosci. Remote Sens.* **2010**, *48*, 4285–4295. [CrossRef]
41. Samadi, S.; Çetin, M.; Masnadi-Shirazi, M.A. Sparse representation-based synthetic aperture radar imaging. *IET Radar Sonar Navig.* **2011**, *5*, 182–193. [CrossRef]
42. Yang, L.; Zhou, J.; Hu, L.; Xiao, H. A Perturbation-Based Approach for Compressed Sensing Radar Imaging. *IEEE Antennas Wirel. Propag. Lett.* **2017**, *16*, 87–90. [CrossRef]
43. Zhang, L.; Qiao, Z.; Xing, M.; Li, Y.; Bao, Z. High-Resolution ISAR imaging with sparse stepped-frequency waveforms. *IEEE Trans. Geosci. Remote Sens.* **2011**, *49*, 4630–4651. [CrossRef]
44. Wang, H.; Quan, Y.; Xing, M.; Zhang, S. ISAR Imaging via sparse probing frequencies. *IEEE Geosci. Remote Sens. Lett.* **2011**, *8*, 451–455. [CrossRef]
45. Du, X.; Duan, C.; Hu, W. Sparse representation based autofocusing technique for ISAR images. *IEEE Trans. Geosci. Remote Sens.* **2013**, *51*, 1826–1835. [CrossRef]
46. Zhu, X.X.; Bamler, R. Tomographic SAR inversion by L1-norm regularization-the compressive sensing approach. *IEEE Trans. Geosci. Remote Sens.* **2010**, *48*, 3839–3846. [CrossRef]
47. Zhu, X.X.; Bamler, R. Super-resolution power and robustness of compressive sensing for spectral estimation with application to spaceborne tomographic SAR. *IEEE Trans. Geosci. Remote Sens.* **2012**, *50*, 247–258. [CrossRef]
48. Zhu, X.X.; Ge, N.; Shahzad, M. Joint Sparsity in SAR Tomography for Urban Mapping. *IEEE J. Sel. Top. Signal Process.* **2015**, *9*, 1498–1509. [CrossRef]
49. Budillon, A.; Evangelista, A.; Schirinzi, G. Three-Dimensional SAR Focusing from Multipass Signals Using Compressive Sampling. *IEEE Trans. Geosci. Remote Sens.* **2011**, *49*, 488–499. [CrossRef]
50. Xilong, S.; Anxi, Y.; Zhen, D.; Diannong, L. Three-dimensional SAR focusing via compressive sensing: The case study of angel stadium. *IEEE Geosci. Remote Sens. Lett.* **2012**, *9*, 759–763. [CrossRef]
51. Aguilera, E.; Nannini, M.; Reigber, A. Multisignal compressed sensing for polarimetric SAR tomography. *IEEE Geosci. Remote Sens. Lett.* **2012**, *9*, 871–875. [CrossRef]
52. Kajbaf, H. Compressed Sensing for 3D Microwave Imaging Systems. Ph.D. Thesis, Missouri University of Science and Technology, Rolla, MO, USA, 2012.
53. Zhang, S.; Dong, G.; Kuang, G. Superresolution downward-looking linear array three-dimensional SAR imaging based on two-dimensional compressive sensing. *IEEE J. Sel. Top. Appl. Earth Obs. Remote Sens.* **2016**, *9*, 2184–2196. [CrossRef]
54. Tian, H.; Li, D. Sparse flight array SAR downward-looking 3-D imaging based on compressed sensing. *IEEE Geosci. Remote Sens. Lett.* **2016**, *13*, 1395–1399. [CrossRef]
55. Lin, Y.G.; Zhang, B.C.; Hong, W.; Wu, Y.R. Along-track interferometric SAR imaging based on distributed compressed sensing. *Electron. Lett.* **2010**, *46*, 858–860. [CrossRef]
56. Prünte, L. Compressed sensing for joint ground imaging and target indication with airborne radar. In Proceedings of the 4th Workshop on Signal Processing with Adaptive Sparse Structured Representations, Edinburgh, UK, 27–30 June 2011.
57. Wu, Q.; Xing, M.; Qiu, C.; Liu, B.; Bao, Z.; Yeo, T.S. Motion parameter estimation in the SAR system with low PRF sampling. *IEEE Geosci. Remote Sens. Lett.* **2010**, *7*, 450–454. [CrossRef]
58. Stojanovic, I.; Karl, W.C. Imaging of moving targets with multi-static SAR using an overcomplete dictionary. *IEEE J. Sel. Top. Signal Process.* **2010**, *4*, 164–176. [CrossRef]
59. Önhon, N.Ö.; Çetin, M. SAR moving target imaging in a sparsity-driven framework. In Proceedings of the SPIE Optics + Photonics Symposium, Wavelets and Sparsity XIV, San Diego, CA, USA, 21–25 August 2011; Volume 8138, pp. 1–9.
60. Prünte, L. Detection performance of GMTI from SAR images with CS. In Proceedings of the 2014 European Conference on Synthetic Aperture Radar, Berlin, Germany, 3–5 June 2014.
61. Prünte, L. Compressed sensing for removing moving target artifacts and reducing noise in SAR images. In Proceedings of the European Conference on Synthetic Aperture Radar, Hamburg, Germany, 6–9 June 2016.
62. Suksmono, A.B.; Bharata, E.; Lestari, A.A.; Yarovoy, A.G.; Ligthart, L.P. Compressive stepped-frequency continuous-wave ground-penetrating radar. *IEEE Geosci. Remote Sens. Lett.* **2010**, *7*, 665–669. [CrossRef]
63. Gurbuz, A.C.; McClellan, J.H.; Scott, W.R. A compressive sensing data acquisition and imaging method for stepped frequency GPRs. *IEEE Trans. Signal Process.* **2009**, *57*, 2640–2650. [CrossRef]

64. Tuncer, M.A.C.; Gurbuz, A.C. Ground reflection removal in compressive sensing ground penetrating radars. *IEEE Geosci. Remote Sens. Lett.* **2012**, *9*, 23–27. [CrossRef]
65. Lagunas, E.; Amin, M.G.; Ahmad, F.; Nájar, M. Joint wall mitigation and compressive sensing for indoor image reconstruction. *IEEE Trans. Geosci. Remote Sens.* **2013**, *51*, 891–906. [CrossRef]
66. Amin, M.G. *Through-the-Wall Radar Imaging*; CRC Press: Boca Raton, FL, USA, 2010.
67. Ahmad, F.; Amin, M.G. Through-the-wall human motion indication using sparsity-driven change detection. *IEEE Trans. Geosci. Remote Sens.* **2013**, *51*, 881–890. [CrossRef]
68. Strohmer, T. Radar and compressive sensing—A perfect couple? In Proceedings of the Keynote Speak of 1st International Workshop on Compressed Sensing Applied to Radar (CoSeRa 2012), Bonn, Germany, 14–16 May 2012.
69. Yang, J.; Thompson, J.; Huang, X.; Jin, T.; Zhou, Z. Random-frequency SAR imaging based on compressed sensing. *IEEE Trans. Geosci. Remote Sens.* **2013**, *51*, 983–994. [CrossRef]
70. Qin, S.; Zhang, Y.D.; Wu, Q.; Amin, M. Large-scale sparse reconstruction through partitioned compressive sensing. In Proceedings of the International Conference on Digital Signal Processing, Hong Kong, China, 20–23 August 2014; pp. 837–840.
71. Yang, J.; Thompson, J.; Huang, X.; Jin, T.; Zhou, Z. Segmented reconstruction for compressed sensing SAR imaging. *IEEE Trans. Geosci. Remote Sens.* **2013**, *51*, 4214–4225. [CrossRef]
72. Duarte, M.F.; Davenport, M.A.; Takhar, D.; Laska, J.N.; Sun, T.; Kelly, K.F.; Baraniuk, R.G. Single-pixel imaging via compressive sampling. *IEEE Signal Process. Mag.* **2008**, *25*, 83–91. [CrossRef]
73. Çetin, M.; Karl, W.C.; Willsky, A.S. Feature-preserving regularization method for complex-valued inverse problems with application to coherent imaging. *Opt. Eng.* **2006**, *45*, 017003.
74. Çetin, M.; Önhon, N.Ö.; Samadi, S. Handling phase in sparse reconstruction for SAR: Imaging, autofocusing, and moving targets. In Proceedings of the EUSAR 2012—9th European Conference on Synthetic Aperture Radar, Nuremberg, Germany, 23–26 April 2012.
75. Varshney, K.R.; Çetin, M.; Fisher, J.W., III; Willsky, A.S. Sparse representation in structured dictionaries with application to synthetic aperture radar. *IEEE Trans. Signal Process.* **2008**, *56*, 3548–3560. [CrossRef]
76. Abolghasemi, V.; Gan, L. Dictionary learning for incomplete SAR data. In Proceedings of the CoSeRa 2012, Bonn, Germany, 14–16 May 2012.
77. Halman, J.; Burkholder, R.J. Sparse expansions using physical and polynomial basis functions for compressed sensing of frequency domain EM scattering. *IEEE Antennas Wirel. Propag. Lett.* **2015**, *14*, 1048–1051. [CrossRef]
78. Van Den Berg, E.; Friedlander, M.P. Probing the Pareto frontier for basis pursuit solutions. *SIAM J. Sci. Comput.* **2008**, *31*, 890–912. [CrossRef]
79. Yang, J.; Jin, T.; Huang, X. Compressed sensing radar imaging with magnitude sparse representation. *IEEE Access* **2019**, *7*, 29722–29733. [CrossRef]
80. Yang, J.; Jin, T.; Huang, X.; Thompson, J.; Zhou, Z. Sparse MIMO array forward-looking GPR imaging based on compressed sensing in clutter environment. *IEEE Trans. Geosci. Remote Sens.* **2014**, *52*, 4480–4494. [CrossRef]
81. Önhon, N.Ö.; Çetin, M. A sparsity-driven approach for joint SAR imaging and phase error correction. *IEEE Trans. Antennas Propag.* **2012**, *21*, 2075–2088.
82. Kelly, S.I.; Yaghoobi, M.; Davies, M.E. Auto-focus for compressively sampled SAR. In Proceedings of the CoSeRa 2012, Bonn, Germany, 14–16 May 2012.
83. Wei, S.J.; Zhang, X.L.; Shi, J. An autofocus approach for model error correction in compressed sensing SAR imaging. In Proceedings of the IEEE International Geoscience and Remote Sensing Symposium (IGARSS 2012), Munich, Germany, 22–27 July 2012.
84. Yang, J.; Huang, X.; Thompson, J.; Jin, T.; Zhou, Z. Compressed sensing radar imaging with compensation of observation position error. *IEEE Trans. Geosci. Remote Sens.* **2014**, *52*, 4608–4620. [CrossRef]

Article

Compressive Sensing-Based Bandwidth Stitching for Multichannel Microwave Radars

Paul Berry [1,*], Ngoc Hung Nguyen [2] and Hai-Tan Tran [1]

1 Intelligence, Surveillance and Space Division, Defence Science and Technology Group, Edinburgh, SA 5111, Australia; haitan.tran@dst.defence.gov.au
2 Maritime Division, Defence Science and Technology Group, Edinburgh, SA 5111, Australia; ngoc.nguyen@dst.defence.gov.au
* Correspondence: paul.berry@dst.defence.gov.au

Received: 18 November 2019; Accepted: 19 January 2020; Published: 24 January 2020

Abstract: The problem of obtaining high range resolution (HRR) profiles for non-cooperative target recognition by coherently combining data from narrowband radars was investigated using sparse reconstruction techniques. If the radars concerned operate within different frequency bands, then this process increases the overall effective bandwidth and consequently enhances resolution. The case of unknown range offsets occurring between the radars' range profiles due to incorrect temporal and spatial synchronisation between the radars was considered, and the use of both pruned orthogonal matching pursuit and refined l_1-norm regularisation solvers was explored to estimate the offsets between the radars' channels so as to attain the necessary coherence for combining their data. The proposed techniques were demonstrated and compared using simulated radar data.

Keywords: radar signal processing techniques; radar imaging; multiband processing; compressive sensing; sparse reconstruction; bandwidth stitching

1. Introduction

The construction of high range resolution profiles (HRRP) of targets is a precursor to feature extraction for automatic target recognition (ATR), and normally requires the employment of a high-bandwidth waveform following detection by a lower resolution radar mode. Examples of recent papers in the non-cooperative target recognition (NCTR) literature focusing on feature extraction for ATR following HRRP construction are [1–3]. This paper considers the problem of HRRP construction, but using low resolution radars operating in different frequency bands for the purpose of combining their signals to achieve a higher resolution, and examines the problem of their data not being mutually coherent.

The ability to acquire high resolution range profiles of targets has improved over time as hardware capability has developed, with higher resolution being achieved by increasing the time-bandwidth product. In the early approaches, for narrowband radars with very limited instantaneous bandwidth, stepped-frequency waveforms were used with a single I,Q (that is, baseband quadrature) signal sample received after each frequency step. The set of samples is effectively used for the Fourier transform of the slant range profile, enabling the range response to be obtained simply by implementing an inverse Fourier transform (see, e.g., [4]). The greater the frequency range, the higher the range resolution, but the downside is that the burst of pulses can be so long that a scatterer may migrate between range cells, causing smearing of the range profile, and therefore requiring range compensation. An example of a recent paper involving the use of stepped-frequency waveforms is [5], and recent papers which have investigated the effects of target motion and aspect sensitivity are [6,7].

An advance on the stepped-frequency approach is to increase the time-bandwidth product using stretch processing, whereby a wideband LFM waveform is transmitted, and pulse-compression is achieved in hardware by mixing the received signal with an extended replica of the transmitted waveform. A point target at a particular range will manifest itself as a single frequency which is proportional to its range. The result, which is digitally sampled in time in order to facilitate the identification of the frequency components, is therefore a superposition of discrete frequencies, each corresponding to a point target at a different range. An application of the inverse Fourier transform again recovers the range profile (see, e.g., [4,8,9]).

A spectral analysis technique to improve range resolution was proposed in [10,11] based on autoregressive linear prediction. The main idea is to combine the mutually coherent signals received from multiple waveforms transmitted sequentially or concurrently, which have widely separate carrier frequencies or may even occupy entirely different frequency bands. Viewed in the spectral domain, the received signals from individual waveforms may be seen to occupy discrete wavebands which are separate or contiguous. If contiguous, then they can potentially be coherently combined to synthesise the signals that would have been received from a single wider bandwidth waveform in the manner presented in [12]. If separate, then presumably this coherent combination of signals would still be feasible, as would be the interpolation of the frequency response in the gaps between the bands under the a priori assumption that the signals are returned from discrete scatterers using spectral estimation techniques (see, e.g., [13]). Alternatively, the signals from different frequency bands can be jointly processed without explicitly filling the gaps between the bands. Since no new synthetic frequency band is actually constructed, this approach can be referred to as bandwidth stitching to distinguish it from bandwidth interpolation and extrapolation. The main challenge of this approach lies in the presence of phase errors in different frequency bands resulting from post-processed motion compensation which is often carried out separately for each frequency band. In this paper, we focus on the problem of bandwidth stitching for radar high-resolution range profiling and explore the use of sparse reconstruction to deal with the phase error problem.

It is convenient to formulate these ideas in the spectral domain, within which point scatterers appear as discrete sinusoids and which are amenable to analysis by spectral estimation techniques such as autoregression, as presented in [10,11]. Compressive sensing and sparse reconstruction, however, provide for the possibility of alternative signal representations, potentially allowing for greater flexibility and discriminating between signals of physical origin and receiver noise [13–20]. Instances of non-sinusoidal signals are waveforms in fast-time and signals returned from rotating objects when the angle of rotation is large. Compressive sensing and sparse reconstruction can also handle the situation corresponding to data being non-uniformly sampled in time or space, such as non-uniform PRF (pulse repetition frequency) waveforms and random sparse arrays.

Compressive sensing and sparse reconstruction were exploited in [17,19,21,22] to address the problem of gaps in the data both in slow-time and in frequency for inverse synthetic aperture radar (ISAR) imaging. However, these works assumed that the data were coherent across different sub-bands and that there were no model uncertainties. The work [23] took account of the possible lack of mutual coherence between the radars operating on the different sub-bands arising from incorrect timing synchronisation, or, equivalently, errors in antenna phase's centre-relative locations. This is achieved by fitting an ultra-wideband all-pole signal model to the mutually-coherent sub-bands, which is then used for bandwidth interpolation and extrapolation prior to recovering the range profile by means of an inverse Fourier transform. This paper, however, proposes the use of compressive sensing and sparse reconstruction to deal with the non-coherence problem between different sub-bands.

To address the sub-band non-coherence problem, two different approaches were explored: (i) greedy pursuit and (ii) l_1-norm regularisation. In the first approach, pruned orthogonal matching pursuit (POMP) [24], which was originally developed for micro-Doppler parameter estimation, is adopted to deal with the dictionary mismatch which is due to the phase errors in each sub-band resulting from the motion-compensation post processing. The main idea is to parameterise the dictionary as a function of the phase errors and to construct multiple realisations of the dictionary. A selective learning process is then used to discard the dictionaries which correspond to incorrect values of phase errors. Since a straight application of the POMP algorithm to the problem under consideration would have been computationally expensive, we first applied the POMP algorithm pairwise to sub-bands in order to estimate the phase errors, and then utilised the conventional OMP algorithm to determine the range profile based on the estimated phase error values. In the second approach, an l_1-norm regularisation problem can be solved jointly both for the range profile vector and the phase errors [25]. The work [25] offers two solutions for joint synthetic aperture radar (SAR) imaging and phase error correction. The first solution is not applicable to the problem under consideration because a constant phase error in each sub-band is assumed. On the other hand, the second solution considers general arbitrary phase errors, and thus can be applied to our problem. Since the phase errors within each sub-band are a linear function of the range error coming from post-processing motion compensation, we also present refined variants of the second solution of [25] to take into account this underlying structure of the phase error.

The paper is organised as follows. Section 2 formulates the problem of bandwidth stitching for HRRP in the presence of phase errors. The POMP algorithm is applied in Section 3 to the bandwidth stitching problem under consideration. Section 4 presents l_1-norm regularisation solvers. Numerical performance comparisons are provided in Section 5 and conclusions are drawn in Section 6.

2. Problem Formulation

Consider a multistatic radar system consisting of M radar channels on different and distinct frequency sub-bands, approximately co-located, and illuminating a common target, such that their radar lines of sight (LoS) coincide but their range profiles are out of alignment. Each channel can individually produce a one dimensional range profile of the target, but with relatively coarse resolution. The bandwidth stitching problem can be briefly stated as follows: for the M generally non-coherent channels, the aim is to coherently combine, or "stitch," the channels together so that they can effectively produce a single range profile with resolution corresponding to the combined overall signal bandwidth.

Let $f_{m,n}$ ($n = 1, \ldots, N_m$) denote the nth frequency bin of the mth channel ($m = 1, \ldots, M$). Here, N_m is the number of frequency bins in the mth channel. We base the formulation on the point-scatterer model and assume that the target can be defined as consisting of K scattering centres at local line-of-sight coordinates x_k (or local "down ranges") and having complex-valued reflectivity coefficients α_k, which are also assumed frequency-independent. The down-converted, pulse-compressed, motion-compensated signals received in each channel, in the frequency domain, can be written as

$$S_m = [\ldots, S_{m,n}, \ldots]^T_{n=1,\ldots,N_m}, \tag{1}$$

where superscript T denotes the transpose operation, and

$$S_{m,n} = |A(f_{m,n})|^2 \exp\left\{-\frac{4\pi j f_{m,n}}{c}\Delta R_m\right\} \sum_{k=1}^{K} \alpha_k \exp\left\{-\frac{4\pi j f_{m,n}}{c} x_k\right\}. \tag{2}$$

Here, $A(f_{n,m})$ represents the transmit radar waveform, the squared amplitude resulting from pulse compression processing; constant c denotes the speed of light; and ΔR_m accounts for the range errors in the motion-compensation processing. Bandwidth stitching in this context amounts to estimating these phase errors as accurately as possible.

The signals S_m in (1) can be rewritten in a more compact form as

$$\mathbf{S}_m = \mathbf{\Lambda}_m \mathbf{F}_m^{\dagger} \boldsymbol{\alpha}^{\dagger}, \tag{3}$$

where

$$\Lambda_m = \text{diag}\left\{\ldots, \exp\left\{-\frac{4\pi j f_{m,n}}{c}\Delta R_m\right\}, \ldots\right\}_{n=1,\ldots,N_m} \quad (4)$$

$$F_m^\dagger = [\ldots, F_{m,k}, \ldots]_{k=1,\ldots,K} \quad (5)$$

$$F_{m,k}^\dagger = \left[\ldots, \exp\left\{-\frac{4\pi j f_{m,n}}{c}x_k\right\}, \ldots\right]^T_{n=1,\ldots,N_m} \quad (6)$$

$$\alpha^\dagger = [\ldots, \alpha_k, \ldots]^T_{k=1,\ldots,K} \quad (7)$$

Here, "diag" denotes a diagonal matrix; Λ_m is referred to as the phase error matrix, of dimension $N_m \times N_m$; $F_m^\dagger$ and $\alpha^\dagger$ are respectively, dimensions $N_m \times K$ and $K \times 1$; S_m is a column vector of dimension $N_m \times 1$; and the dagger symbol † refers to the K actual scatterers on the target.

To apply the sparse representation techniques of compressive sensing, we discretise the target's local range coordinate x using a regularly-spaced range grid $\{x_l\}$ for $l = 1, \ldots, L_x$, with $L_x \gg K$, and construct the $N_m \times L_x$ dictionary matrices

$$F_m = [\ldots, F_{m,l}, \ldots]_{l=1,\ldots,L_x}, \quad (8)$$

where

$$F_{m,l} = \left[\ldots, \exp\left\{-\frac{4\pi j f_{m,n}}{c}x_l\right\}, \ldots\right]^T_{n=1,\ldots,N_m} \quad (9)$$

Are the "atoms" of dictionary F_m in the frequency domain. The corresponding range profile vector

$$\alpha = [\ldots, \alpha_l, \ldots]^T_{l=1,\ldots,L_x} \quad (10)$$

Spans over the range grid $\{x_l\}$. The received signal S_m can also be written as

$$S_m = \Lambda_m F_m \alpha. \quad (11)$$

Since the target usually contains only a small number of dominant scattering centres relative to the total number of range resolution cells, the range profile α can be considered sparse (i.e., containing a small number of non-zero elements).

In the presence of unknown noise, (11) becomes

$$\tilde{S}_m = \Lambda_m F_m \alpha + n_m. \quad (12)$$

where n_m is the additive noise for channel m. Stacking up the individual channel signals $\tilde{S}_m$, $m = 1, \ldots, M$, gives

$$\tilde{S} = \Lambda F \alpha + n, \quad (13)$$

where

$$\tilde{S} = [\ldots, \tilde{S}_m^T, \ldots]^T_{m=1,\ldots,M} \quad (14)$$

$$\Lambda = \text{diag}\{\ldots, \Lambda_m, \ldots\}_{m=1,\ldots,M} \quad (15)$$

$$F = [\ldots, F_m^T, \ldots]^T_{m=1,\ldots,M} \quad (16)$$

$$n = [\ldots, n_m^T, \ldots]^T_{m=1,\ldots,M}. \quad (17)$$

Note that $\tilde{S}$ and n are column vectors of size $(\sum_m N_m) \times 1$; diagonal phase error matrix Λ is of size $(\sum_m N_m) \times (\sum_m N_m)$; dictionary matrix F is $(\sum_m N_m) \times L_x$; and the range profile α is again a column vector of size $L_x \times 1$. Stacking the received signals amounts to a vertical stacking of the dictionary matrices from all channels and a diagonal concatenation of the corresponding phase error matrices.

The stacking of multiple channels in this manner can improve the estimation accuracy for $\boldsymbol{\alpha}$, as will be demonstrated later in the paper.

A *problem statement* can thus be expressed as follows: given $\tilde{S}$ as the measured signal,

$$\text{find } \boldsymbol{\alpha} \text{ and } \boldsymbol{\Lambda}, \text{ subject to } \begin{cases} \tilde{S} \approx \boldsymbol{\Lambda} \boldsymbol{F} \boldsymbol{\alpha}, \\ \boldsymbol{\alpha} \text{ is sparse} \end{cases}. \tag{18}$$

The estimation of $\boldsymbol{\alpha}$ over $\{x_l\}$ is the process of range profiling, giving the main desired output, whereas the estimation of the phase error matrix $\boldsymbol{\Lambda}$ is really only a necessary intermediate result; it is a function of $\Delta R_1, \Delta R_2, \ldots$, and ΔR_M (recall that ΔR_m is the range estimation error resulting from the motion-compensation process for channel m). Furthermore, since these errors arise from a lack of precise knowledge of the relative locations of the radar channels' phase centres and are small relative to a range resolution cell, we may assume, without loss of generality, that $\Delta R_1 = 0$.

3. Greedy Pursuit Solutions

In this section, we adopt the pruned OMP (POMP) technique, which was originally proposed for micro-Doppler parameter estimation, [24], for the problem of bandwidth stitching for range profiling. We start with the simplest case of two channels and then generalise it to the multiple channel case.

3.1. The Two-Channel Case

For this case, the signal model in (13) can be expressed as

$$\tilde{S} = \boldsymbol{\Lambda}(\Delta R_2) \boldsymbol{F} \boldsymbol{\alpha} + \boldsymbol{n}, \tag{19}$$

where $\boldsymbol{\Lambda}(\Delta R_2)$ is a function of the single unknown relative range error ΔR_2,

$$\boldsymbol{\Lambda}(\Delta R_2) = \text{diag}\{\boldsymbol{I}_{N_1}, \boldsymbol{\Lambda}_2(\Delta R_2)\}, \tag{20}$$

with

$$\boldsymbol{\Lambda}_2(\Delta R_2) = \text{diag}\left\{\ldots, \exp\left\{-\frac{4\pi j f_{2,n}}{c} \Delta R_2\right\}, \ldots\right\}_{n=1,\ldots,N_2}. \tag{21}$$

In addition to the sparse range profile vector $\boldsymbol{\alpha}$, ΔR_2 is the only additional unknown parameter to be estimated. Let us rewrite (19) as

$$\tilde{S} = \boldsymbol{\Phi}(\Delta R_2)\, \boldsymbol{\alpha} + \boldsymbol{n} \tag{22}$$

where

$$\boldsymbol{\Phi}(\Delta R_2) = \boldsymbol{\Lambda}(\Delta R_2)\, \boldsymbol{F}.$$

In this form, the problem can be viewed as a joint sparse reconstruction and parameter estimation problem with the parametric dictionary $\boldsymbol{\Phi}(\Delta R_2)$ itself a function of the parameter ΔR_2. This can be considered as a special dictionary learning problem where the objective is to solve simultaneously for both the sparse solution of $\boldsymbol{\alpha}$ and the range error ΔR_2.

To solve this problem, we adopt the POMP technique [24], which embeds a pruning operation into the iterative process of OMP. The main idea of POMP is to construct multiple realisations of the dictionary $\boldsymbol{\Phi}$ based on a number L_Δ of candidate values of ΔR_2; the OMP algorithm is applied to each dictionary realisation to find the atom which correlates most strongly with the current residual for that dictionary, and to recompute the residual with that atom's contribution to the residual removed. To overcome possible excessive computations arising from outlier candidate values of ΔR_2, a pruning operation is performed to exclude the half of the dictionaries which yield the largest residual errors, until a single dictionary realisation remains. The OMP iterations for the remaining dictionary are continued until a termination criterion is satisfied. The candidate value of ΔR_2 corresponding to this dictionary gives the final estimate of ΔR_2. The basic POMP algorithm is summarised in Table 1 and its computational cost is shown in Appendix C to be of the order of $L_\Delta N_m L_x$.

Table 1. The pruned orthogonal matching pursuit (POMP) algorithm (M = 2).

INPUT:
- Noisy signal data vector $\tilde{S}$
- Candidate dictionaries $\mathbf{\Phi}_1, \mathbf{\Phi}_2, \ldots, \mathbf{\Phi}_{L_\Delta}$, corresponding to L_Δ candidate values of ΔR_2

PROCEDURE:
- Initialization:
  - set the initial indexes of active dictionaries to $\Theta_1 = \{1, \ldots, L_\Delta\}$;
  - set the corresponding residual vectors to $\boldsymbol{r}_1 = \cdots = \boldsymbol{r}_{L_\Delta} = \tilde{S}$;
  - set the initial support Λ to $\emptyset$, the null set;
- **for** $i = 1; i := i + 1$ until $|\Theta_i| == 1$ (the cardinality of Θ_i) and $|\boldsymbol{r}_l| < \epsilon$ for $l \in \Theta_i$,

  for every $l \in \Theta_i$, perform OMP as follows
  - Identify:

    $\boldsymbol{c}_l = \mathbf{\Phi}_l^H \boldsymbol{r}_l$

    $j_l = \arg\max_j |\boldsymbol{c}_j|$
  - Merge supports:

    $\Lambda_l = \Lambda_l \cup j_l$
  - Update*:

    $\hat{\boldsymbol{\alpha}}_{l,\Lambda_l} = \left(\mathbf{\Phi}_{l,\Lambda_l}^H \mathbf{\Phi}_{l,\Lambda_l}\right)^{-1} \mathbf{\Phi}_{l,\Lambda_l}^H \tilde{S}$

    $\boldsymbol{r}_l = \tilde{S} - \mathbf{\Phi}_{l,\Lambda_l} \hat{\boldsymbol{\alpha}}_{l,\Lambda_l}$

  end for

  if $|\Theta_i| > 1$

  Remove indices of $\lceil |\Theta_i|/2 \rceil$ candidate dictionaries that correspond to $\lceil |\Theta_i|/2 \rceil$ largest residual errors from $|\Theta_i|$

  end if

  end for

OUTPUT:
- The range profile estimate $\hat{\boldsymbol{\alpha}}_{l^\star,\Lambda_{l^\star}}$ where $l^\star$ is the last remaining element of Θ_i
- The estimate of ΔR_2 is the value of ΔR_2 corresponding to $\mathbf{\Phi}_{l^\star}$

* $\mathbf{\Phi}_{l,\Lambda_l}$ consists of the columns of $\mathbf{\Phi}_l$ with indices belonging to Λ_l and $\hat{\boldsymbol{\alpha}}_{\Lambda_l}$ consists of the elements of $\hat{\boldsymbol{\alpha}}_l$ with indices belonging to Λ_l.

3.2. The Multi-Channel Case

For this case, the noisy signal model (19) becomes

$$\tilde{S} = \mathbf{\Phi}(\Delta R_2, \ldots, \Delta R_M)\boldsymbol{\alpha} + \boldsymbol{n} \tag{23}$$

where

$$\mathbf{\Phi}(\Delta R_2, \ldots, \Delta R_M) = \mathrm{diag}\{\boldsymbol{I}_{N_1}, \mathbf{\Lambda}_2(\Delta R_2), \ldots, \mathbf{\Lambda}_M(\Delta R_M)\}\, \boldsymbol{F}. \tag{24}$$

Here, the dictionary matrix is a function of the $(M-1)$ unknowns $\Delta R_2, \Delta R_3, \ldots$ and ΔR_M.

The POMP algorithm could be extended to multiple channels by computing candidate dictionaries based on a multi-dimensional grid of candidate values for $\Delta R_2, \Delta R_3, \ldots,$ and ΔR_M. The grid would consist of a total of $(M-1)$ dimensions, where the mth dimension corresponds to the unknown range error ΔR_{m+1} of the $(m+1)$ channel. Note that only a one-dimensional grid for ΔR_2 is required for the case of two channels. However, the cardinality of the dictionary set is exponentially dependent on the number of available channels; i.e., $V^{(M-1)}$, where V denotes the number of grid points in each parameter dimension. As a result, although this extension would be simple and straightforward, it is computationally expensive.

To alleviate this computational burden, we instead apply the POMP algorithm pairwise to channels in order to estimate the range errors $\Delta R_2, \dots, \Delta R_M$ relative to the first channel, and then utilise the conventional OMP algorithm to determine the range profile vector $\boldsymbol{\alpha}$ based on these estimated values of $\Delta R_2, \dots, \Delta R_M$. The procedure is summarised as below:

STEP 1: Estimation of range errors.

- For each pair between the 1st and m^{th} channel, $m \in \{2, \dots, M\}$:
    - Calculate input signal:

$$\tilde{\boldsymbol{S}}_{1m} = [\tilde{\boldsymbol{S}}_1^T, \tilde{\boldsymbol{S}}_m^T]^T \tag{25a}$$

$$\boldsymbol{F}_{1m} = [\boldsymbol{F}_1^T, \boldsymbol{F}_m^T]^T. \tag{25b}$$

    - Construct candidate dictionaries based on a grid of L candidate values of $\Delta R_m^{(l)}$ $(l = 1, \dots, L)$:

$$\boldsymbol{\Phi}_{1m,l} = \text{diag}\{\boldsymbol{I}_{N_1}, \boldsymbol{\Lambda}_m(\Delta R_m^{(l)})\}\boldsymbol{F}_{1m}. \tag{25c}$$

    - Perform POMP given $\tilde{\boldsymbol{S}}_{1m}$ and $\boldsymbol{\Phi}_{1m,1}, \dots, \boldsymbol{\Phi}_{1m,L}$ to obtain an estimate of ΔR_m (denoted as $\widehat{\Delta R}_m$).

  End for.

STEP 2: Estimation of range profile vector.

- Compute signal and dictionary.

$$\tilde{\boldsymbol{S}} = [\dots, \tilde{\boldsymbol{S}}_m^T, \dots]^T_{m=1,\dots,M} \tag{26}$$

$$\boldsymbol{F} = [\dots, \boldsymbol{F}_m^T, \dots]^T_{m=1,\dots,M} \tag{27}$$

$$\boldsymbol{\Lambda} = \text{diag}\{\dots, \boldsymbol{\Lambda}_m(\widehat{\Delta R}_m), \dots\}_{m=1,\dots,M} \tag{28}$$

$$\boldsymbol{\Phi} = \boldsymbol{\Lambda}\boldsymbol{F}. \tag{29}$$

- Estimate $\boldsymbol{\alpha}$ using OMP given $\tilde{\boldsymbol{S}}$ and $\boldsymbol{\Phi}$.

The computational cost of the general POMP algorithm is shown in Appendix C to be of the order of $(M-1)L_\Delta N_m L_x$.

4. L_1-Norm Regularisation Approach

The sparse reconstruction problem (18) can be solved via the following l_1 regularised optimisation:

$$\min_{\boldsymbol{\alpha},\boldsymbol{\Lambda}} \left\{ \|\boldsymbol{S} - \boldsymbol{\Lambda}\boldsymbol{F}\boldsymbol{\alpha}\|_2^2 + \mu\|\boldsymbol{\alpha}\|_1 \right\}, \tag{30}$$

where μ is a regularisation parameter. It should be emphasized that this is not a conventional l_1 regularisation formulation because of the unknown phase error matrix $\boldsymbol{\Lambda}$ resulting from the estimation error of the motion-compensation phase. Therefore, $\boldsymbol{\Lambda}$ must be jointly estimated with $\boldsymbol{\alpha}$:

$$\{\hat{\boldsymbol{\alpha}}, \hat{\boldsymbol{\Lambda}}\} = \arg\min_{\boldsymbol{\alpha},\boldsymbol{\Lambda}} \left\{ \|\boldsymbol{S} - \boldsymbol{\Lambda}\boldsymbol{F}\boldsymbol{\alpha}\|_2^2 + \mu\|\boldsymbol{\alpha}\|_1 \right\}. \tag{31}$$

Two solutions for this joint estimation problem were presented in [25]. The first solution assumes that the phase error matrix for the mth sub-band is modelled as

$$\boldsymbol{\Lambda}_m = \exp\{j\phi_m\}\, \boldsymbol{I}_{N_m \times N_m}. \tag{32}$$

In other words, the phase errors for different frequency bins of a particular sub-band are identical. However, this assumption is invalid in the problem under consideration because the phase error is a function of frequency and thus has different values for different frequency bins. Therefore, that solution is not applicable in this case. The second solution considers a general phase error matrix

$$\boldsymbol{\Lambda}_m = \text{diag}\{\ldots, \exp\{j\phi_{m,n}\}, \ldots\}_{n=1,\ldots,N_m} \tag{33}$$

where the phase errors $\phi_{m,n}$ can be arbitrary. Although this solution can be used, it does not exploit the underlying structure of the phase errors; i.e., $\phi_{m,n} = -\frac{4\pi f_{m,n}}{c}\Delta R_m$. In what follows, we will also present other refined versions, building on the second solution of [25], while exploiting prior knowledge of the structure of the phase error.

The l_1 norm can be approximated as [26–29]:

$$\|\boldsymbol{\alpha}\|_1 \approx \sum_{l=1}^{L}\left(|\alpha_l|^2 + \delta\right)^{1/2} \tag{34}$$

In order to overcome the nondifferentiably of the l_1 norm at the origin. Here, δ is a small non-negative parameter. Using this approximation, the minimisation problem in (31) becomes

$$\{\hat{\boldsymbol{\alpha}}, \hat{\boldsymbol{\Lambda}}\} = \arg\min_{\boldsymbol{\alpha},\boldsymbol{\Lambda}}\left\{\|\boldsymbol{S} - \boldsymbol{\Lambda}\boldsymbol{F}\boldsymbol{\alpha}\|_2^2 + \lambda\sum_{l=1}^{L}\left(|\alpha_l|^2 + \delta\right)^{1/2}\right\}. \tag{35}$$

The solution of (35) tends to the solution of (31) as δ approaches zero. Therefore, a small value of δ should be used to ensure the validity of this approximation. The quasi-Newton approach can be adopted to solve the modified l_1 regularised optimisation (35), as below.

The gradient of the objective function of (35) is given by

$$\nabla(\boldsymbol{\alpha}) = \boldsymbol{H}(\boldsymbol{\alpha})\boldsymbol{\alpha} - 2\,\boldsymbol{F}^H\boldsymbol{\Lambda}^H\boldsymbol{S}, \tag{36}$$

where the superscript ^{H} denotes the Hermitian transpose operation. Here, $\boldsymbol{H}$ is the Hessian matrix given by

$$\boldsymbol{H}(\boldsymbol{\alpha}) = 2\,\boldsymbol{F}^H\boldsymbol{\Lambda}^H\boldsymbol{\Lambda}\boldsymbol{F} + \lambda\boldsymbol{W}(\boldsymbol{\alpha}) = 2\boldsymbol{F}^H\boldsymbol{F} + \lambda\boldsymbol{W}(\boldsymbol{\alpha}), \tag{37}$$

where

$$\boldsymbol{W}(\boldsymbol{\alpha}) = \text{diag}\left\{\ldots, \left(|\alpha_l|^2 + \delta\right)^{-1/2}, \ldots\right\}. \tag{38}$$

Since the Hessian matrix is a function of the unknown $\boldsymbol{\alpha}$, the minimisation (35) is solved iteratively. Given the estimates $\hat{\boldsymbol{\alpha}}(i)$ and $\hat{\boldsymbol{\Lambda}}(i)$ from a previous iteration i, the new solutions at iteration $i+1$ are obtained in the following two steps.

1. Calculate $\hat{\boldsymbol{\alpha}}(i+1)$ by setting $\nabla(\boldsymbol{\alpha}) = \boldsymbol{0}$ given $\boldsymbol{H}(\hat{\boldsymbol{\alpha}}(i))$ and $\hat{\boldsymbol{\Lambda}}(i)$:

$$\begin{aligned}\hat{\boldsymbol{\alpha}}(i+1) &= 2\,(\boldsymbol{H}(\hat{\boldsymbol{\alpha}}(i)))^{-1}\boldsymbol{F}^H(\hat{\boldsymbol{\Lambda}}(i))^H\boldsymbol{S} \\ &= \left(\boldsymbol{F}^H\boldsymbol{F} + \frac{1}{2}\lambda\boldsymbol{W}(\hat{\boldsymbol{\alpha}}(i))\right)^{-1}\boldsymbol{F}^H(\hat{\boldsymbol{\Lambda}}(i))^H\boldsymbol{S}.\end{aligned} \tag{39}$$

2. Calculate $\hat{\boldsymbol{\Lambda}}(i+1)$ given $\hat{\boldsymbol{\alpha}}(i+1)$. The phase error matrix $\hat{\boldsymbol{\Lambda}}(i+1)$ is obtained by solving:

$$\hat{\boldsymbol{\Lambda}}(i+1) = \arg\min_{\boldsymbol{\Lambda}}\|\boldsymbol{S} - \boldsymbol{\Lambda}\boldsymbol{F}\hat{\boldsymbol{\alpha}}(i+1)\|_2^2 \tag{40}$$

or equivalently

$$\hat{\boldsymbol{\Lambda}}_m(i+1) = \arg\min_{\boldsymbol{\Lambda}_m}\|\boldsymbol{S}_m - \boldsymbol{\Lambda}_m\boldsymbol{F}_m\hat{\boldsymbol{\alpha}}(i+1)\|_2^2 \tag{41}$$

for $m = 2, \ldots, M$. Note that $\mathbf{\Lambda}_1 = \boldsymbol{I}_{N_1}$ (since $\Delta R_1 = 0$); thus, no estimation is required for $\mathbf{\Lambda}_1$.

The algorithm may be halted when the objective function falls below a threshold, or when a maximum number of iterations is reached, or when the relative change in the objective function falls below a threshold.

Various methods for calculating the phase error matrix $\hat{\mathbf{\Lambda}}(i+1)$ in Step 2 are given in the following sections.

4.1. Unstructured Approach

Ignoring the underlying structure of the phase errors, i.e., $\phi_{m,n} = -\frac{4\pi f_{m,n}}{c}\Delta R_m$, $\hat{\mathbf{\Lambda}}_m$ can be considered as a diagonal matrix with arbitrary elements $\phi_{m,n}$:

$$\mathbf{\Lambda}_m = \mathrm{diag}\{\ldots, \exp\{j\phi_{m,n}\}, \ldots\}_{n=1,\ldots,N_m}. \tag{42}$$

Therefore, $\phi_{m,n}$ can be estimated as [25]

$$\hat{\phi}_{m,n}(i+1) = \tan^{-1}\frac{\Im\{S_{m,n}\hat{Y}^*_{m,n}(i+1)\}}{\Re\{S_{m,n}\hat{Y}^*_{m,n}(i+1)\}} \tag{43}$$

where $\Im\{\cdot\}$ and $\Re\{\cdot\}$ denote operations to extract the imaginary and real parts of a complex number, and $\tan^{-1}$ stands for a four-quadrant arctangent operation. Here, $\hat{Y}_{m,n}(i+1)$ is the nth element of $\hat{\boldsymbol{Y}}_m(i+1)$ which is defined as $\hat{\boldsymbol{Y}}_m(i+1) = \boldsymbol{F}_m\hat{\boldsymbol{\alpha}}(i+1)$. As a result, we obtain:

$$\hat{\mathbf{\Lambda}}_m(i+1) = \mathrm{diag}\{\ldots, \exp\{j\hat{\phi}_{m,n}(i+1)\}, \ldots\}_{n=1,\ldots,N_m}. \tag{44}$$

4.2. Gauss–Newton Approach

Taking into account the underlying structure of the phase errors, $\mathbf{\Lambda}_m(\Delta R_m)$ is a function of ΔR_m, and the minimisation (41) can be re-expressed as

$$\widehat{\Delta R}_m(i+1) = \underset{\Delta R_m}{\arg\min} \|\boldsymbol{S}_m - \mathbf{\Lambda}_m(\Delta R_m)\boldsymbol{F}_m\hat{\boldsymbol{\alpha}}(i+1)\|_2^2. \tag{45}$$

By letting $\boldsymbol{e}_m = \boldsymbol{S}_m - \mathbf{\Lambda}_m(\Delta R_m)\boldsymbol{F}_m\hat{\boldsymbol{\alpha}}(i+1)$, we have

$$\boldsymbol{e}_m = [\ldots, e_{m,n}, \ldots]^T_{n=1,\ldots,N_m} \tag{46}$$

where

$$e_{m,n} = S_{m,n} - U_{m,n}\exp\left\{-\frac{4\pi j f_{m,n}}{c}\Delta R_m\right\}\sum_{l=1}^{L}\hat{\alpha}_l(i+1)\exp\left\{-\frac{4\pi j f_{m,n}}{c}x_l\right\} \tag{47}$$

and

$$U_{m,n} = \exp\{j\phi_{m,n}\}. \tag{48}$$

Here $\hat{\alpha}_l(i+1)$ is the lth element of $\hat{\boldsymbol{\alpha}}(i+1)$. As we are estimating real quantities, it is more convenient to reformulate the problem as the minimisation of a real function in order to apply the Gauss–Newton. The details of the Gauss–Newton algorithm for updating $\widehat{\Delta R}_m(i+1)$ are given in Appendix A.

Using $\widehat{\Delta R}_m(i+1)$, we obtain the phase error matrix as

$$\hat{\mathbf{\Lambda}}_m(i+1) = \mathrm{diag}\left\{\ldots, \exp\left\{-\frac{4\pi j\widehat{\Delta R}_m(i+1)}{c}f_{m,n}\right\}, \ldots\right\}_{n=1,\ldots,N_m}. \tag{49}$$

4.3. Linear Regression-Based Approach

By noting that $\phi_{m,n} = -\frac{4\pi\Delta R_m}{c} f_{m,n}$, the gradient $-\frac{4\pi\Delta R_m}{c}$ can be calculated via a linear least squares estimator using $\hat{\phi}_{m,n}$ obtained from (43) [30]. Specifically, we have

$$\left[-\frac{4\pi\widehat{\Delta R}_m(i+1)}{c}, \hat{\phi}_m^{\dagger}(i+1)\right]^T = (\boldsymbol{A}_m^T \boldsymbol{A}_m)^{-1} \boldsymbol{A}_m^T \boldsymbol{b}_m(i+1) \tag{50}$$

where

$$\boldsymbol{A}_m = [\ldots, \boldsymbol{A}_{m,n}, \ldots]_{n=1,\ldots,N_m}^T, \text{ with } \boldsymbol{A}_{m,n} = [f_{m,n}, 1] \tag{51a}$$

$$\boldsymbol{b}_m(i+1) = [\ldots, \hat{\phi}_{m,n}^{\text{unwrapped}}(i+1), \ldots]_{n=1,\ldots,N_m}^T. \tag{51b}$$

Note that $\hat{\phi}_{m,n}^{\text{unwrapped}}(i+1)$ is the unwrapped version of $\hat{\phi}_{m,n}(i+1)$ and $\hat{\phi}_m^{\dagger}(i+1)$ is an estimate for the initial phase $\phi_m^{\dagger}(i+1)$ which results from the unwrapping process. From (50), $\widehat{\Delta R}_m(i+1)$ is obtained and can then be used for computing $\hat{\boldsymbol{\Lambda}}_m(i+1)$ as in (49).

4.4. Differenced-Phase-Based Approach

Subtracting the estimated phase errors of two successive frequency bins, we obtain:

$$-\frac{4\pi(f_{m,n+1} - f_{m,n})}{c}\Delta R_m = \phi_{m,n+1} - \phi_{m,n}. \tag{52}$$

Therefore, ΔR_m can be estimated as [30]

$$\widehat{\Delta R}_m(i+1) = -\frac{c}{4\Delta f_m(N_m - 1)} \sum_{n=1}^{N_m-1} \Delta\phi_{m,n}(i+1) \tag{53}$$

where

$$\Delta\phi_{m,n}(i+1) = \tan^{-1}\frac{\sin(\hat{\phi}_{m,n+1}(i+1) - \hat{\phi}_{m,n}(i+1))}{\cos(\hat{\phi}_{m,n+1}(i+1) - \hat{\phi}_{m,n}(i+1))}. \tag{54}$$

Note that the four-quadrant arctangent has been used here to handle the phase wrapping. An estimate of the phase error matrix $\hat{\boldsymbol{\Lambda}}_m(i+1)$ is now obtained as in (49) using $\widehat{\Delta R}_m(i+1)$ in (53).

5. Simulation and Discussion

Numerical simulations are presented in this section to evaluate the performance of the methods described in previous sections.

5.1. Scenario 1: Two Sub-Bands

We consider a synthetic scenario with two sub-bands at carrier frequencies of $f_1 = 6$ GHz and $f_2 = 8$ GHz, each having a bandwidth of $B = 300$ MHz and 64 frequency steps (i.e., $N = N_1 = N_2 = 64$). The range profile is discretised over a grid with a length of $(N-1)c/(2B) = 31.5$ m and a grid step of $\Delta_{\text{Grid}} = c/(10B) = 0.1$ m. We consider a far-field target consisting of six point scatterers which are aligned with the grid. Figure 1 plots the true range profile of the target. We set $\Delta R_2 = 2.78\Delta_{\text{Grid}}$ for the case of existing phase errors. The signal-to-noise ratio is set to 10 dB.

Figure 2 compares the reconstructed range profiles obtained by the conventional OMP algorithm without and with the presence of phase errors. The OMP is terminated when the signal residual reaches the noise level or after 15 iterations have been carried out. We observe that OMP successfully reconstructs the range profile of the target by correctly identifying the scatterers of the target with accurate range and coefficient estimates when no phase errors exist. However, OMP provides unsatisfactory results in the presence of phase errors, where the reconstructed image is observed as exhibiting many spurious scatterers. Similar observations are obtained for the results obtained by the conventional l_1-norm regularised optimisation solver (without phase error correction), as shown in Figure 3. Here, we set $\delta = 10^{-5}$ and $\lambda = 0.001 \max |c_l|$ where c_l is the lth element of $\boldsymbol{c} = \boldsymbol{F}^H \tilde{\boldsymbol{S}}$. The l_1-norm regularised optimisation solver is stopped if the relative change in the l_2-norm of the range profile vector $\boldsymbol{\alpha}$ falls below 10^{-5} or after it reaches 500 iterations. The performance degradation of these conventional sparse reconstruction techniques is not unexpected, since they were not originally developed to cope with dictionary mismatch arising from the presence of the phase errors.

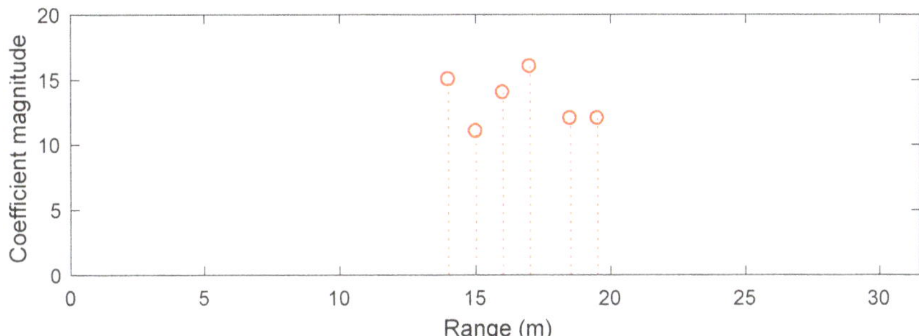

Figure 1. True range profile of synthetic target under consideration.

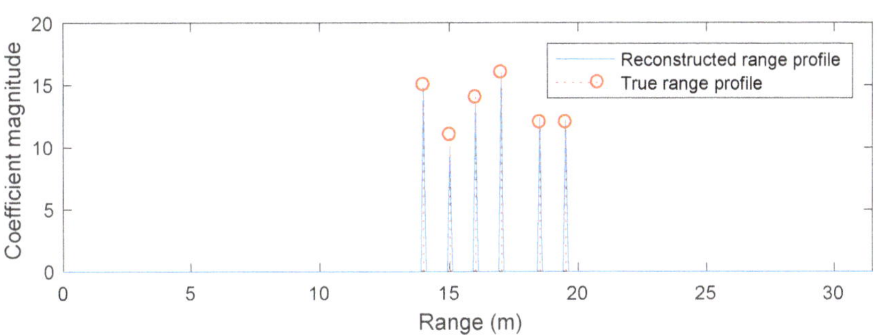

(**a**) Without the presence of phase errors.

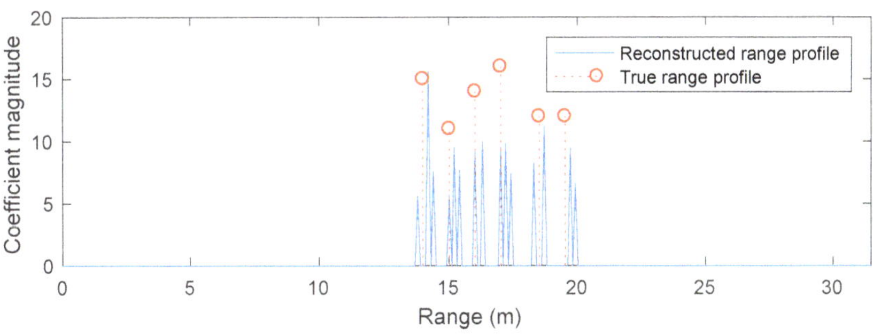

(**b**) With the presence of phase errors.

Figure 2. Performance of conventional OMP.

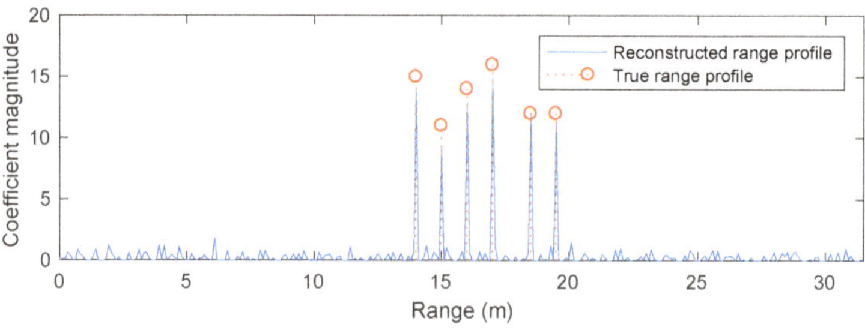

(a) Without the presence of phase errors.

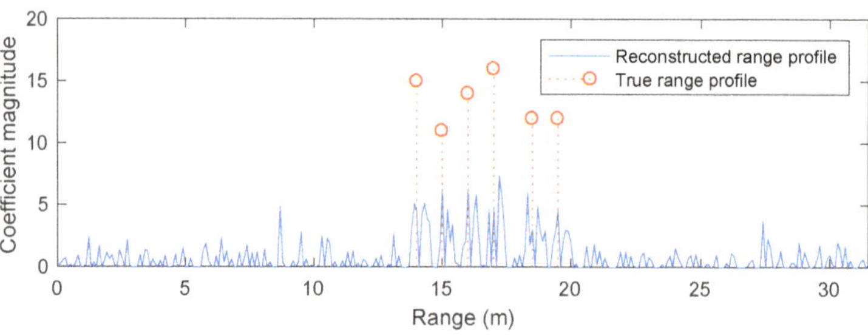

(b) With the presence of phase errors.

Figure 3. Performance of conventional l_1-norm regularised optimisation solver.

Figure 4 shows the reconstructed range profile obtained by the POMP. POMP constructs candidate dictionaries based on a grid of ΔR_2 with a grid step size of $\Delta_{\text{Grid}}/100$. The same stopping criteria of OMP is used for POMP. We observe that POMP produces a range profile which is almost identical to the ground truth, thereby demonstrating the effectiveness of POMP in terms of dealing with the phase errors between different sub-bands.

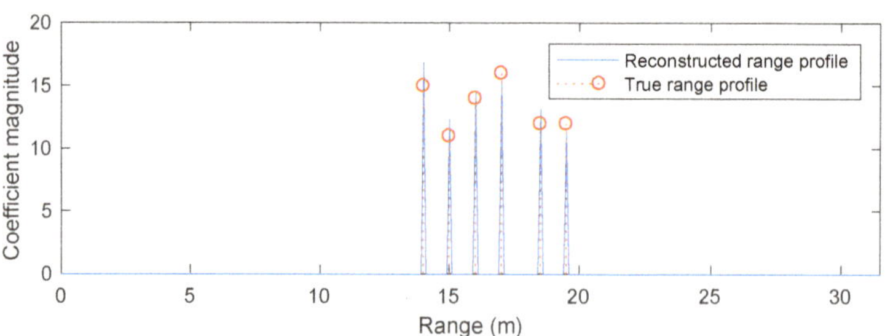

Figure 4. Performance of POMP in the presence of phase errors.

Figure 5 shows the results obtained by different l_1-norm regularised optimisation solvers with phase error correction, as presented in Section 4. The same parameters and stopping criteria of the conventional l_1-norm regularised optimisation solver as described above are used in the simulations. Although these algorithms exhibit some improvements over the conventional l_1-norm regularisation (i.e., without phase error correction), they provide poorer results compared to that of POMP. Specifically, the peaks of the reconstructed range profiles obtained by these algorithms only appear close to but not exactly at the true scatterer positions. In addition, the magnitudes of the peaks are much smaller than the ground truth values.

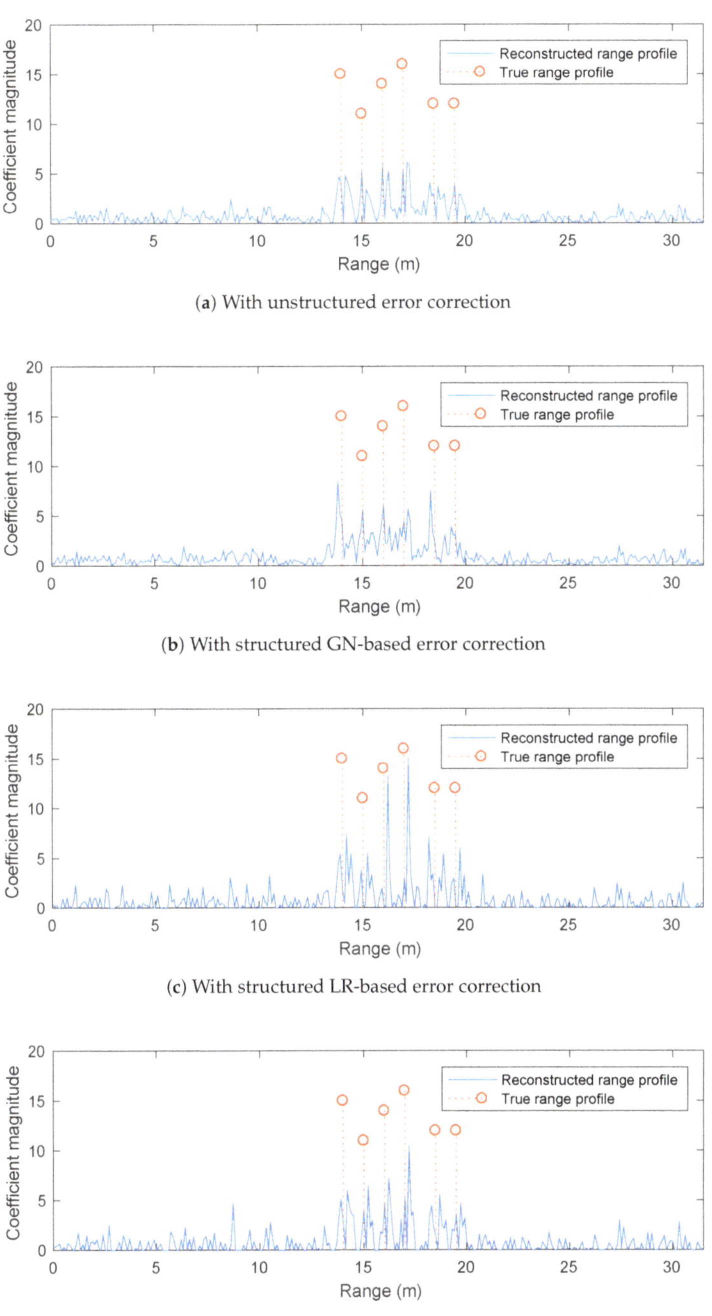

(**a**) With unstructured error correction

(**b**) With structured GN-based error correction

(**c**) With structured LR-based error correction

(**d**) With structured DP-based error correction

Figure 5. Performance of l_1-norm regularised optimisation solver with phase error correction.

The inferior performances of these algorithms can be explained by noting that the l_1-norm regularised optimisation in (31) is a nonconvex problem due to the phase error matrix $\mathbf{\Lambda}$. Figure 6 plots the objective function of (31) as a function of ΔR_2 assuming that $\boldsymbol{\alpha}$ is perfectly known. We observe that this objective function has many local maxima and minima, confirming the non-convexity of the l_1-norm regularised optimisation in (31). The reason for this non-convexity is explained in Appendix B. Due to this nonconvexity, the iterative solvers presented in Section 4 are prone to converge to local minima; thus, limiting the effectiveness of this approach.

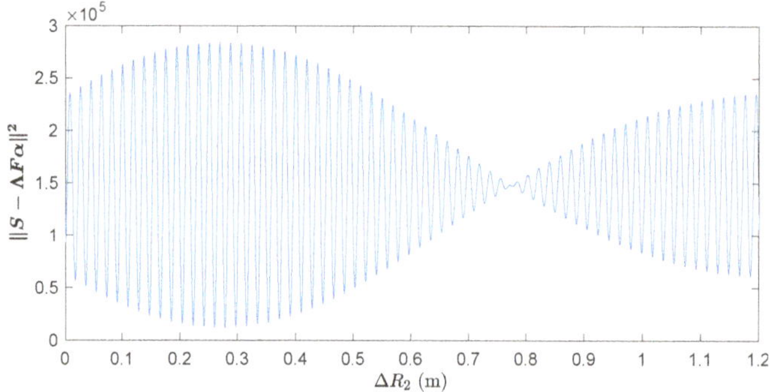

Figure 6. Illustration of the nonconvexity of the l_1-norm regularised optimisation problem (31). The objective function of (31) is plotted against ΔR_2 assuming that $\boldsymbol{\alpha}$ is perfectly known.

The performance of the OMP, POMP, and l_1-norm regularised optimisation methods are now compared using the earth mover's distance (EMD) between the true and reconstructed range profiles. EMD [31] is a metric estimating the distance between two distributions or equivalently the minimal amount of work required to transform one distribution to the other. Figures 7 and 8 show the EMD performance of the OMP, POMP, and l_1-norm regularised optimisation methods, averaged over 100 Monte Carlo runs, versus different levels of SNR (signal-to-noise ratio) and phase error, respectively. It is observed that the POMP method yields the smallest EMD values amongst all algorithms considered. Since a smaller value of EMD corresponds to a higher level of similarity between the true and reconstructed range profiles, this observation indicates that the reconstructed range profile obtained by POMP is closer to the ground-truth range profiles than those obtained from the OMP and l_1-norm regularised optimisation methods. This verifies the performance advantage of the POMP method from a statistical point of view.

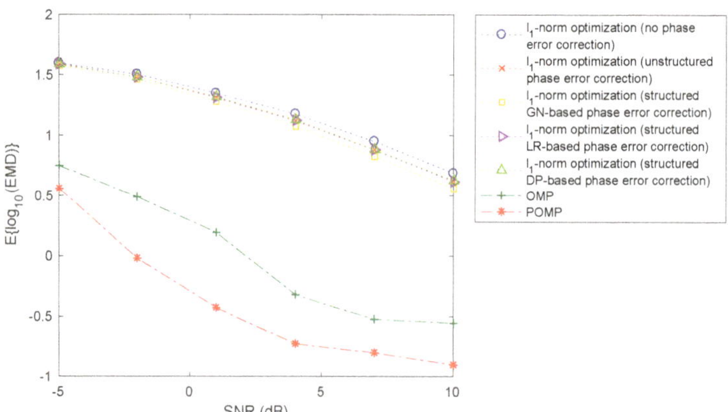

Figure 7. Earth mover's distance (EMD) performance of the OMP, POMP, and l_1-norm regularised optimisation methods versus various of SNRs ($\Delta R_2 = 2.78\Delta_{\text{Grid}}$).

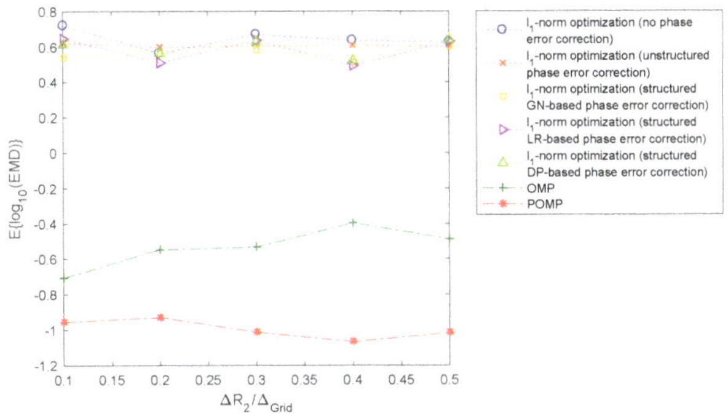

Figure 8. EMD performance of the OMP, POMP, and l_1-norm regularised optimisation methods versus various levels of phase error (SNR = 10 dB).

5.2. Scenario 2: Four Sub-Bands

We now consider another scenario with four sub-bands at carrier frequencies of $f_1 = 6$ GHz, $f_2 = 8$ GHz, $f_3 = 10$ GHz, and $f_4 = 12$ GHz, each having a bandwidth of $B = 300$ MHz and 64 frequency steps (i.e., $N = N_1 = N_2 = N_3 = N_4 = 64$). The range errors are set to $\Delta R_2 = 2.78\Delta_{\text{Grid}}$, $\Delta R_3 = 1.33\Delta_{\text{Grid}}$, and $\Delta R_4 = 3.69\Delta_{\text{Grid}}$. Other simulation parameters and the true range profile of the target remain unchanged as in the previous simulation example.

Figure 9 compares the reconstructed range profiles obtained by the conventional OMP algorithm and the POMP algorithm presented in Section 3.2. OMP results in an unsatisfactorily reconstructed range profile with many spurious peaks, as expected, because it ignores the phase errors between different sub-bands. In contrast, the POMP is capable of reconstructing the true range profile with a high accuracy thanks to the use of dictionary learning with a pruning process. Note that, given the inferior performance of the l_1-norm regularised optimisation approach compared to POMP, as demonstrated in the previous simulation scenario, this approach is excluded from the comparison here.

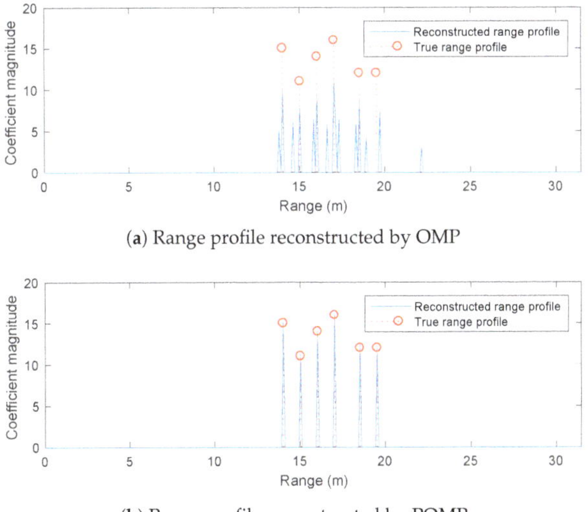

(**a**) Range profile reconstructed by OMP

(**b**) Range profile reconstructed by POMP

Figure 9. Performance comparison between OMP and POMP for Simulation Scenario 2 (with four sub-bands).

6. Conclusions

This paper explores the use of the POMP algorithm and l_1-norm regularisation solvers for the problem of sparsity-driven HRRP with bandwidth stitching in the presence of phase errors. We observe that the l_1-norm regularisation solvers do not provide significant performance improvement over the conventional sparse reconstruction algorithms due to the nonconvexity of l_1-norm regularised optimisation when phase errors exists. In contrast, POMP is observed to be capable of effectively dealing with the phase errors and thus be able to reconstruct the range profile of the target with high accuracy. Simulation results show a significant performance improvement by POMP over OMP and the conventional and refined l_1-norm regularisation. In future work, we propose using experimental data for a more general scenario where the true scatterers constituting the target are located in off-grid positions with respect to the dictionary grid, and the true range errors have off-grid values. We shall also consider the more general case of frequency-dependence of scatterer RCS (radar cross-section). A potential approach for this is to exploit the framework of spectral compressive sensing [32,33].

Author Contributions: Conceptualisation, P.B., N.H.N., and H.-T.T.; methodology, P.B., N.H.N., and H.-T.T.; simulation, N.H.N.; validation, P.B., N.H.N., and H.-T.T.; writing—original draft preparation, N.H.N.; writing—review and editing, P.B., N.H.N., and H.-T.T. All authors have read and agreed to the published version of the manuscript.

Funding: This research was funded by Defence Science and Technology Group, Australia.

Conflicts of Interest: The authors declare no conflict of interest.

Appendix A. Derivation of Gauss–Newton Algorithm for Estimation of $\widehat{\Delta R}_m(i+1)$

We rewrite $e_{m,n}$ as

$$e_{m,n} = S_{m,n} - \exp\left\{-\frac{4\pi j f_{m,n}}{c}\Delta R_m\right\} Z_{m,n} \tag{A1}$$

with

$$Z_{m,n} = U_{m,n}\sum_{l=1}^{L}\hat{\alpha}_l(i+1)\exp\left\{-\frac{4\pi j f_{m,n}}{c}x_l\right\}. \tag{A2}$$

By noting that

$$\begin{aligned}
&\exp\left\{-\frac{4\pi j f_{m,n}}{c}\Delta R_m\right\} Z_{m,n}\\
&= Z^R_{m,n}\cos\left\{\frac{4\pi f_{m,n}}{c}\Delta R_m\right\} + Z^I_{m,n}\sin\left\{\frac{4\pi f_{m,n}}{c}\Delta R_m\right\}\\
&\quad + j\left(Z^I_{m,n}\cos\left\{\frac{4\pi f_{m,n}}{c}\Delta R_m\right\} - Z^R_{m,n}\sin\left\{\frac{4\pi f_{m,n}}{c}\Delta R_m\right\}\right)
\end{aligned} \tag{A3}$$

where $Z^R_{m,n}$ and $Z^I_{m,n}$ are the real and imaginary components of $Z_{m,n}$, we can decouple and stack the real and imaginary components of $\boldsymbol{e}_m$ to form a real-valued vector as

$$\boldsymbol{\epsilon}_m = \left[\left(\boldsymbol{e}^R_m\right)^T, \left(\boldsymbol{e}^I_m\right)^T\right]^T \tag{A4}$$

where

$$\boldsymbol{e}^R_m = \left[\ldots, e^R_{m,n}, \ldots\right]^T_{n=1,\ldots,N_m} \tag{A5}$$

$$\boldsymbol{e}^I_m = \left[\ldots, e^I_{m,n}, \ldots\right]^T_{n=1,\ldots,N_m} \tag{A6}$$

and

$$e_{m,n}^R = Z_{m,n}^R \cos\left\{\frac{4\pi f_{m,n}}{c}\Delta R_m\right\} + Z_{m,n}^I \sin\left\{\frac{4\pi f_{m,n}}{c}\Delta R_m\right\} \tag{A7}$$

$$e_{m,n}^I = Z_{m,n}^I \cos\left\{\frac{4\pi f_{m,n}}{c}\Delta R_m\right\} - Z_{m,n}^R \sin\left\{\frac{4\pi f_{m,n}}{c}\Delta R_m\right\}. \tag{A8}$$

Using (46)–(A8), the minimization (45) becomes

$$\widehat{\Delta R}_m(i+1) = \arg\min_{\Delta R_m} \|\boldsymbol{\epsilon}(\Delta R_m)\|_2^2. \tag{A9}$$

By adopting the Gauss–Newton algorithm, $\widehat{\Delta R}_m(i+1)$ can be computed from $\widehat{\Delta R}_m(i)$ as

$$\widehat{\Delta R}_m(i+1) = \widehat{\Delta R}_m(i) - \left(\boldsymbol{J}_m^T(i)\boldsymbol{J}_m(i)\right)^{-1} \boldsymbol{J}_m^T(i)\boldsymbol{\epsilon}(\widehat{\Delta R}_m(i)) \tag{A10}$$

where $\boldsymbol{\epsilon}(\widehat{\Delta R}_m(i))$ is an estimated version of $\boldsymbol{\epsilon}$ computed from $\widehat{\Delta R}_m(i)$ and $\boldsymbol{J}_m(i)$ is the Jacobian of $\boldsymbol{\epsilon}(\Delta R_m)$ with respect to ΔR_m evaluated at $\Delta R_m = \widehat{\Delta R}_m(i)$.

The expression for the Jacobian $\boldsymbol{J_m}$ of $\boldsymbol{\epsilon}(\Delta R_m)$ with respect to ΔR_m is given by

$$\boldsymbol{J}_m = \left[\left(\boldsymbol{J}_m^R\right)^T, \left(\boldsymbol{J}_m^I\right)^T\right]^T \tag{A11}$$

where

$$\boldsymbol{J}_m^R = \left[\ldots, J_{m,n}^R, \ldots\right]_{n=1,\ldots,N_m}^T \tag{A12}$$

$$\boldsymbol{J}_m^I = \left[\ldots, J_{m,n}^I, \ldots\right]_{n=1,\ldots,N_m}^T \tag{A13}$$

and

$$J_{m,n}^R = \frac{4\pi f_{m,n}}{c}\left(-Z_{m,n}^R \sin\left\{\frac{4\pi f_{m,n}}{c}\Delta R_m\right\} + Z_{m,n}^I \cos\left\{\frac{4\pi f_{m,n}}{c}\Delta R_m\right\}\right) \tag{A14}$$

$$J_{m,n}^I = -\frac{4\pi f_{m,n}}{c}\left(Z_{m,n}^I \sin\left\{\frac{4\pi f_{m,n}}{c}\Delta R_m\right\} + Z_{m,n}^R \cos\left\{\frac{4\pi f_{m,n}}{c}\Delta R_m\right\}\right). \tag{A15}$$

Appendix B. Analysis of the Nonconvexity of the l_1-Norm Regularised Optimisation Problem (31)

From Equation (31)

$$\{\hat{\boldsymbol{\alpha}}, \hat{\boldsymbol{\Lambda}}\} = \arg\min_{\boldsymbol{\alpha},\boldsymbol{\Lambda}} \left\{\|\boldsymbol{S} - \boldsymbol{\Lambda F \alpha}\|_2^2 + \mu\|\boldsymbol{\alpha}\|_1\right\}, \tag{A16}$$

or

$$\{\hat{\boldsymbol{\alpha}}, \hat{\boldsymbol{\Lambda}}\} = \arg\min_{\boldsymbol{\alpha}} \arg\min_{\boldsymbol{\Lambda}} \left\{\|\boldsymbol{S} - \boldsymbol{\Lambda F \alpha}\|_2^2 + \mu\|\boldsymbol{\alpha}\|_1\right\}. \tag{A17}$$

$$\|\boldsymbol{S} - \boldsymbol{\Lambda F \alpha}\|_2^2 = \{\boldsymbol{S} - \boldsymbol{\Lambda F \alpha}\}^H \{\boldsymbol{S} - \boldsymbol{\Lambda F \alpha}\} \tag{A18}$$

$$\|\boldsymbol{S} - \boldsymbol{\Lambda F \alpha}\|_2^2 = \boldsymbol{S}^H\boldsymbol{S} - 2\Re e[(\boldsymbol{F}^H\boldsymbol{\Lambda}^H\boldsymbol{S})^H\boldsymbol{\alpha}] + \boldsymbol{\alpha}^H\boldsymbol{F}^H\boldsymbol{F\alpha}. \tag{A19}$$

Only the term $(\boldsymbol{F}^H\boldsymbol{\Lambda}^H\boldsymbol{S})^H\boldsymbol{\alpha}$ is dependant on ΔR_2, so the objective function is minimised with respect to ΔR_2 when the correlation between $\boldsymbol{F}^H\boldsymbol{\Lambda}^H\boldsymbol{S}$ and $\boldsymbol{\alpha}$ is maximised. Now

$$\boldsymbol{F}^H\boldsymbol{\Lambda}^H\boldsymbol{S} = \boldsymbol{F}_1^H\boldsymbol{S}_1 + \boldsymbol{F}_2^H\boldsymbol{\Lambda}_2^H\boldsymbol{S}_2 \tag{A20}$$

And the terms $\boldsymbol{F}_1^H\boldsymbol{S}_1$ and $\boldsymbol{F}_2^H\boldsymbol{\Lambda}_2^H\boldsymbol{S}_2$ represent the conventional pulse compression (i.e., transformation from frequency domain to range domain) for each of the radars with a range offset

ΔR_2. These are non-sparse and low resolution due to the oversampling in range. So $F^H \Lambda^H(\Delta R_2) S$ represents the linear superposition of the two conventional range profiles. When these range profiles from the two radars are correctly aligned in range, they will better correlate with the true range profile α. Also if the range offset ΔR_2 is such that a scatterer for one radar is superimposed upon a different scatterer for the other radar, a local minimum will occur. That explains Figure 6 and the reason for its non-convexity.

Appendix C. Analysis of Operation Count for POMP

Consider first the case of $M = 2$. Let the size of the grid for the parameter ΔR_2 be $L_\Delta = 2^{N_\Delta}$, so that a dictionary is constructed for each of these 2^{N_Δ} values of ΔR_2. OMP is implemented with the number of dictionaries halved at each stage; hence, the name "pruned" OMP. At stage k of OMP, the number of complex multiplications and divisions for a single dictionary is denoted $C_{omp}(k)$. Here $k = 1, \ldots, N_\Delta + 1$ with $2^{N_\Delta - k+1}$ dictionaries considered at stage k. The purpose of this section is to estimate the dependence of the computational cost of POMP on the size of the dictionary L_x and the grid size L_Δ for the parameter ΔR_2.

With reference to Table 1, the significant costs for POMP are associated with the Identify and Update steps. At each stage of OMP for a given dictionary, the Identify step performs atom/residual correlations which require $\sim O(N_m L_x)$ complex multiplications. The Update step performs a linear least squares estimation requiring Gaussian elimination which, at stage k of OMP, has an operation count $\sim O(k^3)$.

Due to the halving of the number of dictionaries at each stage, the total operational count required until only one dictionary is left (although more OMP steps may be required for that dictionary until the residual is sufficiently small) is therefore of order

$$\sum_{k=1}^{N_\Delta+1} (N_m L_x + k^3) 2^{N_\Delta - k+1} \tag{A21}$$

or

$$(2^{N_\Delta+1} - 1) N_m L_x + (N_\Delta + 1)^3 + 2^{N_\Delta+1} \sum_{k=1}^{N_\Delta} k^3 z^k \tag{A22}$$

with $z = \frac{1}{2}$.

The finite sum $\sum_{k=1}^{n} k^3 z^k$ is referred to as a low-order polylogarithm, for which a formula may be derived [34]. This formula can be shown to have a leading term of order $n^3 z^{n+3}$ so that the overall operation count for POMP is $\sim O((2^{N_\Delta+1} - 1) N_m L_x + N_\Delta^3)$. As the size of the grid for ΔR_2 is $L_\Delta = 2^{N_\Delta}$, $N_\Delta = \frac{\log L_\Delta}{\log 2}$, and the operation count in terms of L_Δ is $\sim O\left(L_\Delta N_m L_x + \left(\frac{\log L_\Delta}{\log 2}\right)^3\right)$. We see that to leading order the computational cost is proportional to the size of the grid for ΔR_2.

For the case of general M this cost is multiplied by $(M-1)$.

References

1. Ai, X.; Li, Y.; Wang, X.; Xiao, S. Some results on characteristics of bistatic high-range resolution profiles for target classification. *IET Radar Sonar Navig.* **2012**, *6*, 379–388. [CrossRef]
2. Ai, X.; Zou, X.; Liu, J.; Li, Y.; Xiao, S. Bistatic high range resolution profiles of precessing cone-shaped targets. *IET Radar Sonar Navig.* **2013**, *7*, 615–622. [CrossRef]
3. Liu, H.; Chen, F.; Du, L.; Bao, Z. Robust radar automatic target recognition algorithm based on HRRP signature. *Front. Electr. Electron. Eng.* **2012**, *7*, 49–55. [CrossRef]
4. Tait, P. *Introduction to Radar Target Recognition*; IET: London, UK, 2005.
5. Mao, Z.; Wei, Y. Interpulse-frequency-agile and intrapulse-phase-coded waveform optimisation for extend-range correlation sidelobe suppression. *IET Radar Sonar Navig.* **2017**, *11*, 1530–1539. [CrossRef]

6. Bai, X.; Zhou, F.; Bao, Z. High-resolution radar imaging of space targets based on HRRP series. *IEEE Trans. Geosci. Remote. Sens.* **2013**, *52*, 2369–2381. [CrossRef]
7. Ajorloo, A.; Hadavi, M.; Bastani, M.H.; Nayebi, M.M. Radar HRRP modeling using dynamic system for radar target recognition. *Radioengineering* **2014**, *23*, 121–127.
8. Rihaczek, A.W. *Principles of High-Resolution Radar*; Artech House: Norwood, MA, USA, 1996.
9. Wehner, D.R. *High Resolution Radar*; Artech House, Inc.: Norwood, MA, USA, 1987; 484p.
10. Cuomo, K. *A Bandwidth Extrapolation Technique for Improved Range Resolution Of Coherent Radar Data*; Project Report CJP-60, Rev. 1; Massachusetts Inst of Tech: Lexington, MA, USA, 1992.
11. Borison, S.; Bowling, S.B.; Cuomo, K.M. Super-resolution methods for wideband radar. *Linc. Lab. J.* **1992**, *5*, 441–461.
12. French, A. Improved high range resolution profiling of aircraft using stepped-frequency waveforms with an S-band phased array radar. In Proceedings of the 2006 IEEE Conference on Radar, Verona, NY, USA, 24–27 April 2006; pp. 69–75.
13. Eldar, Y.C.; Kutyniok, G. (Eds.) *Compressed Sensing: Theory and Applications*; Cambridge University Press: New York, NY, USA, 2012.
14. Amin, M. (Ed.) *Compressive Sensing for Urban Radar*; CRC Press: Boca Raton, FL, USA, 2017.
15. Tropp, J.A.; Wright, S.J. Computational methods for sparse solution of linear inverse problems. *Proc. IEEE* **2010**, *98*, 948–958. [CrossRef]
16. Potter, L.C.; Ertin, E.; Parker, J.T.; Cetin, M. Sparsity and compressed sensing in radar imaging. *Proc. IEEE* **2010**, *98*, 1006–1020. [CrossRef]
17. Tomei, S.; Bacci, A.; Giusti, E.; Martorella, M.; Berizzi, F. Compressive sensing-based inverse synthetic radar imaging imaging from incomplete data. *IET Radar Sonar Navig.* **2016**, *10*, 386–397. [CrossRef]
18. Nguyen, N.H.; Dogancay, K.; Tran, H.T.; Berry, P. An image focusing method for sparsity-driven radar imaging of rotating targets. *Sensors* **2018**, *18*. [CrossRef] [PubMed]
19. Qiu, W.; Giusti, E.; Bacci, A.; Martorella, M.; Berizzi, F.; Zhao, H.; Fu, Q. Compressive sensing-based algorithm for passive bistatic ISAR with DVB-T signals. *IEEE Trans. Aerosp. Electron. Syst.* **2015**, *51*, 2166–2180. [CrossRef]
20. Nguyen, N.H.; Dogancay, K.; Tran, H.; Berry, P. Nonlinear least-squares post-processing for compressive radar imaging of a rotating target. In Proceedings of the International Conference on Radar (RADAR), Brisbane, Australia, 27–30 August 2018; pp. 1–6.
21. Winkler, V.; Edrich, M. Multiband radar signal processing. In Proceedings of the 2015 16th International Radar Symposium (IRS), Dresden, Germany, 24–26 June 2015; pp. 108–113.
22. Giusti, E.; Tomei, S.; Bacci, A.; Martorella, M.; Berizzi, F. Autofocus for CS based ISAR imaging in the presence of gapped data. In Proceedings of the 2nd International Workshop on Compressed Sensing Applied to Radar (CoSeRa), Bonn, Germany, 17–19 September 2013.
23. Cuomo, K.M.; Piou, J.E.; Mayhan, J.T. Ultra-wideband coherent processing. *Linc. Lab. J.* **1997**, *10*, 203–222.
24. Li, G.; Varshney, P.K. Micro-Doppler parameter estimation via parametric sparse representation and pruned orthogonal matching pursuit. *IEEE J. Sel. Top. Appl. Earth Obs. Remote. Sens.* **2014**, *7*, 4937–4948. [CrossRef]
25. Onhon, N.O.; Cetin, M. A sparsity-driven approach for joint SAR imaging and phase error correction. *IEEE Trans. Image Process.* **2012**, *21*, 2075–2088. [CrossRef]
26. Cetin, M.; Karl, W.C. Feature-enhanced synthetic aperture radar image formation based on nonquadratic regularization. *IEEE Trans. Image Process.* **2001**, *10*, 623–631. [CrossRef]
27. Zhang, L.; j. Qiao, Z.; Xing, M.; Li, Y.; Bao, Z. High-resolution ISAR imaging with sparse stepped-frequency waveforms. *IEEE Trans. Geosci. Remote. Sens.* **2011**, *49*, 4630–4651. [CrossRef]
28. Vogel, C.R.; Oman, M.E. Fast, robust total variation-based reconstruction of noisy, blurred images. *IEEE Trans. Image Process.* **1998**, *7*, 813–824. [CrossRef]
29. Acar, R.; Vogel, C.R. Analysis of bounded variation penalty methods for ill-posed problems. *Inverse Probl.* **1994**, *10*, 1217–1229. [CrossRef]
30. Liao, Y. Phase and Frequency Estimation: High-Accuracy and Low-Complexity Techniques. Master's Thesis, Worcester Polytechnic Institute, Worcester, MA, USA, 2011.
31. Rubner, Y.; Tomasi, C.; Guibas, L.J. The Earth Mover's Distance as a Metric for Image Retrieval. *Int. J. Comput. Vis.* **2000**, *40*, 99–121. [CrossRef]

32. Duarte, M.F.; Baraniuk, R.G. Spectral compressive sensing. *Appl. Comput. Harmon. Anal.* **2013**, *35*, 111–129. [CrossRef]
33. Liu, S.; Zhang, Y.D.; Shan, T.; Tao, R. Structure-aware Bayesian compressive sensing for frequency-hopping spectrum estimation with missing observations. *IEEE Trans. Signal Process.* **2018**, *66*, 2153–2166. [CrossRef]
34. Olver, F.W.J.; Olde Daalhuis, A.B.; Lozier, D.W.; Schneider, B.I.; Boisvert, R.F.; Clark, C.W.; Miller, B.R.; Saunders, B.V.; Cohl, H.S.; McClain, M.A. (Eds.). *NIST Digital Library of Mathematical Functions;* Release 1.0.25 of 2019-12-15; National Institute of Standards and Technology: Gaithersburg, MD, USA, 2019. Available online: http://dlmf.nist.gov/ (accessed on 23 January 2020).

Article

Compressive Sensing for Tomographic Imaging of a Target with a Narrowband Bistatic Radar

Ngoc Hung Nguyen [1,*], **Paul Berry** [2] **and Hai-Tan Tran** [2]

[1] Maritime Division, Defence Science and Technology Group, Edinburgh 5111, SA, Australia
[2] Intelligence, Surveillance and Space Division, Defence Science and Technology Group, Edinburgh 5111, SA, Australia; paul.berry@dst.defence.gov.au (P.B.); haitan.tran@dst.defence.gov.au (H.-T.T.)
* Correspondence: ngoc.nguyen@dst.defence.gov.au

Received: 12 November 2019; Accepted: 9 December 2019; Published: 13 December 2019

Abstract: This paper introduces a new approach to bistatic radar tomographic imaging based on the concept of compressive sensing and sparse reconstruction. The field of compressive sensing has established a mathematical framework which guarantees sparse solutions for under-determined linear inverse problems. In this paper, we present a new formulation for the bistatic radar tomography problem based on sparse inversion, moving away from the conventional k-space tomography approach. The proposed sparse inversion approach allows high-quality images of the target to be obtained from limited narrowband radar data. In particular, we exploit the use of the parameter-refined orthogonal matching pursuit (PROMP) algorithm to obtain a sparse solution for the sparse-based tomography formulation. A key important feature of the PROMP algorithm is that it is capable of tackling the dictionary mismatch problem arising from off-grid scatterers by perturbing the dictionary atoms and allowing them to go off the grid. Performance evaluation studies involving both simulated and real data are presented to demonstrate the performance advantage of the proposed sparsity-based tomography method over the conventional k-space tomography method.

Keywords: radar tomography; compressive sensing; sparse reconstruction; bistatic radar; radar imaging; parameter-refined orthogonal matching pursuit (PROMP); orthogonal matching pursuit (OMP); k-space tomography; narrowband radar; off-grid compressive sensing

---

1. Introduction

Radar imaging has received much attention for several decades, having a wide range of applications in both civilian and military domains [1–3]. In principle, to obtain high-resolution radar images, a wide bandwidth of radar waveform is required for a fine resolution in the range direction, while a large antenna aperture is required for a fine resolution in the cross-range direction. To overcome the physical constraints of the radar aperture size, a synthesized aperture with a much larger size can be formed by exploiting the relative motion between the radar and target. This is, in fact, the main idea behind the synthetic aperture radar (SAR) and inverse SAR (ISAR) [1]. In recent years, there has been an increasing demand on the radio-frequency (RF) electromagnetic spectrum due to rapid advances in radar and communications, and the radar has to compete for spectrums with many different services, including radio and television broadcasting, communications, and radio-navigation [4]. As a result, the constraints on spectrum availability may present severe limits on signal bandwidth, prompting the need for high-resolution imaging techniques using narrowband radars.

As a consequence, Doppler tomography has been considered for narrowband radar imaging [5–12], which is also called "Doppler radar tomography" (DRT). The main idea of DRT is to utilize the information given by the Doppler frequencies induced from the relative radar-target rotational motion to construct an image of the target, which can be conveniently formulated in the slow-time k-space [8]. Imaging can also be formulated in the more traditional fast-time k-space, the support for which is created by sweeping out the complex samples of the received signal in the angular direction. For a particular transmit frequency, the complex samples for all available aspect angles form a circular arc in the spatial frequency space. The traditional range-Doppler ISAR imaging can be considered as a special case of this fast-time k-space technique when a wideband signal is available and the total rotation angle is small enough that the support region can be approximated as being rectangular. The inversion process for image formation has evolved from traditional tools, such as filtered back projection, to the more modern non-uniform fast Fourier transform (NUFFT). The k-space radar tomography was also considered in bistatic settings [13,14]. The bistatic radar offers several advantages over a monostatic radar, including higher performance against stealth targets, less vulnerability to jamming, and its covertness.

The main objective of this paper is to present a new tomographic imaging technique for a narrowband bistatic radar based on the framework of compressive sensing and sparse reconstruction. The field of compressive sensing has established a mathematical framework which guarantees sparse solutions for underdetermined linear inverse problems that occur across numerous engineering and mathematical science fields. In particular, this framework has found applications in various radar imaging problems, ranging from moving target indication, ISAR imaging, coherence imaging, multichannel imaging, micro-Doppler imaging, to through-the-wall radar imaging (see, e.g., [15–26]). The key contributions of this paper are summarized as follows.

- A new formulation for radar tomography based on sparse inversion is introduced. The main idea is to construct a dictionary of signal prototypes by discretizing the illuminated scene of interest into a grid of discrete points. In this formulation, the received radar signal vector becomes linear to the unknown reflection vector to be estimated. This effectively casts the radar tomography problem under consideration to a sparse linear inverse problem, given that the illuminated scene of interest only contains a small number of dominant scatterers, as is often the case in practice. Such a formulation allows a high-quality image of the target to be obtained under the compressive sensing framework.
- A technical challenge for the tomography formulation based on sparse inversion is the dictionary mismatch problem, resulting from the fact that the true scatterers almost always do not coincide exactly with the dictionary grid. This dictionary mismatch problem has been known in the literature to significantly degrade the performance of conventional sparse reconstruction techniques [27,28]. To overcome this problem, we tried exploiting the use of the parameter-refined orthogonal matching pursuit (PROMP) algorithm [29] to solve the sparsity-based tomographic formulation. Compared to other conventional sparse reconstruction techniques, like the orthogonal matching pursuit (OMP) and convex optimization (see e.g., [30,31], and the references therein), PROMP has the advantage of being capable of dealing with the dictionary mismatch arising from off-grid scatterers by perturbing the dictionary atoms and allowing them to go off the grid. PROMP belongs to the greedy pursuit family which identifies the support of the solution in an iterative manner based on the level of correlation between the input data and the dictionary atoms. As a result, PROMP is computationally efficient and thus suitable for real-time operation.
- Performance evaluation studies involving both simulated and real data are presented to demonstrate the superior performance of the proposed sparsity-based tomography method over the conventional k-space tomography technique.

The remainder of this paper is organized as follows. Section 2 describes the signal model for bistatic radar tomography. Section 3 formulates bistatic radar tomographic imaging as a sparse inversion problem. Section 4 derives the sparse solution based on the PROMP algorithm. Performance studies with simulated and real data are presented in Section 5. The paper ends in Section 6 with some concluding remarks.

2. Signal Model

Figure 1 shows the geometry for the problem of bistatic radar tomographic imaging under consideration. The transmitter Tx and receiver Rx are located in the far field of the target of interest. The bistatic angle between the transmitter and receiver with respect to the target is denoted as β. The transmitter is narrowband, transmitting a continuous waveform at a single frequency f (i.e., the wavelength is $\lambda = c/f$, where c is the speed of signal propagation). A local target coordinate frame $\mathcal{T}(x_1, x_2, x_3)$, which is fixed and rotated with the target, is chosen as the reference frame. Here, both the transmitter and receiver lie on the image plane $\mathcal{X}(x_1, x_2)$ of the target, and the origin of the frame $\mathcal{T}$ is placed at the target rotation centre. It is also assumed that the target rotational speed Ω is constant over the coherent processing interval (CPI) and is accurately estimated a priori.

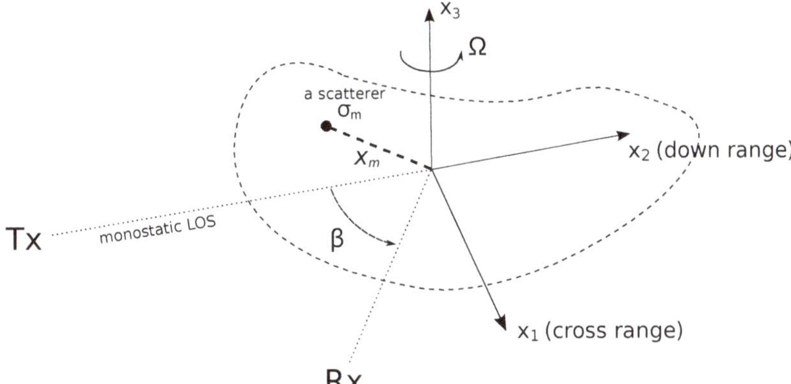

Figure 1. The radar-target geometry of the considered bistatic radar tomographic imaging problem.

The receiver takes one complex-valued scattered signal sample for each rotation angle $\theta_n = \Omega t_n$ of the target (with respect to the axis x_2 of the frame $\mathcal{T}$) at time t_n. The expression of the scattered signal sample collected at time t_n is given by [13,14]

$$s(t_n) = \int_{x_1}\int_{x_2} \sigma(\boldsymbol{x}) \exp\left\{ -j\frac{2\pi}{c} f \left[R(t_n) + 2\cos\left(\frac{1}{2}|\beta|\right) \boldsymbol{x}\cdot\boldsymbol{u}(t_n) \right] \right\} dx_1 dx_2, \tag{1}$$

where $\boldsymbol{x} = [x_1, x_2]^T$, R is the total bistatic range between the target centre (or focus point) and the transmitter and receiver, and $\sigma(\boldsymbol{x})$ is the scatterer reflectivity distribution projected onto the image plane. Note that the total bistatic range R is, in general, a function of time t_n because of the target translational motion. In (1), $\boldsymbol{u} = [u_1, u_2]^T$ denotes the unit vector along the bisector of the bistatic angle β. It is also noted that the radar tomographic imaging problem under interest is considered in the rotating local frame $\mathcal{T}$, and the signal model (1) has already taken into account the rotational motion of the target by the rotation of the unit vector $\boldsymbol{u}(t_n)$ relative to $\mathcal{T}$.

Since x_1 and x_2 in the image domain are discrete variables in practical radar imaging applications, the reflectivity function $\sigma(\boldsymbol{x})$ is commonly discretized over the image plane $\mathcal{X}(x_1, x_2)$ onto a grid of points $\boldsymbol{x}_m$ for $m \in \{1, \ldots, M\}$, as

$$\sigma(\boldsymbol{x}) = \sum_{m=1}^{M} \sigma_m \delta(\boldsymbol{x} - \boldsymbol{x}_m). \tag{2}$$

Substituting (2) into (1) and assuming that translational motion compensation is accurately accomplished by additional pre-processing, we obtain

$$s(t_n) = \sum_{m=1}^{M} \sigma_m \exp\left\{ -j\frac{4\pi}{c} f \cos\left(\frac{1}{2}|\beta|\right) \boldsymbol{x}_m \cdot \boldsymbol{u}(t_n) \right\}. \tag{3}$$

A compact vector form of (3) is given by

$$\boldsymbol{s} = \boldsymbol{\Phi}\boldsymbol{\sigma}, \tag{4}$$

where

$$\boldsymbol{s} = [s(t_1), \ldots, s(t_N)]^T \tag{5}$$

$$\boldsymbol{\sigma} = [\sigma_1, \ldots, \sigma_M]^T \tag{6}$$

and

$$\boldsymbol{\Phi} = [\boldsymbol{\phi}(\boldsymbol{x}_1), \ldots, \boldsymbol{\phi}(\boldsymbol{x}_M)] \tag{7a}$$

$$\boldsymbol{\phi}(\boldsymbol{x}_m) = [\phi(\boldsymbol{x}_m, t_1), \ldots, \phi(\boldsymbol{x}_m, t_N)]^T \tag{7b}$$

$$\phi(\boldsymbol{x}_m, t_n) = \exp\left\{ -j\frac{4\pi}{c} f \cos\left(\frac{1}{2}|\beta|\right) \boldsymbol{x}_m \cdot \boldsymbol{u}(t_n) \right\}. \tag{7c}$$

In practice, where noise is presented, the radar received signal becomes

$$\tilde{\boldsymbol{s}} = \boldsymbol{s} + \boldsymbol{n} = \boldsymbol{\Phi}\boldsymbol{\sigma} + \boldsymbol{n}, \tag{8}$$

where $\boldsymbol{n} = [n(t_1), \ldots, n(t_N)]^T$, with $n(t_n)$ denoting the complex-valued noise term at time t_n.

3. Sparse Inversion Formulation of Bistatic Radar Tomography

The objective of any target imaging problem is to construct a spatial reflectivity map of the target from the backscattered radar signal. Specifically, the ultimate objective is to estimate the unknown reflection vector $\boldsymbol{\sigma}$ from the noisy received signal vector $\tilde{\boldsymbol{s}}$ by solving (8). Since the number of signal samples received, as often is the case in practice, is much smaller than the number of grid points in the reflectivity map (i.e., $N \ll M$), solving (8) is essentially an underdetermined linear inverse problem which requires additional regularization constraints to obtain meaningful solutions.

Typical target images captured by microwave radar signals has been known in the literature to contain a few dominant scattering centers (see, e.g., [15–17,19–23]). As a result, the reflection vector $\boldsymbol{\sigma}$ only has a small number of non-zero elements, thus enjoying a sparse characteristic. Such a sparse characteristic of $\boldsymbol{\sigma}$ can be utilized as a regularizing constraint to solve the underdetermined inverse problem (8), that is,

$$\text{find sparse } \boldsymbol{\sigma} \text{ such that } \tilde{\boldsymbol{s}} \approx \boldsymbol{\Phi}\boldsymbol{\sigma}. \tag{9}$$

This sparse inversion problem can be effectively solved under the compressive sensing framework using sparse reconstruction algorithms. Note that in the compressive sensing context, the matrix $\boldsymbol{\Phi}$ is commonly referred to as the dictionary, and the columns of $\boldsymbol{\Phi}$ are called the atoms, each representing the theoretical scattered signal component of a hypothetical scatterer residing on a grid point of the reflectivity map.

Compressive sensing and sparse reconstruction have been extensively studied in the last two decades, with various techniques proposed. Comprehensive surveys of the state-of-the-art on this topic can be found in [22,23,30,31]. The objective of this paper is to apply the sparse reconstruction approach to the bistatic radar tomographic imaging problem.

The main challenge for this work is that the true scatterers constituting the target do not coincide exactly with the grid which is used to construct the dictionary, leading to dictionary mismatch problems which in turn significantly degrade the performance of conventional sparse reconstruction techniques [27,28]. Several methods have been presented in the literature to address the off-grid dictionary mismatch problems based on the ideas of joint-sparse recovery [32], dictionary perturbation [33,34], sparse Bayesian learning [35,36], and parameter perturbation [29,37]. In this paper, we will exploit the use of the PROMP method [29,37], that is, a parameter perturbation method, to solve the sparse inversion problem (9). The main motivations of using PROMP are twofold. Firstly, PROMP is capable of tackling the dictionary mismatch problem by perturbing the dictionary atoms and allowing them to go off the grid. Secondly, PROMP is computationally efficient and thus suitable for real-time operation because it belongs to the greedy pursuit family which identifies the support of the solution in an iterative manner based on the level of correlation between the input data and the dictionary atoms.

4. Parameter-Refined Orthogonal Matching Pursuit

Table 1 summarizes the overall structure of the PROMP algorithm. As a variant of the greedy pursuit technique, PROMP solves the sparse inversion problem (9) by identifying the support of σ in an iterative greedy manner. In particular, it starts with an empty support set $\Lambda^{[0]} = \emptyset$ and sets the signal $\tilde{\mathbf{s}}$ as the initial signal residual $\boldsymbol{r}^{[0]}$. Like other greedy techniques, one column of $\boldsymbol{\Phi}$ (corresponding to one atom of the dictionary) that produces the largest correlation with the current signal residual $\boldsymbol{r}^{[i]}$ is chosen and added to the support set $\Lambda^{[i]}$ in each iteration. However, a unique feature of PROMP is that it allows the dictionary atoms to go off the grid by perturbing their parameters, thus it can overcome the off-grid dictionary mismatch problem. Specifically, the updated step of PROMP, as different to that of other greedy pursuit techniques, not only estimates the coefficients $\sigma_k^{[i]}$ but also determines the positions $\boldsymbol{x}_k^{[i]} = \left[x_{1,k}^{[i]}, x_{2,k}^{[i]}\right]^T, k = 1, \ldots, i$, of the scatterers associated with the current support set $\Lambda^{[i]}$ via the least-square sense as

$$\begin{aligned} \left\{\hat{\sigma}_k^{[i]}, \hat{\boldsymbol{x}}_k^{[i]}\right\}_{k=1,\ldots,i} &= \arg\min \left\| \tilde{\mathbf{s}} - \sum_{k=1}^{i} \sigma_k^{[i]} \boldsymbol{\phi}\left(\boldsymbol{x}_k^{[i]}\right) \right\|_2 \\ \text{subject to} & \\ \left\| \hat{x}_{1,k}^{[i]} - \bar{x}_{1,k}^{[i]} \right\| &\leq \zeta \text{ and } \left\| \hat{x}_{2,k}^{[i]} - \bar{x}_{2,k}^{[i]} \right\| \leq \zeta, \end{aligned} \tag{10}$$

which is in fact a nonlinear least-square (NLS) estimation problem. Here, $\bar{\boldsymbol{x}}_k^{[i]} = \left[\bar{x}_{1,k}^{[i]}, \bar{x}_{2,k}^{[i]}\right]^T, k = 1, \ldots, i$ are the positions of the dictionary atoms in the current support set $\Lambda^{[i]}$. Note that the constraint in (10) ensures the position estimate for each scatterer staying within the vicinity of the corresponding dictionary atom. A nominal resolution of $\lambda/2$ can be used to set the value of ζ.

Table 1. Summary of PROMP.

INPUT: $\tilde{\mathbf{s}}, \mathbf{\Phi}$.

- Initialization: $\mathbf{r}^{[0]} = \tilde{\mathbf{s}}$, $\Lambda^{[0]} = \emptyset$
- **for** iteration $i = 1; i := i + 1$ until stopping criterion is met **do**
    - Identify:
      $\mathbf{c}^{[i]} = \mathbf{\Phi}^H \mathbf{r}^{[i-1]}$
      $j^{[i]} = \arg\max_j |\mathbf{c}_j^{[i]}|$
      quit the iteration **if** $j^{[i]} \in \Lambda^{[i-1]}$
    - Record selected supports:
      $\Lambda^{[i]} = \Lambda^{[i-1]} \cup j^{[i]}$
    - Nonlinear least-squares estimation:

$$\left\{\hat{\sigma}_k^{[i]}, \hat{\mathbf{x}}_k^{[i]}\right\}_{k=1,\dots,i} = \arg\min \left\| \tilde{\mathbf{s}} - \sum_{k=1}^{i} \sigma_k^{[i]} \boldsymbol{\phi}(\mathbf{x}_k^{[i]}) \right\|_2$$

subject to

$$\left\| \hat{x}_{1,k}^{[i]} - \bar{x}_{1,k}^{[i]} \right\| \leq \zeta \text{ and } \left\| \hat{x}_{2,k}^{[i]} - \bar{x}_{2,k}^{[i]} \right\| \leq \zeta$$

    - Update residual:
      $\mathbf{r}^{[i]} = \tilde{\mathbf{s}} - \sum_{k=1}^{i} \hat{\sigma}_k^{[i]} \boldsymbol{\phi}(\hat{\mathbf{x}}_k^{[i]})$

end for.

OUTPUT: $\left\{\hat{\sigma}_k^{[i]}, \hat{\mathbf{x}}_k^{[i]}\right\}_{k=1,\dots,i}$

The superscript ^{H} stands for the Hermitian transpose.

Since the NLS problem in (10) does not admit a closed-form solution, in what follows we will derive an iterative solution based on the Gauss–Newton (GN) approach [38]. Note that the scatterer reflection coefficient is a complex-valued variable, while the scatterer position is a real-valued variable. For the sake of convenience, we transform (10) into a NLS problem purely in the real-valued domain. In particular, we re-express the cost function in (10) as

$$\left\| \tilde{\mathbf{s}} - \sum_{k=1}^{i} \sigma_k^{[i]} \boldsymbol{\phi}(\mathbf{x}_k^{[i]}) \right\|_2 = \left\| \begin{bmatrix} \text{Real}\{\tilde{\mathbf{s}}\} \\ \text{Imag}\{\tilde{\mathbf{s}}\} \end{bmatrix} - \sum_{k=1}^{i} \begin{bmatrix} \text{Real}\left\{\sigma_k^{[i]} \boldsymbol{\phi}(\mathbf{x}_k^{[i]})\right\} \\ \text{Imag}\left\{\sigma_k^{[i]} \boldsymbol{\phi}(\mathbf{x}_k^{[i]})\right\}, \end{bmatrix} \right\|_2 \tag{11}$$

where explicit expressions of $\text{Real}\left\{\sigma_k^{[i]} \boldsymbol{\phi}(\mathbf{x}_k^{[i]})\right\}$ and $\text{Imag}\left\{\sigma_k^{[i]} \boldsymbol{\phi}(\mathbf{x}_k^{[i]})\right\}$ are given by

$$\text{Real}\left\{\sigma_k^{[i]} \boldsymbol{\phi}(\mathbf{x}_k^{[i]})\right\} = \left[\dots, \left(\text{Real}\{\sigma_k^{[i]}\} \cos\theta_k^{[i]}(t_n) - \text{Imag}\{\sigma_k^{[i]}\} \sin\theta_k^{[i]}(t_n)\right), \dots\right]_{n=1,\dots,N'}^T \tag{12a}$$

$$\text{Imag}\left\{\sigma_k^{[i]} \boldsymbol{\phi}(\mathbf{x}_k^{[i]})\right\} = \left[\dots, \left(\text{Real}\{\sigma_k^{[i]}\} \sin\theta_k^{[i]}(t_n) + \text{Imag}\{\sigma_k^{[i]}\} \cos\theta_k^{[i]}(t_n)\right), \dots\right]_{n=1,\dots,N'}^T \tag{12b}$$

and

$$\theta_k^{[i]}(t_n) = -\frac{4\pi f}{c} \cos\left(\frac{|\beta|}{2}\right) \left(x_{1,k}^{[i]} u_1(t_n) + x_{2,k}^{[i]} u_2(t_n)\right). \tag{13}$$

Now we define

$$\tilde{\mathbf{z}} = \begin{bmatrix} \text{Real}\{\tilde{\mathbf{s}}\} \\ \text{Imag}\{\tilde{\mathbf{s}}\} \end{bmatrix}, \quad \mathbf{z} = \sum_{k=1}^{i} \begin{bmatrix} \text{Real}\left\{\sigma_k^{[i]} \boldsymbol{\phi}(\mathbf{x}_k^{[i]})\right\} \\ \text{Imag}\left\{\sigma_k^{[i]} \boldsymbol{\phi}(\mathbf{x}_k^{[i]})\right\} \end{bmatrix} \tag{14}$$

and

$$\boldsymbol{\xi}^{[i]} = [\boldsymbol{\xi}_1^{[i]T}, \ldots, \boldsymbol{\xi}_i^{[i]T}]^T \text{with } \boldsymbol{\xi}_k^{[i]} = [\sigma_{R,k}^{[i]}, \sigma_{I,k}^{[i]}, x_{1,k}^{[i]}, x_{2,k}^{[i]}]^T. \tag{15}$$

Here, $\sigma_{R,k}^{[i]} = \text{Real}\{\sigma_k^{[i]}\}, \sigma_{I,k}^{[i]} = \text{Imag}\{\sigma_k^{[i]}\}$. Noting that z is a function of $\boldsymbol{\xi}^{[i]}$, (10) is equivalent to

$$\hat{\boldsymbol{\xi}}^{[i]} = \arg\min_{\boldsymbol{\xi}^{[i]}} \left\| \tilde{z} - z(\boldsymbol{\xi}^{[i]}) \right\|_2. \tag{16}$$

This is a NLS problem solely in the real-valued domain, and its solution can be obtained via the following GN iteration [38]

$$\hat{\boldsymbol{\xi}}^{[i]}(h+1) = \hat{\boldsymbol{\xi}}^{[i]}(h) + \left(\boldsymbol{\Gamma}^T(h)\boldsymbol{\Gamma}(h)\right)^{-1}\boldsymbol{\Gamma}^T(h)\left(\tilde{z} - z(\hat{\boldsymbol{\xi}}^{[i]}(h))\right) \tag{17}$$

for $h = 0, 1, \ldots,$ where $\boldsymbol{\Gamma}(h) = \boldsymbol{\Gamma}(\hat{\boldsymbol{\xi}}^{[i]}(h))$ is the Jacobian matrix of z with respect to $\boldsymbol{\xi}^{[i]}$ evaluated at $\boldsymbol{\xi}^{[i]} = \hat{\boldsymbol{\xi}}^{[i]}(h)$ and $z(\hat{\boldsymbol{\xi}})$ is an estimate of z calculated at $\boldsymbol{\xi}^{[i]} = \hat{\boldsymbol{\xi}}^{[i]}(h)$.

The expression of the Jacobian matrix $\boldsymbol{\Gamma}(\boldsymbol{\xi}^{[i]})$ is given by

$$\boldsymbol{\Gamma} = \left[\ldots, \boldsymbol{\Gamma}_k, \ldots\right]_{k=1,\ldots,i} \tag{18a}$$

$$\boldsymbol{\Gamma}_k = \begin{bmatrix} \boldsymbol{\Gamma}_k^{(1)} & \boldsymbol{\Gamma}_k^{(2)} & \boldsymbol{\Gamma}_k^{(3)} & \boldsymbol{\Gamma}_k^{(4)} \\ \boldsymbol{\Gamma}_k^{(5)} & \boldsymbol{\Gamma}_k^{(6)} & \boldsymbol{\Gamma}_k^{(7)} & \boldsymbol{\Gamma}_k^{(8)} \end{bmatrix} \tag{18b}$$

$$\boldsymbol{\Gamma}_k^{(1)} = \left[\ldots, \cos\theta_k^{[i]}(t_n), \ldots\right]_{n=1,\ldots,N}^T \tag{18c}$$

$$\boldsymbol{\Gamma}_k^{(5)} = \left[\ldots, \sin\theta_k^{[i]}(t_n), \ldots\right]_{n=1,\ldots,N}^T \tag{18d}$$

$$\boldsymbol{\Gamma}_k^{(2)} = \left[\ldots, -\sin\theta_k^{[i]}(t_n), \ldots\right]_{n=1,\ldots,N}^T \tag{18e}$$

$$\boldsymbol{\Gamma}_k^{(6)} = \left[\ldots, \cos\theta_k^{[i]}(t_n), \ldots\right]_{n=1,\ldots,N}^T \tag{18f}$$

$$\boldsymbol{\Gamma}_k^{(3)} = \left[\ldots, \frac{4\pi f}{c}\cos\left(\frac{|\beta|}{2}\right)u_1(t_n)\left(\sigma_{R,k}^{[i]}\sin\theta_k^{[i]}(t_n) + \sigma_{I,k}^{[i]}\cos\theta_k^{[i]}(t_n)\right), \ldots\right]_{n=1,\ldots,N}^T \tag{18g}$$

$$\boldsymbol{\Gamma}_k^{(7)} = \left[\ldots, \frac{4\pi f}{c}\cos\left(\frac{|\beta|}{2}\right)u_1(t_n)\left(-\sigma_{R,k}^{[i]}\cos\theta_k^{[i]}(t_n) + \sigma_{I,k}^{[i]}\sin\theta_k^{[i]}(t_n)\right), \ldots\right]_{n=1,\ldots,N}^T \tag{18h}$$

$$\boldsymbol{\Gamma}_k^{(4)} = \left[\ldots, \frac{4\pi f}{c}\cos\left(\frac{|\beta|}{2}\right)u_2(t_n)\left(\sigma_{R,k}^{[i]}\sin\theta_k^{[i]}(t_n) + \sigma_{I,k}^{[i]}\cos\theta_k^{[i]}(t_n)\right), \ldots\right]_{n=1,\ldots,N}^T \tag{18i}$$

$$\boldsymbol{\Gamma}_k^{(8)} = \left[\ldots, \frac{4\pi f}{c}\cos\left(\frac{|\beta|}{2}\right)u_2(t_n)\left(-\sigma_{R,k}^{[i]}\cos\theta_k^{[i]}(t_n) + \sigma_{I,k}^{[i]}\sin\theta_k^{[i]}(t_n)\right), \ldots\right]_{n=1,\ldots,N}^T. \tag{18j}$$

The following decision logic is then applied at the end of each GN iteration to ensure the constraint in (10) is met:

$$\begin{aligned} &\textbf{if } \left(\hat{x}_{1,k}^{[i]}(h+1) - \bar{x}_{1,k}^{[i]}\right) \gtrless \pm\zeta \textbf{ set } \hat{x}_{1,k}^{[i]}(h+1) = \bar{x}_{1,k}^{[i]} \pm \zeta, \\ &\textbf{if } \left(\hat{x}_{2,k}^{[i]}(h+1) - \bar{x}_{2,k}^{[i]}\right) \gtrless \pm\zeta \textbf{ set } \hat{x}_{2,k}^{[i]}(h+1) = \bar{x}_{2,k}^{[i]} \pm \zeta. \end{aligned} \tag{19}$$

When the constraint is in effect, a re-estimation of the reflection coefficients is performed as

$$[\hat{\sigma}_1^{[i]}(h+1), \ldots, \hat{\sigma}_i^{[i]}(h+1)]^T = \left(\mathbf{Y}^H\mathbf{Y}\right)^{-1}\mathbf{Y}^H\tilde{\mathbf{s}} \tag{20}$$

with $\mathbf{Y} = \left[\boldsymbol{\phi}(\hat{\boldsymbol{x}}_1^{[i]}(h+1)), \ldots, \boldsymbol{\phi}(\hat{\boldsymbol{x}}_i^{[i]}(h+1))\right]$. The GN iteration can be halted after a fixed number of iterations or if the l_2 norm of the updating term falls below a given threshold.

The GN iteration is initialized to the solution $\hat{\xi}^{[i-1]}$ obtained from the previous PROMP iteration $i-1$ and the newly selected atom:

$$\hat{\xi}^{[i]}(0) = [\hat{\xi}^{[i-1]T}, \text{Real}\{\bar{\sigma}_{j[i]}\}, \text{Imag}\{\bar{\sigma}_{j[i]}\}, \bar{x}_{j[i]}^{T}]^{T} \tag{21}$$

where $\bar{x}_{j[i]}$ is the position of the newly selected atom at index $j^{[i]}$ within the dictionary and $\bar{\sigma}_{j[i]} = \left(\boldsymbol{\phi}^{H}(\bar{x}_{j[i]})\boldsymbol{\phi}(\bar{x}_{j[i]})\right)^{-1}\boldsymbol{\phi}^{H}(\bar{x}_{j[i]})\boldsymbol{r}^{[i-1]}$ is the corresponding initial coefficient estimate for this atom.

5. Results

In this section, we demonstrate the performance superiority of the proposed sparsity-based tomography method based on the PROMP algorithm over the conventional k-space tomography method via results using both simulated and real data. The result comparison also includes the performance of the OMP algorithm to illustrate the off-grid dictionary mismatch problem of the sparsity-based tomography formulation and to verify the effectiveness of PROMP in dealing with this issue.

5.1. Results with Simulated Data

We consider two synthetic targets with two and eight scatterers, respectively, as depicted in Figure 2. In this simulation, the target rotational speed is set to $\Omega = 37.70$ rad/s and the signal frequency is set to $f = 9.96$ GHz. The constraint value ζ is set to $\zeta = \lambda/2$ for PROMP. The dictionary matrix is constructed using a regularly spaced grid in Cartesian coordinates with the grid step size of $\Delta = \lambda/5$ on each x- and y-axis. The sampling frequency at the receiver is set to 2.16 kHz. PROMP and OMP iterations are halted if the signal residual reaches the noise level.

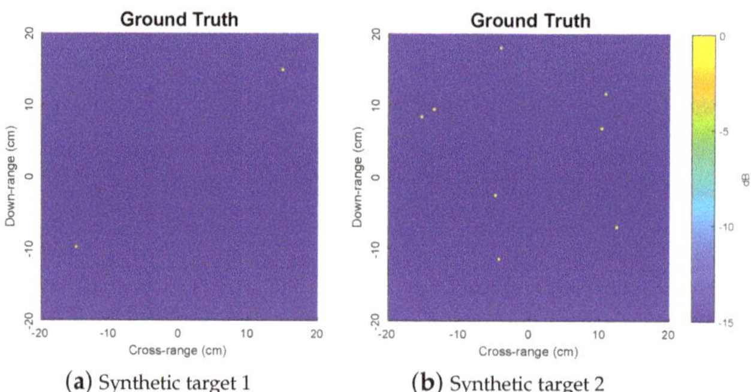

(**a**) Synthetic target 1 (**b**) Synthetic target 2

Figure 2. Ground truth images of two synthetic targets under consideration.

Figure 3 compares the reconstructed images for synthetic target 1 obtained by the k-space, OMP and PROMP algorithms for various noise levels. Here, the number of data samples is $N = 360$ (i.e., the CPI approximately being one full rotation cycle of the target). Note that the SNR is defined by $\text{SNR} = 20\log_{10}(\|\boldsymbol{s}\|/\|\boldsymbol{n}\|)$. We observe that the k-space technique produces images with two main peaks corresponding to the true target scatterers. However, along with these two main peaks, the images obtained by the k-space technique also contain other sidelobes with many spurious peaks. Specifically, the higher the noise is, the poorer the performance of the k-space technique (i.e., yielding a larger numbers of spurious peaks). In contrast, such a problem associated with spurious peaks does not appear in the OMP and PROMP images, thus demonstrating the performance advantage of the sparsity-based tomography approach over the conventional k-space approach. It is observed that one of the scatterers is split into multiple peaks in the OMP images. This observation can be explained by the fact the OMP solution relies on the fixed dictionary which is built based on a grid of atoms while the true scatterers of the target do not coincide with this dictionary grid, thereby demonstrating the dictionary mismatch problem. On the other hand, by perturbing the dictionary atoms and allowing them to go off the grid, the PROMP algorithm can effectively overcome the dictionary mismatch problem by exhibiting a clean image with only two peaks corresponding to the true scatterers. More importantly, the locations of the peaks in the PROMP images almost exactly match the locations of the true scatterers.

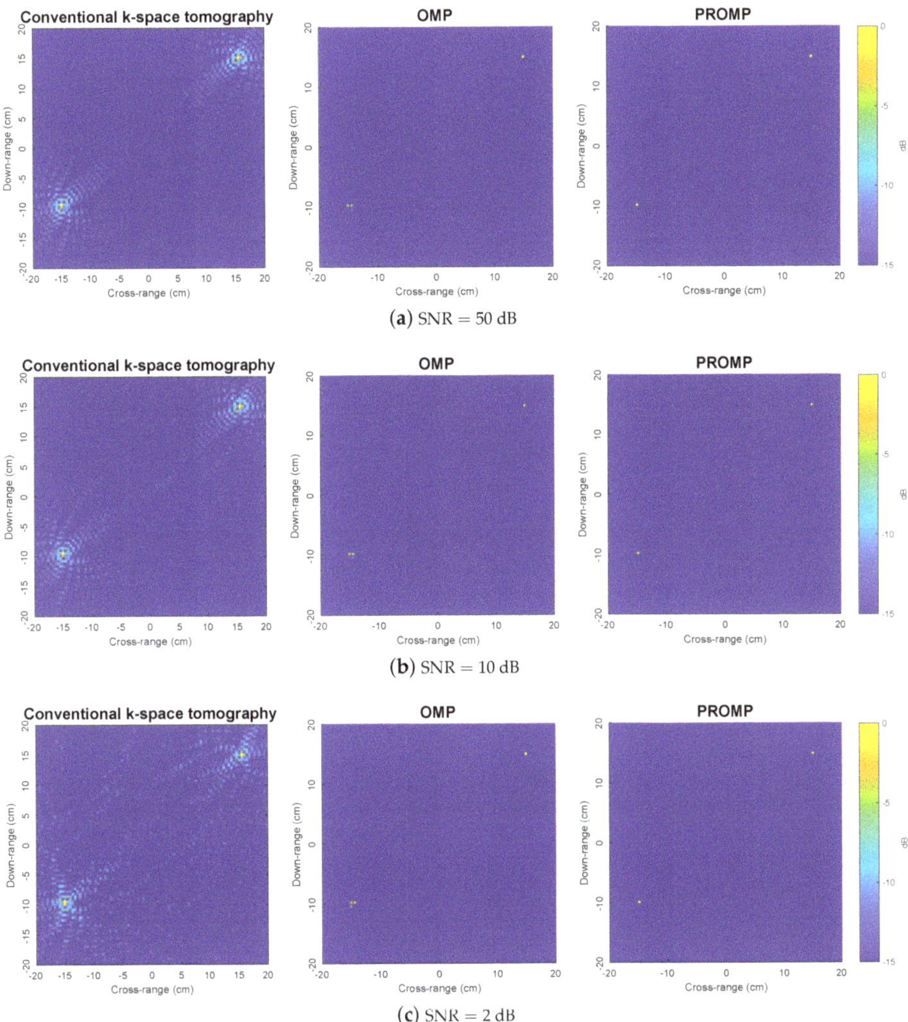

Figure 3. Reconstructed images obtained by the *k*-space, OMP and PROMP algorithms for synthetic target 1.

Figure 4 shows the results for synthetic target 2. This is a more challenging scenario because target 2 contains much more scatterers than target 1. We observer that the conventional *k*-space method is struggling to produce reliable image results because of the interaction between the sidelobes of different main peaks, especially in large noise scenarios. Such an interaction leads to some strong spurious peaks which have similar magnitudes to the correct peaks that correspond to the true scatterers, thus making the resulting images severely distorted. On the other hand, compared to the k-space method, OMP results in much more satisfactory images. However, the OMP performance is significantly affected by the dictionary mismatch problem arising from off-grid scatterers. As a result, the true scatterers are split into multiple peaks in the OMP images, and some spurious peaks also appear. In contrast, PROMP produces clean and clear images which are almost identical to the ground truth target image, even at large noise levels.

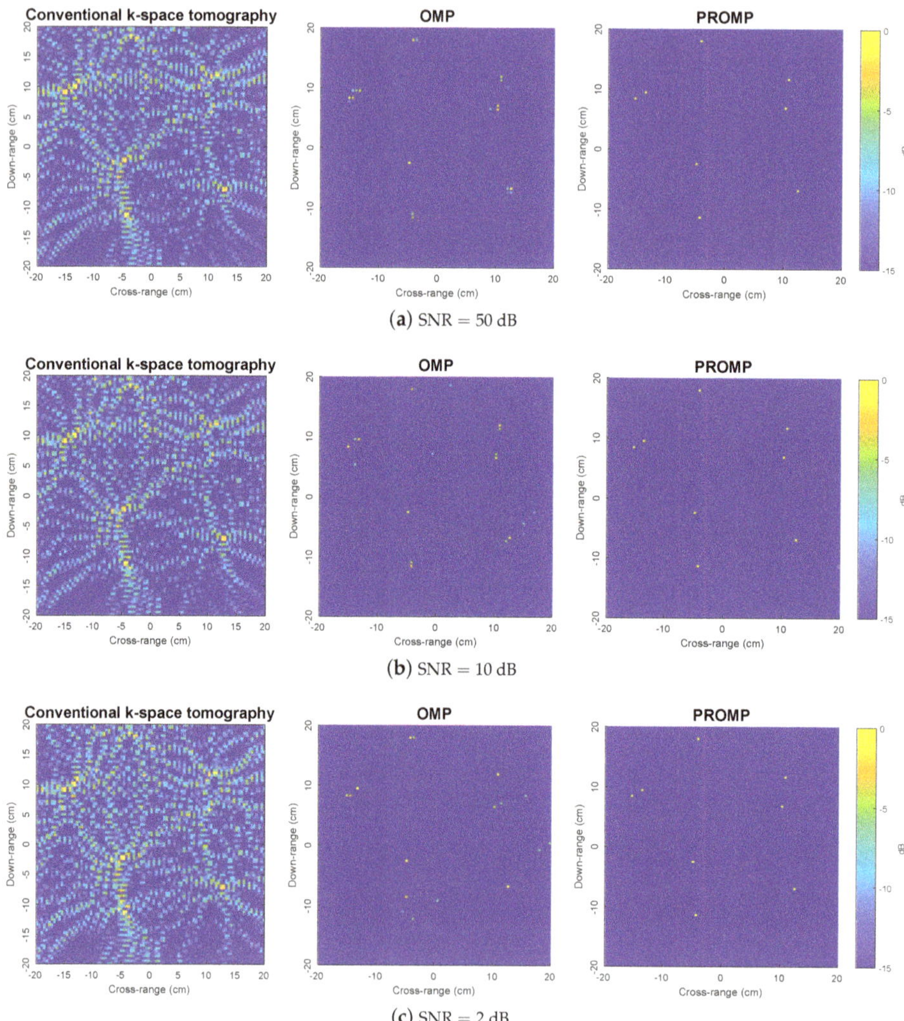

Figure 4. Reconstructed images obtained by the k-space, OMP and PROMP algorithms for synthetic target 2.

To further demonstrate the superior performance of PROMP, the reconstructed images obtained from less data samples (i.e., with CPI $= 2/3$ and $1/3$ target rotation cycle) are shown in Figure 5. With a limited number of data samples, the k-space method results in unsatisfactory images with incorrect peaks, while OMP and PROMP are observed to retain their good performance. In addition, similar to the observations in Figure 4, PROMP outperforms OMP and provides a better image of the target thanks to its ability to deal with off-grid scatterers.

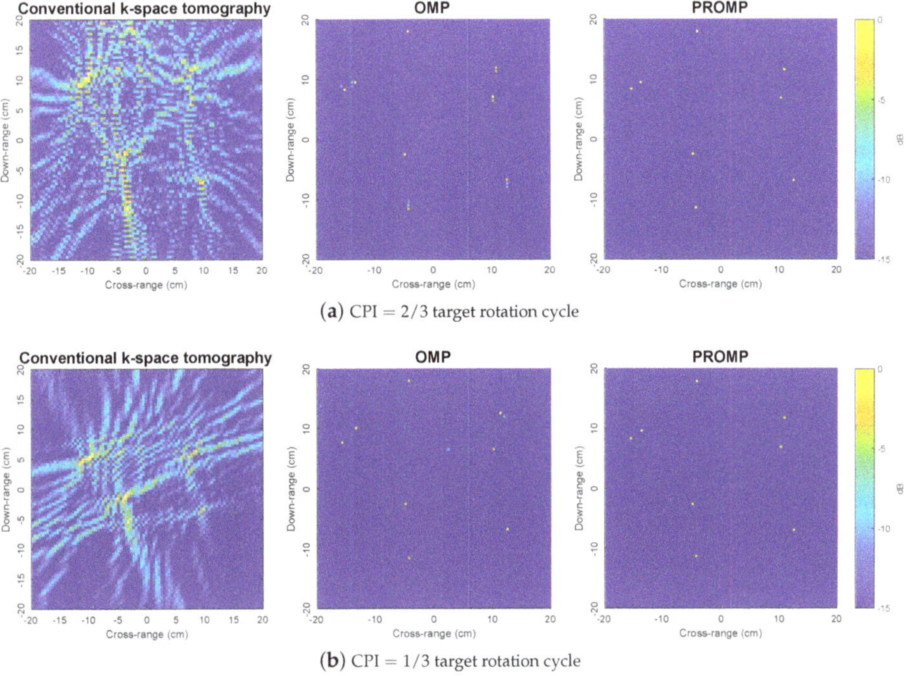

(**a**) CPI = 2/3 target rotation cycle

(**b**) CPI = 1/3 target rotation cycle

Figure 5. Reconstructed images obtained by the *k*-space, OMP and PROMP algorithms for synthetic target 2, given less data samples.

Note that, to satisfy the sparsity condition, the number of dominant scatterers constituting the target must be sufficiently small compared to the number of grid points on the reflectivity map. The required sparsity level in general depends on several factors, including the number of data samples, the noise level, as well as the level of coherence between the atoms of the dictionary. In compressive sensing, the restricted isometry property and the mutual incoherence property establish theoretical connections between those factors required for the effectiveness of sparse reconstruction [23]. However, these analytical metrics are overly-conservative and do not reflect the average performance which is often of interest from the practical point of view [39].

We now compare the performance of the *k*-space, OMP, and PROMP methods using the earth mover's distance (EMD) between the true and reconstructed images. EMD [40] is a widely-used metric to compare the similarity between different images. In principle, EMD is an estimate of the distance between two distributions which is equivalent to the minimal amount of work required for one distribution to be transformed to the other [40]. Figure 6 shows the EMD performance of the *k*-space, OMP, and PROMP methods, averaged from 1000 Monte Carlo runs, against various levels of SNR for the synthetic target 2 and CPI = 1 target rotation cycle. We observe that the PROMP method exhibits an EMD much smaller than those of the *k*-space and OMP methods. This indicates that, from a statistical point of view, the image obtained by PROMP is much closer to the ground-truth image than those obtained by the *k*-space and OMP methods, thus verifying the performance superiority of the PROMP method.

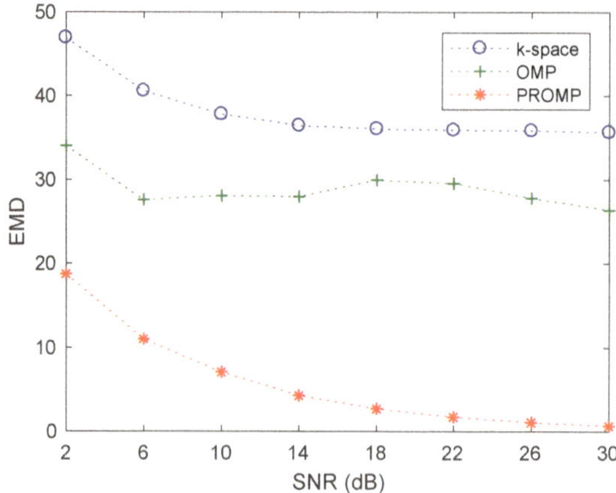

Figure 6. EMD performance of the k-space, OMP and PROMP algorithms versus SNR for synthetic target 2 and CPI = 1 target rotation cycle.

Table 2 compares the runtimes of the k-space, OMP, and PROMP algorithm for the image results shown in Figure 4b. For a fair comparison, all methods were implemented in MATLAB on the same Intel Core i7 3.40 GHz CPU with 16 GHz RAM. We observe that the k-space method is much slower than the OMP and PROMP methods. The reason for this is that the k-space method requires the non-uniform fast Fourier transform to be performed, thus being computationally more demanding compared to OMP and PROMP. On the other hand, the OMP and PROMP methods are computationally fast, thanks to the fact that they belong to the greedy pursuit family. In addition, each iteration of PROMP is about 7.3 times slower than each iteration of OMP because PROMP incorporates a NLS solver, rather than a linear least-squares solver as in OMP. However, the overall timerun of PROMP is only 2.4 times slower than that of OMP because PROMP requires many fewer iterations than OMP to make the signal residual reach the noise level.

Table 2. Complexity comparison *.

Algorithm	k-Space	OMP	PROMP
Averaged runtime (second)	0.1561	0.0331	0.0799
Averaged number of iterations	–	27.65	8.96
Averaged per-iteration runtime (second)	–	0.0012	0.0088

* All methods are implemented in MATLAB on an Intel Core i7 3.40 GHz CPU with 16 GHz RAM, for the image results shown in Figure 4b.

5.2. Results with Real Data

The experimental data used in this paper was collected in the Mumma Radar Laboratory at the University of Dayton, Ohio, USA. Although this paper focuses on narrowband tomographic imaging, a wideband waveform at X-band with stepped frequency pulses over 101 regular frequency steps from 8 GHz to 12 GHz was used in the experiment. The aim of using wideband data was only for the removal of extraneous clutter components existing in the lab environment, as described in [13]. After that, only the measured data from one discrete frequency is actually used for algorithm performance evaluation for the problem of narrowband tomographic imaging under consideration.

Figure 7 shows photos of the experimental system configuration. The experimental setup involves transmitting and receiving horn antennas mounted on separate robotic arms in a controlled laboratory environment with radar-absorbing material to reduce the radar reflections from the floor and walls.

These robotic arms could be oriented and positioned with high precision. During the experiment, the antennas were kept stationary while the target was rotated over 1° steps through 360°, where the stepped-frequency waveform was transmitted and sampled (one sample for each frequency). Recall from above that only one frequency sample set was used for tomographic imaging purposes. The experimental target was comprised of two vertical metallic rods with a 19 cm separation to emulate two point scatterers which rotate around a vertical pedestal. Figure 8 shows the wideband k-space tomographic image obtained from monostatic data with all 101 frequency steps. This will be used as a reference benchmark for the performance evaluation of narrowband bistatic imaging presented in this section.

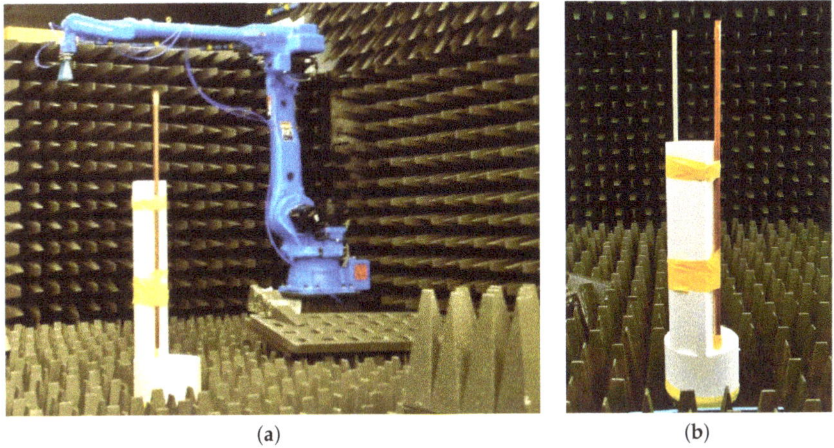

(a) (b)

Figure 7. Photos of experimental system configuration: (**a**) an antenna mounted on a robotic arm, and (**b**) two vertical metallic rods secured to a rotating pedestal.

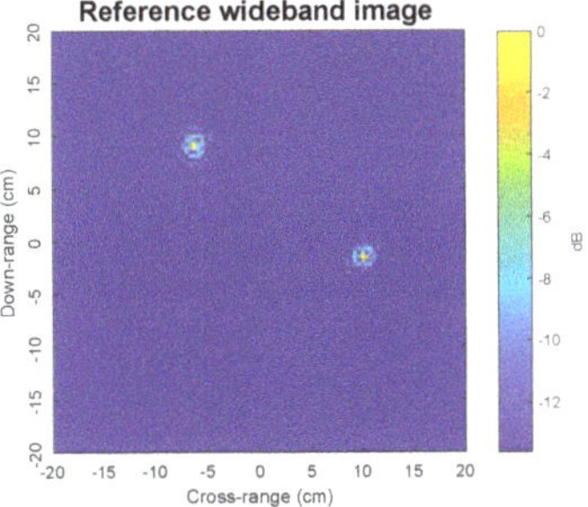

Figure 8. A benchmark experimental target image using wideband signals received by the monostatic receiver for all 101 frequency steps between 8 GHz and 12 GHz.

Figure 9 shows the experimental narrowband images obtained by the k-space, OMP, and PROMP algorithms using a narrowband signal received by the bistatic receiver with $\beta = 86°$ at a single frequency of 8.8 GHz for various values of CPI. Since the SNR is unknown, the OMP and PROMP iterations are halted when the change in the signal residual norm falls below 1% of the input signal norm. Compared to the benchmark image in Figure 8, the k-space method produces images with a much lower quality when only a narrowband signal from a single frequency is available. We observe that the images obtained by the k-space method are distorted with numerous spurious sibelode peaks. In particular, the number of spurious sibelode peaks increases significantly for shorter CPI.

Moreover, the main lodes corresponding to the true scatterers are also spread when the CPI is reduced. In contrast, the PROMP images contain two clear peaks at the locations very close to the peaks of the reference image in Figure 8, even when the CPI is reduced to one third of the target rotation cycle. This observation demonstrates the performance superiority of the sparsity-based tomography approach over the conventional *k*-space tomography approach. Figure 9 also shows the images obtained by OMP to illustrate the dictionary mismatch problem associated with off-grid scatterers, where each true scatterer is split into multiple peaks; thus, verifying the effectiveness of PROMP in terms of tackling the dictionary mismatch problem.

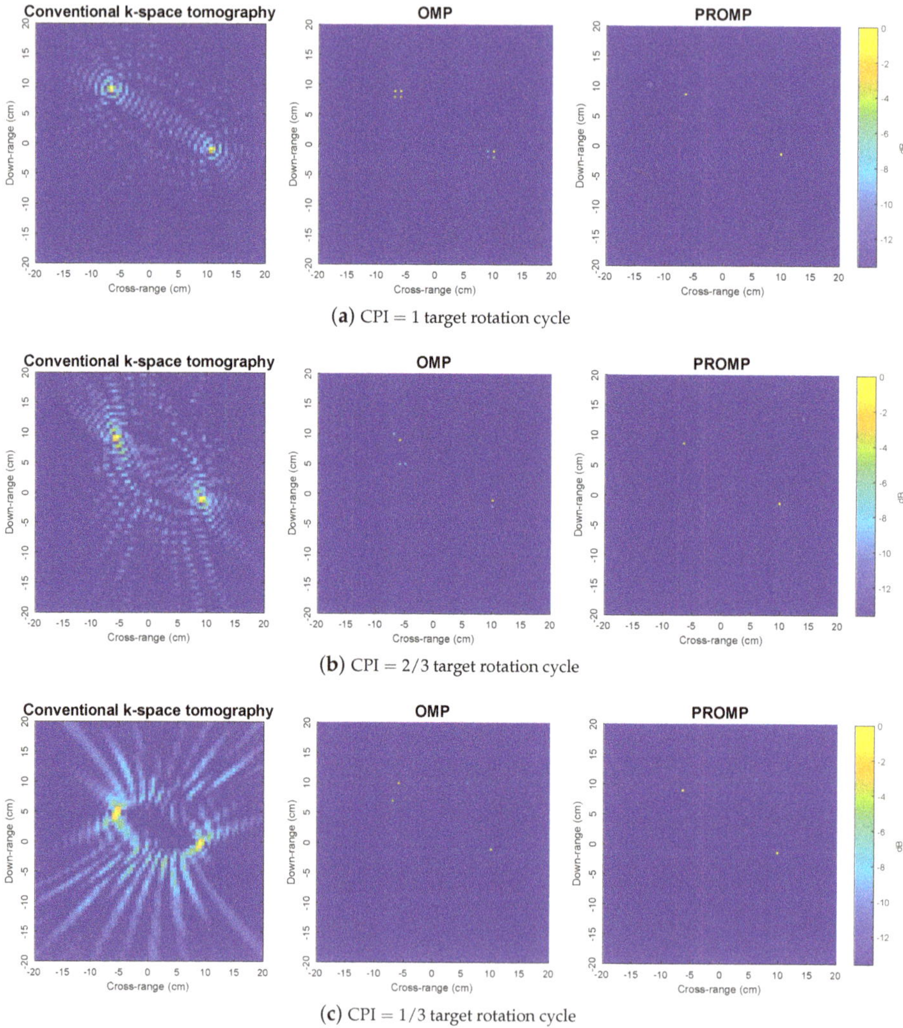

Figure 9. Experimental narrowband images obtained by the *k*-space, OMP, and PROMP algorithms using a narrowband signal received by the bistatic receiver with $\beta = 86^\circ$ at a single frequency of 8.8 GHz, for various values of CPI.

Figure 10 shows the experimental results where the data is downsampled. Here, we observe a similar relative performance comparison to Figures 3–5 and 9, once again confirming the performance advantages of the proposed sparsity-based tomographic imaging method based on the PROMP algorithm.

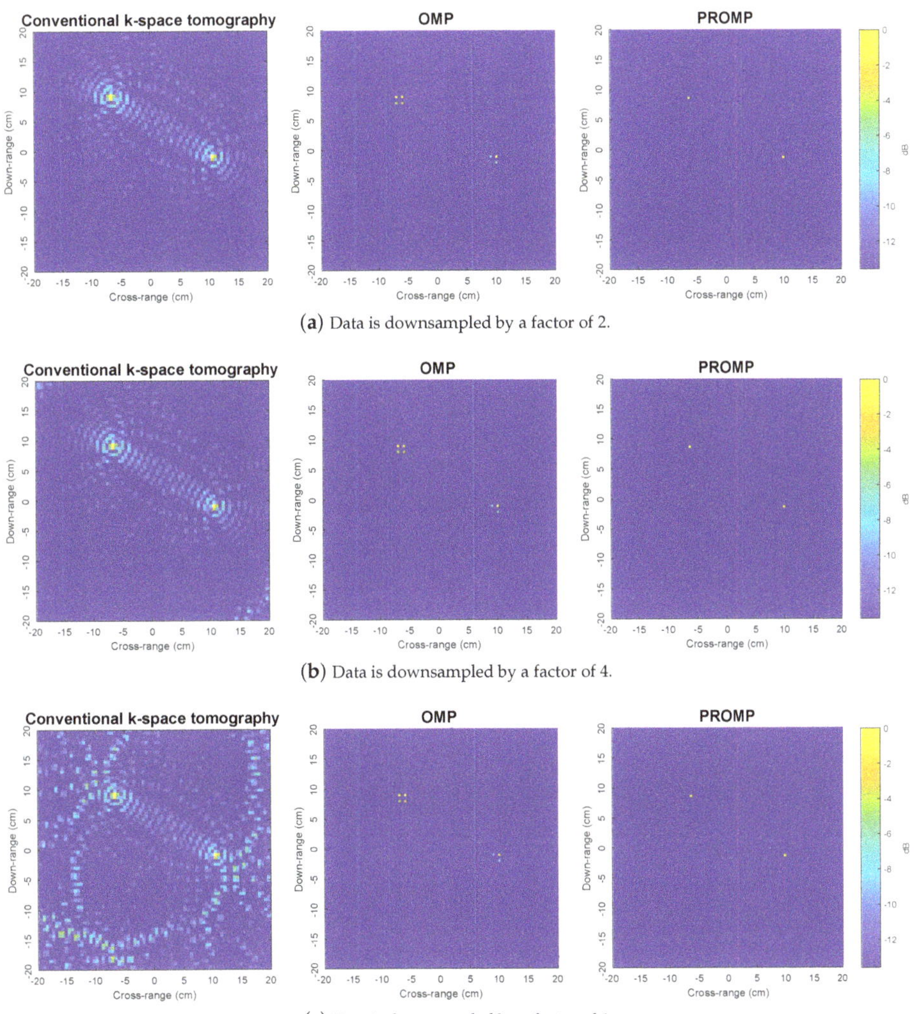

(a) Data is downsampled by a factor of 2.

(b) Data is downsampled by a factor of 4.

(c) Data is downsampled by a factor of 6.

Figure 10. Experimental narrowband images obtained by the k-space, OMP, and PROMP algorithms for downsampled data (CPI = 1 target rotation cycle).

6. Conclusions

In this paper, we have proposed a new sparsity-based bistatic radar tomographic imaging method exploiting the use of the PROMP algorithm. A new formulation for radar tomography building on the framework of compressive sensing and sparse reconstruction was presented, moving away from conventional k-space tomography which is prone to sidelobe responses and their interference. The PROMP algorithm was adopted to obtain a sparse solution for the resulting sparsity-based tomography formulation. By perturbing the dictionary atoms and allowing the estimated scatterers to go off the grid, PROMP is capable of tackling the dictionary mismatch problem arising from off-grid scatterers. The performance advantages of the proposed sparsity-based tomography method over the conventional k-space tomography method were demonstrated via numerical studies involving both simulated and real data.

Author Contributions: Conceptualization, N.H.N., H.-T.T. and P.B.; Methodology, N.H.N., H.-T.T. and P.B.; Simulation, N.H.N.; Validation, N.H.N., H.-T.T. and P.B.; Writing—Original Draft Preparation, N.H.N.; Writing—Review & Editing, N.H.N., H.-T.T. and P.B.

Funding: This research was funded by Defence Science and Technology Group, Australia.

Acknowledgments: Preliminary results of this work were reported in [41], where the feasibility and benefits of using compressive sensing and sparse reconstruction for narrowband bistatic radar tomographic imaging were demonstrated. This paper presents detailed algorithmic design and implementation as well as experimental evaluation of the PROMP-based tomographic imaging method.

Conflicts of Interest: The authors declare no conflict of interest.

References

1. Chen, V.C.; Martorella, M. *Inverse Synthetic Aperture Radar Imaging: Principles, Algorithms and Applications*; Institution of Engineering and Technology: Edison, NJ, USA, 2014.
2. Amin, M.G. (Ed.) *Through-the-Wall Radar Imaging*; CRC Press: Boca Raton, FL, USA, 2017.
3. Chen, V.C.; Ling, H. *Time-Frequency Transforms for Radar Imaging and Signal Analysis*; Artech House: Norwood, MA, USA, 2002.
4. Griffiths, H. Where has all the spectrum gone? In Proceedings of the International Conference on Radar, Adelaide, Australia, 9–12 September 2013; pp. 1–5, [CrossRef]
5. Mensa, D.L.; Halevy, S.; Wade, G. Coherent Doppler tomography for microwave imaging. *Proc. IEEE* **1983**, *71*, 254–261. [CrossRef]
6. Coetzee, S.L.; Baker, C.J.; Griffiths, H.D. Narrow band high resolution radar imaging. In Proceedings of the IEEE Conference on Radar, Verona, NY, USA, 24–27 April 2006; pp. 622–625, [CrossRef]
7. Sun, H.; Feng, H.; Lu, Y. High resolution radar tomographic imaging using single-tone CW signals. In Proceedings of the IEEE Radar Conference, Washington, DC, USA, 10–14 May 2010; pp. 975–980, [CrossRef]
8. Tran, H.T.; Melino, R. The Slow-Time k-Space of Radar Tomography and Applications to High-Resolution Target Imaging. *IEEE Trans. Aerosp. Electron. Syst.* **2018**, 1–12, [CrossRef]
9. Wang, L.; Yazici, B. Bistatic Synthetic Aperture Radar Imaging Using UltraNarrowband Continuous Waveforms. *IEEE Trans. Image Process.* **2012**, *21*, 3673–3686. [CrossRef] [PubMed]
10. Sego, D.J.; Griffiths, H.; Wicks, M.C. Radar tomography using Doppler-based projections. In Proceedings of the IEEE Radar Conference (RADAR), Kansas City, MO, USA, 23–27 May 2011; pp. 403–408. [CrossRef]
11. Tran, H.T.; Melino, R. *Application of the Fractional Fourier Transform and S-Method in Doppler Radar Tomography*; Technical Report, DSTO-RR-0357; DST Group: Edinburgh, Australia, 2010.
12. Borden, B.; Cheney, M. Synthetic-aperture imaging from high-Doppler-resolution measurements. *Inverse Probl.* **2005**, *21*, 1–11. [CrossRef]
13. Tran, H.T.; Heading, E.; Monte, L.L.; Alfaisali, N. An experimental study of radar tomographic imaging in a multi-bistatic scenario. In Proceedings of the IEEE Radar Conference (RadarConf), Seattle, WA, USA, 8–12 May 2017; pp. 419–424. [CrossRef]
14. Tran, H.T.; Heading, E.; Ng, B. Multi-Bistatic Doppler Radar Tomography for Non-Cooperative Target Imaging. In Proceedings of the International Conference on Radar (RADAR), Brisbane, Australia, 27–30 August 2018.
15. Cetin, M.; Stojanovic, I.; Onhon, O.; Varshney, K.; Samadi, S.; Karl, W.C.; Willsky, A.S. Sparsity-Driven Synthetic Aperture Radar Imaging: Reconstruction, autofocusing, moving targets, and compressed sensing. *IEEE Signal Process. Mag.* **2014**, *31*, 27–40. [CrossRef]
16. Leigsnering, M.; Ahmad, F.; Amin, M.; Zoubir, A. Multipath exploitation in through-the-wall radar imaging using sparse reconstruction. *IEEE Trans. Aerosp. Electron. Syst.* **2014**, *50*, 920–939. [CrossRef]
17. Nguyen, L.H.; Tran, T.; Do, T. Sparse models and sparse recovery for ultra-wideband SAR applications. *IEEE Trans. Aerosp. Electron. Syst.* **2014**, *50*, 940–958. [CrossRef]

18. Feng, C.; Xiao, L.; Wei, Z.H. Compressive Sensing Inverse Synthetic Aperture Radar Imaging Based on Gini Index Regularization. *Int. J. Autom. Comput.* **2014**, *11*, 441–448. [CrossRef]
19. Giusti, E.; Cataldo, D.; Bacci, A.; Tomei, S.; Martorella, M. ISAR Image Resolution Enhancement: Compressive Sensing Versus State-of-the-Art Super-Resolution Techniques. *IEEE Trans. Aerosp. Electron. Syst.* **2018**, *54*, 1983–1997. [CrossRef]
20. Nguyen, N.H.; Dogancay, K.; Tran, H.T.; Berry, P. An Image Focusing Method for Sparsity-Driven Radar Imaging of Rotating Targets. *Sensors* **2018**, *18*. [CrossRef]
21. Tomei, S.; Bacci, A.; Giusti, E.; Martorella, M.; Berizzi, F. Compressive sensing-based inverse synthetic radar imaging imaging from incomplete data. *IET Radar Sonar Navig.* **2016**, *10*, 386–397. [CrossRef]
22. Amin, M. (Ed.) *Compressive Sensing for Urban Radar*; CRC Press: Boca Raton, FL, USA, 2017.
23. Potter, L.C.; Ertin, E.; Parker, J.T.; Cetin, M. Sparsity and compressed sensing in radar imaging. *Proc. IEEE* **2010**, *98*, 1006–1020. [CrossRef]
24. Nguyen, N.H.; Dogancay, K.; Tran, H.; Berry, P. Nonlinear Least-Squares Post-Processing for Compressive Radar Imaging of a Rotating Target. In Proceedings of the 2018 International Conference on Radar (RADAR), Brisbane, Australia, 27–30 August 2018; pp. 1–6. [CrossRef]
25. Qiu, W.; Giusti, E.; Bacci, A.; Martorella, M.; Berizzi, F.; Zhao, H.; Fu, Q. Compressive sensing-based algorithm for passive bistatic ISAR with DVB-T signals. *IEEE Trans. Aerosp. Electron. Syst.* **2015**, *51*, 2166–2180. [CrossRef]
26. Nguyen, N.H.; Dogancay, K.; Berry, P.; Tran, H. *Convex Relaxation Methods: A Review and Application to Sparse Radar Imaging of Rotating Targets*; Technical Report, DST-Group-RR-0444; DST Group: Edinburgh, Australia, 2017.
27. Chi, Y.; Scharf, L.L.; Pezeshki, A.; Calderbank, A.R. Sensitivity to Basis Mismatch in Compressed Sensing. *IEEE Trans. Signal Process.* **2011**, *59*, 2182–2195. [CrossRef]
28. Tang, G.; Bhaskar, B.N.; Shah, P.; Recht, B. Compressed Sensing Off the Grid. *IEEE Trans. Inf. Theory* **2013**, *59*, 7465–7490. [CrossRef]
29. Nguyen, N.H.; Dogancay, K.; Tran, H.; Berry, P.E. Parameter-Refined OMP for Compressive Radar Imaging of Rotating Targets. *IEEE Trans. Aerosp. Electron. Syst.* **2019**. [CrossRef]
30. Eldar, Y.C.; Kutyniok, G. (Eds.) *Compressed Sensing: Theory and Applications*; Cambridge University Press: New York, NY, USA, 2012.
31. Tropp, J.A.; Wright, S.J. Computational methods for sparse solution of linear inverse problems. *Proc. IEEE* **2010**, *98*, 948–958. [CrossRef]
32. Tan, Z.; Yang, P.; Nehorai, A. Joint sparse recovery method for compressed sensing with structured dictionary mismatches. *IEEE Trans. Signal Process.* **2014**, *62*, 4997–5008. [CrossRef]
33. Teke, O.; Gurbuz, A.C.; Arikan, O. Perturbed Orthogonal Matching Pursuit. *IEEE Trans. Signal Process.* **2013**, *61*, 6220–6231. [CrossRef]
34. Arablouei, R. Fast reconstruction algorithm for perturbed compressive sensing based on total least-squares and proximal splitting. *Signal Process.* **2017**, *130*, 57–63. [CrossRef]
35. Yang, Z.; Xie, L.; Zhang, C. Off-Grid Direction of Arrival Estimation Using Sparse Bayesian Inference. *IEEE Trans. Signal Process.* **2013**, *61*, 38–43. [CrossRef]
36. Liu, S.; Wu, H.; Huang, Y.; Yang, Y.; Jia, J. Accelerated Structure-Aware Sparse Bayesian Learning for Three-Dimensional Electrical Impedance Tomography. *IEEE Trans. Ind. Inform.* **2019**, *15*, 5033–5041. [CrossRef]
37. Teke, O.; Gurbuz, A.C.; Arikan, O. A robust compressive sensing based technique for reconstruction of sparse radar scenes. *Digit. Signal Process.* **2014**, *27*, 23–32. [CrossRef]
38. Ljung, L.; Soderstrom, T. *Theory and Practice of Recursive Identification*; MIT Press: Cambridge, MA, USA, 1983.
39. Blumensath, T.; Davies, M.E. Stagewise Weak Gradient Pursuits. *IEEE Trans. Signal Process.* **2009**, *57*, 4333–4346. [CrossRef]

40. Rubner, Y.; Tomasi, C.; Guibas, L.J. The Earth Mover's Distance as a Metric for Image Retrieval. *Int. J. Comput. Vis.* **2000**, *40*, 99–121. [CrossRef]
41. Nguyen, N.H.; Tran, H.; Berry, P.; Dogancay, K. Experimental Demonstration of Bistatic Radar Tomography using Parameter-Refined OMP. In Proceedings of the International Radar Conference, Toulon, France, 23–27 September 2019.

Article

Two-Dimensional Augmented State–Space Approach with Applications to Sparse Representation of Radar Signatures

Kejiang Wu and Xiaojian Xu *

School of Electronics and Information Engineering, Beihang University, Beijing 100191, China; wukejiang88@buaa.edu.cn
* Correspondence: xiaojianxu@buaa.edu.cn; Tel.: +86-10-8231-6065

Received: 16 September 2019; Accepted: 21 October 2019; Published: 24 October 2019

Abstract: In this work, we focus on sparse representation of two-dimensional (2-D) radar signatures for man-made targets. Based on the damped exponential (DE) model, a 2-D augmented state–space approach (ASSA) is proposed to estimate the parameters of scattering centers on complex man-made targets, i.e., the complex amplitudes and the poles in down-range and aspect dimensions. An augmented state–space approach is developed for pole estimation of down-range dimension. Multiple-range search strategy, which applies one-dimensional (1-D) state–space approach (SSA) to the 1-D data for each down-range cell, is used to alleviate the pole-pairing problem occurring in previous algorithms. Effectiveness of the proposed approach is verified by the numerical and measured inverse synthetic aperture radar (ISAR) data.

Keywords: damped exponential (DE) model; inverse synthetic aperture radar (ISAR); radar signatures; state–space approach (SSA); sparse representation

---

1. Introduction

Sparse representation of two-dimensional (2-D) radar signatures has been widely used in many applications, such as super-resolution radar imaging, data compression, and target identification [1–4]. 2-D radar signatures can be reconstructed with fewer data, where a set of parameters including the locations, amplitudes, and damping factors are used to represent the returned signals from spatial distributed scattering centers. Moreover, 2-D ultra-wideband images of radar targets can be obtained through parameter interpolation or extrapolation [5].

Over the years, several model-based spectral estimation approaches have been developed for sparse representation of 2-D radar signatures, in which the common idea of these approaches is the development of a parametric model based on electromagnetic scattering mechanisms. In this way, a set of parameters may be found to represent the original signatures. Please note that most sparse representation approaches are considered to have super-resolution capabilities because they are supposed to estimate the parameters of scattering centers that cannot be distinguished by standard processing [6–21]. Typically, these approaches include the amplitude and phase estimation of a sinusoid (APES) algorithm [6], fast Fourier transform (FFT)-based technique (CLEAN) [7,8], the compressed sensing (CS)-based procedure [9–11] and the subspace-based approach [12–21]. The APES algorithm could provide accurate estimation of the complex amplitudes whereas it does not estimate the locations of scattering centers. The 2-D CLEAN technique is a computationally efficient procedure which uses the undamped exponential model and deconvolution algorithm for optimization [7,22]. The main process of the CS-based procedure is to represent the signatures using an over-complete dictionary and the corresponding coefficients [9,10]. The performance of this approach depends on the initial coefficients and the threshold used in terminating iterations.

Another important group of 2-D sparse representation technique is the subspace-based approaches. The representative algorithms of this kind include the 2-D multiple signal classification (MUSIC) [12], matrix enhancement and matrix pencil (MEMP) [13,14], 2-D total least squares (TLS) Prony [15], algebraically coupled matrix pencils (ACMP) [16], 2-D estimation of signal parameters via rotational invariance techniques (2-D ESPRIT) [17] and 2-D system realization technique [18,19]. The rationale and pole-pairing scheme of these algorithms are listed in Table 1.

Table 1. Synopsis of the 2-D subspace-based algorithms.

Technique	Rationale	Pole [(1)] Pairing Scheme
2-D MUSIC	Signal-noise subspace decomposition	No need for pole-pairing
MEMP	Enhanced matrix decomposition	Maximizing a certain criterion
2-D TLS Prony	TLS-based Prony model	Minimizing a certain distance
ACMP	Signal subspace decomposition	Rank-restoration scheme
2-D ESPRIT	Shift-invariance structure of the signal subspace	Joint diagonalization scheme
2-D system realization	State–space model	Algebraic method

[(1)]. Pole denotes a transfer function of the scattering center collected on target. Usually, the scattering centers are characterized via pairing the complex amplitudes and the poles in down-range and aspect dimensions.

While these subspace-based algorithms are demonstrated to be useful for the signatures of targets or simulated scattering centers, there are still some technique challenges, especially for sparse representation of the wideband radar signatures collected on complex man-made targets. First, the pairing schemes used by many existing algorithms cannot provide correct pole pairs in certain circumstances. i.e., MEMP, ACMP and 2-D system realization meet with the pole-pairing problem when there are numerous repeated poles in either the down-range or aspect dimension [17,20,23]. Second, the model order, which directly affects the result of parameter estimation such as 2-D MUSIC, MEMP, and 2-D ESPRIT et al., is difficult to be determined in the measurement environment [8]. Third, the compromises between the computational complexity and accuracy should be considered in the application of these algorithms [12]; for instance, the 2-D TLS Prony has smaller computational complexity than MEMP but with loss in accuracy [15].

In this paper, we focus on developing an approach for sparse representation of 2-D radar signatures collected on man-made targets. An augmented state–space approach is proposed for pole estimation of the down-range dimension. Multiple-range search strategy is then applied to estimate the pairing poles and corresponding amplitudes along the aspect dimension. Compared to the existing methods [14,17,18], the advantages of the 2-D augmented state–space approach (ASSA) are: (1) Computational complexity of the algorithm is much reduced since the newly defined Hankel matrix and several time-saving operations are adopted, whereas the pole-estimation accuracy is still at the same level; (2) The pole-pairing problem can be alleviated because all the poles are adaptively paired by using the multiple-range search strategy; And (3) an eigenvalue sequences transform algorithm is proposed, which could provide fast model order selection.

The remainder of the paper is organized as follows. Section 2 briefly presents the damped exponential models for 1-D and 2-D signals. In Section 3, a two-step procedure for 2-D ASSA is developed. Results for numerical and measured inverse synthetic aperture radar (ISAR) data representation are demonstrated to validate the effectiveness of the proposed procedure in Section 4. We conclude the paper in Section 5. Appendices are given to show more mathematical details of the proposed approach.

2. Data Model

2.1. 1-D Damped Exponential Model

As described in DE model [22,24], the 1-D radar signatures $y(f_n)$ can be expressed as a summation of K scattering centers corrupted with noise $w(n)$.

$$\begin{aligned} y(f_n) &= \sum_{k=1}^{K} A_k \exp\left[\left(\beta_k + j2\pi \tfrac{r_k}{c}\right) f_n\right] + w(n) \\ &= \sum_{k=1}^{K} a_k p_k^n + w(n) \end{aligned} \quad (1)$$

where n = 1,2, ... ,N and N denotes the number of pulses, A_k is the complex amplitude of the k-th scattering center; β_k is the damping factor with respect to frequency; r_k denotes the relative range; The parameter f_n denotes the radar frequency $f_n = f_c + (n-1-N2)\Delta f$ where f_c is the center frequency and $N2 = ceil(N/2)$ denotes the smallest integer less than or equal to $N/2$; a_k represents the amplitude of the k-th scattering center in pole form, the pole $p_k = \exp[(\beta_k + j2\pi r_k/c)\Delta f]$ represents the transfer function of the k-th scattering center; $c = 3 \times 10^8$ m/s is the propagation velocity. It worth noting that the data models used in this paper are all considered in a stepped frequency radar [25,26].

According to the discrete-time control theory and auto-regressive moving average (ARMA) model, Piou and Naishadham proposed a one-dimensional state–space approach (1-D SSA) which use a state–space description to the 1-D radar signatures in (1) [22].

$$x(n+1) = \mathbf{A}x(n) + \mathbf{B}u(n) \quad (2)$$

$$y(n) = \mathbf{C}x(n) + u(n) \quad (3)$$

where $x(n) \in \mathbb{C}^{K\times 1}$ is the state vector, $y(n)$ denotes the signal sequence $y(f_n)$, $u(n)$ is the input vector, $\mathbf{A} \in \mathbb{C}^{K\times K}$ represents the open-loop matrix, $\mathbf{B} \in \mathbb{C}^{K\times 1}$ and $\mathbf{C} \in \mathbb{C}^{1\times K}$ are the constant matrices. Thus, the 1-D noiseless radar signatures same as (1) can be expressed as

$$\begin{bmatrix} \widetilde{y}(1) & \widetilde{y}(2) & \dots & \widetilde{y}(N) \end{bmatrix} = \begin{bmatrix} \mathbf{CB} & \mathbf{CAB} & \dots & \mathbf{CA}^{N-1}\mathbf{B} \end{bmatrix} \quad (4)$$

As described in [22], 1-D SSA could precisely estimate the state matrices **A**, **B**, and **C**. Once these three state matrices are computed, the model parameters in (1) can be estimated by using the eigen-decomposition technique.

2.2. 2-D Damped Exponential Model

As a 2-D extension of 1-D DE model, the radar signatures obtained from different aspect angles are considered to be the summation of a finite number of dispersive scattering centers [13–18]. Typically, it is applicable to modeling the 2-D radar signatures with small aspect ranges [27].

$$\begin{aligned} y(\theta_m, f_n) &= \sum_{k=1}^{K} A_k \exp\left[\left(\tfrac{j2\pi r_{1k}}{c} + \beta_{1k}\right) f_c \theta_m\right] \exp\left[\left(\tfrac{j2\pi r_{2k}}{c} + \beta_{2k}\right) f_n\right] + w(m,n) \\ &= \sum_{k=1}^{K} a_k s_k^m p_k^n + w(m,n) \end{aligned} \quad (5)$$

in matrix notation, the 2-D radar signatures in (5) can be expressed as:

$$\mathbf{Y} = \begin{bmatrix} y(1,1) & y(1,2) & \dots & y(1,N) \\ y(2,1) & y(2,2) & \dots & y(2,N) \\ \vdots & \vdots & \dots & \vdots \\ y(M,1) & y(M,2) & \dots & y(M,N) \end{bmatrix} \quad (6)$$

where $m = 1,\ldots,M$ and $n = 1,\ldots,N$; $\{r_{1k}, r_{2k}\}$ give the relative locations of the *k-th* scattering center; $\{\beta_{1k}, \beta_{2k}\}$ characterize the frequency and aspect dependence of scattering; $\theta_m = \theta_0 + (m-1)\Delta\theta$ denotes the *m-th* aspect angle where θ_0 is the starting angle and $\Delta\theta$ represents the angle interval; $w(m,n)$ is the Gaussian noise with zero-mean; $y(m,n)$ denotes the signal sequences $y(\theta_m, f_n)$; $\{s_k, p_k\}$ refer to poles of the down-range and aspect dimension.

$$s_k = \exp\left[\left(\frac{j2\pi r_{1k}}{c} + \beta_{1k}\right) f_c \Delta\theta\right] \tag{7}$$

$$p_k = \exp\left[\left(\frac{j2\pi r_{2k}}{c} + \beta_{2k}\right) \Delta f\right] \tag{8}$$

3. Two-Dimensional Augmented State–Space Approach

From the 2-D DE model in (5), the vector $\widetilde{\mathbf{Y}}(n)$, which represents the n-th column of the noiseless signature matrix $\widetilde{\mathbf{Y}} = \mathbf{Y} - \mathbf{W}$ (where $\mathbf{W}$ is the noise matrix), can be decomposed as.

$$\widetilde{\mathbf{Y}}(n) = \begin{bmatrix} a_1 s_1^1 p_1 & a_2 s_2^1 p_2 & \vdots & a_K s_K^1 p_K \\ a_1 s_1^2 p_1 & a_2 s_2^2 p_2 & \vdots & a_K s_K^2 p_K \\ \vdots & \vdots & \vdots & \vdots \\ a_1 s_1^M p_1 & a_2 s_2^M p_2 & \vdots & a_K s_K^M p_K \end{bmatrix} \begin{bmatrix} p_1 & 0 & \vdots & 0 \\ 0 & p_2 & \vdots & 0 \\ \vdots & \vdots & \vdots & \vdots \\ 0 & 0 & \vdots & p_K \end{bmatrix}^{n-1} \mathbf{l}_K \tag{9}$$

where K denotes the number of scattering centers, $\mathbf{l}_K$ indicates the column vector of ones with length K.

Actually, we often meet the one-to-multiple matching situation, i.e., one pole p_k is corresponding to more than one poles $s_{k1}, s_{k2}, \ldots$. These repeated poles can be merged and (9) could be rewritten as.

$$\widetilde{\mathbf{Y}}(n) = \begin{bmatrix} \sum\limits_{t=1}^{l_1} a_{1t} s_{1t} p_1 & \sum\limits_{t=1}^{l_2} a_{2t} s_{2t} p_2 & \cdots & \sum\limits_{t=1}^{l_{K_1}} a_{K_1 t} s_{K_1 t} p_{K_1} \\ \sum\limits_{t=1}^{l_1} a_{1t} s_{1t}^2 p_1 & \sum\limits_{t=1}^{l_2} a_{2t} s_{2t}^2 p_2 & \cdots & \sum\limits_{t=1}^{l_{K_1}} a_{K_1 t} s_{K_1 t}^2 p_{K_1} \\ \vdots & \vdots & \vdots & \vdots \\ \sum\limits_{t=1}^{l_1} a_{1t} s_{1t}^M p_1 & \sum\limits_{t=1}^{l_2} a_{2t} s_{2t}^M p_2 & \vdots & \sum\limits_{t=1}^{l_{K_1}} a_{K_1 t} s_{K_1 t}^M p_{K_1} \end{bmatrix} \begin{bmatrix} p_1 & 0 & \vdots & 0 \\ 0 & p_2 & \vdots & 0 \\ \vdots & \vdots & \vdots & \vdots \\ 0 & 0 & \vdots & p_K \end{bmatrix}^{n-1} \mathbf{l}_{K_1} \tag{10}$$

where K_1 represents the number of non-repeated poles in the down-range dimension. $l_1, l_2, \ldots l_{K_1}$ denote the number of repeated poles, respectively, for the corresponding poles $p_1, p_2, \ldots p_{K_1}$.

We define the matrix $\mathbf{P}$ as.

$$\mathbf{P} = \mathbf{Q} \begin{bmatrix} p_1 & 0 & \vdots & 0 \\ 0 & p_2 & \vdots & 0 \\ \vdots & \vdots & \vdots & \vdots \\ 0 & 0 & \vdots & p_{K_1} \end{bmatrix} \mathbf{Q}^* \tag{11}$$

where * denotes the Hermitian operator; $\mathbf{Q}$ is a unitary matrix which satisfies $\mathbf{Q}^*\mathbf{Q} = \mathbf{Q}\mathbf{Q}^* = \mathbf{E}$, $\mathbf{E}$ is the identity matrix.

Two constant matrices, i.e., $\mathbf{S}$ and $\mathbf{D}$, are defined to simplify the expression in (10).

$$\mathbf{S} = \begin{bmatrix} \sum_{t=1}^{l_1} a_{1t}s_{1t}p_1 & \sum_{t=1}^{l_2} a_{2t}s_{2t}p_2 & \cdots & \sum_{t=1}^{l_{K_1}} a_{K_1t}s_{K_1t}p_{K_1} \\ \sum_{t=1}^{l_1} a_{1t}s_{1t}^2p_1 & \sum_{t=1}^{l_2} a_{2t}s_{2t}^2p_2 & \cdots & \sum_{t=1}^{l_{K_1}} a_{K_1t}s_{K_1t}^2p_{K_1} \\ \vdots & \vdots & \vdots & \vdots \\ \sum_{t=1}^{l_1} a_{1t}s_{1t}^Mp_1 & \sum_{t=1}^{l_2} a_{2t}s_{2t}^Mp_2 & \vdots & \sum_{t=1}^{l_{K_1}} a_{K_1t}s_{K_1t}^Mp_{K_1} \end{bmatrix} \mathbf{Q}_* \quad (12)$$

$$\mathbf{D} = \mathbf{Q}\mathbf{l}_{K_1} \quad (13)$$

Using (11)–(13), expression in (10) is simplified as.

$$\widetilde{\mathbf{Y}}(n) = \mathbf{S}\mathbf{P}^{n-1}\mathbf{D} \quad (14)$$

Thus, the noiseless signatures $\widetilde{\mathbf{Y}}$ can be written as.

$$\widetilde{\mathbf{Y}} = \begin{bmatrix} \mathbf{SD} & \mathbf{SPD} & \ldots & \mathbf{SP}^{N-1}\mathbf{D} \end{bmatrix} \quad (15)$$

Comparing (15) with (4), it can be found that these two equations have extremely similar structures. Thus, the parameters a_k, s_k and p_k can be estimated by 2-D augmentation of 1-D SSA, which is called 2-D augmentation state–space approach (2-D ASSA).

Here, the 2-D ASSA consists of two steps. An augmented state–space approach is applied to estimating the pole p_k in down-range dimension. Then multiple-range search strategy is used for estimating the matching pole s_k as well as the corresponding amplitude a_k. Details are as follows.

3.1. Pole Estimation of the Down-Range Dimension

First, we introduce a single augmented Hankel matrix $\mathbf{H}$ of size $M(N-L+1) \times L$ by analogy of the enhanced Hankel matrix in 1-D SSA:

$$\mathbf{H} = \begin{bmatrix} \mathbf{Y}(1) & \mathbf{Y}(2) & \ldots & \mathbf{Y}(L) \\ \mathbf{Y}(2) & \mathbf{Y}(3) & \ldots & \mathbf{Y}(L+1) \\ \vdots & \vdots & \ldots & \vdots \\ \mathbf{Y}(N-L+1) & \mathbf{Y}(N-L+2) & \ldots & \mathbf{Y}(N) \end{bmatrix} \quad (16)$$

where each element $\mathbf{Y}(n)$, first mentioned in (9), represents the n-th column of the matrix $\mathbf{Y}$; L is the step size of the correlation window, which is heuristically set to be L=$N2$ [22].

Next, do the singular value decomposition (SVD) of the Hankel matrix.

$$\mathbf{H} = \begin{bmatrix} \mathbf{U}_{sn} & \mathbf{U}_n \end{bmatrix} \begin{bmatrix} \Re_{sn} & \\ & \Re_n \end{bmatrix} \begin{bmatrix} \mathbf{V}_{sn}^* \\ \mathbf{V}_n^* \end{bmatrix} \quad (17)$$

where $\mathbf{U}_{sn}$,$\mathbf{U}_n$,$\mathbf{V}_{sn}^*$ and $\mathbf{V}_n^*$ are unitary matrices; $\Re_{sn}$ and $\Re_n$ are diagonal matrices; The matrices with subscript 'sn' refer to the components of signal space and those matrices with subscript 'n' denote the components of noise space; The rank of the noiseless matrix $\mathbf{U}_{sn}\Re_{sn}\mathbf{V}_{sn}^*$ is known as the model order which has been widely used in [14,22,27].

As proved in [14,17], the model order in the down-range dimension is equal to K_1 and should be less than the number of columns or rows of $\mathbf{H}$ at least. Thus, the number of non-repeated poles K_1 should satisfy the following condition.

$$\begin{cases} MN - ML + M \geq K_1 \\ L \geq K_1 \end{cases} \tag{18}$$

Model order selection is an inevitable problem in modeling the 2-D signatures [14,22,27]. A series of eigenvalue-based criteria, such as Akaike information criterion (AIC), minimum description length (MDL) criterion and the minimum eigenvalue (MEV) criterion, have been proposed for solving this problem [28,29]. Those criteria are useful in model order selection but with time consuming process [30]. Here an eigenvalue sequences transform algorithm is proposed for estimating the model order due to its low computational complexity. This algorithm can provide fast estimation but with loss in robustness to noise. More details of this algorithm are presented in Appendix A.

Based on the linear systems theory [31] and the matrix expression in (14) and (16), the noiseless matrix $\widetilde{\mathbf{H}}$ can be further factorized as:

$$\widetilde{\mathbf{H}} = \mathbf{U}_{sn}\mathfrak{R}_{sn}\mathbf{V}_{sn}^{*} = \mathbf{\Omega}\mathbf{\Gamma} \tag{19}$$

where $\mathbf{\Omega}$ is the $(N-L+1)M \times K_1$ observability matrix which can be further expressed by the matrices $\mathbf{S}$ and $\mathbf{P}$.

$$\mathbf{\Omega} = \mathbf{U}_{sn}\mathfrak{R}_{sn}^{1/2} = \begin{bmatrix} \mathbf{S} & \mathbf{SP} & \ldots & \mathbf{SP}^{N-L} \end{bmatrix}^{T} \tag{20}$$

and $\mathbf{\Gamma}$ is the $K_1 \times L$ controllability matrix which can be expressed by the matrices $\mathbf{P}$ and $\mathbf{D}$.

$$\mathbf{\Gamma} = \mathfrak{R}_{sn}^{1/2}\mathbf{V}_{sn}^{*} = \begin{bmatrix} \mathbf{D} & \mathbf{PD} & \ldots & \mathbf{P}^{L-1}\mathbf{D} \end{bmatrix} \tag{21}$$

Considering that computational complexity of (19)–(21) is enlarged observably for large data sets, operations (22)–(24) are used for alternative steps which have lower computational load but with minimal calculation error.

$$\widetilde{\mathbf{H}}^{*}\widetilde{\mathbf{H}} = \mathbf{V}_{sn}\mathfrak{R}_{sn}^{2}\mathbf{V}_{sn}^{*} = \mathbf{V}_{sn}\mathbf{Z}\mathbf{V}_{sn}^{*} \tag{22}$$

$$\mathbf{\Omega} \approx \mathbf{H}\mathbf{V}_{sn}\mathbf{Z}^{-1/4} \tag{23}$$

$$\mathbf{\Gamma} = \mathbf{Z}^{-1/4}\mathbf{V}_{sn}^{*} \tag{24}$$

where $\mathbf{Z} = \mathfrak{R}_{sn}^{2}$.

As an analogy of the open-loop matrix in 1-D SSA, the augmented open-loop matrix $\mathbf{P}$ can be derived from the observability matrix by using least square.

$$\mathbf{P} = \left(\mathbf{\Omega}_{rf}^{*}\mathbf{\Omega}_{rf}\right)^{-1}\mathbf{\Omega}_{rf}^{*}\mathbf{\Omega}_{rl} \tag{25}$$

Or it can be computed by the controllability matrix $\mathbf{\Gamma}$.

$$\mathbf{P} = \mathbf{\Gamma}_{cl}\mathbf{\Gamma}_{cf}^{*}\left(\mathbf{\Gamma}_{cf}\mathbf{\Gamma}_{cf}^{*}\right)^{-1} \tag{26}$$

where $\mathbf{\Omega}_{rf}$ is the first $(N-L)M$ rows of $\mathbf{\Omega}$ and $\mathbf{\Omega}_{rl}$ denotes the last $(N-L)M$ rows of $\mathbf{\Omega}$; $\mathbf{\Gamma}_{cf}$ represents the first $(L-1)$ columns of $\mathbf{\Gamma}$ and $\mathbf{\Gamma}_{cl}$ is the last $(L-1)$ columns of $\mathbf{\Gamma}$. From (20) and (21), these matrices can be rewritten as

$$\begin{bmatrix} \mathbf{S} & \mathbf{SP} & \ldots & \mathbf{SP}^{N-L-1} \end{bmatrix}^{T} = \mathbf{\Omega}_{rf} \tag{27}$$

$$\begin{bmatrix} \mathbf{SP} & \mathbf{SP}^{2} & \ldots & \mathbf{SP}^{N-L} \end{bmatrix}^{T} = \mathbf{\Omega}_{rl} \tag{28}$$

$$\left[\begin{array}{cccc} \mathbf{D} & \mathbf{PD} & \ldots & \mathbf{P}^{L-2}\mathbf{D} \end{array} \right] = \mathbf{\Gamma}_{cf} \tag{29}$$

$$\left[\begin{array}{cccc} \mathbf{PD} & \mathbf{P}^2\mathbf{D} & \ldots & \mathbf{P}^{L-1}\mathbf{D} \end{array} \right] = \mathbf{\Gamma}_{cl} \tag{30}$$

More details of the derivation of the matrices **S**, **P**, and **D** are listed in Appendix B.

According to (11), the vector $\left[\begin{array}{cccc} p_1 & p_2 & \ldots & p_{K_1} \end{array} \right]$ is obtained by performing the SVD.

$$\mathbf{\Lambda} = \mathbf{Q}^*\mathbf{PQ} \tag{31}$$

$$diag(\mathbf{\Lambda}) = \left[\begin{array}{cccc} p_1 & p_2 & \ldots & p_{K_1} \end{array} \right] \tag{32}$$

where $diag(\mathbf{\Lambda})$ denotes the diagonal element of matrix $\mathbf{\Lambda}$.

3.2. Pole Estimation of the Aspect Dimension

In this step, multiple-range search, which applies one-dimensional (1-D) state–space approach (SSA) to the 1-D data for each down-range cell, is used for pole estimation of the aspect dimension and pole adaptive pairing. Details are as follows.

We introduce the Vandermonde sub-matrix **O**.

$$\mathbf{O} = \begin{bmatrix} p_1^1 & p_1^2 & \vdots & p_1^N \\ p_2^1 & p_2^2 & \vdots & p_2^N \\ \vdots & \vdots & \vdots & \vdots \\ p_{K_1}^1 & p_{K_1}^2 & \vdots & p_{K_1}^N \end{bmatrix} \tag{33}$$

From (5), the noiseless signatures $\widetilde{\mathbf{Y}}$ can be factorized as a product of the pairing matrix **G** and **O**, where each column of **G** is associated with each row of **O**.

$$\widetilde{\mathbf{Y}} = \begin{bmatrix} \sum\limits_{t=1}^{l_1} a_{1t}s_{1t} & \sum\limits_{t=1}^{l_2} a_{2t}s_{2t} & \cdots & \sum\limits_{t=1}^{l_{K_1}} a_{K_1t}s_{K_1t} \\ \sum\limits_{t=1}^{l_1} a_{1t}s_{1t}^2 & \sum\limits_{t=1}^{l_2} a_{2t}s_{2t}^2 & \cdots & \sum\limits_{t=1}^{l_{K_1}} a_{K_1t}s_{K_1t}^2 \\ \vdots & \vdots & \vdots & \vdots \\ \sum\limits_{t=1}^{l_1} a_{1t}s_{1t}^M & \sum\limits_{t=1}^{l_2} a_{2t}s_{2t}^M & \vdots & \sum\limits_{t=1}^{l_{K_1}} a_{K_1t}s_{K_1t}^M \end{bmatrix} \mathbf{O} = \mathbf{GO} \tag{34}$$

Thus, the pairing matrix **G** in (34) can be calculated by using least square, i.e.

$$\mathbf{G} = \widetilde{\mathbf{Y}}\mathbf{O}^*(\mathbf{O}\mathbf{O}^*)^{-1} \tag{35}$$

The k-th column of the pairing matrix **G** is presented as

$$\mathbf{G}(k) = \left[\begin{array}{cccc} \sum\limits_{t=1}^{l_k} a_{kt}s_{kt}^1 & \sum\limits_{t=1}^{l_k} a_{kt}s_{kt}^2 & \cdots & \sum\limits_{t=1}^{l_k} a_{kt}s_{kt}^M \end{array} \right]^T \tag{36}$$

where $k = 1, 2, \ldots, K_1$.

Using (36), each column of the pairing matrix is constructed in a same structure to 1-D DE model defined in (1). This indicates that the pole s_{kt} and a_{kt}, which correspond to the *k-th* pole p_k in (32), can be solved by using 1-D SSA [22] to each column of the pairing matrix **G**. For example, for the first column of the pairing matrix **G** corresponds to p_1, the parameter matrices (**C**, **A**, **B**) can be obtained by using 1-D SSA in (4).

$$\mathbf{G}(1) = \left[\begin{array}{cccc} \mathbf{CB} & \mathbf{CAB} & \ldots & \mathbf{CA}^{M-1}\mathbf{B} \end{array} \right]^T \tag{37}$$

The eigenvalue decomposition of the open-loop matrix **A** leads to.

$$\left[\begin{array}{cccc} \widetilde{y}(1) & \widetilde{y}(2) & \ldots & \widetilde{y}(N) \end{array}\right] = \left[\begin{array}{cccc} \mathbf{CB} & \mathbf{CAB} & \ldots & \mathbf{CA}^{N-1}\mathbf{B} \end{array}\right] \tag{38}$$

The matching pole s_{1t} and the corresponding amplitude a_{1t} are computed as.

$$\left[s_{11}, s_{12}, \ldots, s_{1l_1}\right] = \left[\begin{array}{cccc} \mathbf{\Lambda}_1(1,1) & \mathbf{\Lambda}_1(2,2) & \ldots & \mathbf{\Lambda}_1(l_1, l_1) \end{array}\right] \tag{39}$$

$$\left[a_{11}, a_{12}, \ldots, a_{1l_1}\right] = (\mathbf{CM}_1)^{*} \cdot \left(\mathbf{M}_1^{-1}\mathbf{B}\right) \tag{40}$$

Thus, the pairing matrix **G** can be searched column by column until all poles s_{kt} and the corresponding amplitudes a_{kt} are estimated. No extra pairing scheme is required because the poles $[s_{k1}, s_{k2}, \ldots, s_{kl_k}]$ and the amplitudes $[a_{k1}, a_{k2}, \ldots, a_{kl_k}]$ have already been adaptively paired to the poles p_k in (34). In addition, considering that the model order may be overestimated sometime, the searched pole pairs are usually checked again according to their amplitudes. Finally, locations and damping factors of all the scattering points can be obtained from (41) and (42).

$$(r_{1k}, r_{2k}) = \left(\frac{\mathrm{Arg}\{s_k\}}{4\pi f_c \Delta\theta / c}, \frac{\mathrm{Arg}\{p_k\}}{4\pi \Delta f / c}\right) \tag{41}$$

$$(\beta_{1k}, \beta_{2k}) = \left(\frac{\ln(|s_k|)}{f_c \Delta\theta}, \frac{ln(|p_k|)}{\Delta f}\right) \tag{42}$$

4. Results and Discussion

In this Section, three examples are presented to demonstrate the usefulness of the proposed procedure, i.e., the numerical signatures with 14 point scattering centers, the numerical signatures of a sphere tipped cone-cylinder-frustum combination model, the measured ISAR data for an aircraft model. The results obtained by 2-D ESPRIT are used for comparison since it is one of the very few techniques which have been used in real radar applications [17,32].

4.1. Numerical Signatures with Point Scattering Centers

In this example, the noisy signatures composed of 14-point scattering centers are considered to be as follows:

$$y(m,n) = \sum_{k=1}^{14} a_k \exp[(j2\pi r_{1k}/c + \beta_{1k}) f_c \theta_n] \exp[(j2\pi r_{2k}/c + \beta_{2k}) f_m] + w(m,n) \tag{43}$$

where $(M, N) = (41, 41)$; a_k is set to be 1; $f_c = 12\text{GHz}$ and $\Delta f = 150\text{MHz}$; $\theta_0 = 0$ rad and $\Delta\theta = 0.0125$ rad; The setting of coupled ranges are shown in Figure 1; All the damping factors including β_{1k} and β_{2k} are set as $-0.05/(f_c\Delta\theta)$ and $-0.05/\Delta f$ except the far left scattering points $(\beta_{11}, \beta_{21}) = (-0.02/(f_c\Delta\theta), -0.02/\Delta f)$; $w(m,n)$ denotes the additive white Gaussian noise.

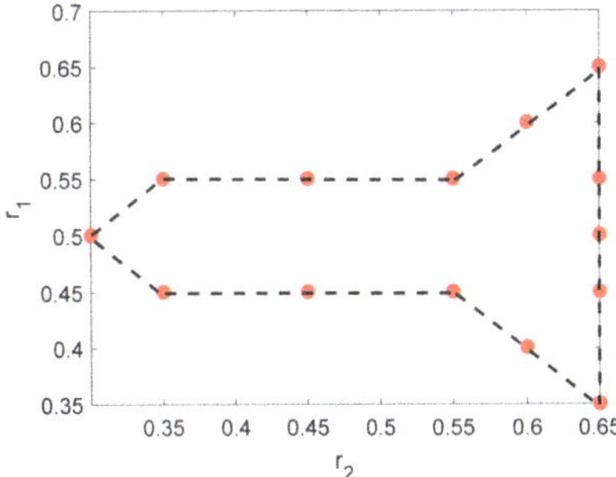

Figure 1. Spatial distribution of the simulated poles.

As shown in Figure 1, these scattering points form a missile-like shape in Cartesian coordinates. It contains 13 pole pairs which have repeated poles in either the down-range or aspect dimensions. The noisy signatures in space domain by Fourier transform-based imaging algorithm (add Taylor window) are displayed in Figure 2a,b. As can be seen, the positions and decay rates of these scattering points are consistent with the corresponding ranges and damping factors. To sparse representation of these noisy signatures, the parameter for 2-D ASSA is chosen to be $L = 20$; the parameters for 2-D ESPRIT are set as: $(P, Q) = (20, 20)$, $\beta = 0.8$, which are suggested in [17]. For each signal to noise ratio (SNR), the number of Monte Carlo simulation is 200. A series of range estimation results in different SNRs are presented in Figure 3. Please note that the model order in 2-D ESPRIT is pre-specified because the singular values-based criterion [28,29] cannot be used when there are repeated poles in either the down-range or aspect dimensions. Because of different pairing strategies, K_1 and K in 2-D ASSA could be estimated by the eigenvalue sequences transform algorithm (Appendix A) when SNR > 10 dB. However, the numbers K_1 and K in 2-D ASSA should be pre-specified or use the other criterion [30] when the noise level is higher (SNR $\leq$ 10 dB).

As can be seen in Figure 3a,b, the positions of scattering points estimated by 2-D ESPRIT are generally according to the preset poles in Figure 1, except for some missing or incorrect points (shown by the black rectangular). The possible reason of this problem is that the pairing procedure in 2-D ESPRIT may provide incorrect pole pairs when there are repeated poles in either the down-range or aspect dimension, i.e., six pole pairs (s_1, p_1), (s_1, p_2), (s_1, p_3), (s_2, p_1), (s_2, p_2), (s_2, p_3). In contrast, the results estimated by 2-D ASSA showed higher accuracy than 2-D ESPRIT for different pairing strategies.

The estimation accuracy of cross/down ranges and damping factors are displayed in Figure 4. Root mean square error (RMSE) is used as the evaluating indicator which is defined in (44). As we can see, estimation accuracy of these two algorithms are basically at the same level although the size of Hankel matrix used by 2-D ASSA is smaller than the block-Hankel matrix defined by 2-D ESPRIT.

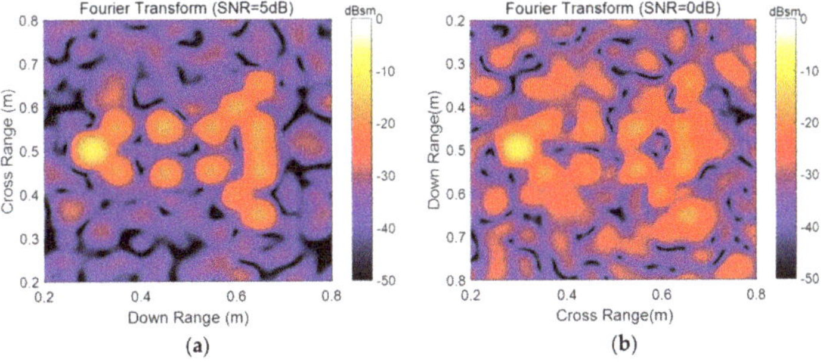

Figure 2. Simulated radar image with different SNR by Fourier transform. (**a**) Simulated radar image with SNR = 5 dB; (**b**) Simulated radar image with SNR = 0 dB.

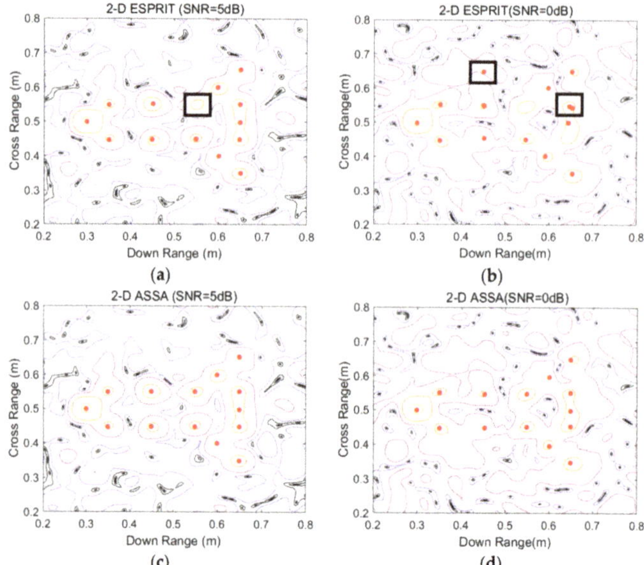

Figure 3. Estimation of the coupled ranges by 2-D ASSA and 2-D ESPRIT with different SNR. (**a**) Estimation by 2-D ESPRIT with SNR = 5 dB; (**b**) Estimation by 2-D ESPRIT with SNR = 0 dB; (**c**) Estimation by 2-D ASSA with SNR = 5 dB; (**d**) Estimation by 2-D ASSA with SNR = 0 dB.

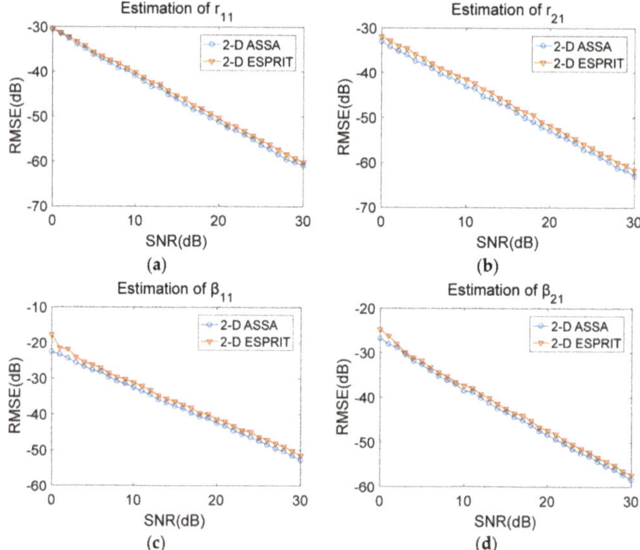

Figure 4. Estimation accuracy of r_{11}, r_{21}, β_{11} and β_{21} with different SNR. (**a**) Estimation of r_{11} with different SNR; (**b**) Estimation of r_{21} with different SNR; (**c**) Estimation of β_{11} with different SNR; (**d**) Estimation of β_{21} with different SNR.

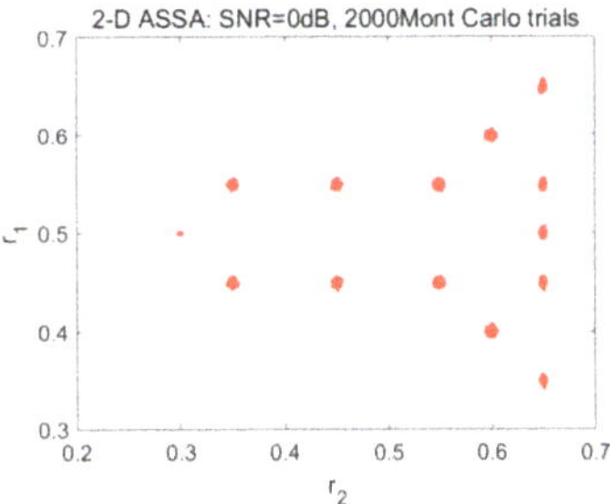

Figure 5. Estimation of the coupled ranges by 2-D ASSA with 2000 Monte Carlo runs (SNR = 0 dB).

$$\delta_{RMSE} = 10\log_{10}\sqrt{\frac{1}{M_c}\sum_{t=1}^{M_c}[X_{\text{est}} - X_{\text{real}}]^2} \tag{44}$$

where $M_c = 2000$ denotes the number of Monte Carlo runs, X_{est} and X_{real} are the esimated and real parameters.

Figure 5 also presents the statistic result with 2000 Monte Carlo runs (SNR = 0 dB). The result demonstrates that the multiple-range search strategy used by 2-D ASSA is robust in pole-pairing for low SNR.

4.2. Numerical Signatures of Computer-Aided Design (CAD) Model

The numerical signatures are obtained using method of moment (MOM), where the target is from a computer-aided design (CAD) model of a sphere tipped cone-cylinder-frustum combination, (shown in Figure 6). The data was calculated from 8–12 GHz in 10 MHz frequency step size, view angle ranging from −5 to 5 deg with an increment of 0.25 deg. Figure 7a presents 2-D radar image processed using 2-D FFT. As it can be seen, strong scattering points can be observed at differential discontinuities, such as the base-edge, the body groove, and nosetip. To sparsely represent the numerical data, the parameters of this example are set as follows. For 2-D ASSA, L is set as $N2$. For 2-D ESPRIT, $(P, Q) = (M2, N2)$, where $M2 = ceil(M/2)$ denotes the smallest integer less than or equal to $M/2$, and $\beta = 0.8$.

Location estimation of key scattering points for these two algorithms are shown in Figure 7b–e. As can be seen, when the number of scattering centers is set to be 14, all key scattering points are accurately estimated by these two algorithms except for some minor differences. However, when K is set to be 18, 2-D ESPRIT encountered the pole-pairing problem and could not provide the right estimation. Relative reconstruction error (RRE) δ_{RRE} (defined in (45)) of these two algorithms are shown in Figure 8. We can see that the RRE of 2-D ASSA has been falling when K increased from 10 to 30, whereas the RRE of 2-D ESPEIT stopped falling after $K = 14$. The result shows that 2-D ASSA tends to be more robust in pole-pairing for different numbers of scattering centers.

$$\delta_{RRE} = \frac{1}{MN}\sum_{m=1}^{M}\sum_{n=1}^{N}\left|20\log_{10}[\mathbf{Y}_{recon}(m,n)/\mathbf{Y}_{real}(m,n)]\right| \tag{45}$$

where $\mathbf{Y}_{recon}$ denotes the signatures reconstructed by the estimated parameters. $\mathbf{Y}_{real}$ is the original signatures.

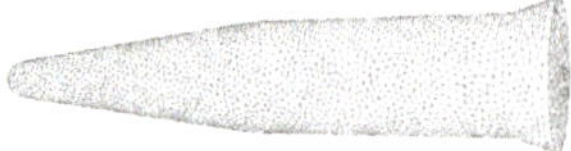

Figure 6. CAD model of the sphere tipped cone-cylinder-frustum combination.

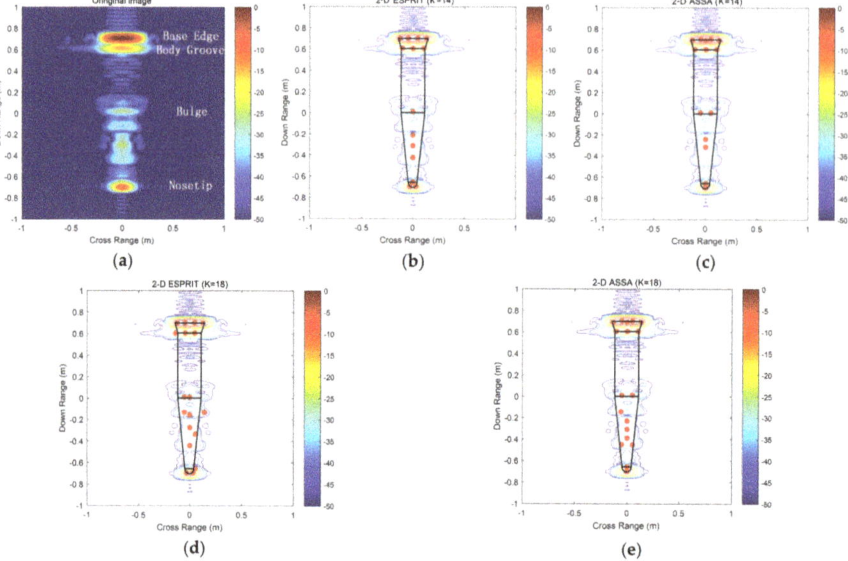

Figure 7. 2-D radar image of the CAD model and location estimation of main scattering points result by 2-D ESPRIT and 2-D ASSA. (**a**) 2-D radar image in aspect angle from −5 to 5 deg; (**b**) Pole estimation by 2-D ESPRIT(K = 14); (**c**) Pole Estimation by 2-D ASSA(K = 14); (**d**) Pole estimation by 2-D ESPRIT(K = 18); (**e**) Pole Estimation by 2-D ASSA(K = 18).

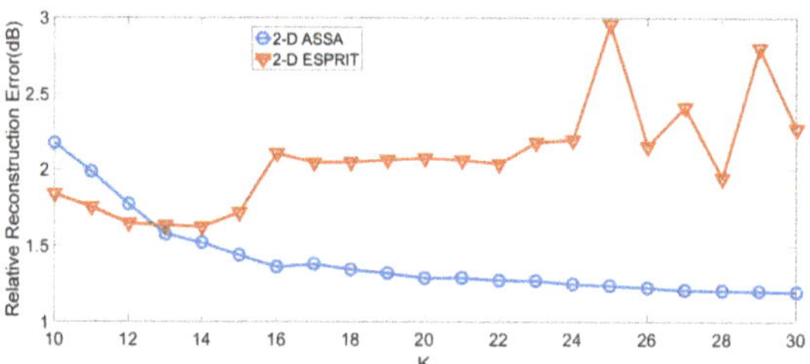

Figure 8. Relative reconstruction error of the numerical signatures of CAD model with different numbers of scattering centers.

Figure 9a displays the sub-band data image (8–10 GHz, from −5 to 5 deg) which was extracted from the full-band data. Compared to the full-band data image in Figure 7a, scattering centers of the base-edge and the body groove cannot be distinguished when the signal bandwidth is only 2 GHz. Figure 9b presents the scattering points estimated by 2-D ASSA and Figure 9c shows the full-band data image extrapolated by these extracted pole pairs. As can be seen, scattering centers of the base-edge and the body groove are distinguished from the extracted scattering points. The full-band data image extrapolated by 2-D ASSA is in keeping with the main scattering centers distributed in the original full-band data image (Figure 7a). It is clear from these figures that 2-D ASSA results in higher-resolution images than traditional 2-D FFT imagery.

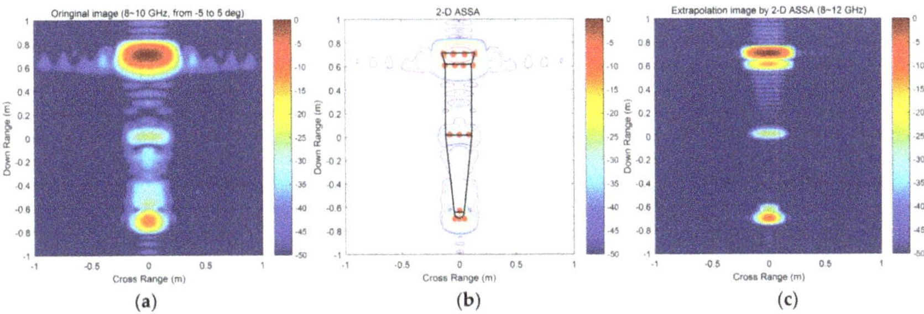

Figure 9. Sub-band data image and extrapolated data image by 2-D ASSA. (**a**) Sub-band data image by 2-D FFT (8–10 GHz); (**b**) Scattering points estimated by 2-D ASSA; (**c**) Full-band data image extrapolated by 2-D ASSA (8–12 GHz).

4.3. Measured ISAR Signatures

In the third example, the measured ISAR signatures are acquired in an indoor test range where a mocked aircraft model is used. The data was collected using S-band radar with center frequency 3 GHz and 1.5 GHz bandwidth. The range of aspect angle is from −4° to 4° with an interval of 0.1°. Figure 10a shows the ISAR image of the aircraft model generated by 2-D FFT. As it can be seen, the scattering distribution of this model is more complex than the previous model, i.e., many scattering points are densely distributed in the fuselage. The estimation parameter for 2-D ASSA is set to be $L = N2$; For 2-D ESPRIT, $(P, Q) = (M2, N2)$ and $\beta = 0.8$.

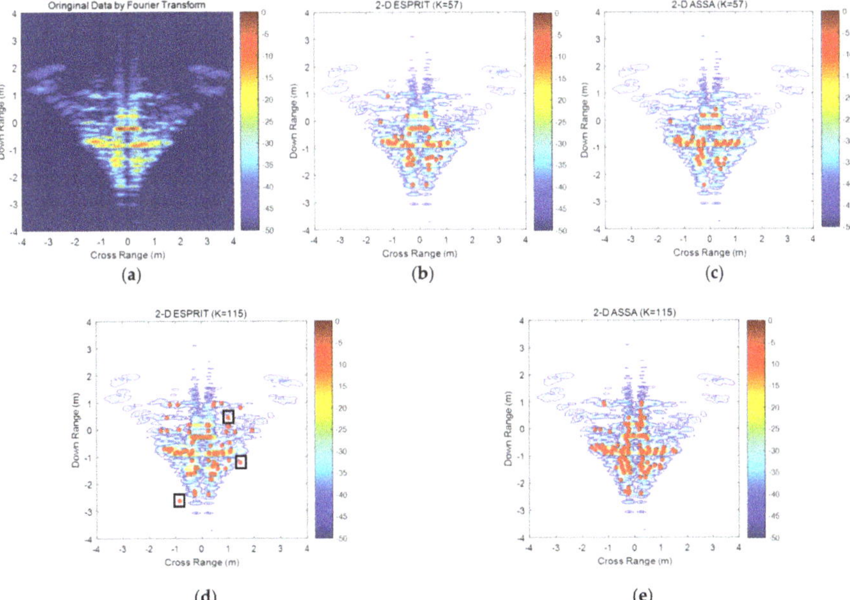

Figure 10. Pole-estimation result of the measured data with different K. (**a**) Measured ISAR image of the aircraft model by using 2-D FFT (**b**) Pole estimation by 2-D ESPRIT ($K = 57$); (**c**) Pole estimation by 2-D ASSA ($K = 57$); (**d**) Pole estimation by 2-D ESPRIT($K = 115$); (**e**) Pole estimation by 2-D ASSA ($K = 115$).

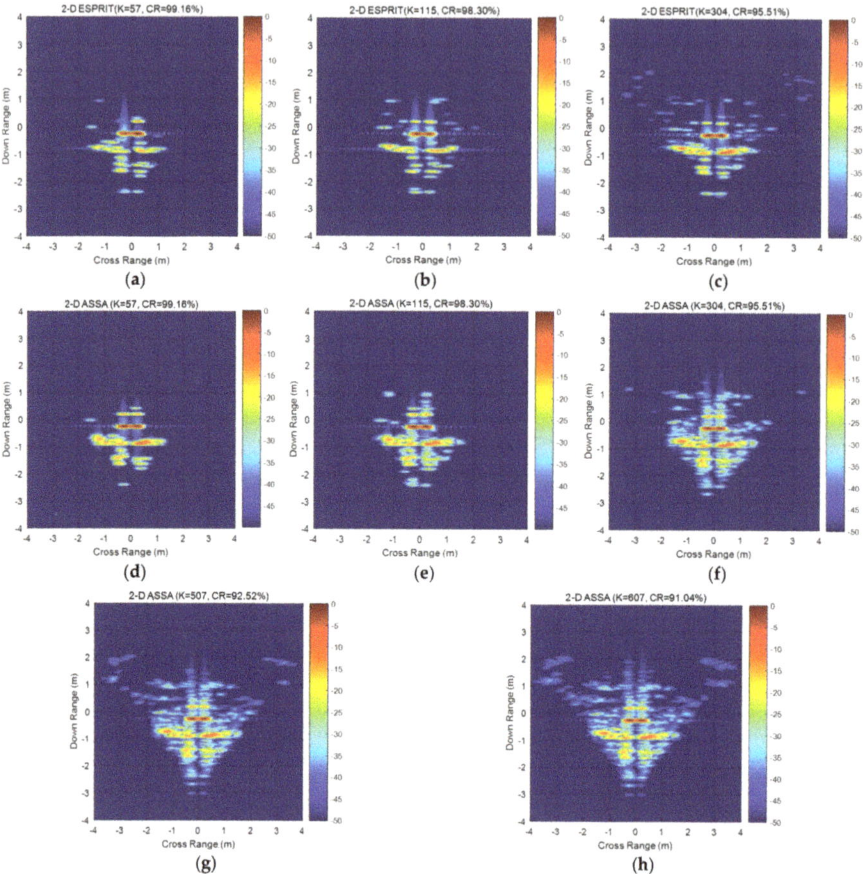

Figure 11. Reconstruction result of the measured data with different K. (**a**) Reconstruction result by 2-D ESPRIT($K = 57$); (**b**) Reconstruction result by 2-D ESPRIT ($K = 115$); (**c**) Reconstruction result by 2-D ESPRIT ($K = 304$); (**d**) Reconstruction result by 2-D ASSA($K = 57$);(**e**) Reconstruction result by 2-D ASSA($K = 115$); (**f**) Reconstruction result by 2-D ASSA($K = 304$); (**g**) Reconstruction result by 2-D ASSA($K = 507$); (**h**) Reconstruction result by 2-D ASSA($K = 607$).

Location estimation of main scattering centers for measured data are shown in Figure 10b–e. As we can see, both 2-D ESPRIT and 2-D ASSA could estimate the right positions of strong scattering centers. The scattering points estimated by 2-D ASSA are more densely distributed around the strong scattering centers than 2-D ESPRIT's for different pole-pairing strategies. A few relatively weak scattering points, i.e., the red point in rectangle box, are wrongly estimated by 2-D ESPRIT when the number of scattering points is set to be 115, and more wrongly estimated pole pairs can be seen in $K > 115$ which are not displayed in Figure 10.

Figure 11a–h display a set of reconstructed 2-D images by the estimated pole pairs. Please note that the number of scattering points K in 2-D ESPRIT should be equal to the model order whereas K in 2-D ASSA could be larger than the model order K_1 (the number of non-repeated poles in the down-range dimension) when there are one-to-multiple pairing poles. That is the reason K could be set to 607 which is already exceed the limitation of model order in [17]. As we can see from the figures, all those two algorithms can reconstruct those strong scattering centers of target which are in consistent with the estimated poles in Figure 10a–d. As displayed in Figure 11f–h, more and more relatively weak scattering centers have been reconstructed by 2-D ASSA with the increasing number of K and the reconstruction result with compression ratio of 91.04% is extremely similar to the original data in Figure 10a.

Evaluation of the reconstructed results by 2-D ASSA are listed in Table 2. Compression ratio (CR) ε_{CR} and image similarity degree (ISD) are defined in (46) and (47). The RRE δ_{RRE} is used for evaluating the reconstructed result in frequency-domain, whereas the ISD γ_{ISD} is for the result in image-domain. From the table, we can see that the numbers of scattering points in 2-D ASSA are negatively correlated with RRE (or ISD). In contrast, 2-D ESPRIT performs higher RRE than 2-D ASSA when the number of scattering centers remains the same. Please note that part of evaluation results by 2-D ESPRIT are abnormal because 2-D ESPRIT cannot do the right reconstruction when $K > 307$. The possible reason is that for 2-D ESPRIT, all amplitudes factors $\begin{bmatrix} a_1 & a_2 & \dots & a_K \end{bmatrix}$ (defined in (3)) are estimated together by using least square technique after all the pole pairs are confirmed. Thus, numerous incorrect pairing poles estimated by 2-D ESPRIT will lead to entirely wrong reconstruction. In comparison, 2-D ASSA performs better robustness for the measurement data.

$$\varepsilon_{CR} = \frac{M \times N - n_P \times K}{M \times N} \times 100\% \tag{46}$$

$$\gamma_{ISD} = \frac{\sum \mathbf{I}_{recon}\mathbf{I}_{real}}{\sqrt{\sum \mathbf{I}_{recon}^2 \sum \mathbf{I}_{real}^2}} \tag{47}$$

where $n_P = 3$ represents parameter number of each paired pole (a_k, s_k, p_k), K is the number of scattering points, $M \times N$ denotes size of the original signatures. $\mathbf{I}_{recon}$, $\mathbf{I}_{real}$ represent the reconstructed and original 2-D images-domain data.

Table 2. Evaluation of the reconstruction result of the measured data by 2-D ESPRIT and 2-D ASSA.

	2-D ASSA		2-D ESPRIT	
Number of Scattering	δ_{RRE}**(dB)**	γ_{ISD}	δ_{RRE}**(dB)**	γ_{ISD}
$K = 57$ ($\varepsilon_{CR} = 99.16\%$)	3.26	0.90	3.93	0.88
$K = 115$ ($\varepsilon_{CR} = 98.30\%$)	2.60	0.92	3.20	0.90
$K = 156$ ($\varepsilon_{CR} = 97.70\%$)	2.53	0.92	3.11	0.90
$K = 207$ ($\varepsilon_{CR} = 96.95\%$)	2.47	0.93	2.96	0.91
$K = 256$ ($\varepsilon_{CR} = 96.22\%$)	2.27	0.94	2.94	0.91
$K = 304$ ($\varepsilon_{CR} = 95.51\%$)	2.07	0.94	2.90	0.91
$K = 407$ ($\varepsilon_{CR} = 93.99\%$)	1.76	0.95	41.28	0.26
$K = 507$ ($\varepsilon_{CR} = 92.52\%$)	1.60	0.96	68.69	0.12
$K = 607$ ($\varepsilon_{CR} = 91.04\%$)	1.51	0.96	43.57	0.24

Running time of the processed examples are listed in Table 3. All the results are carried out by Matlab (R2016b) with same hardware platform: Intel Core i7-6900k 3.2 GHz and 128 G memory. Please note that the running time of 2-D ESPRIT does not contain the estimation of model order. According to the table, 2-D ASSA performs higher efficiency than 2-D ESPRIT in processing the same data with the following differences.

(1) The size of the augmented Hankel matrix used by 2-D ASSA is approximate $1/(M/4)$ of the block-Hankel matrix used by 2-D ESPRIT when $L = Q = N2$ and $P = M2$. Moreover, the data size of the Hankel matrix could be further reduced by using the operation (22)–(24);

(2) No extra pairing scheme is required by using multiple-range search strategy which can finish the pairing process once the pole estimation of the aspect dimension is confirmed. Moreover, calculation

of the block-Hankel matrix for the aspect dimension in 2-D ESPRIT is simplified as calculating multiple small Hankel matrices with size $M2 \times (M - M2 + 1)$.

Table 3. Running time of the numerical radar signatures and the measured ISAR data by 2-D ESPRIT and 2-D ASSA.

	Method	K	Hankel Matrix (p_k)	Hankel Matrix (s_k)	Running Time
Numerical data (41 × 401)	2-D ESPRIT	14	4000 × 4444	4000 × 4444	34.81 s
	2-D ASSA	14	8282 × 200	$K_1 \times 20 \times 22$	0.94 s
Measured ISAR data (81 × 251)	2-D ESPRIT	57	5000 × 5334	5000 × 5334	53.27 s
		115	5000 × 5334	5000 × 5334	94.18 s
	2-D ASSA	57	10287 × 125	$K_1 \times 40 \times 42$	1.04 s
		115	10287 × 125	$K_1 \times 40 \times 42$	1.19 s

5. Conclusions

In this paper, we have presented a two-dimensional augmented state–space approach for sparse representation of 2-D wideband radar signatures collected on man-made targets. To do this, a two-step procedure, i.e., an augmented state–space approach followed by multiple-range search strategy, is proposed to estimate the complex amplitudes and poles in down-range and aspect dimensions. In general, there are mainly two contributions provided in this paper. First, we develop a computationally efficient approach by adopting several time-saving operations, whereas the pole-estimation accuracy is still at the same level; second, the proposed approach can apparently alleviate the pole-pairing problem by using the multiple-range search strategy.

Numerical as well as measured ISAR data are processed to validate the proposed approach. Experimental results demonstrate that 2-D ASSA is robust and accurate in pole-paring for different SNRs, and is applicable for sparse representation of 2-D wideband radar signatures collected on man-made targets with low computational cost.

Future works are considered to be as follows. On the one hand, physical meanings of the extracted parameters of scattering centers by 2-D ASSA might be further studied to the possible applications in automatic target recognition (ATR). On the other hand, the extension of the proposed approach to 3-D radar signatures obtained from different azimuth and elevation angles would also be an interesting study.

Author Contributions: Methodology, K.W.; validation, K.W. and X.X.; writing—original draft preparation, K.W.; writing—review and editing, K.W. and X.X.

Funding: This research was funded by National Natural Science Foundation of China: 61371005.

Conflicts of Interest: The authors declare no conflict of interest.

Appendix A

In the appendix, we provide an eigenvalue sequences transform algorithm to estimate the model order K_1. Details are as follows.

Suppose that the size of the input matrix $\mathbf{H}$ is $m \times n (m \geq n)$, do the SVD of the matrix $\mathbf{H}^*\mathbf{H}$.

$$\mathbf{H}^*\mathbf{H} = \mathbf{V}\Re^2\mathbf{V}^* \quad \text{(A1)}$$

where $\mathbf{V}$ is unitary matrix; $\Re^2$ is the diagonal matrix.

$\Lambda = [\lambda_1, \lambda_2, , \ldots \lambda_n]$ is the diagonal vector of $\Re$. It can be normalized as:

$$w_i = \frac{\lambda_i - \min(\Lambda)}{\max(\Lambda) - \min(\Lambda)} \quad \text{(A2)}$$

where $i = 1, 2, \ldots, n$; $\max(\Lambda)$ and $\min(\Lambda)$ denote the maximum and minimum elements of Λ.

The coordinate form of the normalized eigenvalue vector is:

$$[Q_1, Q_2, \ldots, Q_n] = [(1, w_1), (2, w_2), \ldots, (n, w_n)] \tag{A3}$$

where Q_n denotes the n-th points.

Based on information theory and principal factor analysis [28], the separable signals and noises in **H** often have the following hypothesis.

$$\begin{array}{c} w_1 > \ldots > w_{K_1} > w_{K_1+1} \ldots > w_n \\ \text{signal space}|\text{noisespace} \end{array} \tag{A4}$$

$$w_{K_1} - w_{K_1+1} > w_{K_1+1} - w_{K_1+2} \approx \ldots \approx w_{n-1} - w_n \tag{A5}$$

where K_1 is the number of separable signals which also denotes the model order.

From (A5), it can be deduced that the point Q_{K_1} belonging to $[Q_1, Q_2, \ldots, Q_n]$ has the longest distance to line Q_1D which has the following condition.

$$k\left(\overrightarrow{Q_i Q_{K_1+1}}\right) > k\left(\overrightarrow{Q_1 D}\right) > k\left(\overrightarrow{Q_{K_1+1} Q_n}\right) \tag{A6}$$

where i = $1, 2, \ldots, K_1$, $k\left(\overrightarrow{Q_i Q_{K_1+1}}\right)$ denotes the slope of line $\overrightarrow{Q_i Q_{K_1+1}}$, D is a point which satisfies the slope condition (A6).

Here, an approximate calculation is used to estimate the slop of $\overrightarrow{Q_1 D}$.

$$k\left(\overrightarrow{Q_1 D}\right) \approx k\left(\overrightarrow{Q_{p-1} Q_n}\right) \tag{A7}$$

$$g(p) = \max[g(1), g(2), \ldots, .g(n)] \tag{A8}$$

$$g(i) = \frac{\left|k\left(\overrightarrow{Q_1 Q_n}\right)(i-1) - (w_i - w_1)\right|}{\sqrt{k\left(\overrightarrow{Q_1 Q_n}\right)^2 + 1}} \tag{A9}$$

where Q_p represents the point in $[Q_1, Q_2, \ldots, Q_n]$ which has the longest distance to line Q_1Q_n; $g(i)$ denotes the distance from point Q_i to $\overrightarrow{Q_1 Q_n}$; $i = 1, 2, \ldots, n$.

Thus, Q_{K_1} can be confirmed by searching the longest distance from point Q_i to line $\overrightarrow{Q_1 D}$.

$$g'(i) = \frac{\left|k\left(\overrightarrow{Q_1 D}\right)(i-1) - (w_i - w_1)\right|}{\sqrt{k\left(\overrightarrow{Q_1 D}\right)^2 + 1}} \tag{A10}$$

$$g'(K_1 + 1) = \max[g'(1), g'(2), \ldots, .g'(n)] \tag{A11}$$

where $i = 1, 2, \ldots, n$; $g'(i)$ denotes the distance from point Q_i to $\overrightarrow{Q_1 D}$.

Appendix B

From the observability matrix $\mathbf{\Omega}$ given by (20), controllability matrix $\mathbf{\Gamma}$ given by (21) and the matrices $\mathbf{\Omega}_{rf}$, $\mathbf{\Omega}_{rl}$, $\mathbf{\Gamma}_{cf}$,$\mathbf{\Gamma}_{cl}$ in (27),(28),(29) and (30), it is not difficult to deduce that the augmented open-loop matrix $\mathbf{P}$ satisfies the following matrix equations.

$$\mathbf{\Omega}_{rf}\mathbf{P} = \mathbf{\Omega}_{rl} \tag{A12}$$

$$\mathbf{P}\boldsymbol{\Gamma}_{cf} = \boldsymbol{\Gamma}_{cl} \tag{A13}$$

Then the augmented open-loop matrix **P** can be obtained by least squares.

$$\mathbf{P} = \left(\boldsymbol{\Omega}_{rf}^{*}\boldsymbol{\Omega}_{rf}\right)^{-1}\boldsymbol{\Omega}_{rf}^{*}\boldsymbol{\Omega}_{rl} \tag{A14}$$

or

$$\mathbf{P} = \boldsymbol{\Gamma}_{cl}\boldsymbol{\Gamma}_{cf}^{*}\left(\boldsymbol{\Gamma}_{cf}\boldsymbol{\Gamma}_{cf}^{*}\right)^{-1} \tag{A15}$$

Moreover, here are two ways to compute the corresponding constant matrices **S** and **P**. The first way is for matrix **P** computed by (A14), the corresponding matrix **S** is defined as the first M rows of the observability matrix $\boldsymbol{\Omega}$.

$$\mathbf{S} = \begin{bmatrix} \boldsymbol{\Omega}(1,1) & \cdots & \boldsymbol{\Omega}(1,K_1) \\ \vdots & \vdots & \vdots \\ \boldsymbol{\Omega}(M,1) & \cdots & \boldsymbol{\Omega}(M,K_1) \end{bmatrix} \tag{A16}$$

Using (A14) and (A16), we calculate the matrix $\widetilde{\boldsymbol{\Omega}}$

$$\widetilde{\boldsymbol{\Omega}} = \begin{bmatrix} \mathbf{S} & \mathbf{SP} & \ldots & \mathbf{SP}^{N-1} \end{bmatrix} \tag{A17}$$

From (15) and (A17), the original 2-D signatures **Y** can be written as

$$\mathbf{Y} = \widetilde{\boldsymbol{\Omega}}\mathbf{D} \tag{A18}$$

By using least square again, the matrix **D** is

$$\mathbf{D} = \left(\widetilde{\boldsymbol{\Omega}}^{*}\widetilde{\boldsymbol{\Omega}}\right)^{-1}\widetilde{\boldsymbol{\Omega}}^{*}\mathbf{Y} \tag{A19}$$

The second way is for matrix **P** computed by (A15), the corresponding matrix **D** is defined as

$$\mathbf{D} = \begin{bmatrix} \boldsymbol{\Gamma}(1,1) \\ \vdots \\ \boldsymbol{\Gamma}(K_1,1) \end{bmatrix} \tag{A20}$$

Using (A15) and (A20), we calculate the matrix $\widetilde{\boldsymbol{\Gamma}}$

$$\widetilde{\boldsymbol{\Gamma}} = \begin{bmatrix} \mathbf{D} & \mathbf{PD} & \ldots & \mathbf{P}^{N-1}\mathbf{D} \end{bmatrix} \tag{A21}$$

Similar to (A18), the original 2-D signatures **Y** can also be rewritten as

$$\mathbf{Y} = \mathbf{S}\widetilde{\boldsymbol{\Gamma}} \tag{A22}$$

By using least square, the matrix **S** is

$$\mathbf{S} = \mathbf{Y}\widetilde{\boldsymbol{\Gamma}}^{*}\left(\widetilde{\boldsymbol{\Gamma}}\widetilde{\boldsymbol{\Gamma}}^{*}\right)^{-1} \tag{A23}$$

In this appendix, we provide two ways to estimate the parameter matrices (**S**, **P**, **D**). For consideration of the robustness to noise, (A23), (A14) (or (A15)) and (A19) are preferred. It is also worth mentioning that 2-D ASSA has two ways, which include the matrix form (**S**, **P**, **D**) and the pole form (a_k, s_k, p_k), to reconstruct the original signatures **Y**. The matrix form could provide more accurate reconstruction result than the pole form but with much more parameters.

References

1. Wang, L.; Zhao, L.F.; Bi, G.; Wan, C. Sparse Representation-Based ISAR Imaging Using Markov Random Fields. *IEEE J. Sel. Topics Appl. Earth Observ. Remote Sens.* **2015**, *8*, 3941–3953. [CrossRef]
2. Lasserre, M.; Bidon, S.; Chevalier, F.L. New Sparse-Promoting Prior for the Estimation of a Radar Scene with Weak and Strong Targets. *IEEE Trans. Signal Process.* **2016**, *64*, 4634–4643. [CrossRef]
3. Tseng, N. A very Efficient RCS Data Compression and Reconstruction Technique. Master's Thesis, The Ohio State University, Columbus, OH, USA, 1992.
4. Dong, G.G.; Wang, N.; Kuang, G.Y. Sparse Representation of Monogenic Signal: With Application to Target Recognition in SAR Images. *IEEE Signal Process. Lett.* **2014**, *21*, 952–956.
5. Xu, X.J.; He, F.Y. An Iterative Procedure for Ultra-Wideband Imagery of Space Objects from Distributed Multi-Band Radar Signatures. *Proc. SPIE.* **2014**, 9227. [CrossRef]
6. Li, J.; Stoica, P. An Adaptive Filtering Approach to Spectral Estimation and SAR Imaging. *IEEE Trans. Signal Process.* **1996**, *44*, 1469–1484.
7. Graaf, S.R.D. Parametric Estimation of Complex 2-D Sinusoids. In Proceedings of the IEEE Fouth Annual ASSP Workshop on Spectrum Estimation and Modeling, Minneapolis, MN, USA, 3–5 August 1988; IEEE: Piscataway, NJ, USA; pp. 391–396.
8. Choi, I.S.; Kim, H.T. Two-Dimensional Evolutionary Programming-Based CLEAN. *IEEE Trans. Aerosp. Electron. Syst.* **2003**, *39*, 373–382. [CrossRef]
9. Varshney, K.R.; Cethin, M.; Fisher, J.W.; Willsky, A.S. Sparse Representation in Structured Dictionaries With Application to Synthetic Aperture Radar. *IEEE Trans. Signal Process.* **2008**, *56*, 3548–3561. [CrossRef]
10. Potter, L.C.; Erthin, E.; Parker, J.T.; Cethin, M. Sparsity and Compressed Sensing in Radar Imaging. *Proc. IEEE.* **2010**, *98*, 1006–1020. [CrossRef]
11. Liu, H.C.; Jiu, B.; Liu, H.W.; Bao, Z. Superresolution ISAR Imaging Based on Sparse Bayesian Learning. *IEEE Trans. Geosci. Remote Sens.* **2014**, *52*, 5005–5013.
12. Odendaal, J.W.; Barnard, E.; Pistorius, C.W.I. Two-Dimensional Superresolution Radar Imaging Using the MUSIC Algorithm. *IEEE Trans. Antennas Propag.* **1994**, *42*, 1386–1391. [CrossRef]
13. Hua, Y.; Sarkar, T.K. Matrix Pencil Method for Estimating Parameters of Exponentially Damped/undamped Sinusoids in Noise. *IEEE Trans. Signal Process.* **1990**, *36*, 814–824. [CrossRef]
14. Hua, Y. Estimating Two-Dimensional Frequencies by Matrix Enhancement and Matrix Pencil. *IEEE Trans. Signal Process.* **1992**, *40*, 2267–2280. [CrossRef]
15. Sacchini, J.J.; Steedly, W.M.; Moses, R.L. Two-Dimensional Prony Modeling and Parameter Estimation. *IEEE Trans. Signal Process.* **1993**, *41*, 3127–3137. [CrossRef]
16. Vanpoucke, F.; Moonen, M.; Berthoumieu, Y. An Efficient Subspace Algorithm for 2-D Harmonic Retrieval. In Proceedings of the ICASSP '94. IEEE International Conference on Acoustics, Speech and Signal Processing, Adelaide, SA, Australia, 19–22 April 1994; pp. 461–464.
17. Rouquette, S.; Najim, M. Estimation of Frequencies and Damping Factors by Two-Dimensional ESPRIT Type Methods. *IEEE Trans. Signal Process.* **2001**, *49*, 237–245. [CrossRef]
18. Piou, J.E. System Realization Using 2-D Output Measurements. In Proceedings of the 2004 American Control Conference (ACC), Boston, MA, USA, 30 June–2 July 2004.
19. Piou, J.E.; Dumanian, A.J. Application of the Fornasini-Marchesini First Model to Data Collected on a Complex Target Model. In Proceedings of the 2014 American Control Conference (ACC), Portland, OR, USA, 4–6 June 2014.
20. Chen, F.J.; Fung, C.C.; Kok, C.W.; Kwong, S. Estimation of Two-Dimensional Frequencies Using Modified Matrix Pencil Method. *IEEE Trans. Signal Process.* **2007**, *55*, 718–724. [CrossRef]
21. Qian, C.; Huang, L.; So, H.C.; Sidiropoulos, N.D.; Xie, J. Unitary PUMA Algorithm for Estimating the Frequency of a Complex Sinusoid. *IEEE Trans. Signal Process.* **2015**, *63*, 5358–5368. [CrossRef]
22. Naishadham, K.; Piou, J.E. A Robust State-Space Model for the Characterization of Extended Returns in Radar Target Signatures. *IEEE Trans. Antennas Propag.* **2008**, *56*, 1742–1750. [CrossRef]
23. Ying, C.J.; Chiang, H.C.; Moses, R.L.; Potter, L.C. Complex SAR Phase History Modeling Using Two-dimensional Parametric Estimation Techniques. *Proc. SPIE.* **1996**, *2757*, 174–185. [CrossRef]
24. Xu, X.J.; He, F.Y. Fast and Accurate RCS Calculation for Squat Cylinder Calibrators. *IEEE Antennas Propagation Mag.* **2015**, *57*, 33–40. [CrossRef]

25. Augusto, A.; Vincenzo, C.; Antonio, D.M.; Luca, P. High Range Resolution Profile Estimation via a Cognitive Stepped Frequency Technique. *IEEE Trans. Aerosp. Electron. Syst.* **2019**, *1*, 444–458.
26. Pia, A.; Augusto, A.; Antonio, D.M.; Luca, P.; Silvia, L.U. HRR profile estimation using SLIM. *IET Radar Sonar Navig.* **2019**, *4*, 512–521.
27. Potter, L.C.; Moses, R.L. Attributed Scattering Centers for SAR ATR. *IEEE Trans. Image Process.* **1997**, *6*, 79–91. [CrossRef] [PubMed]
28. Wax, M.; Kailath, T. Detection of Signals by Information Theoretic Criteria. *IEEE Trans. Signal Process.* **1995**, *33*, 387–392. [CrossRef]
29. Wax, M.; Ziskind, I. Detection of the Number of Coherent Signals by the MDL Principle. *IEEE Trans. Signal Process.* **1989**, *27*, 1190–1196. [CrossRef]
30. Aksasse, B.; Radouane, L. Two-Dimensional Autoregressive (2-D AR) Model Order Estimation. *IEEE Trans. Signal Process.* **1999**, *47*, 2072–2077. [CrossRef]
31. Kung, S.Y.; Arun, K.S.; Rao, D.V.B. State-space and Singular-Value Decomposition based Approximation Methods for the Harmonic Retrieval Problem. *J. Opt. Soc. Am.* **1983**, *73*, 1799–1811. [CrossRef]
32. Michael, L.B. Two-Dimensional ESPRIT With Tracking for Radar Imaging and Feature Extraction. *IEEE Trans. Antennas Propag.* **2004**, *2*, 524–532.

Article

On the Slow-Time k-Space and its Augmentation in Doppler Radar Tomography

Hai-Tan Tran [1,*], Emma Heading [1,2] and Brian W.-H. Ng [2]

1 Intelligence, Surveillance, and Space Division, Defence Science and Technology Group, Edinburgh, SA 5111, Australia; emma.heading@dst.defence.gov.au

2 Faculty of Engineering Computer & Mathematical Sciences, School of Electrical and Electronic Engineering, The University of Adelaide, North Terrace Campus, Adelaide, SA 5005, Australia; brian.ng@adelaide.edu.au

* Correspondence: haitan.tran@dst.defence.gov.au

Received: 16 November 2019; Accepted: 11 January 2020; Published: 16 January 2020

Abstract: Doppler Radar Tomography (DRT) relies on spatial diversity from rotational motion of a target rather than spectral diversity from wide bandwidth signals. The slow-time k-space is a novel form of the spatial frequency space generated by the relative rotational motion of a target at a single radar frequency, which can be exploited for high-resolution target imaging by a narrowband radar with Doppler tomographic signal processing. This paper builds on a previously published work and demonstrates, with real experimental data, a unique and interesting characteristic of the slow-time k-space: it can be augmented and significantly enhance imaging resolution by signal processing. High resolution can reveal finer details in the image, providing more information to identify unknown targets detected by the radar.

Keywords: slow-time k-space; spatial frequency; Doppler radar tomography; radar imaging; k-space augmentation; high-resolution narrowband radar

---

1. Introduction

Tomography is a general imaging technique that is based on lower-dimensional projections of an object from different spatial aspects, which are then processed using the projection-slice theorem [1] to reconstruct an image of the object. Radar tomography uses reflective scattering phenomenology and radar waveforms for the measurements, which may be wideband or narrowband. Wideband waveforms exploit *spectral* diversity as system resources to facilitate radar imaging and have probably been the most exploited resources in practical applications in the last few decades. The well-known synthetic aperture radar (SAR) and inverse SAR (ISAR) imaging techniques may be described as two special forms of wideband tomography, in which another system resource—*spatial* diversity—is exploited only minimally [2]. Range-Doppler ISAR imaging, and stripmap SAR in particular, typically involve aspect angle changes of a few degrees [3–5]. This constraint of small rotation angles in the linear phase regimes allows the image inversion processing to take advantage of the computationally efficient fast Fourier transform (FFT) without needing signal interpolation onto rectangular grids.

Spotlight SAR makes use of wider angles [6], while circular SAR [7] may coherently process up to a complete cycle of target aspect rotation, with sophisticated and precise motion compensation in range. More notably, in the associated spatial frequency spaces, also known as k-spaces [2], traditionally intensive interpolation processing prior to image inversion processing may be necessary. Nevertheless, these forms of SAR and ISAR rely on the bandwidth resource to achieve high down-range resolution, and so can be considered as belonging to the category of 'wideband radar tomography'.

Radar tomographic imaging with ultra-narrowband or single-frequency waveforms relies on spatial diversity as the only system resource for image formation [8–11]. Spatial diversity may be realized by: (i) having a radar with multiple receivers looking at the target from diverse angular locations, the received signals from which are processed coherently, or (ii) using a single receiver looking at a target undergoing relative rotational motion, i.e., changing target aspect. Both cases widen the angular extents of the measurement support of the received signal in the k-spaces.

Previous work [2] showed that narrowband radar tomography can be most effectively formulated in the *slow-time* k-space in conjunction with the classical Doppler processing and Doppler radar tomography (DRT) [12,13]. The DRT algorithm applies the projection-slice theorem in which the inputs of the target's cross-range projections are formed from Doppler profiles. The slow-time k-space is not only convenient for describing the DRT algorithm, it is also a natural tool to formulate high-resolution DRT imaging with an *augmentation* of its measurement support. Augmentation is the process of significantly enlarging the support of the slow-time k-space by using longer coherent processing intervals (CPI) in the DRT algorithm and correcting for nonlinear phase effects due to strong rotational motion. This 'augmentability' is a unique characteristic of the slow-time k-space.

The introduction of nonlinear phase terms in the k-space augmentation causes a blurring effect in the resulting image. Spectral compression techniques for chirped signals can be used to address this problem using bilinear transforms such as the Wigner-Ville Distribution (WVD), the Cohen's class and the time-frequency distribution series (TFDS) as discussed in [14]. The problem with these techniques is the presence of undesirable cross terms when instantaneous component frequencies may overlap, which is the case for DRT imaging [13]. The combination of the fractional Fourier transform and S-method was used to overcome the problem of cross terms in [13], which demonstrated the slow-time k-space augmentation with DRT. The current work is extended to a more novel technique based on the orthogonal matching pursuit (OMP) technique, inspired by related work in compressive sensing.

Radar imaging naturally is suited to compressive sensing techniques, given that real targets often resemble a sparse collection of discrete point scatterers [5,15]. OMP is fundamentally a technique for parameter estimation by matching a given signal to a dictionary of possible elemental functions spanning a finite parameter space. The dictionary is designed for the particular application and has been applied to the area of DRT imaging in varying contexts [16–18]. The particular application in this work is to estimate the non-linear phase term in the radar signal to reduce the image blurring for improved resolution. The main contribution of the paper is two-fold: to highlight the augmentability of the slow-time k-space as a fundamentally useful characteristic for narrowband radar imaging, and to present a novel application of the OMP technique to such augmentation processing.

The slow-time k-space processing technique as presented in this paper provides a complimentary approach to traditional high-resolution ISAR imaging. The dependence of wide bandwidth signals for high resolution in ISAR imaging is not always readily achievable within the confines of the available spectrum and limitations at lower frequency bands [9,19]. The proposed high-resolution imaging scheme can lead to improved target recognition despite an absence of wide bandwidth signals, provided sufficient spatial diversity is available. This is an important capability of great interest to the radar research community [20].

The rest of the paper is organized as follows. The next Section summarizes the fundamental theory: system geometry and signal model, cross-range bandwidth and resolution, and DRT. Section 3 describes the slow-time k-space and its augmentation with OMP processing. Section 4 describes the experimental setup using simple point scatterers on a rotating turntable with imaging results for both standard and augmented DRT processing. The final Section presents some relevant discussion points and concluding remarks.

2. Background

This Section defines the signal model, the fundamental concept of cross-range bandwidth and resolution, and summarizes the known theory of Doppler radar tomography (DRT) in its standard version.

2.1. Signal Model

Consider a monostatic radar system geometry as illustrated in Figure 1. Without loss of generality, an inertial local target reference frame, denoted as $\boldsymbol{T}_x$ with origin O at the target's nominal centre of rotation, is chosen to have the the x_2 ('down-range') axis aligned with the radar line of sight (LOS), with x_1 axis denoting cross-range. The plane (x_1, x_2) is often known as the image projection plane (IPP, or just 'image plane'). The axis orthogonal to the IPP is denoted as the x_3-axis (sometimes referred to as 'height'). The target's *effective* rotation vector $\boldsymbol{\Omega}_e$ is defined as the projection of the target's total rotational velocity vector $\boldsymbol{\Omega}$ along the x_3-axis.

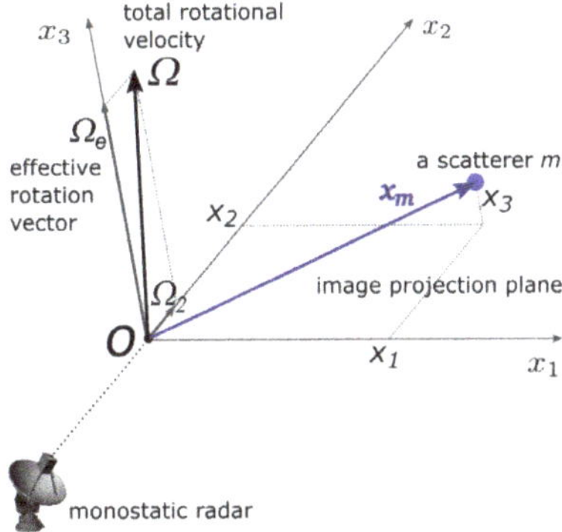

Figure 1. Imaging system geometry: The (x_1, x_2, x_3) coordinates are defined in the local target frame, $\boldsymbol{T}_x$. For clarity, a single point scatterer is shown at $\boldsymbol{x}_m$, which rotates around origin O with velocity Ω. Targets are modeled as a discrete, distributed collection of similar point scatterers.

Using the definition above, the total rotational velocity vector can be written in $\boldsymbol{T}_x$ as

$$\boldsymbol{\Omega} = (0, \Omega_2, \Omega_e). \tag{1}$$

Physically, Ω_e introduces cross-range dependent Doppler shifts in the radar backscatter and is the principal reason that motion-based target imaging is possible. In comparison, Ω_2 has minimal (sometimes deleterious) impact on radar imaging. For non-cooperative targets, neither Ω_e nor Ω_2, or the orientation of the IPP itself, are known a priori. In this paper, we further assume that $\Omega_2 = 0$, and Ω_e is approximately constant during a coherent processing interval (CPI).

For this paper, we use an idealized point-scatterer model for the target: it is adequately modeled as an ensemble of M point scatterers with reflectivity coefficients σ_m, located in the far field of the radar. The approximate range to the mth point-scatterer on the target with position vector $\boldsymbol{x}_m$, defined relative to O, can be defined as

$$R(\boldsymbol{x}_m) \approx \boldsymbol{R}(\boldsymbol{x}_m) \cdot \boldsymbol{i}_{LOS} = R_0 + r_m, \tag{2}$$

in which $\boldsymbol{R}(\boldsymbol{x}_m)$ is the range vector to the mth scatterer, R_0 is the radar range to O, the scatterer's local down range is

$$r_m = \boldsymbol{x}_m \cdot \boldsymbol{i}_{LOS}, \tag{3}$$

and $\boldsymbol{i}_{LOS} = (0, 1, 0)$ is the unit vector along the radar LOS in the $\boldsymbol{T}_x$ frame.

Formulation in the $\boldsymbol{T}_x$ frame is appropriate in traditional ISAR imaging where the change of aspect is small (a few degrees), or for signal analysis within a relatively short CPI. In contrast, radar tomography exploits spatial diversity through wide changes of target aspect. For this formulation, a second, dynamic local target frame denoted as $\boldsymbol{T}_y$, is needed. This frame rotates with the target and coincides with $\boldsymbol{T}_x$ at a reference time, usually assumed to be $t_k = 0$. By the $\Omega_2 = 0$ assumption, it follows also that T_x and T_y share the same x_3-axis. The reason for choosing T_y frame is that its axes are aligned with those of the underlying k-spaces and thus preserves angles across the T_y frame and the k-spaces.

Let

$$s_T(t_k;f) \propto \exp\{j2\pi f t_k\}$$

denote the simple transmit continuous waveform at a single frequency f, where only the slow time t_k is involved; there are no pulses and hence no 'fast time' spanning a pulse. The slow-time index is $k = 0, 1, 2, \ldots, K-1$, and we assume a total of K time samples in a CPI. The received signal $s_R(t_k;f)$ is a delayed version of $s_T(t_k;f)$, summed over all scatterers,

$$s_R(t_k;f) \propto \exp\left\{-j4\pi f\frac{R_0(t_k)}{c}\right\} \sum_{m=1}^{M} \sigma_m \exp\left\{-j\frac{4\pi f}{c} r_m(t_k)\right\}. \tag{4}$$

Here, we have also assumed that radar hardware perfectly removes the carrier frequency term $\exp\{j2\pi f t_k\}$. The first factor in (4) describes translational motion of the target as a whole; the second factor captures the target geometry and scattering reflectivities to be processed for imaging.

Furthermore, we shall assume a linear translational motion model for the target,

$$R_0(t_k) = R_0(0) + v\, t_k, \tag{5}$$

where v is the velocity, assumed known prior to DRT processing, and $R_0(0)$ is target range at a reference time $t_k = 0$.

2.2. Cross-Range Bandwidth and Resolution

The position of each scatterer executing rotational motion with rotation vector $\boldsymbol{\Omega}$ is described to a second order approximation by

$$\boldsymbol{x}(t_k) = \boldsymbol{x}_0 + (\boldsymbol{\Omega} \times \boldsymbol{x}_0)t_k - \frac{1}{2}[\Omega^2 \boldsymbol{x}_0 - (\boldsymbol{\Omega} \cdot \boldsymbol{x}_0)\boldsymbol{\Omega}]t^2,$$

where $\boldsymbol{x}_0 \equiv \boldsymbol{x}(0)$ for convenience. Relative to T_x, the local down range r_m in (4) can be expressed as

$$r_m(t_k) = x_{m_2} + x_{m_1}\,\Omega_e\, t_k - \frac{1}{2} x_{m_2}\,\Omega_e^2\, t_k^2 + \cdots, \tag{6}$$

where x_{m_1}, x_{m_2} are the initial ($t_k = 0$) cross range and range, respectively, of the m-th scatterer in T_x. The CPI duration is denoted by T_{CPI}. As has been thoroughly discussed in [2], although x_2 (dropping the subscript m for brevity) cannot be directly estimated with a zero-bandwith signal, the *first*-order term of (6) suggests that a so-called *cross-range bandwidth*,

$$B_\perp = f\,\Omega_e T_{CPI} = f\,\Delta\theta, \tag{7}$$

can be used to estimate cross range x_1. In other words, the target's rotation generates an effective bandwidth which allows for the resolving cross-range measurements, as long as the rotation angle through T_{CPI},

$$\Delta\theta = \Omega\, T_{CPI},$$

is sufficiently small such that higher-order terms (quadratic and above) in (6) can be ignored. In practice, the $\Delta\theta$ is limited to a few degrees, which is consistent with wideband ISAR imaging. Note that the presence of the (unknown) zeroth-order term x_{m_2} means x_1 cannot be directly estimated from the time-domain signal. Doppler tomography, as formulated below, overcomes such constraints to achieve target imaging.

Consider a segmented CPI of the received signal $s_R(t_k, l)$ as illustrated in Figure 2. Taking a Fourier transform over t_k produces a Doppler profile $S_R(f_d) = \mathcal{F}\{s_R(t_k, l)\}$, with zero Doppler ($f_d = 0$) corresponding to the centre of rotation at O (ignoring any residual translational motion after preprocessing). For a segment of duration T_{CPI}, the achievable Doppler resolution is

$$\Delta f_d = \frac{1}{T_{CPI}} = \frac{f\,\Omega_e}{B_\perp}. \tag{8}$$

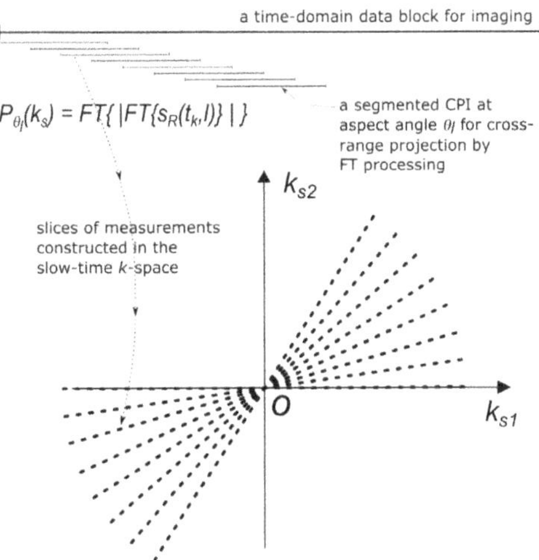

Figure 2. Illustration of the DRT narrowband imaging algorithm. k_{s2}- and k_{s1} are the components of the slow-time k-vector $\mathbf{k}_s$ aligned with the target's initial range and cross-range directions, respectively. Each radial line represents the slow-time k-space samples obtained from one segmented CPI.

Each Doppler profile contains contributions from all scatterers, with the down ranges coordinates x_{m_2} encoded as constant phase terms. Since the cross-range of a scatterer is directly proportional to its Doppler frequency f_d, namely

$$x_1 = \frac{\lambda}{2\,\Omega_e} f_d, \tag{9}$$

it follows that the *magnitude* of the Doppler profile,

$$p_\theta(x) = |S_R(f_d)|,$$

represents a cross-range projection of the target's reflectivity function at angle θ, the average aspect angle over the CPI. The achievable *cross-range resolution* is

$$\Delta x_1 = \frac{\lambda}{2\Omega_e} \Delta f_d = \frac{c}{2B_\perp}. \tag{10}$$

This expression is exactly analogous to the down-range resolution $\Delta x_2 = c/2B$ for wideband imaging with spectral bandwidth B.

2.3. Doppler Radar Tomography (DRT)

The *Projection-Slice Theorem* (PST) states that the Fourier transform $P_\theta(f_{s\perp})$ of projection $p_\theta(x)$ is a slice of the 2D FT of the target's reflectivity function at aspect angle θ. This theorem can be used to invert the cross-range profiles accumulated from a range of aspect angles θ_l to recover the target reflectivity function, i.e., estimate the scatterer coordinates x_{m_1} and x_{m_2} in T_y frame. For this to be effective, the target's rotation must subtend a significant change in aspect angles; the 1D cross-range projections are computed in the frequency domain as discussed above, after which the target reflectivity function (image) can be reconstructed by a 2D inverse FT.

2.3.1. The Monostatic DRT Algorithm

To perform radar imaging using the DRT method, it is necessary to populate the slow-time k-space from the radar backscatter. The algorithm to generate the slow-time k-space samples consists of the following steps:

a. *Data segmentation:* Partition the N samples of the received signal $s_R(t_n)$ into L overlapping CPIs of K samples, $s_R(t_k, l), k = 0, 1, \ldots, K-1; l = 1, 2, \ldots, L$. These are referred to as 'segmented CPIs' below. Denote the overlap factor η with $0 \leq \eta < 1$. At the midpoint of each segment, the target aspect angle (relative to $\boldsymbol{T}_x$) is denoted as θ_l;

b. *Translational motion compensation (TMC):* this step shifts the Doppler component induced by translational motion to zero Doppler frequency by modulating the segmented CPI by $\exp(j2\pi \nu t_k)$, where ν is the target's translational velocity as noted in (5). This quantity is assumed to be known or estimated by other methods. A discrete Fourier transform is then applied to the modulated segments to obtain the Doppler spectrum. The magnitude of the output,

$$p_{\theta_l}(x) = |\mathcal{F}\{s_R(t_k, l)\, \exp(j2\pi \nu t_k)\}|\,, \tag{11}$$

is the cross-range (which is proportional to Doppler) profile for the target at an angle θ_l from its original orientation. Accumulate all such cross-range profiles for all the corresponding aspect angles θ_l, i.e., for all L segmented CPIs.

c. *Populating the k-space:* The spatial Fourier transform of $p_{\theta_l}(x)$

$$P_{\theta_l}(f_{s\perp}) = \mathcal{F}\{p_{\theta_l}(x)\} \tag{12}$$

at target aspect angle θ_l are then used as the 'measurement samples' in the slow-time k-space. As the target rotates, the measurements sweep out a region of support in slow-time k-space as indicated in Figure 2. Due to our choice of reference frames, the measurement population always starts close to the k_{s1}-axis because $p_{\theta_1}(x)$ is the initial cross-range profile.

d. *Image inversion:* An inverse Fourier transform is applied to the populated support of the k-space to yield the target image. Other works have either used filtered back projection, or interpolated the samples onto a rectangular grid to utilise a standard 2D inverse Fourier transform, for this task applied [12,13]. In this paper, we use the non-uniform Fast Fourier transform (NUFFT) [21–24].

It is worth noting that the image resolution is inversely proportional to the diameter of the span of the k-space samples which is dependent on the cross range bandwidth $B_\perp$ as defined in (10). The resulting supportable size of the image is then determined from the image resolution cell multiplied by a factor of K being the number of samples spanning the diameter of the k-space. Although limited amounts of target rotation can reduce image resolution in the sparsely populated direction, here we focus on the case where a half cycle of the target scatterers is visible to the radar to completely populate the k-space. Under this assumption, the angular sampling density of the k-space samples drives the image contrast and is a trade off with computational cost [25].

2.3.2. Standard DRT

By standard DRT, we refer to the case where the input cross-range profiles, as defined by (11), are Doppler migration free (DMF), and the rotation angle corresponding to each profile formed under this condition is said to be within the linear limit (of phase variation). The DMF condition can be satisfied when the segmented CPI lengths are sufficiently short such that the nonlinear phase terms in (6) are negligible and hence compensation is not necessary, or when $|\boldsymbol{x}_m|$ is small. The former case is particularly sensitive for scatterers at larger radial distances from the centre of rotation, while the latter case applies more to scatterers sufficiently close to the centre of rotation whose Doppler frequencies are small and Doppler migration effects (if any) are also small.

As derived in the Appendix A, the standard DRT constraint on CPI rotation angle is

$$\Delta\theta \leq \min\{\Delta\theta_{DM}, \Delta\theta_{LM}\}, \quad \text{(rad)} \tag{13}$$

where $\Delta\theta_{DM} = (\lambda/2\,r_{\max})^{1/2}$ is an effective rotation angle required to induce a Doppler migration (DM) of one bin, $\Delta\theta_{LM}$ is the 'linear limit', while DRT image resolution, in both range and cross range, is

$$\Delta x_1 = \Delta x_2 \geq \left(\frac{\lambda\, r_{\max}}{2}\right)^{1/2}. \quad \text{(m)} \tag{14}$$

Here, r_{max} is the maximum radial dimension of the target. Note that $\Delta\theta$ and Δx are independent of rotation rate and signal sampling rate, but only on radar wavelength and the dimension of the target (through maximum radial dimension $r_{\max}$ to any scatterer). $\Delta\theta_{LM}$ is roughly 10 degrees; Equations (13) and (14) can be used as a guide to predict the expected imaging performance or applicability of standard DRT for a specific radar wavelength and target size.

The limitations imposed by these nonlinear effects at wider rotation angles can be compensated by a processing technique described in the next Section. For differentiation from standard DRT, such cases are referred to as 'Augmented DRT'.

3. The Slow-Time *k*-Space and Its Augmentation

While it is possible to formulate the problem and solution entirely in terms of the spatial frequency space of $f_{s\perp}$, we shall keep up with tradition and formulate it in terms of a '*k*-space', with

$$k_s = 2\pi f_{s\perp}.$$

3.1. The Slow-Time k-Space

In basic Fourier analysis, for signal with a pulse repetition interval PRI, the Doppler frequency extent of the signal is $PRF = 1/PRI$, which spans the interval $(-PRF/2, PRF/2)$. Analogously, from the spatial (cross-range) resolution Δx_1 as given in (10), the values of spatial frequency $f_{s\perp}$ spans the interval $(-B_\perp/c, B_\perp/c)$. It follows from (7) that the interval for k_s is

$$k_s \in \left(-\frac{2\pi f}{c}\Delta\theta, \frac{2\pi f}{c}\Delta\theta\right).$$

These limits are illustrated by extents of the radial dashed lines in Figure 2. Since the cross-range profiles are computed from FFT, both the discretized time and frequency domain vectors have K samples. That is, the slow-time k-vectors $\boldsymbol{k}_s$ corresponding to each cross-range projection contains samples given by

$$\boldsymbol{k}_s = k'\frac{2\pi f}{c}(\Omega_e\, T_{PRI})\,\boldsymbol{i}_\perp, \tag{15}$$

where $k' = -K, -K+2, \ldots, -2, 0, 2, \ldots, K-2$; $T_{PRI} = T_{CPI}/K$, and $\boldsymbol{i}_\perp$ is the cross-range unit vector (perpendicular to LOS) along the x_1-axis of the $\boldsymbol{T}_y$ frame.

The slow-time k-space arises naturally out of DRT: its *radial support* determined by the cross-range bandwidth $B_\perp$ and its *populating samples* are $P_{\theta_l}(k_s)$ given by (12); as the target rotates, the slow-time k-space support is swept out in fan-like shapes around the k-space origin. Also, for a given $B_\perp$, the number of $\boldsymbol{k}_s$ points is a processing design parameter not necessarily fixed to K; its chosen value however would affect only the sidelobes of the impulse response, and hence image contrast, not image resolution.

An important and useful characteristic of the slow-time k-space is it can be *augmented*. As implied by (7), $B_\perp$ can be increased by using a wider rotation angle $\Delta\theta$, providing processing can effectlively correct for the nonlinear terms in the phase function of (6). In Section 3.2 below, we discuss one typical technique to correct for the *second*-order term, i.e., linear chirp components. In other words, augmentation of the k-space enhances resolution by permitting the CPI to be lengthened to the limit where rotational motion of all point scatterers can be modelled as linear chirps.

By comparing a standard CPI $T_{CPI}^{(s)}$ and corresponding rotation angle $\Delta\theta^{(s)}$ in the conventional linear limit of narrowband imaging to a longer CPI we define an 'augmentation factor'

$$\kappa = \frac{\Delta\theta}{\Delta\theta^{(s)}} = \frac{T_{CPI}}{T_{CPI}^{(s)}}, \tag{16}$$

where T_{CPI} is the lengthened CPI and corresponding larger rotation angle $\Delta\theta$. The augmentation factor of κ describes the expansion of the cross range bandwidth $B_\perp$ or equivalently the radial span of the slow-time k-space described in (7) and (15). The DRT image resolution is inversely proportional to $B_\perp$ defined in (9), hence, an improvement in resolution can be achieved with adequate compensation of the linear chirps which is described further in Section 3.2. The concept of the slow-time k-space is illustrated in Figure 3. The DRT algorithm based on an augmented k-space is called augmented DRT.

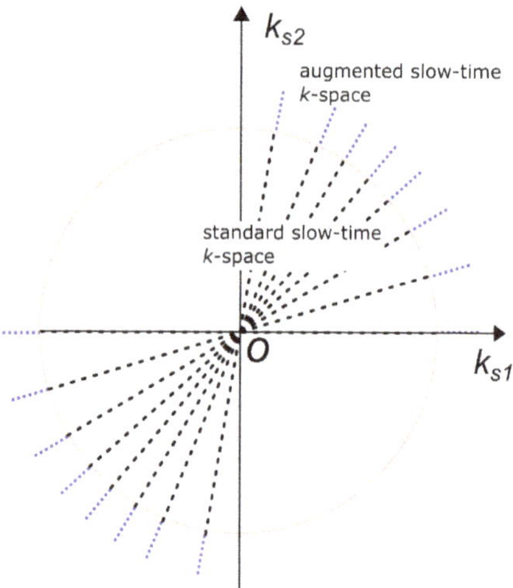

Figure 3. An illustration of the augmentation of the slow-time k-space to generate longer segmented CPIs for cross-range profile formation, which compensates for nonlinear effects of rotation arising from wider angles. The circle indicates the boundary of support in standard DRT imaging.

3.2. *Augmented DRT with Orthogonal Matching Pursuit (OMP)*

This technique shares the same objective as the FrFTS-based technique [13] but instead makes use of a popular tool in the more modern approach of sparse signal approximation, OMP. Again, TMC is assumed to have been perfectly processed prior to this processing.

3.2.1. Sparse Representation

With reference to (4) and (6), the segmented CPI signal received is represented in vector form as

$$s_R = \boldsymbol{\Psi}\sigma + \epsilon, \tag{17}$$

where $\mathbf{\Psi}$ is the dictionary matrix of size $K \times N_\sigma$; $\boldsymbol{\sigma}$ is a length-N_σ column vector of (complex-valued) atom coefficients; and $\boldsymbol{\epsilon}$ is a length-K column vector of noise and/or clutter components. The columns of $\mathbf{\Psi}$ are the *chirp atoms*, of the form

$$g(k) = \exp\left\{-j2\pi\left(f_g t_k + \frac{1}{2}c_g t_k^2\right)\right\}, \tag{18}$$

where $k = 0, 1, \ldots, K-1$, and the parameters

$$f_g = \frac{2\,x_1\,\Omega_e}{\lambda}, \quad \text{and} \quad c_g = -\frac{2\,x_2\,\Omega_e^2}{\lambda} \tag{19}$$

respectively represent the Doppler frequency and chirp rate of a scatterer due to rotation, at reference time $t_k = 0$ of the current segmented CPI, which define the atom $g(t_k)$. Furthermore, let f_g and c_g, or equivalently x_1 and x_2, be discretized as vectors of expected or possible values, of lengths N_f and N_c respectively, then $N_\sigma = N_f N_c$.

Different options for discretizing (f_g, c_g) lead to different definitions of the dictionary $\mathbf{\Psi}$. The above option in terms of (x_1, x_2) uses rectangular scatterer coordinates. Another option is by polar coordinates (d, α) with

$$x_1 = d\cos(\alpha), \quad x_2 = d\sin(\alpha), \tag{20}$$

which may be useful when prior knowledge about the expected scatterer locations is available. It is desirable to use a coordinate grid for (f_g, c_g) in such a way that the grid points efficiently spans the target while keeping the total number of grid points (the dictionary size) to a minimum. A demonstration of these options is shown in Section 4.2.4.

The OMP algorithm itself is well-known, hence will not be described here (see [26] for example). In fact, OMP is only one of several sparse approximation techniques that could be used in this algorithm.

3.2.2. The OMP-Based Augmented DRT Algorithm

The augmented DRT algorithm is modified from standard DRT by simply lengthening the segmented (and overlapping) CPIs with an augmentation factor κ, as defined by (16); the target signal in each CPI can then be represented as a sum of linear chirp components. The aim is then to estimate such a representation and to correct for the chirps, i.e., focusing the range profile, before applying them to remaining steps of the DRT algorithm.

For each augmented CPI, suppose the output of the OMP processing is a (sparse) representation $\{f_g^{(m)}, c_g^{(m)}\}$ of size M with corresponding atoms $\{g_m(t_k)\}$ and coefficients $\{\sigma_m\}$, then a *dechirped* version for the segmented receive signal is

$$\tilde{s}_R(t_k) = \sum_{m=1}^{M} \sigma_m\, g_m(t_k) \to \sum_{m=1}^{M} \sigma_m\, \tilde{g}_m(t_k). \tag{21}$$

The right arrow $\to$ above denotes a *replacement* of the $g_m(t_k)$ atom with a corresponding *monotone* signal

$$\tilde{g}_m(t_k) = \exp\left\{-j2\pi f_g^{(\mathrm{mid})} t_k\right\}, \tag{22}$$

with Doppler frequency

$$f_g^{(mid)} = f_g^{(m)} + \frac{1}{2}c_g^{(m)} t_{\mathrm{mid}}, \tag{23}$$

so defined as the instantaneous frequency at t_{mid}–the middle time of the segmented CPI. The operation

$$p_{\theta_l}(x) = |\mathcal{F}\{\tilde{s}_R(t_k)\}| \tag{24}$$

then would give a focused cross-range projection for tomographic processing.

The augmentation algorithm, applied to each CPI of the augmented DRT algorithm, can thus be summarized as follows.

0. *Initialize:*
   - define or select expected intervals of Doppler frequency f_g and chirp rate c_g;
   - define the corresponding chirp atoms and set up the dictionary $\mathbf{\Psi}$;
   - input segmented CPI data $s_R(t_k)$;
1. Compute the OMP-based sparse solution;
2. Replace all chirp atoms in the sparse solution with single-tone sinusoid functions with Doppler frequency at the mid-point of the segmented CPI;
3. Compute the focused cross-range profile p_{θ_l} as given by (24).
4. Compute NUFFT on the populated slow-time k space to produce the output image.

4. Experimental Results

We present two different datasets using simple point-like scatterers on a rotating turntable to represent a target. This scenario is analagous to rotating components on a target such as a helicopter rotor blade tips [20,25,27]. The first dataset is a target with a small dimension and small scatterers to showcase improvements in resolution. The second dataset is representative of a much larger target which highlights the effect of blurring in the image that we aim to remove for improved image resolution.

4.1. Small Target

4.1.1. Experimental Setup

The data was collected in the Mumma Radar Laboratory at the University of Dayton, Ohio, USA. Although the aim of the study is narrowband imaging, a wideband waveform at X-band was used with stepped-frequency pulses between 8 GHz and 12 GHz, over 101 regular frequency steps. Only the measured data from one of the available discrete frequencies f_k was used to study narrowband tomographic radar imaging.

The transmit and receive horn antennas were mounted on separate robotic arms which could be oriented and positioned with high precision. The measurements were conducted in a controlled laboratory environment with some Radar Absorbing Material (RAM) reducing the radar reflections from the floor and walls. The experimental target consisted of two vertical metallic rods, separated by 19 cm (approximately), emulating two point scatterers which rotated around a vertical pedestal, as illustrated in Figure 4. The maximum radial distance is 11 cm. The antennas were kept stationary whilst the target was rotated through 360°, at 0.1° steps. At each step, the stepped-frequency waveform was transmitted and sampled, one sample for each frequency.

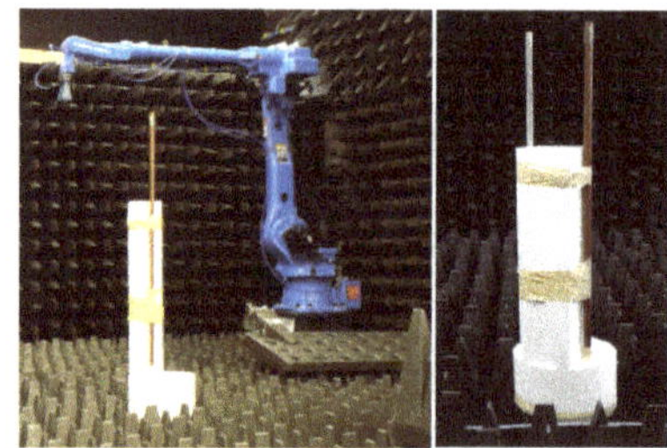

Figure 4. An example of an antenna mounted on a robotic arm at the Mumma Radar Laboratory with the two vertical metallic rods secured to the rotating pedestal.

4.1.2. System Requirements

Successful imaging is not dependent on shifts in relative velocity of the target from pulse to pulse, in fact the target could completely stop at each sampled rotation angle [28]. This is the case when a target rotates on a turntable with a very slow rotation rate during which the Doppler frequency is derived from the change in phase in time from the different target perspectives. Therefore we describe the system parameters such as target rotational speed and radar sampling rate based on the angular sampling rate.

While the full theoretical details are included in the Appendix A, the key requirements are summarized as follows.

- f_k: we choose the lowest and highest frequencies available in this experiment, 8 and 12 GHz, corresponding to $\lambda = 3.75$ or 2.5 cm. With $r_{\max} \approx 0.11$ m, $\Delta\theta_{DM} \approx 23.7^\circ$ or 19.3°, respectively. We also choose $\Delta\theta_{LM} = 10^\circ$; the system is thus $\Delta\theta_{LM}$-limited and poor imaging performance can be expected from standard DRT;
- Inequality (A1) is the Doppler ambiguity free condition; PRF should be designed such that the angular sampling rate $PRF_a = PRF/\omega$ (in samp/rad) is greater than $(4\,r_{\max}/\lambda)$ (11.7 or 17.6 for this setup), but with as small a margin as possible, to ease hardware requirement.
- The angular sampling interval of 0.1° per sample in the experiment translates to a $PRF_a =$ 573 samp/rad. Over the chosen $\Delta\theta_{LM}$ value, 100 samples are available. To reduce computational cost while retaining a reasonable FFT length and satisfying the Doppler ambiguity free condition, we use a *down sampling* ratio of 3:1, leading to $K = 33$ samples per CPI, and $PRF_a \approx 191$ (samp/rad). This choice also automatically satisfies the constraint in (A8).

Realistic values for PRF and ω can also be chosen such that $PRF/\omega = 191$, however, this is not necessary for DRT processing.

For each of the selected frequencies, an elliptic filter with a very narrow stop band is also applied to the signal as a pre-processing for clutter removal. Results are shown in Sections 4.2.3 and 4.2.4.

4.1.3. Standard DRT Imaging

To demonstrate k-space augmentation and the usefulness of sparse signal approximation, some typical results of standard DRT imaging is now shown. Figure 5 shows a spectrogram of the signal at 8 GHz and Figure 6 shows the corresponding slow-time k-space support and standard DRT image. We have used an overlapping factor η of 0.99 in the segmentation step to provide a very smooth angular coverage of the k-space. However, standard DRT imaging performance is poor; the k-space support is small; the two scatterers (metallic rods) are not distinguishable in the image.

If longer CPIs are used with the standard DRT algorithms, image blurring occurs. Suppose κ as defined by (16) is set to 6, the Doppler resolution in the spectrogram of the signal becomes higher, as shown in Figure 7. When the corresponding cross-range profiles (with Doppler bin migration effects present) are applied to standard DRT, the resulting image is in Figure 8.

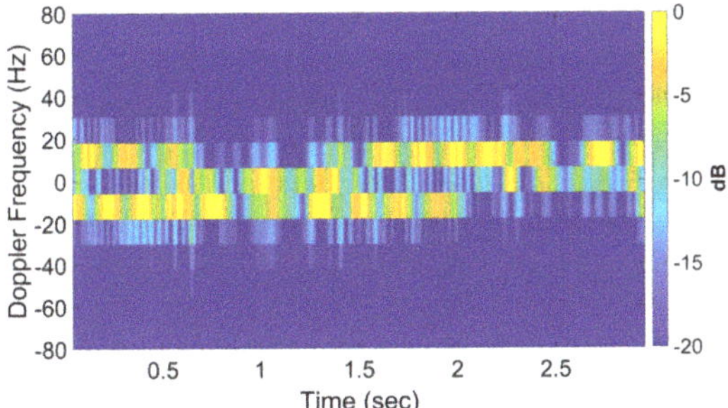

Figure 5. Spectrogram using standard DRT processing for $f = 8$ GHz.

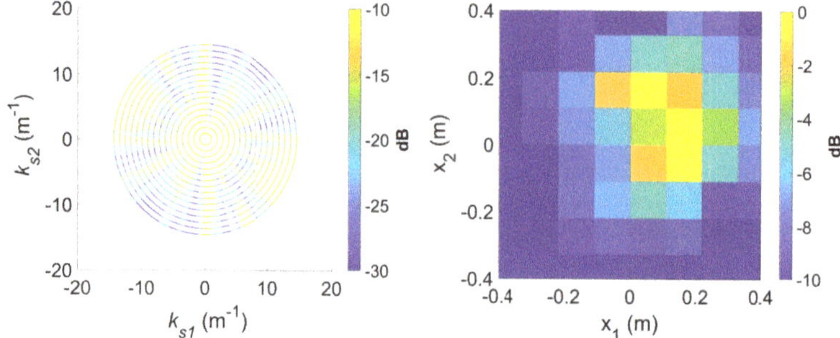

Figure 6. The slow-time k-space support (left) and corresponding standard DRT image (right), at $f = 8$ GHz.

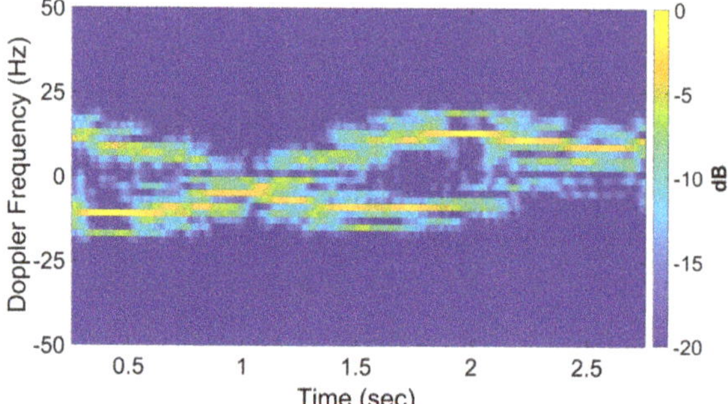

Figure 7. Signal spectrogram with augmented CPIs, $\kappa = 6$, at $f = 8$ GHz.

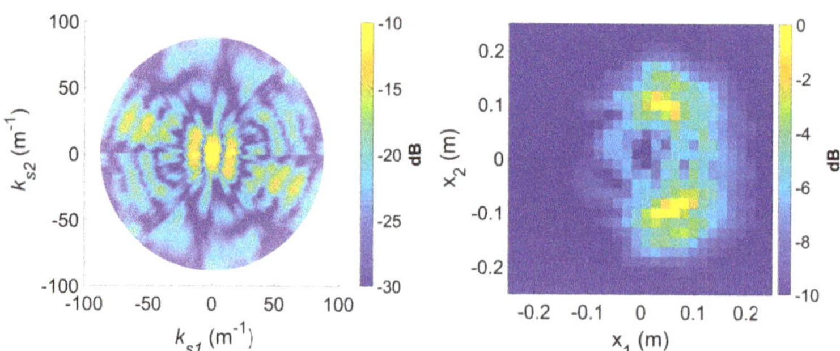

Figure 8. The slow-time k-space support (left) and image (right) for standard DRT with $\kappa = 6$, at $f = 8$ GHz.

For a better insight into the electromagnetic scattering effects in this experiment, similar results using the highest frequency (12 GHz) available are shown in Figures 9 and 10. From Figures 7 and 9 it is clear that in addition to direct (specular) scattering off the inner side of a metallic rod, creeping waves around the rods are the most likely cause of the twin sinusoidal traces for each of the rods [29]. The effects are more pronounced with the shorter wavelength of 2.5 cm, which is more comparable to the rod diameter of approximately 2 cm. The double scattering effects are highlighted in the DRT image.

Note that image blurring in standard DRT imaging is only in the azimuthal direction; image focusing is still generally achieved in the radial direction.

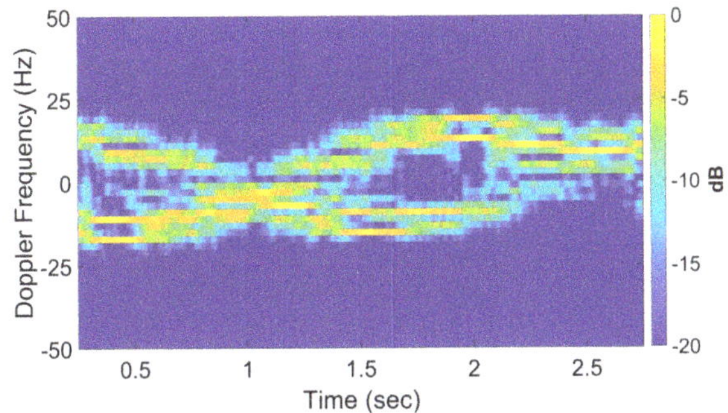

Figure 9. Signal spectrogram with augmented CPIs, $\kappa = 6$, at $f = 12\,\text{GHz}$.

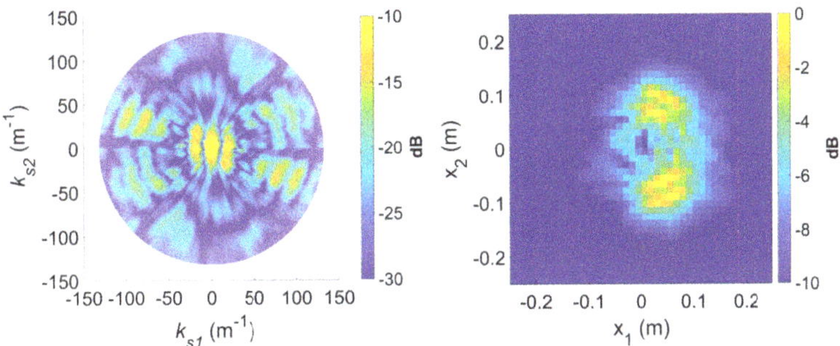

Figure 10. The slow-time k-space support (left) and image (right) for standard DRT with $\kappa = 6$, at $f = 12\,\text{GHz}$.

4.1.4. Augmented DRT Imaging with OMP

To apply OMP for image focusing, the dictionary $\boldsymbol{\Psi}$ is set up with chirp atoms as defined in (18) and (19). As mentioned in Section 3.2.1, two coordinate options for spatial scatterer grids are possible: rectangular in (x_1, x_2) or polar in (d, α). In either case, prior knowledge can be used from the standard DRT processing to constrain the parameter span for the dictionary.

A rectangular scatterer grid was chosen spanning $\pm 0.4\,\text{m}$ with a nominal spacing proportional to $\lambda/2$. As for the standard DRT demonstration, the two frequencies of 8 GHz ($\lambda = 3.75\,\text{cm}$) and 12 GHz ($\lambda = 2.5\,\text{cm}$) are used. Over the selected discretization interval, the number N_σ of atoms was 1849 (for 8 GHz) or 4096 (for 12 GHz).

The coefficient magnitudes of the first 20 atoms extracted from the 8 GHz signal show a clear convergence, as shown in Figure 11.

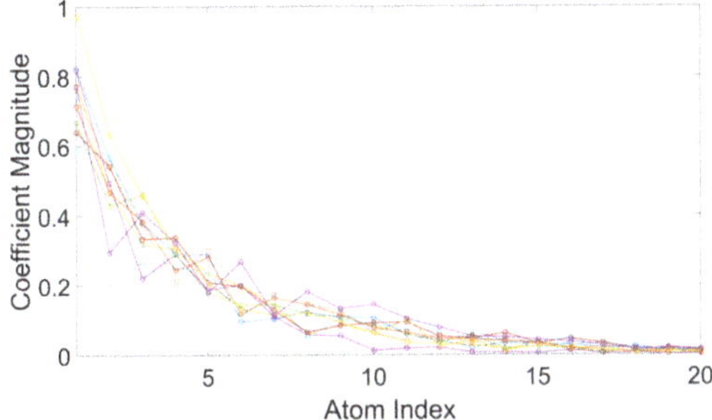

Figure 11. Magnitude of atom coefficients at $f = 8\,\text{GHz}$ for first 20 atoms, shown for a subset of the total number of CPIs.

To reduce signal processing noise effects in the resulting image, a simple thresholding method can be used to control the number of atoms to keep in the sparse representation: in each CPI, stop the OMP iteration when atom coefficient magnitude falls below 20% of the maximum magnitude, as an example. This criterion can also save on computational cost, as less atoms need to be extracted in the processing.

The spectrogram of the reconstructed and 'OMP-focused' signal is shown in Figure 12 which clearly shows more resolvable sinusoidal traces compared to Figure 7.

An example is shown in Figure 13 for the 8 GHz signal. Compared to Figure 8, this is a clearly significantly more focused image where the scatterers are more easily resolvable.

For completeness, we also show results in Figure 14 for the 12 GHz signal, which also resolve the scatterers significantly better using the OMP processing as compared to Figure 10 for standard DRT. The double scattering effects resulting in 'double rods' are also enhanced.

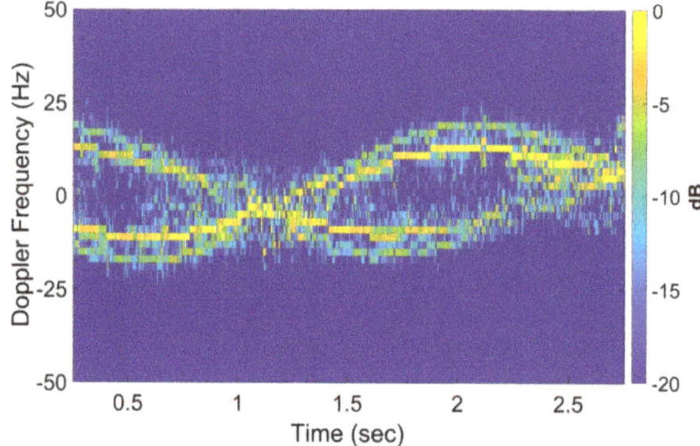

Figure 12. Spectrogram reconstructed, OMP-focused signal with first 20 atoms, at $f = 8\,\text{GHz}$ (compared to Figure 7).

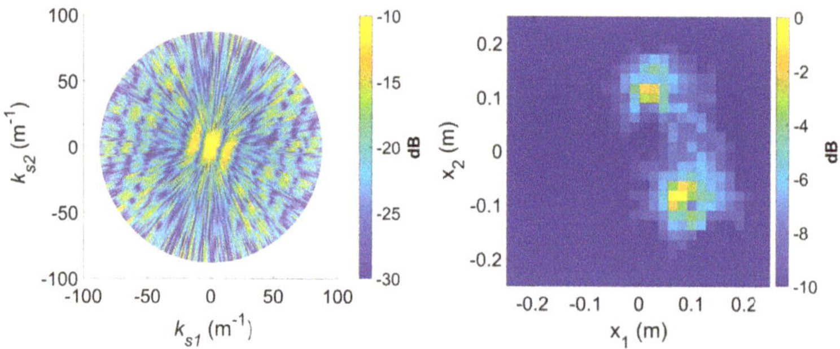

Figure 13. The slow time k-space support (left) and image (right) after OMP processing using a 20% coefficient magnitude threshold at $f = 8\,\text{GHz}$ (compared to Figure 8).

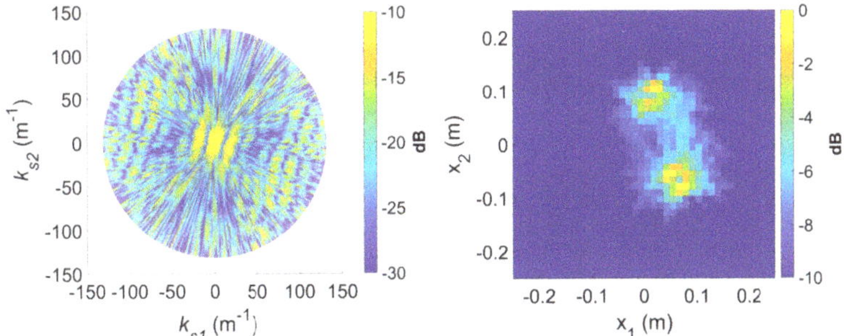

Figure 14. The slow time k-space support (left) and image (right) after OMP processing using a 20% coefficient magnitude threshold at $f = 12$ GHz (compared to Figure 10).

4.2. Large Target

4.2.1. Experimental Setup for Large Target

The experiment was carried out on the turntable at the RAAF Edinburgh airbase, which has a diameter of 17 m. The test target consists of three metallic cylinders as shown in Figure 15 with physical specification listed in Table 1. The experimental X-band radar employed a vertically polarised pulsed stepped frequency waveform starting at 9 GHz with 4 MHz steps, spanning a total of 256 frequencies. The turntable was rotated at approximately one revolution per 15 minutes with a receiver sampling rate of $PRF = 20$ Hz at each frequency, which translates to an angular sampling interval of 0.02°.

Figure 15. The turntable at the RAAF Edinburgh airbase with three metallic cylinders as a test target.

Table 1. Metallic cylinder configuration.

Cylinder	r_m (m)	Diameter (m)	Height (m)
1	2.5	0.15	0.30
2	5	0.38	0.18
3	8	0.21	0.46

4.2.2. System Requirements

As previously described in Section 4.1.2 we designed the system requirements such that the backscattered radar signal is dependent on the angular sampling rate as follows.

- We choose $f_k = 9$ GHz, corresponding to $\lambda = 3.0$ cm. With $r_{\max} \approx 8.0$ m, $\Delta\theta_{DM} \approx 2.48°$; this is well below the typical linear limit of several degrees. The system is thus $\Delta\theta_{DM}$-limited;
- $PRF_a = 1067$ for this experiment which satisfies inequality (A1) for ambiguity free Doppler frequency.

The angular sampling interval of 0.02° per sample in the experiment translates to a $PRF_a = 2864\,\text{samp/rad}$, which satisfies (A1).

Similar to the previous data set, an elliptic filter with a very narrow stop band is applied to the signal for clutter removal. Results are shown in Sections 4.2.3 and 4.2.4.

4.2.3. Standard DRT Imaging

Figure 16 shows a spectrogram of the signal at 9 GHz, featuring three distinct sinusoidal traces corresponding to the three cylinders. Figure 17 shows the corresponding slow-time k-space support and standard DRT image. The standard DRT imaging performance is poor due to the small diameter of the k-space support, with the cylinder locations represented by coarsely granulated pixels.

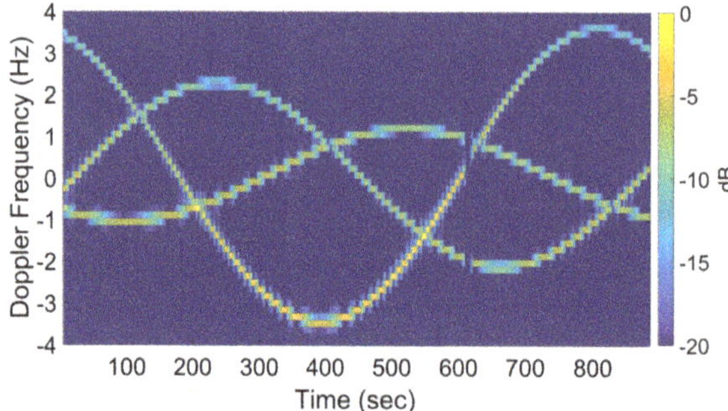

Figure 16. Spectrogram using standard DRT processing for $f = 9\,\text{GHz}$. (The small gap near 600 sec is due to an antenna pointing error during the measurements.)

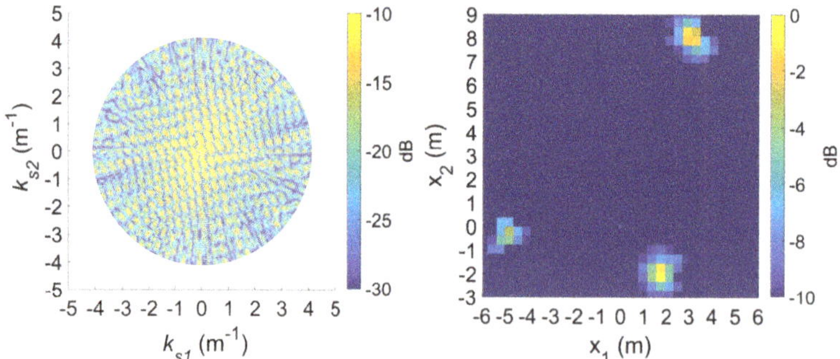

Figure 17. Slow-time k-space support (left) and image (right) for standard DRT processing; $f = 9\,\text{GHz}$.

When longer CPIs are used with the standard DRT algorithms; for example, when κ in (16) is set to 6, the Doppler resolution in the spectrogram of the signal becomes higher, as evident in Figure 18; the corresponding cross-range profiles (with Doppler bin migration effects present) applied to standard DRT result in Figure 19.

Again it is shown that image blurring in standard DRT imaging is only in the azimuthal direction; image focusing is still generally achieved in the radial direction. The blurring effect in the image is more severe for scatterers at larger radial distances which travel along greater arc lengths within a given angular rotation angle (i.e., larger Doppler effects) and hence more severe Doppler bin migration.

4.2.4. Augmented DRT Imaging with OMP

We choose the polar grids representation for this dataset as defined in Section 3.2.1. This approach is useful when some prior knowledge about the radial coordinate of the major scatterers is available from the standard DRT processing.

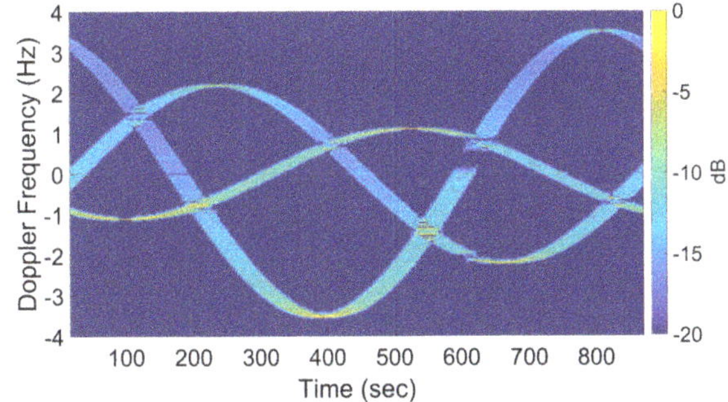

Figure 18. Signal spectrogram with augmented CPIs, $\kappa = 6$ at $f = 9\,\text{GHz}$.

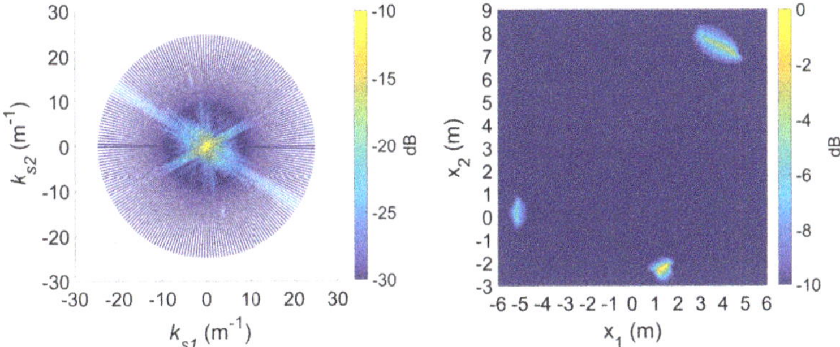

Figure 19. The slow time k-space support (left) and image (right) for standard DRT with $\kappa = 6$, at $f = 9\,\text{GHz}$.

A relatively narrow window is used for discretization of the d-dimension derived from the Doppler information of the scatterers in Figure 16 with a $\lambda/2$ spacing. The full 360° with 1° spacing is used for α. Twenty atoms were extracted in each CPI from the OMP process giving a reconstructed spectrogram that is virtually identical to that in Figure 18, affirming the sufficient accuracy of the sparse representation. After the de-chirping operation, the scatterer locations are much more focussed as shown in Figure 20 compared to the same scatterers in Figure 19 with the same augmentation factor. The technique shows some degradation with the furthermost cylinder which exhibited the most blurring.

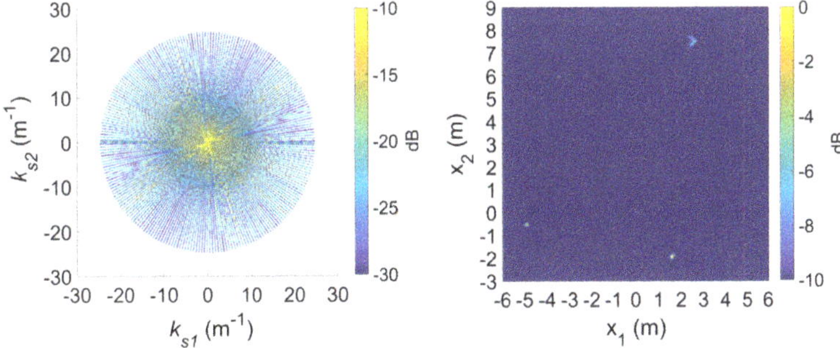

Figure 20. The slow time k-space support (left) and image (right) after OMP processing using a 20% coefficient magnitude threshold at 9 GHz (compared to Figure 19).

Our current study is focused more on imaging performance rather than computational cost; nevertheless, to give some idea on computational cost, we ran the algorithm on the high performance computer called 'Phoenix' at the University of Adelaide which took approximately 1 hour to run on 16 CPUs using 64 GB RAM.

5. Further Discussion

This paper is an expansion to the work reported earlier in [2], demonstrating high-resolution DRT imaging with real experimental data. As this is not a real moving and rotating target in a typical operational scenario, a number of issues could be noted.

Firstly, the target's translational velocity is exactly zero for the entire data collection. Nevertheless, this is not expected to be a sensitive factor. For most real moving targets, translational velocity can be readily compensated by shifting the 'body Doppler' line to zero Doppler. Sensitive propagation phases, as in the case of fast-time k-spaces, do not enter the slow-time k-spaces.

Secondly, the measured data were collected at precise angular sampling rates PRF_a, which can only be estimated in typical operational scenarios. Errors in PRF_a or Ω_e would translate into errors of the locations of populated samples as well as image scaling factor. Hence both image focusing and image scaling could be affected. We have not fully addressed these issues in this work.

The experimental data does reveal interesting electromagnetic phenomenology, highlighting the limiting simplicity of the ideal point-scatterer assumption; creeping waves and nonlinear scattering effects do exist, which are not taken into account in the current DRT theory.

On application of the OMP algorithm, what this work has demonstrated its feasibility: techniques such as OMP can be used for slow-time k-space augmentation. Other alternative sparse approximation techniques can possibly be used to yield higher performance. Numerous other aspects can also be considered, such as dictionary 'learning': how to select an optimum spatial scatterer grid for the best focusing performance while keeping computational cost at manageable levels? Or how to deal with the off-grid/mismatched scatterer problem [30]. Many open questions remain, some of which will be addressed in future publications.

6. Concluding Remarks

We have demonstrated, with two datasets, the ability to improve image resolution using a rotating target with an ultra-narrowband radar. The enabling signal processing technique presented was a combination of Doppler radar tomography and a sparse reconstruction technique such as OMP, with a unifying mathematical framework based on the slow-time k-space. We have shown that closely spaced scatterers can be resolved by illustrating the creeping wave effect when the scatterer size is similar to the radar wavelength. The technique also performed well addressing the adverse effect of blurring in the image with scatterers at larger radial distances to the centre of rotation. By compensating for the blurred scatterer locations in the image, the ability to resolve closely spaced scatterers is improved providing finer details for target recognition.

Although the demonstration of this technique is effective, the application to a real complex target with many non-ideal scatterers may present additional challenges including discontinuous scattering effects, larger dictionaries affecting computational cost and inaccuracies due to signal mismatch with finite dictionary elements. In future work, we aim at investigating the use of multiple widely separated radar receivers to reduce the requirement on large target rotation angles for DRT imaging, where the direct application of OMP may not scale efficiently for large amounts of data. The increase in data may require a modified approach such as dictionary learning to help reduce the computational cost.

Author Contributions: Conceptualization, H.-T.T.; Methodology, E.H., H.-T.T. and B.W.-H.N.; Simulation, E.H.; Validation, B.W.-H.N., H.-T.T.; Writing—Original Draft Preparation, H.-T.T. and E.H.; Writing—Review & Editing, B.W.-H.N., E.H., and H.-T.T. All authors have read and agreed to the published version of the manuscript.

Funding: This research was funded by Defence Science and Technology Group, Australia.

Acknowledgments: We sincerely thank Lorenzo Lo Monte and Nihad Afaisali from the University of Dayton and Mark Ingham, John Senior, Peter Drake and Shane Hatty from DST Group for their technical expertise and professionalism in performing the radar measurements presented in this manuscript.

Conflicts of Interest: The authors declare no conflict of interest.

Appendix A. Standard DRT: System Parameters and Image Resolution

Figure A1 illustrates the spectral composition of a segmented CPI in the DRT algorithm. The sinusoidal traces depicts the instantaneous Doppler frequencies of scattererers as the target rotates, which are generally chirp signals. The chirps are approximately linear for short CPIs.

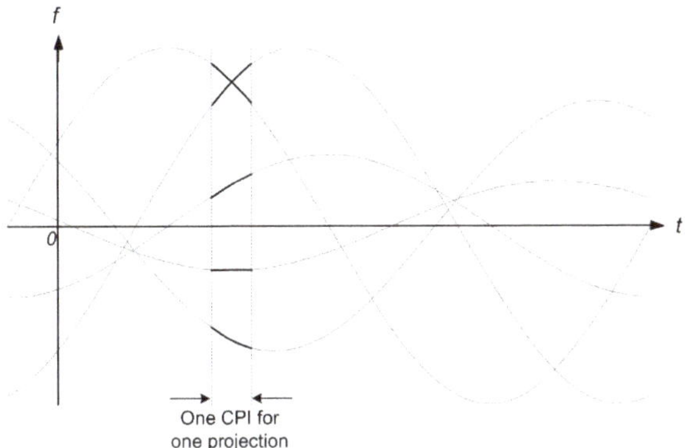

Figure A1. Instantaneous Doppler traces of point scatterers on a rotating target.

There are three main constraints on system parameters for standard DRT to be applicable. The first one is Doppler ambiguity free condition: the sampling rate PRF must be at least two times the largest Doppler components in the received signal–the well-known Nyquist criterion. With $r_{\max}$ denoting a largest radial distance of scatterers on the target, this constraint can be written as

$$PRF \geq \frac{4\,\omega\, r_{\max}}{\lambda}, \quad \text{or} \quad PRF_a \geq \frac{4\, r_{\max}}{\lambda}, \tag{A1}$$

where $PRF_a = PRF/\omega$ is the angular sampling rate (in units of samples/rad).

The second constraint is: Doppler migration-free (DMF), i.e., variation of the instantaneous Doppler frequency of any scatterer is less than a Doppler bin size. This constraint may be derived as follows: a scatterer's cross range is given by $x_1 = r\cos\theta$, hence the differential change in x_1 is

$$dx_1 = -r\omega \sin\theta \, dt.$$

Maximum cross range migration occurs near $\theta = n\pi/2$ for odd integral values of n. If dt represents a CPI time, here denoted as T_{CPI}, then the DMF requirement translates to having

$$|dx_1| \equiv r\omega T_{CPI} \tag{A2}$$

to be not larger than a cross range bin size, at all range bins.

A cross range bin size Δx_1 is related to Doppler filter size Δf through the well-known relationship $\Delta f = (2\,\omega/\lambda)\Delta x_1$ for monostatic radars, hence

$$\Delta x_1 = \frac{\lambda}{2\omega}\Delta f = \frac{\lambda}{2\omega}\frac{PRF}{K}. \tag{A3}$$

Here, K denotes both the number of samples spanning the CPI and the FFT length; and therefore

$$T_{CPI} = \frac{K}{PRF}. \tag{A4}$$

Using (A2), (A3) and (A4) in the DMF requirement $|dx_1| \leq \Delta x_1$ then leads to the condition

$$\Delta\theta = \frac{\omega K}{PRF} \leq \left(\frac{\lambda}{2\, r_{\max}}\right)^{1/2} \equiv \Delta\theta_{DM}, \tag{A5}$$

where $\Delta\theta$ is the (segmented) CPI rotation angle (or angular extent), and $\Delta\theta_{DM}$ is an effective angle the target would need to rotate to induce a Doppler migration (DM) through one frequency bin. Note that $\Delta\theta_{DM}$ depends only on radar wavelength λ and maximum radial extent $r_{\max}$, not rotation speed. Another useful related expression is

$$T_{CPI} = \frac{K}{PRF} \leq \frac{\Delta\theta_{DM}}{\omega}, \tag{A6}$$

for the corresponding CPI time.

The third constraint is the so-called 'linear limit': the maximum rotation angle at which the Taylor expansion in (6) up to the first order in time remains valid. In other words,

$$\Delta\theta < \Delta\theta_{LM}, \tag{A7}$$

where $\Delta\theta_{LM} \approx 10^\circ$.

Combining the three constraints above, the system constraints on PRF and K are (A1) and

$$K \leq PRF_a \min\{\Delta\theta_{DM}, \Delta\theta_{LM}\}. \tag{A8}$$

As an example, suppose $\Delta\theta_{LM} = 8^\circ$ is used; and $\lambda = 5$ cm, $r_{\max} = 1$ m and $\omega = 300$ RPM, then (A5) gives $\Delta\theta_{DM} \approx 9.1^\circ > \Delta\theta_{LM}$, and the choice of $PRF = 15$ kHz and $K = 64$ would satisfy all constraints.

The result in (A8) also highlights a useful comparison between the DMF condition and the linear limit $\Delta\theta_{LM}$: the CPI length K may be limited by either of the two factors; it is however more desirable to be DMF-limited, i.e, without strong dependence on wide rotation angles. Indeed, shorter radar wavelengths and larger target dimensions would induce more pronounced Doppler effects which are required for the applicability of the DRT imaging algorithm itself.

Image resolution for standard DRT can be derived as follows. The minimum PRF for unambiguous Doppler effects, given by (A1), means a cross-range profile given by (11) exactly spans the cross-range extent of the target. Larger values would lead to outer range bins of the profile containing noise only, while those range bins spanning the target remain the same, in both bin size and number. For the case of minimum unambiguous PRF, the maximum selectable value of K, given by (A6), is

$$K_{STD} = \left(\frac{8\, r_{\max}}{\lambda}\right)^{1/2}, \tag{A9}$$

or smaller if K is limited by $\Delta\theta_{LM}$ in (A8). The resolution of a cross-range profile is therefore

$$\Delta x_{RB} = \frac{2\, r_{\max}}{K_{STD}} = \left(\frac{\lambda\, r_{\max}}{2}\right)^{1/2}. \tag{A10}$$

The same result can be obtained from (10) and (7) by using the equality in (A5) for $\Delta\theta$. The result in (A10) is also the expected image resolution in standard DRT imaging.

References

1. Kak, A.C.; Slaney, M. *Principles of Computerized Tomographic Imaging*; Society of Industrial and Applied Mathematics: Philadelphia, PA, USA, 2001.
2. Tran, H.; Melino, R. The Slow-Time k-Space of Radar Tomography and Applications to High-Resolution Target Imaging. *IEEE Trans. Aerosp. Electron. Syst.* **2018**, *54*, 3047–3059. [CrossRef]
3. Walker, J.L. Range-Doppler Imaging of Rotating Objects. *IEEE Trans. Aerosp. Electron. Syst.* **1980**, *AES-16*, 23–52. [CrossRef]
4. Ausherman, D.A.; Kozma, A.; Walker, J.L.; Jones, H.M.; Poggio, E.C. Developments in Radar Imaging. *IEEE Trans. Aerosp. Electron. Syst.* **1984**, *AES-20*, 363–400. [CrossRef]
5. Chen, V.C.; Martorella, M. *Inverse Synthetic Aperture Radar*; Scitech Publishing: Edison, NJ, USA, 2014.
6. Jakowatz, C.V.; Wahl, D.E.; Eichel, P.H.; Ghiglia, D.C.; Thompson, P. *Spotlight-Mode Synthetic Aperture Radar: A Signal Processing Approach*; Springer: Berlin/Heidelberg, Germany, 1996.
7. Soumekh, M. Reconnaissance with slant plane circular SAR imaging. *IEEE Trans. Image Process.* **1996**, *5*, 1252–1265. [CrossRef] [PubMed]
8. Wang, L.; Yazici, B. Bistatic Synthetic Aperture Radar Imaging Using UltraNarrowband Continuous Waveforms. *IEEE Trans. Image Process.* **2012**, *21*, 3673–3686. [CrossRef] [PubMed]
9. Coetzee, S.L.; Baker, C.J.; Griffiths, H.D. Narrow band high resolution radar imaging. In Proceedings of the 2006 IEEE Conference on Radar, Verona, NY, USA, 24–27 April 2006.
10. Sun, H.; Feng, H.; Lu, Y. High resolution radar tomographic imaging using single-tone CW signals. In Proceedings of the 2010 IEEE Radar Conference, Washington, DC, USA, 10–14 May 2010; pp. 975–980.
11. Sego, D.J.; Griffiths, H.; Wicks, M.C. Radar tomography using Doppler-based projections. In Proceedings of the 2011 IEEE RadarCon (RADAR), Kansas City, MO, USA, 23–27 May 2011; pp. 403–408.
12. Mensa, D.L.; Halevy, S.; Wade, G. Coherent Doppler tomography for microwave imaging. *Proc. IEEE* **1983**, *71*, 254–261. [CrossRef]
13. Tran, H.T.; Melino, R. *Application of the Fractional Fourier Transform and S-Method in Doppler Radar Tomography*; Technical Report DSTO-RR-0357; The Defence Science and Technology Organisation: Edinburgh, Australia, 2010.
14. Chen, V.; Ling, H. *Time-Frequency Transforms for Radar Imaging and Signal Analysis*; Artech House: Norwood, MA, USA, 2003.
15. Potter, L.C.; Ertin, E.; Parker, J.T.; Cetin, M. Sparsity and Compressed Sensing in Radar Imaging. *Proc. IEEE* **2010**, *98*, 1006–1020. [CrossRef]
16. Kodituwakku, S.; Melino, R.; Berry, P.; Tran, H. Tilted-wire scatterer model for narrowband radar imaging of rotating blades. *IET Radar Sonar Navig.* **2017**, *11*, 640–645. [CrossRef]
17. Melino, R.; Kodituwakku, S.; Tran, H. Orthogonal matching pursuit and matched filter techniques for the imaging of rotating blades. In Proceedings of the 2015 IEEE Radar Conference, Johannesburg, South Africa, 27–30 October 2015; pp. 1–6.
18. Tran, H.; Heading, E.; Melino, R. OMP-based translational motion estimation for a rotating target by narrowband radar. *IET Radar Sonar Navig.* **2017**, *11*, 854–860. [CrossRef]
19. Baker, C.J.; Griffiths, H.D. Bistatic and Multistatic Radar Sensors for Homeland Security. In *Advances in Sensing with Security Applications*; Springer: Dordrecht, The Netherlands, 2006; pp. 1–22.
20. Cilliers, A.; Nel, W.A.J. Helicopter parameter extraction using joint time-frequency and tomographic techniques. In Proceedings of the 2008 International Conference on Radar, Adelaide, SA, Australia, 2–5 September 2008; pp. 598–603.
21. Dutt, A.; Rokhlin, V. Fast Fourier Transforms for Nonequispaced Data, II. *Appl. Comput. Harmon. Anal.* **1995**, *2*, 85–100. [CrossRef]
22. Nguyen, N.; Liu, Q.H. The Regular Fourier Matrices and Nonuniform Fast Fourier Transform. *SIAM J. Sci. Comput.* **1999**, *21*, 283–293. [CrossRef]
23. Greengard, L.; Lee, J. Accelerating the Nonuniform Fast Fourier Transform. *SIAM Rev.* **2004**, *46*, 443–454. [CrossRef]
24. Ferrara, M. AFOSR Lab Task 'Moving-Target Radar Feature Extraction'. 2009. Available online: https://au.mathworks.com/matlabcentral/fileexchange/25135-nufft--nfft--usfft/all_files (accessed on 7 November 2010).

25. Tran, H.; Heading, E.; Ng, B. Multi-Bistatic Doppler Radar Tomography for Non-Cooperative Target Imaging. In Proceedings of the 2018 International Conference on Radar (RADAR), Brisbane, Australia, 27–30 August 2018; pp. 1–6.
26. Elad, M. *Sparse and Redundant Representations: From Theory to Applications in Signal and Image Processing*; Springer: Berlin/Heidelberg, Germany, 2010.
27. Fliss, F. Tomographic Radar Imaging of Rotating Structures. In *Synthetic Aperture Radar*; International Society for Optics and Photonics: Los Angeles, CA, USA, 1992; Volume 1630.
28. Munson, D.C.; O'Brien, J.D.; Jenkins, W.K. A tomographic formulation of spotlight-mode synthetic aperture radar. *Proc. IEEE* **1983**, *71*, 917–925. [CrossRef]
29. Knott, E.F.; Shaeffer, J.F.; Tuley, M.T. *Radar Cross Section*, 2nd ed.; SciTech Publishing Inc: Rayleigh, NC, USA, 2004.
30. Nguyen, N.H.; Dogancay, K.; Tran, H.; Berry, P.E. Parameter-Refined OMP for Compressive Radar Imaging of Rotating Targets. *IEEE Trans. Aerosp. Electron. Syst.* **2019**, *55*, 3561–3577. [CrossRef]

Article

Target Doppler Rate Estimation Based on the Complex Phase of STFT in Passive Forward Scattering Radar

Karol Abratkiewicz *, Piotr Krysik, Zbigniew Gajo and Piotr Samczyński

Institute of Electronic Systems, Faculty of Electronics and Information Technology, Warsaw University of Technology, 00-665 Warsaw, Poland
* Correspondence: k.abratkiewicz@elka.pw.edu.pl

Received: 19 July 2019; Accepted: 17 August 2019; Published: 20 August 2019

Abstract: This article presents a novel approach to the estimation of motion parameters of objects in passive forward scattering radars (PFSR). In such systems, most frequency modulated signals which are used have parameters that depend on the geometry of a radar scene and an object's motion. Worth noting is that in bistatic (or multistatic) radars forward scattering geometry is present thus in this case only Doppler measurements are available while the range measurement is unambiguous. In this article the modulation factor, also called the Doppler rate, was determined based on the chirp rate (equivalent Doppler rate) estimation concept in the time-frequency (TF) domain. This approach utilizes the idea of the complex phase of the short-time Fourier transform (STFT) and its modification known from the literature. Mathematical dependencies were implemented and verified and the simulation results were described. The accuracy of the considered estimators were also verified using the Cramer-Rao lower bound (CRLB) to which simulated data for the considered estimators was compared. The proposed method was validated using a real-life signal collected from a radar operating in PFSR geometry. The Doppler rate provided by a car crossing the baseline between the receiver and the GSM transmitter was estimated. Finally, the concept of using CR estimation, which in the case of PFSR can be understood as Doppler rate, was confirmed on the basis of both simulated and real-life data.

Keywords: passive forward scattering radar; chirp rate estimation; passive radar; forward scattering radar; radar measurements; time-frequency analysis

---

1. Introduction

Passive forward scattering radars are a special class of passive bistatic radars (PBR), in which the bistatic angle β between the non-cooperating transmitter, the target and the receiver is $\beta \approx 180^\circ$ [1–3]. In such kinds of passive radars, the illuminating signal can be a wave from a commercial transmitter of popular systems such as FM, DAB, DVB-T, GSM, and so forth [4–7]. The simplified passive forward scattering radar (PFSR) geometry is presented in Figure 1.

Unlike typical PBRs, the PFSR is characterized by the fact that objects cross the baseline, which has certain consequences. In such a case the PBR radar using classical PCL (Passive Coherent Location) processing is blind, as a target is crossing the line of sight between a receiver and a transmitter, and for PBR in this geometry there is no existing range resolution. Additionally, the target disturbs the reference signal, thus based on PCL principles it is difficult to detect the target at the line of sight to the transmitter as only the reference antenna is pointed in this direction, and surveillance beams are pointed in other directions. In bistatic radars a target moving at a velocity V provides the Doppler shift expressed as follows: [8]:

$$f_d = \frac{2V}{\lambda}\cos(\alpha)\cos(\beta/2). \tag{1}$$

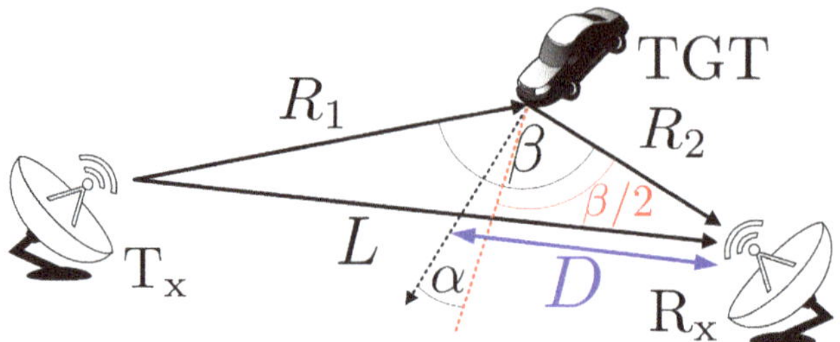

Figure 1. Simplified passive forward scattering radar (PFSR) geometry. β—bistatic angle, T_x—non-cooperative transmitter, R_x—radar receiver, TGT—target, L—baseline, R_1—distance from the transmitter to the target, R_2—distance from the target to the receiver, D—distance from the receiver to the crossing point.

As previously mentioned, in PFSR $\beta \approx 180°$, which makes $f_d \approx 0$ (Hz). However, observing the object in a slightly wider range, that is $\beta \in (180° - \Delta, 180° + \Delta)$, where Δ is a certain angle, it is possible to measure the Doppler rate (or equivalently the chirp rate (CR)). This parameter describes the kinematic properties of the measured object.

The literature describes some methods of Doppler rate estimation in PFSR. Ustalli et al. proposed in Reference [9] a four-step processing technique for the extraction of kinematic motion parameters of targets in forward scattering radar (FSR). The method is based on multiple matched filtration of the signal with simultaneous time-frequency analysis, resulting in the precise estimation of the motion parameters of a single object near the intersection of the baseline. However, a problem may be the analysis of several objects that intersect the baseline at the same time. In addition, due to multiple matched filtration and other processing operations, the complexity of the method is significant. The same authors developed this approach and described it in Reference [10]. Another solution is to use the Radon transformation to calculate the phase acceleration [11]. After transforming the signal into a two-dimensional distribution in the TF domain, the components responsible for the Doppler rate of objects are found. Analyzing the aforementioned papers, it can be noticed that the analysis in the TF domain is the proper approach to the PFSR signal's considerations. A waveform received by the radar can be treated as a non-stationary frequency modulated signal. In the vicinity of $f_d \approx 0$ (Hz) the signal can be approximated as a linear frequency modulated waveform. This methodology was also presented in References [12–14]. This is due to the fact that the phase of the received signal can be described by the dependency:

$$\phi(t) = -\frac{2\pi}{\lambda}\left[R_1(t) + R_2(t) - L\right], \tag{2}$$

which can be approximated using Taylor expansion into the following formula [15]:

$$\phi(t) \approx \frac{\pi}{\lambda} v_p \left(\frac{1}{L-D} + \frac{1}{D}\right)(t-t_0)^2, \tag{3}$$

where v_p is the velocity component perpendicular to the baseline L, t_0 is the moment when the target crosses the baseline and D is the range from the receiver to the crossing point (see Figure 1). As can be noted, Equation (3) describes the second order polynomial characteristic for the frequency modulated signals.

The above considerations prompted the authors to use the CR estimation in the TF domain to analyze signals from PFSR. This approach is based on short–time Fourier transform (STFT) modification and allows the CR at each point in the TF plane to be determined. This is consistent with the assumption that when the baseline and the trajectory of the object are crossed, the signal appearing in the receiver can be approximated with the linear frequency modulated waveform, and this technique is dedicated for such a problem. In addition, the method is computationally efficient, which reduces the calculation time.

This paper is organized as follows: Section 2 presents the CR estimation theory background, including Cramer–Rao lower bound (CRLB) analysis. Section 3 depicts simulation results, and Section 4 covers real-life signal analysis provided by the GSM PFSR. Discussion and comments close the article.

2. Chirp Rate Estimation

2.1. Theory Background

The pioneer of CR estimation in the TF domain using the complex phase of STFT was Czarnecki, who proposed a method for determining the instantaneous frequency rate a two-dimensional signal distribution [16,17]. In general, the signal described by the following model will be considered:

$$x(t) = A_x \exp(j\Phi_x(t)), \tag{4}$$

for the amplitude A_x, $j = \sqrt{-1}$ and phase described as:

$$\Phi_x(t) = \phi_x + \omega_x t + 2\pi \cdot \alpha t^2/2 = \phi_x + 2\pi t\left(f_0 + \alpha t/2\right), \tag{5}$$

where $\omega_x = 2\pi f_0$ is the angular frequency with the carrier frequency f_0, and α is the CR. $x(t)$ can be transformed into a two-dimensional distribution using STFT, which is given by the formula:

$$F_x^h(t,\omega) = \int_{\mathbb{R}} x(\tau)^* h(t-\tau) e^{-j\omega\tau} \mathrm{d}\tau, \tag{6}$$

where $(\cdot)^*$ is the complex conjugate, and the upper index in the F_x^h expression denotes the analysis window $h(t)$ whereas the lower index expressed the signal under consideration $x(t)$. Energy distribution in the TF domain can be calculated as a squared absolute value of the STFT and is called a spectrogram:

$$S_x^h(t,\omega) = \left|F_x^h(t,\omega)\right|^2, \tag{7}$$

where $|\cdot|$ denotes the absolute value operator. STFT can be presented using the concept of a complex phase in the thought of dependency [16]:

$$F_x^h(t,\omega) = \int_{\mathbb{R}} x(\tau)^* h(t-\tau) e^{-j\omega\tau} \mathrm{d}\tau = A_x^h(t,\omega) e^{j\phi_x^h(t,\omega)} = e^{\Lambda_x^h(t,\omega) + j\phi_x^h(t,\omega)}, \tag{8}$$

where STFT phase is described as $\phi_x^h(t,\omega)$, whereas $A_x^h(t,\omega)$ denotes STFT absolute value ($A_x^h(t,\omega) > 0$). By using the property described in Reference [18], the STFT phase in Equation (8) was transformed into a complex form in which $\Lambda_x^h(t,\omega) = \ln(A_x^h(t,\omega))$ then the complex phase of the STFT is defined as:

$$\Psi_x^h(t,\omega) = \ln\left(F_x^h(t,\omega)\right) = \Lambda_x^h(t,\omega) + j\phi_x^h(t,\omega). \tag{9}$$

such a transformation allows many useful signal parameters in the TF domain to be determined. By calculating the partial derivatives of the real part of the complex phase with respect to time and frequency, the instantaneous bandwidth and the local group delay are obtained respectively. The ratio of these values gives the CR estimator in the manner described as follows:

$$\mathfrak{K}(t,\omega) = -\left(\frac{\partial \Lambda_x^h(t,\omega)}{\partial t} \Big/ \frac{\partial \Lambda_x^h(t,\omega)}{\partial \omega}\right). \tag{10}$$

The graphic interpretation of the estimator is presented in Figure 2.

In Reference [19] it was proposed that the $\mathfrak{K}$ estimator can be calculated more efficiently utilizing the modified analysis window. Additionally, two new estimators were revealed. The uncertainty effect occurring in the $\mathfrak{K}$ estimator has been reduced by the differentiation of the numerator and the denominator with respect to time (giving the $\mathfrak{D}$ estimator) and frequency (giving the $\mathfrak{F}$ estimator) giving the following relationships:

$$\mathfrak{D}(t,\omega) = -\left(\frac{\partial^2 \Lambda_x^h(t,\omega)}{\partial t^2} \Big/ \frac{\partial^2 \Lambda_x^h(t,\omega)}{\partial \omega \partial t} \right), \tag{11}$$

$$\mathfrak{F}(t,\omega) = -\left(\frac{\partial^2 \Lambda_x^h(t,\omega)}{\partial \omega \partial t} \Big/ \frac{\partial^2 \Lambda_x^h(t,\omega)}{\partial \omega^2} \right). \tag{12}$$

This method was tested using different types of signals and is described in the literature. Acoustic signals were processed using this method, which can be found in References [17,20]. Radar applications utilizing this approach are presented in References [21–23]. In this paper, the PFSR application is proposed in order to verify the possibility af assessing the motion parameters of a target. Because the estimators given by Equations (10)–(12) are actually equivalent, further considerations are made using one of them to present the correctness of the concept. This is an example of using the idea in PFSR applications, and each of the estimators should give similar results. The comparison of estimators as well as their limitations and computational complexity have been made in the literature [19,21]. Due to the smaller variance in comparison to the estimator $\mathfrak{K}$ (see Section 2.2) and the lower sensitivity to noise (see Reference [21]), the authors decided to perform the tests using the estimator $\mathfrak{F}$; however, the estimators $\mathfrak{K}$ and $\mathfrak{D}$ can be used in the same way.

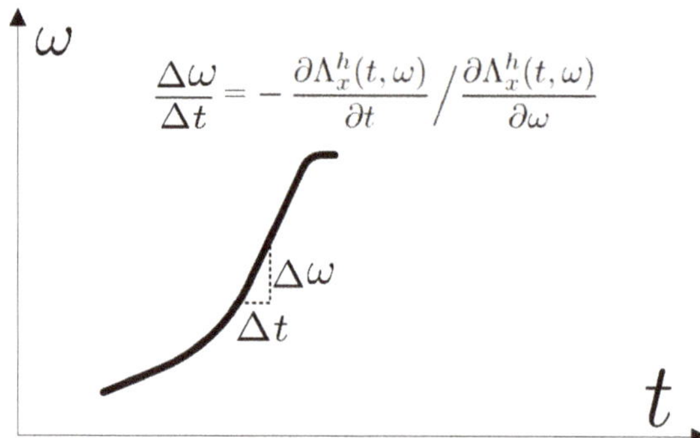

Figure 2. CR estimation in the TF domain—an interpretation.

2.2. Analysis of the Estimation Accuracy

The accuracy of the estimators has been compared in this section to the CRLB for the considered signal model. The analyzed complex chirp signal is given by:

$$x[n] = A_x \exp\left(j\alpha \frac{n^2}{2} \right), \tag{13}$$

where $A_x = 1$, $n \in [0, N-1]$. This signal is merged in a white Gaussian noise $w[n]$ with a variance σ_w^2. Thus, the observation vector $\mathbf{x} = [x[0], x[1], ..., x[N-2], x[N-1]]^T$ is normally distributed:

$$x \sim \mathcal{N}(\boldsymbol{\mu}(\alpha), \mathbf{C}(\alpha)), \tag{14}$$

where

$$\boldsymbol{\mu}(\alpha) = \boldsymbol{x} = \left[A_x, A_x \exp\left(j\alpha\frac{1^2}{2}\right), A_x \exp\left(j\alpha\frac{2^2}{2}\right), ..., A_x \exp\left(j\alpha\frac{(N-1)^2}{2}\right) \right]^T, \tag{15}$$

and $\boldsymbol{C}(\alpha)$ is a covariance matrix of observation vector whereas $(\cdot)^T$ is the matrix transpose. For such a Gaussian observation model depending on the scalar parameter, α the Fisher information matrix (FIM) is a 1 by 1 matrix (scalar) given by [24]:

$$I(\alpha) = 2\Re\left(\left[\frac{\partial\boldsymbol{\mu}(\alpha)}{\partial\alpha}\right]^H C^{-1}(\alpha)\left[\frac{\partial\boldsymbol{\mu}(\alpha)}{\partial\alpha}\right]\right) + \frac{1}{2}\mathfrak{T}\left[\left(C^{-1}(\alpha)\frac{\partial \boldsymbol{C}(\alpha)}{\partial\alpha}\right)^2\right], \tag{16}$$

where $(\cdot)^H$ is the Hermitian transpose and $\mathfrak{T}$ denotes the matrix trace and $\Re$ expresses a real part. Since, the additive noise $w[n]$ is white, the covariance matrix $\boldsymbol{C}(\alpha)$ equals $\sigma_w^2\boldsymbol{I}$ and does not depend on the parameter α. Thus $\frac{\partial \boldsymbol{C}(\alpha)}{\partial\alpha} = 0$ and the second term in Equation (15) vanishes. Moreover, $\boldsymbol{C}^{-1}(\alpha) = \frac{1}{\sigma_w^2}\boldsymbol{I}$. The FIM now takes the following form:

$$I(\alpha) = 2\Re\left(\frac{1}{\sigma_w^2}\left[\frac{\partial\boldsymbol{\mu}(\alpha)}{\partial\alpha}\right]^H\left[\frac{\partial\boldsymbol{\mu}(\alpha)}{\partial\alpha}\right]\right). \tag{17}$$

According to Equation (15) it can be written as:

$$\begin{aligned} I(\alpha) &= 2\Re\left(\frac{1}{\sigma_w^2}\sum_{n=0}^{N-1}\left(\frac{\partial}{\partial\alpha}A_x e^{\left(\frac{-j\alpha n^2}{2}\right)}\right)\left(\frac{\partial}{\partial\alpha}A_x e^{\left(\frac{j\alpha n^2}{2}\right)}\right)\right) = \\ &= 2\Re\frac{A_x^2}{\sigma_w^2}\left(\sum_{n=0}^{N-1} e^{\left(\frac{-j\alpha n^2}{2}\right)}\left(-j\frac{n^2}{2}\right)e^{\left(\frac{j\alpha n^2}{2}\right)}\left(j\frac{n^2}{2}\right)\right) = \frac{A_x^2}{2\sigma_w^2}\sum_{n=0}^{N-1} n^4. \end{aligned} \tag{18}$$

Finally, the CRLB for the estimator variance is given by:

$$\sigma^2(\hat{\alpha}) \geq [I(\alpha)]^{-1} = \frac{2\sigma_w^2}{A_x^2\sum_{n=0}^{N-1} n^4}. \tag{19}$$

In reference to the article describing the estimators used [19], the accuracy of the estimate was investigated and compared to the presented CRLB given by Equation (19). The linear complex chirp was considered. The signal model given by Equation (13) has the following parameters: $N = 250$, $\alpha = 2\pi\frac{0.36}{N}$. In 1000 realizations of noise in the range from -20 to 30 dB, the CR value was estimated and verified at point n_0 on the reference frequency $f_{\text{ref}} = \frac{\alpha_{\text{ref}}}{2\pi}n_0$, where $\alpha_{\text{ref}} = 0.009$. Results for the estimators $\mathfrak{K}$, $\mathfrak{D}$ and $\mathfrak{F}$ are presented in Figure 3.

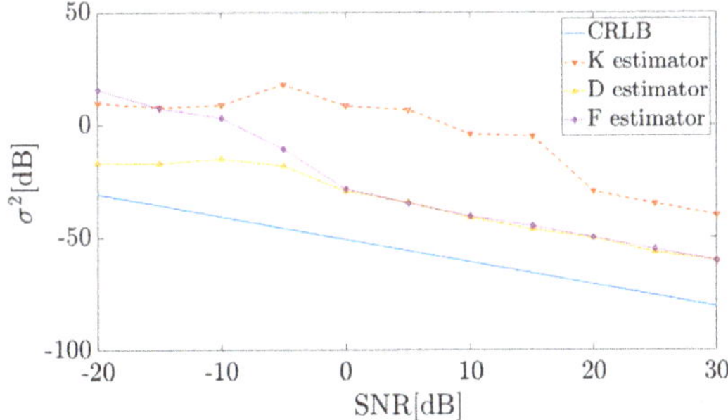

Figure 3. Comparison of the accuracy of the utilized estimators.

As can be seen, for signals with signal–noise ratio (SNR) ≥ 0 dB the $\mathfrak{D}$ and $\mathfrak{F}$ estimator give very similar results. The signals from this range will be considered later in the paper, so choosing one of the two more accurate tools was right. Additionally, according to the authors' experience, this estimator is less sensitive to the changes of the analysis window width. Because of the uncertainty problem in the $\mathfrak{K}$ estimator, the CR was verified in the vicinity of the f_{ref}. For this reason, the variance of this tool is clearly greater than in the case of the other two estimators. Although all of the tools used are characterized by an error in relation to the CRLB, it should be kept in mind that for SNR ≥ 0 dB the variance of the $\mathfrak{D}$ and $\mathfrak{F}$ estimators is satisfactorily low, and in many practical cases is sufficient to estimate CR.

3. Results of the Simulation

In order to verify the proposed method simulations were carried out. Two radar scenes were considered in which one or two targets crossed the baseline. The first of the analyzed cases covers the situation presented in Figure 4.

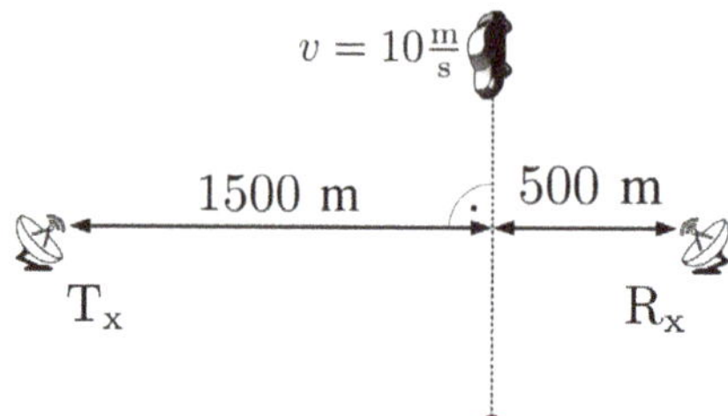

Figure 4. Radar scene for the 1st simulation case.

The continuous wave (CW) transmitter working with the harmonic signal using the carrier frequency $f_c = 900$ MHz is 1500 m away from the target trajectory (at the closest point) which, in turn, is 500 m from the receiver. The point target (dimension is neglected) moves with the velocity $v = 10\frac{\text{m}}{\text{s}}$ perpendicularly to the baseline. The TF distribution of the signal energy in the form of a spectrogram is presented in Figure 5. The signal was merged with the white Gaussian noise, for which SNR = 30 dB. This was due to the fact that in the further part of the article covering the real-life signal analysis, waveforms characterized by SNR > 0 dB are considered, and on such conditions the simulations were focused. STFT parameters during the simulations are as follows:

- N = 1024—amount of points in FFT analysis,
- W = 350—Blackman-Harris window length (in samples),
- H = 1—hop-size (in samples),
- $f_s = 1$ kHz—sampling rate.

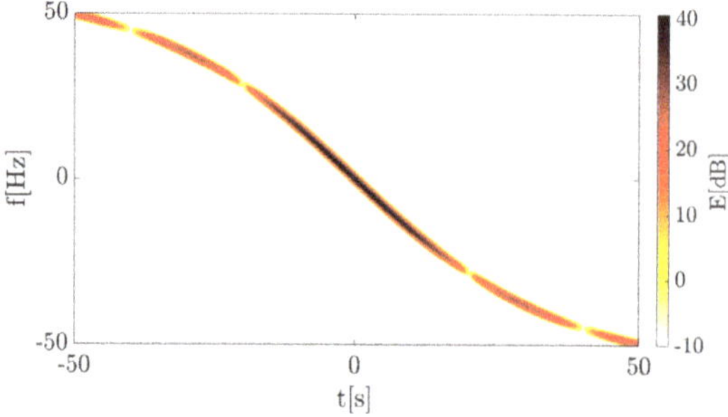

Figure 5. Spectrogram of the 1st simulation case.

The spectrogram shows a characteristic curve, typical for a bistatic geometry utilizing the forward scattering phenomenon. The received signal is defined only by the Doppler frequency (due to the fact that the transmitter works with the CW) expressed by Equation (1), the consequence of which is the frequency modulation of the wave. The Doppler rate, in this case, can be considered as the CR described in the previous section and the physical interpretation of both is the same. Thus, the proposed CR estimation methods were employed to verify their usability in such a context. A calculated accelerogram presenting instantaneous CR for each point in the TF plane is depicted in Figure 6.

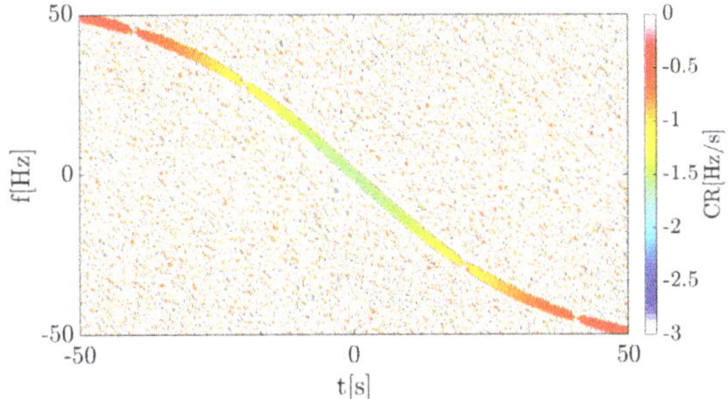

Figure 6. Accelerogram of the 1st simulation case.

By analyzing this distribution, it can be noted that during the crossing of the baseline the target provided a Doppler rate of $\sim -1.5\frac{\text{m}}{\text{s}}$. Additionally, the Doppler rate can be determined for each point for the observation time. This is an advantage of the proposed approach due to the fact that the estimation process automatically provides a set of information about the movement parameters of the target, allowing an unknown trajectory or velocity to be assessed. In comparison to the methods presented in the literature, unique movement signatures are provided by this approach, which allows additional information about the object to be distinguished.

For the purpose of validation, the theoretical Doppler rate, as the first order derivative of the f_d given by Equation (1), was compared with the estimated value. This value was read in the maximum point of the spectrogram in each time frame. Because of the amplitude modulation, the estimated value in the minimum of the envelope is distorted, thus additionally results for the constant signal amplitude were plotted. A comparison of the true value and estimated CR for both cases (amplitude modulation and no amplitude modulation) is presented in Figure 7.

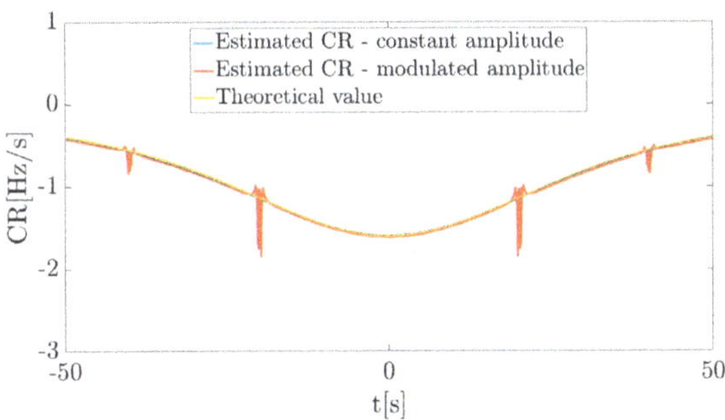

Figure 7. Comparison of the true Doppler rate and estimated CR for two amplitude cases.

As can be noticed, the estimated Doppler rate (CR) for the constant amplitude signal is similar to the theoretical value. In Figure 7 the blue line overlapped completely with the yellow line, which confirms the convergence of the results of the estimation with the theory. However, because of the fact that the signal has an amplitude locally near or equal to zero, the estimation process returns distorted results (see the red line in Figure 7). This is caused by the character of the signal, not by the error of the estimation, which can be seen in the case where the amplitude is constant. In the simulation, the maximum value of the spectrogram in each time portion was extracted, so moments in which signal amplitude is near to zero, the maximum energy value was found in the noise, which caused significant errors.

In order to present the advantages of the proposed method, an extended scenario of the simulation was carried out. In this situation two objects are considered, however, one of them approaches the baseline from a different angle in comparison to the first one, and with slightly higher velocity. The simulation parameters are presented in Figure 8.

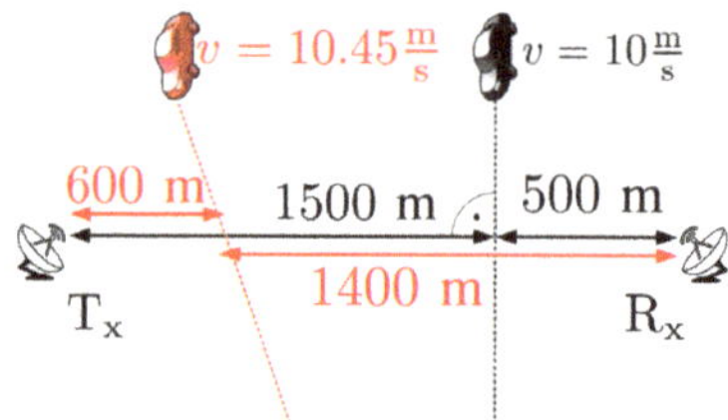

Figure 8. Radar scene for the 2nd simulation case.

A spectrogram presenting the energy distribution of the reflected signal from the two targets is depicted in Figure 9. The additional target introduces a second curve with a different Doppler rate depending on the velocity and trajectory. Thus, based on the previous results it should be possible to differentiate considered targets on the accelerogram because of the individual Doppler rate seen during the time of analysis.

In the simulated case, it was assumed that both targets cross the baseline approximately the same time in order to verify the possibility of extracting particular targets in PFSR geometry. An accelerogram of the simulated data is presented in Figures 9 and 10.

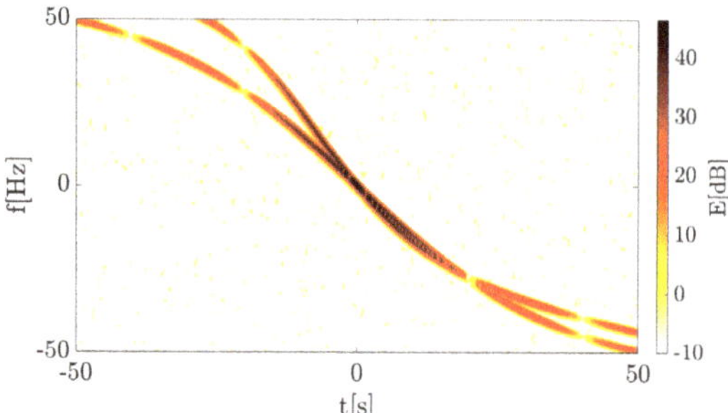

Figure 9. Spectrogram of the 2nd simulation case.

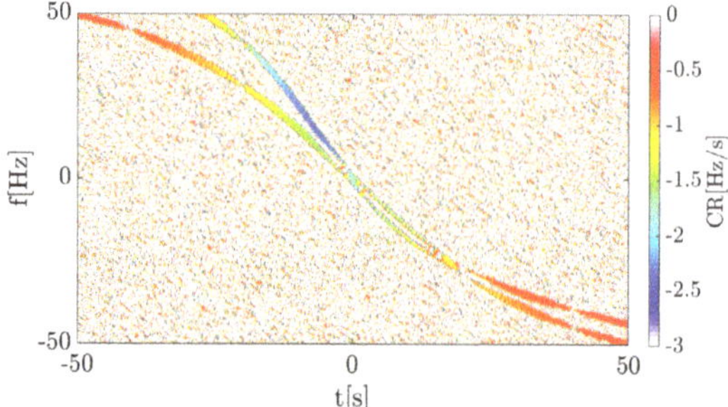

Figure 10. Accelerogram of the 2nd simulation case.

Based on the simulation carried out it was proved that CR estimation may be an effective technique, allowing additional information in PFSR to be extracted. Even if the Doppler rate in the vicinity of $t = 0$ (s) is similar for both targets, the Doppler rate history from the entire observation time provides additional information about the trajectory of the objects. Such data can be used as an extension for PFSR systems. In the next section, the method is tested using real-life data to verify the adaptability and usability of the proposed approach.

4. Real-Life Signal Analysis

4.1. Measurement Campaign

The experimental data were gathered during a measurement campaign using a GSM-based passive radar for monitoring ground moving objects. The source of the illuminating signal was a GSM base transceiver station with an antenna mounted at the height of approximately 50 m. During the trials, a cooperating vehicle was used as an observed target moving with the velocity $v \approx 10\frac{\mathrm{m}}{\mathrm{s}}$ [25]. Measurements were carried out a few times under similar conditions. This was done to confirm the repeatability of results in comparable scenarios, and to verify the possibility of distinguishing characteristic features in the movement.

The location of the radar's surveillance antenna and trajectory of the target were selected in order to ensure that the forward scattering effect of the vehicle occurred at the observation spot. The surveillance antenna was mounted 1.5 m above the ground and was located approximately 1.5 km from the transmitter. The cooperating car crossed the baseline between the transmitter and the receiver within approximately 1 km of the transmitter.

The reference signal was acquired at the same location as where the measurement was performed. In order to reduce the influence of the target echo in the reference channel, an antenna with vertical beamwidth of approximately 25 degrees was mounted on a 12 m mast and tilted upwards.

The measurement equipment consisted of commercial-off-the shelf components, with a two channel receiver based on the National Instruments PXIe-5667 vector signal analyzer, low noise amplifiers and band-pass filters for the GSM900 band (925–960 MHz) connected after the antennas of both the surveillance and reference channels. The measurement scene utilized during the trials is presented in Figure 11, and the general measurement geometry diagram is depicted in Figure 12. More details about the measurement campaign and the signal processing chain are available in Reference [25].

Figure 11. Measurement scene. In white—Range from the GSM transmitter to the receiver, in red—range from the receiver to the intersection point, in blue—the target trajectory.

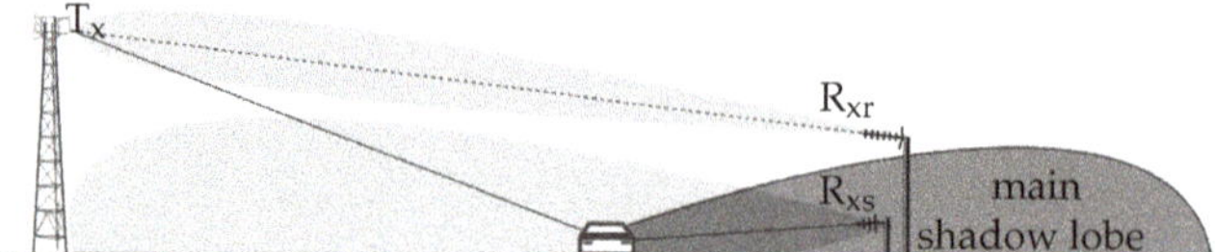

Figure 12. Measurement geometry diagram. T_x—GSM transmitter of opportunity, R_{xr}—the reference antenna, R_{xs}—the surveillance antenna.

4.2. Target Doppler Rate Estimation

The obtained data were examined with one of the analyzed estimators in order to verify whether the proposed tool is effective in analyzing real-life signals from PFSR. The recorded waveform was processed using the cross-correlation of the reference and reflected from the target signal. As with the simulation, the object's dimensions can be neglected. This is due to the fact that the available GSM signal bandwidth is ~200 kHz, giving a bistatic range resolution of ~1500 m. Four cases were considered during which the object examined crossed the baseline between the transmitter and the receiver. During the real-life signal analysis, the following parameters were employed:

- N = 4096—amount of points in FFT analysis,
- W = 1300—Blackman-Harris window length (in samples),
- H = 1—hop-size (in samples),
- $f_s = 1$ kHz—sampling rate (after decimation).

The spectrograms of the analyzed cases are presented in Figure 13.

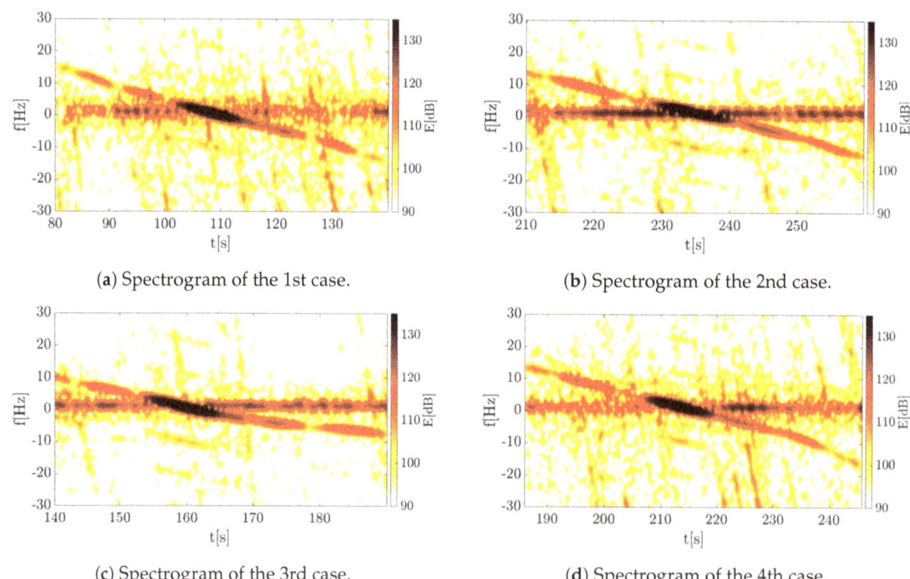

(a) Spectrogram of the 1st case.

(b) Spectrogram of the 2nd case.

(c) Spectrogram of the 3rd case.

(d) Spectrogram of the 4th case.

Figure 13. Spectrograms of all considered cases.

Apart from the signal that is caused by the analyzed phenomenon, that is the signal with frequency modulation, additional components are visible on the spectrograms. They arise for several reasons. The first of these is the presence of a significant number of stationary objects in the radar beam. A strong echo from the surface of the earth, trees or buildings creates a strong permanent component at $f \approx 0$ (Hz), the impact of which can be reduced, as described in Reference [25], for example. The second effect is visible as numerous components with a much higher modulation coefficient.

However, these are only visible for a short time. This phenomenon, in turn, arises as a result of the presence of side lobes, which cause the receiving of a signal reflected from a moving object as well as from other objects in space. In addition, there was a highway near the measuring scene, and the cars moving on it are visible in the spectrograms. However, it is possible to distinguish a significant component, which is the echo coming from the analyzed object.

For such obtained data, accelerograms, that is a local (instantaneous) CR distribution on the TF plane, were determined. The results for the analyzed cases are shown in Figure 14.

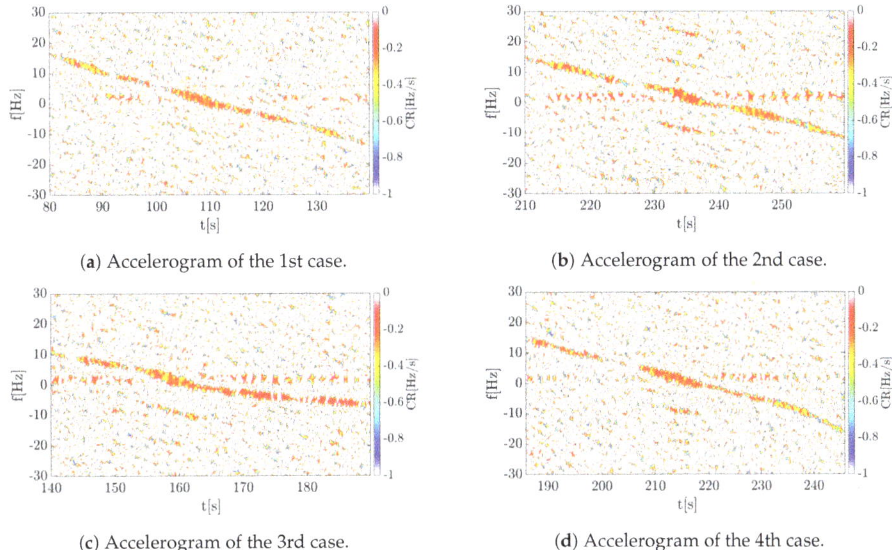

(**a**) Accelerogram of the 1st case. (**b**) Accelerogram of the 2nd case.

(**c**) Accelerogram of the 3rd case. (**d**) Accelerogram of the 4th case.

Figure 14. Accelerograms of all considered cases.

As can be noticed, similar results were obtained for four measurements which confirmed the repeatability of the results. Although the most significant portion of the signal energy is at the moment of changing the frequency sign (around 0 Hz), it is possible, as in the case of the simulation, to determine the CR for the entire recorded segment. Such information can be extremely useful when determining the trajectory or kinematic parameters of the target. In the case of the previously mentioned solutions [9–15], only the Doppler rate was determined at the moment when the baseline is crossed. The proposed method therefore extends the spectrum of possibilities, allowing more information to be extracted. For all cases the estimated CR value at the time of t_0 is CR$\in [-0.3, -0.5]$ (Hz/s), which agrees with the actual value. For example in the first case shown in Figures 13a and 14a, it can be observed that the frequency is reduced by less than 30 Hz within 60 s, which gives an estimated value. In fact, it is difficult to reproduce the vehicle's motion perfectly a few times, which resulted in temporary changes in the signal. It can be observed by comparing the 1st and 2nd cases (corresponding spectrograms Figure 13a,b and accelerograms Figure 14a,b) with 3rd and 4th (corresponding spectrograms Figure 13c,d and accelerograms Figure 14c,d). For the first two cases the car maintains a constant velocity, while for the other two cases the velocity is not retained in the second phase of the move. For the third case it is noticeable that the car moves slowly, which results in a decrease in the chirp rate (absolute) value. In the fourth case, the acceleration of the car is noticeable, which was also observed. In this situation, the accelerogram presents more rapid changes in the range of $-0.5 \sim -0.8$ $\frac{\text{Hz}}{\text{s}}$ in the final part of the observed movement. The presented results confirmed the correctness of the proposed method in the estimation of the object motion parameters observed by the PFSR radar, as well as extraction characteristic features of the object movement. The tested algorithms can help and/or speed up the estimation of the object's parameters, which is especially important due to the fact that information about the distance of the object is lost in the PFSR radar. In this situation, any additional information about the object may be useful.

5. Conclusions

The article has presented a novel approach to the analysis of signals occurring in radars using the forward scattering phenomenon. Based on the concept of the complex STFT phase and the CR estimators known from the literature, the signals from the PFSR radar have been analyzed. In the first part, using a mathematical model, the possibility of applying the proposed method has been tested using simulated signals. A situation has been considered in which one object and two objects with different kinematic parameters intersect the baseline. Simulations have confirmed the applicability of the method, after which the method was verified using real-life data. The role of the illuminator of opportunity has been fulfilled by the GSM transmitter and the transmitter-receiver line was crossed by the car. The analysis allowed the Doppler rate to be determined in the real scenario. In addition, the method gives the Doppler rate estimation, which had not yet been available, not only at the time of the baseline intersection, but also for the entire observed trajectory. A valuable property is the ability to extract temporary changes in velocity, which increases the amount of information describing the observed object.

The accuracy of the considered tools has been verified by statistical analysis and the comparison of results to the CRLB, which mainly showed the differences between the estimators as well as the expected accuracy. In the future, the authors want to verify the accuracy of the estimators for a different analysis window length. This is an important parameter due to the fact that in the classical STFT the length of the analysis window affects the estimation variance and bias, and because the tools used are based on STFT, the bias and the variance are also related to the analysis window's length.

A promising point of further work is the extension of the method with the possibility of classifying objects and estimating the targets tracks. Based on the Doppler rate, not only at the intersection of the baseline but also in the wider observation period, it is possible to classify objects. This can be particularly important in security systems, where the detection of fast and maneuvering missiles is difficult for both active and passive radars. The use of the proposed approach for a transmitter located on the ground and a receiver on the ground or in orbit would allow for detection and classification of dangerous rockets and flying objects.

Author Contributions: Project administration, K.A.; formal analysis, K.A. and Z.G.; investigation, K.A.; methodology, K.A. and P.S.; raw radar data collection, P.K.; software, K.A. and P.K.; supervision, P.S.; validation, K.A.; visualization, K.A. and P.K.; Writing—Original Draft, K.A., P.K. and Z.G.; Writing—Review & Editing, P.S.

Funding: This research received no external funding.

Conflicts of Interest: The authors declare no conflict of interest.

References

1. Cherniakov, M. Geometry of bistatic radars. In *Bistatic Radar: Principles and Practice, Part II*; Wiley: New York, NY, USA, 2007.
2. Willis, N. Forward-scatter fences. In *Bistatic Radar*; SciTech Publishing: Raleigh, NC, USA, 2005.
3. Cherniakov, M. Basic principles of forward-scattering radars. In *Bistatic Radar: Principles and Practice, Part III*; Wiley: New York, NY, USA, 2007.
4. Edrich, M.; Schroeder, A.; Meyer, F. Design and performance evaluation of a mature FM/DAB/DVB-T multi-illuminator passive radar system. *IET Radar Sonar Navig.* **2014**, *8*, 114–122. [CrossRef]
5. Krysik, P.; Kulpa, K.; Bączyk, M.; Maślikowski, L.; Samczyński, P. Ground moving vehicles velocity monitoring using a GSM based passive bistatic radar. In Proceedings of the 2011 IEEE CIE International Conference on Radar, Chengdu, China, 24–27 October 2011; pp. 781–784.
6. O'Hagan, D.W.; Baker, C.J. Passive Bistatic Radar (PBR) using FM radio illuminators of opportunity. In Proceedings of the 2008 New Trends for Environmental Monitoring Using Passive Systems, Hyeres, French Riviera, France, 14–17 October 2008; pp. 1–6.

7. Weiß, M. Compressive sensing for passive surveillance radar using DAB signals. In Proceedings of the 2014 International Radar Conference, Lille, France, 13–17 October 2014; pp. 1–6.
8. Kuschel, H.; Cristallini, D.; Olsen, K.E. Tutorial: Passive radar tutorial. *IEEE Aerosp. Electron. Syst. Mag.* **2019**, *34*, 2–19. [CrossRef]
9. Ustalli, N.; Pastina, D.; Lombardo, P. Kinematic parameters extraction from a single node Forward Scatter Radar configuration. In Proceedings of the 2018 19th International Radar Symposium (IRS), Bonn, Germany, 20–22 June 2018; pp. 1–10.
10. Ustalli, N.; Pastina, D.; Lombardo, P. Target motion parameters estimation in Forward Scatter Radar. *IEEE Trans. Aerosp. Electron. Syst.* **2019**, doi: 10.1109/TAES.2019.2913731 [CrossRef]
11. Ustalli, N.; Pastina, D.; Bongioanni, C.; Lombardo, P. Motion parameters estimation in dual-baseline Forward Scatter Radar configuration. In Proceedings of the International Conference on Radar Systems (Radar 2017), Belfast, UK, 23–26 October 2017; pp. 1–6.
12. Pastina, D.; Contu, M.; Lombardo, P.; Gashinova, M.; de Luca, A.; Daniem, L.; Cherniakov, M. Target motion estimation via multi-node forward scatter radar system. *IET Radar Sonar Navig.* **2016**, *10*, 3–14. [CrossRef]
13. Contu, M.; de Luca, A.; Hristov, S.; Daniel, L.; Stove, A.; Gashinova, M.; Cherniakov, M.; Pastina, D.; Lombardo, P.; Baruzzi, A.; et al. Passive Multifrequency Forward-Scatter Radar Measurements of Airborne Targets Using Broadcasting Signals. *IEEE Trans. Aerosp. Electron. Syst.* **2017**, *53*, 1067–1087. [CrossRef]
14. Arcangeli, A.; Bongioanni, C.; Ustalli, N.; Pastina, D.; Lombardo, P. Passive forward scatter radar based on satellite TV broadcast for air target detection: Preliminary experimental results. In Proceedings of the 2017 IEEE Radar Conference (RadarConf), Seattle, WA, USA, 8–12 May 2017; pp. 1592–1596.
15. Ustalli, N.; di Lello, F.; Pastina, D.; Bongioanni, C.; Rainaldi, S.; Lombardo, P. Two-dimensional filter bank design for velocity estimation in Forward Scatter Radar configuration. In Proceedings of the 2017 18th International Radar Symposium (IRS), Prague, Czech Republic, 28–30 June 2017; pp. 1–10.
16. Czarnecki, K.; Moszyński, M. A novel method of local chirp-rate estimation of LFM chirp signals in the time-frequency domain. In Proceedings of the 2013 36th International Conference on Telecommunications and Signal Processing (TSP), Rome, Italy, 2–4 July 2013; pp. 704–708.
17. Czarnecki, K. The instantaneous frequency rate spectrogram. *Mech. Syst. Signal Process.* **2016**, *66*, 361–373. [CrossRef]
18. Hahn, S.L. On the uniqueness of the definition of the amplitude and phase of the analytic signal. *Signal Process.* **2003**, *83*, 1815–1820. [CrossRef]
19. Fourer, D.; Auger, F.; Czarnecki, K.; Meignen, S.; Flandrin, P. Chirp Rate and Instantaneous Frequency Estimation: Application to Recursive Vertical Synchrosqueezing. *IEEE Signal Process. Lett.* **2017**, *24*, 1724–1728. [CrossRef]
20. Abratkiewicz, K.; Czarnecki, K.; Fourer, D.; Auger, F. Estimation of time-frequency complex phase-based speech attributes using narrow band filter banks. In Proceedings of the 2017 Signal Processing Symposium (SPSympo), Jachranka, Poland, 12–14 September 2017; pp. 1–6.
21. Abratkiewicz, K.; Samczyński, P.; Czarnecki, K. Radar Signal Parameters Estimation Using Phase Accelerogram in the Time-Frequency Domain. *IEEE Sens. J.* **2019**, *19*, 5078–5085. [CrossRef]
22. Abratkiewicz, K.; Samczyński, P. A Block Method Utilizing the Instantaneous Chirp-Rate Estimation for Non-Linear Frequency Modulated Pulse Reconstruction. *IEEE Trans. Aerosp. Electron. Syst.* **2019**, under review.
23. Abratkiewicz, K.; Gromek, D.; Samczyński, P. Chirp Rate Estimation and micro-Doppler Signatures for Pedestrian Security Radar Systems. In Proceedings of the Signal Processing Symposium (SPSympo 2019), Krakow, Poland, 17–19 September 2019.
24. Kay, S.M. *Fundamentals of Statistical Signal Processing: Estimation Theory*; Prentice Hall: Upper Saddle River, NJ, USA, 1997.
25. Krysik, P.; Kulpa, K.; Samczyński, P. GSM based passive receiver using forward scatter radar geometry. In Proceedings of the 2013 14th International Radar Symposium (IRS), Dresden, Germany, 19–21 June 2013; pp. 637–642.

Article

The Use of the Reassignment Technique in the Time-Frequency Analysis Applied in VHF-Based Passive Forward Scattering Radar

Marek Płotka *, Karol Abratkiewicz, Mateusz Malanowski , Piotr Samczyński and Krzysztof Kulpa

Institute of Electronic Systems, Faculty of Electronics and Information Technology, Warsaw University of Technology, 00-665 Warsaw, Poland; k.abratkiewicz@elka.pw.edu.pl (K.A.); M.Malanowski@elka.pw.edu.pl (M.M.); P.Samczynski@elka.pw.edu.pl (P.S.); k.kulpa@elka.pw.edu.pl (K.K.)
* Correspondence: m.plotka@elka.pw.edu.pl

Received: 12 May 2020; Accepted: 15 June 2020; Published: 17 June 2020

Abstract: This paper presents the application of the time-frequency (TF) reassignment technique in passive forward scattering radar (FSR) using Digital Video Broadcasting – Terrestrial (DVB-T) transmitters of opportunity operating in the Very High Frequency (VHF) band. The validation of the proposed technique was done using real-life signals collected by the passive radar demonstrator during a measurement campaign. The scenario was chosen to test detection ranges and the capability of estimating the kinematic parameters of a cooperative airborne target in passive FSR geometry. Additionally, in the experiment the possibility of utilizing FSR geometry in foliage penetration conditions taking advantage of the VHF band of a DVB-T illuminator of opportunity was tested. The results presented in this paper show that the concentrated (reassigned) energy distribution of the signal in the TF domain allows a more precise target Doppler rate to be estimated using the Hough transform.

Keywords: passive forward scattering radar; forward scattering radar; passive radar; radar measurements; time-frequency analysis; time-frequency reassignment

---

1. Introduction

Over the past decades, passive radars have evolved significantly [1–3], which can be seen in numerous demonstrations and works devoted to this topic [4–8]. This has resulted from the advantages of passive radars and the possibility to detect targets which do not have their own emission. In fact, issues related to passive radars are generally widely described in the literature, however, there are still problems that require a specific approach.

The main problem in Passive Coherent Location (PCL) radar technology is that using classical passive radar processing for air target detection [9] does not allow one to detect and localize the target in the direction of the illuminator of opportunity, therefore the radar is "blind" at this particular angle and additionally the target is in the first range cell, which provides unclear detection results. This direction is reserved for the collection of the reference signal, which is used for cross-correlation with signals collected from other receiver channels whose measurement antennas are pointed in other surveillance directions where a target echo is suspected to be received. However, this paper deals with the methodology which allows kinematic parameters of the object to be distinguished even if the range information is lost. Various answers to the problem of passive radar direct path "blindness" can be found in numerous literature positions [10–15].

A possible solution to this issue may be the employment of reference signal reconstruction [10]. However, this technique only works for digital signals and with sufficient signal-to-noise ratio (SNR) values. Using a beamforming technique to reduce direct signal leakage to surveillance channels [11,12] can also be employed. Aubry et al. [13], used a Constrained Least Squares two-dimensional localization algorithm. Its performance, expressed in terms of Root Mean Square Error (RMSE), is even comparable to square root of the Cramer Rao Lower Bound (CRLB) for some of the simulation scenarios presented. A significant disadvantage of this algorithm is a necessity of employment multiple transmitters of opportunity. A different solution to the localization issue can be found in Aubry et al. [14]. Joint target location is based on a PCL and Time Difference of Arrival (TDOA) measurement techniques. However, the TDOA method requires multiple dislocated radar receivers. Another study on the target location accuracy in multistatic scenario is presented by Anastasio et al. [15].

The other solution for this problem, utilizing a single receiver and single transmitter of opportunity only, might be to additionally use FSR methods in the PCL processing chain. The use of FSR geometry allows passive radar to detect and estimate main movement parameters such as target velocity for the targets crossing the T_x–R_x baseline [16,17]. In such a case, data from the FSR module applied in PCL radars might be used as additional information for the radar tracker, and consequently the detection and velocity estimation from the FSR module to the tracker working in the bistatic range-Doppler plane. Such a method will significantly improve the detection and tracking performance in PCL processing. This fact led the authors to study in more detail the possibility of applying FSR geometry in passive radar, and test novel methods for target Doppler frequency rate estimation which might be applied in PCL processing. An additional motivation was using low-frequency DVB-T sources of illumination in passive radars and their ability to perform foliage penetration. The VHF DVB-T operates in the band of 174–230 MHz [18]. As these are relatively low frequencies, they penetrate the foliage well. The authors did one experiment where a VHF DVB-T based passive radar was deployed in a forest on a low mast around 3 m in height, which was much lower than the surrounding trees, and successfully detected the air targets. The results have been described by Plotka et al. [19]. These valuable results also motivated the authors to check how efficiently the VHF DVB-T illuminator of opportunity would be used in FSR geometry, where the reference signal is also received through the transmitter.

This paper has the following structure: Section 2 presents the passive FSR geometry principle that is considered in this work. Section 3 covers the description of the proposed method for the target Doppler rate estimation. In Section 4, the measurement campaign and the numerical results using the real-life signals are depicted. The paper is closed by comments and conclusions.

2. Passive FSR Geometry

The forward scattering phenomenon [16,17] is schematically depicted in Figure 1.

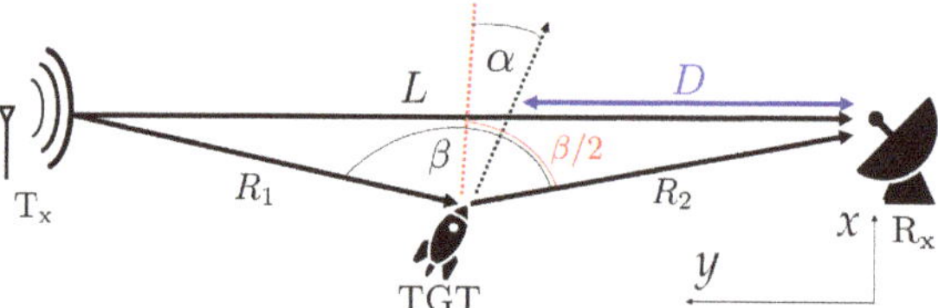

Figure 1. A typical passive forward scattering radar (FSR) geometry. T_x—transmitter, R_x—receiver, TGT—target, L—baseline, R_1—range from the transmitter to the target, R_2—range from the receiver to the target, D—range from the receiver to the crossing point, β—bistatic angle, α—angle between the bistatic bisector and the velocity vector.

In bistatic radars using an electromagnetic wave of a length λ, the Doppler shift f_d produced by the target moving at the velocity V can be expressed as

$$f_d = \frac{2V}{\lambda}\cos(\alpha)\cos(\beta/2), \tag{1}$$

where β is the bistatic angle and α is the angle between the bistatic bisector and the movement vector (see Figure 1). For such a spatial configuration where the target crosses the baseline (when the range information is lost and the Doppler frequency tends to 0 Hz) $\beta \approx 180°$, thus $f_d \approx 0$ Hz, however, the target can be observed in a range $\beta \in (180° - \Delta, 180° + \Delta)$, where Δ is a small angle, which delivers more information about the target trajectory. In the vicinity of $f_d \approx 0$ Hz the signal impinging the receiving antenna can be defined as [20]

$$\phi(t) = \frac{2\pi}{\lambda}\left[R_1(t) + R_2(t) - L\right]. \tag{2}$$

Assuming that the target moves along a linear trajectory at a constant velocity v and the velocity vector (composed of v_x and v_y components corresponding respectively to the x- and y-axis) creates an angle with respect to the x-axis that is normal to the baseline $\alpha = \tan^{-1}\left(\frac{v_y}{v_x}\right)$. If the target crosses the baseline at the point D from the receiver (see Figure 1) then $x(t) = v_x t$ and $y(t) = D + v_y t$ which leads to

$$R_1(t) = \sqrt{x(t)^2 + (L - y(t))^2} \tag{3}$$

and

$$R_2(t) = \sqrt{x(t)^2 + y(t)^2}. \tag{4}$$

Apart from the Doppler history, the target forward radar cross section (FRCS) has a contribution to the signal reaching the receiver in the passive FSR system. Namely, a rectangular target of horizontal l_h and vertical l_v dimension such that $l_h >> \lambda$ and $l_v >> \lambda$, is given by [21]

$$\sigma(t) = \frac{L^2}{R_1(t)R_2(t)}\frac{\cos(\theta_T(t) - \alpha) + \cos(\theta_R(t) + \alpha)}{2}\Pi\left(\frac{l_h}{\lambda}\left(\sin(\theta_T(t) - \alpha) + \sin(\theta_R(t) + \alpha)\right)\right) \tag{5}$$

where the target aspect angle with respect to transmitter is $\theta_T = \tan^{-1}\left(\frac{x(t)}{L - y(t)}\right)$ and with respect to receiver is $\theta_R = \tan^{-1}\left(\frac{x(t)}{y(t)}\right)$ and Π is the function such that $\Pi(x) = \frac{\sin(x)}{x}$. Then, the signal impinging the FSR receiving antenna can be written as

$$s(t) = -\sigma(t)\sin(\phi(t)). \tag{6}$$

Approximating Equation (6) by the third order Taylor polynomial around the crossing point $t = t_0$, Equation (2) becomes

$$\phi(t) \approx \pi\epsilon(t - t_0)^2 + \pi\zeta(t - t_0)^3, \tag{7}$$

where

$$\epsilon = \frac{v_x^2}{\lambda}\left[\frac{1}{L - D} + \frac{1}{D}\right], \tag{8}$$

and

$$\zeta = \frac{v_x^2 v_y}{\lambda}\left[\frac{1}{(L - D)^2} + \frac{1}{D^2}\right], \tag{9}$$

and the latter equation goes to 0 for $v_y = 0$ or $D = L/2$. Finally, the signal phase can be approximated as follows

$$\phi(t) \approx \frac{\pi}{\lambda}v_x\left(\frac{1}{L - D} + \frac{1}{D}\right)(t - t_0)^2, \tag{10}$$

where v_x denotes the velocity component which is perpendicular to the baseline L, D expresses the range from the receiver to the crossing point, and t_0 is the particular moment the target crosses the baseline. In fact, Equation (10) can be considered as a quadratic phase function resulting in the linear frequency-modulated signal. The modulation factor (also known as a chirp rate, frequency rate, frequency slope, etc.) is valuable information describing the target in the situation when the range measurement is ambiguous, which is the case in passive FSR. Thus, any additional characterization of the target movement is significant in this case. In the literature, different approaches are proposed in order to estimate motion parameters, such as spectrogram analysis [22], chirp rate estimation in the time-frequency (TF) domain [23], or the Radon transform [20]. This paper refers to the latter example, and the authors of this work proposed the method known from the literature to improve the resolution of the TF distribution using the TF reassignment technique in order to distinguish the kinematic parameter of the cooperative target using the Hough transform, which can be interpreted as a discrete realization of the Radon transform [24,25]. The obtained outcomes are compared to existing methodology [20,26–29], and the improvement in the estimation precision is shown.

3. Target Doppler Rate Estimation

Typically, the signal received by the passive FSR antenna after initial processing is presented in the TF domain as a quasi-linear frequency-modulated waveform in accordance to Equation (10). Next, by using the Radon transform the Doppler rate is estimated at the crossing point which determines the kinematic parameters of the target [20,26,28,29]. As shown by Toft et al. [24,25], the Hough transform can be used equivalently as a discrete realization of the Radon method, which may be used for the estimation of the frequency slope of the signal in the TF domain. This approach is very fast and can be implemented in real-life systems, however, some details have to be taken into account. Namely, classical TF representations suffer from the limited resolution resulting from the Heisenberg–Gabor uncertainty principle [30], which spoils the estimation accuracy in this case. Especially when the SNR is low or when several objects cross the baseline at the same time, the method proposed by Ustalli et al. may require additional processing steps. Widely speaking, in many practical applications the short-time Fourier transform (STFT) based approach may be insufficient due to the finite resolution of the TF plane. Additionally, the resolution is strongly dependent on the processing parameters, such as window type, window width, overlap, etc. Even a high-SNR signal can be distributed incorrectly over the TF plane if the processing parameters are badly conditioned. One of the popular and widely applied methods for TF resolution enhancement is TF reassignment [30–33]. This method can be implemented through the classical STFT-based method, as well as using a recursive version presented by Fourer et al. [34]. The latter is particularly interesting due to the possibility of its efficient implementation and the fast operation of the energy relocation $(t, \omega) \mapsto (\hat{t}, \hat{\omega})$. As both the recursive and fast Fourier transform (FFT)-based implementations are equivalent, the FFT-based method is used in this paper as an example of the technique.

In general, the STFT of the signal $x(t)$ can be computed as follows:

$$F_x^h(t, \omega) = \int_{\mathbb{R}} x(\tau) h^*(t - \tau) e^{-j\omega\tau} \mathrm{d}\tau = M_x^h(t, \omega) e^{j\phi_x^h(t,\omega)}, \tag{11}$$

where $j = \sqrt{-1}$, $(\cdot)^*$ is the complex conjugate, $\mathbb{R}$ denotes the set of real numbers, $M_x^h(t, \omega)$ is the amplitude, and $\phi_x^h(t, \omega)$ is the phase of the transform. The energy distribution, commonly called a spectrogram, is defined as a squared absolute value of the STFT and is given by

$$E_x^h(t, \omega) = |F_x^h(t, \omega)|^2, \tag{12}$$

where $|\cdot|$ denotes the absolute value operator.

Using the relationship defined by Hahn [35], Equation (11) can be transformed into a complex phase as follows:

$$\Phi_x^h(t,\omega) = \ln\left(F_x^h(t,\omega)\right) = \Lambda_x^h(t,\omega) + j\phi_x^h(t,\omega), \tag{13}$$

where $\Lambda_x^h(t,\omega) = \ln\left(M_x^h(t,\omega)\right)$. The concept of complex phase is widely used in the literature as an effective tool for the estimation of signal parameters in the TF domain, which is also applied in the considered approach. For TF reassignment, the relocation operators have to be estimated, which correspond to vectors for both the time and frequency axes along which the energy has to be moved. In the investigated approach, the reassignment operators may be estimated respectively [31]:

$$\hat{t}(t,\omega) = -\Im\left(\frac{\partial\Phi_x^h(t,\omega)}{\partial\omega}\right) = t - \Re\left(\frac{\partial\Phi_x^h(t,\omega)}{\partial\omega}\right) = t - \Re\left(\frac{F_x^{\mathcal{T}h}(t,\omega)}{F_x^h(t,\omega)}\right), \tag{14}$$

$$\hat{\omega}(t,\omega) = \omega + \Im\left(\frac{\partial\Phi_x^h(t,\omega)}{\partial t}\right) = \omega + \Im\left(\frac{F_x^{\mathcal{D}h}(t,\omega)}{F_x^h(t,\omega)}\right), \tag{15}$$

where $\Re$ is the real and $\Im$ is the imaginary part. $\hat{t}(t,\omega)$ denotes the relocation along the t-axis and $\hat{\omega}(t,\omega)$ expresses the reassignment vector along the frequency axis. In contrast to another energy concentration technique known from the literature, for example TF synchrosqueezing transform [32], the reassignment method allows strong concentration to be obtained, however, this technique is irreversible. In Equation (14), the expression $\mathcal{T}h = th(t)$ is a window multiplied by the linear time ramp with a root in 0, and $\mathcal{D}h$ in Equation (15) denotes the first order derivative of the analyzing window $\mathcal{D}h = \frac{\mathrm{d}h(t)}{\mathrm{d}t}$. In fact, these operations can be interpreted as follows:

$$\partial F_x^h(t,\omega)\big/\partial t = \int x(\tau)\frac{\partial h^*(t-\tau)e^{-j\omega(t-\tau)}}{\partial t}\mathrm{d}\tau = F_x^{\mathcal{D}h}(t,\omega), \tag{16}$$

as well as:

$$\partial F_x^h(t,\omega)\big/\partial\omega = \int x(t-\tau)h^*(\tau)\frac{\partial e^{-j\omega\tau}}{\partial\omega}\mathrm{d}\tau = -\int x(t-\tau)h^*(\tau)j\tau e^{-j\omega\tau}\mathrm{d}\tau = -jF_x^{\mathcal{T}h}(t,\omega), \tag{17}$$

which leads to:

$$\frac{\partial\ln(F_x^h(t,\omega))}{\partial t} = \frac{\partial F_x^h(t,\omega)}{\partial t}\frac{1}{F_x^h(t,\omega)} = \frac{F_x^{\mathcal{D}h}(t,\omega)}{F_x^h(t,\omega)} \tag{18}$$

and

$$\frac{\partial\ln(F_x^h(t,\omega))}{\partial\omega} = \frac{\partial F_x^h(t,\omega)}{\partial\omega}\frac{1}{F_x^h(t,\omega)} = -j\frac{F_x^{\mathcal{T}h}(t,\omega)}{F_x^h(t,\omega)} \tag{19}$$

which can be directly applied in Equations (14) and (15). This means that the reassignment operators can be easily computed through the STFT method using the modified analyzing window, which increases the utility of the method and reduces the computational effort. Equivalently, the method may be implemented using a recursive filter bank, as described in [34].

Finally, the energy relocation using the reassignment method can be expressed as [31]

$$R_x^h(t,\omega) = \iint_{\mathbb{R}^2}|F_x^h(t,\omega)|^2\delta(t-\hat{t}(t,\omega))\delta(\omega-\hat{\omega}(t,\omega))\mathrm{d}t\mathrm{d}\omega, \tag{20}$$

where $\delta(\cdot)$ denotes the Dirac distribution. The distribution given by Equation (20) results in strongly concentrated energy on the TF plane, with an enhanced readability and separated components. In fact, the reassignment method usually does not relocate the maximum of the energy but only attracts the surrounding distribution, hence the signal localization remains stable whilst the readability of the transform is improved.

The TF reassignment is a widely used technique in many applications, e.g., ultrasound signal processing [36], audio signal analysis [37,38], or sonar applications [39]. However, despite the high potential of this method, it is still not very popular in the radar community. Namely, in the literature one can find results for the energy concentration in micro-Doppler signature analysis [40–42], characterization of frequency shift keying (FSK) signals [43], improving the quality of inverse synthetic aperture radar (ISAR) imaging [44] as well as in direction of arrival estimation [45] and Doppler radar tomography imaging [46]. The novelty presented in this paper is to apply the TF reassignment method and combine it with the Hough transform that aims to extract the Doppler rate of the target with the improved accuracy comparing the classical method existing in the literature.

The valuable properties of the TF reassignment method prompted the authors to apply this technique to analyze radar signals in a passive FSR application. The "energy gathering" properties applied in such an application may improve the accuracy of the Doppler rate estimation in passive FSR systems. This, in fact is the novelty proposed in this paper, since to the authors knowledge there is no similar application that uses the TF reassignment method to improve the accuracy of the Doppler rate estimate in passive FSR system. Additionally, the outcomes are compared to the classical approach existing in the literature with particular emphasis on the usability in real-life data processing which is investigated in the next section.

4. Numerical Experiments

4.1. Measurement Campaign

The measurements took place during the APART GAS 2019 (Active PAssive Radar Trials Ground based, Airborne, Sea-borne) trials. The trials were described by Plotka et al. [19], nevertheless some details of the measurement FSR scenario geometry will be examined here. The positions of the receiving station, transmitter of opportunity, aircraft trajectory, and transmitter–receiver baselines are depicted in Figure 2.

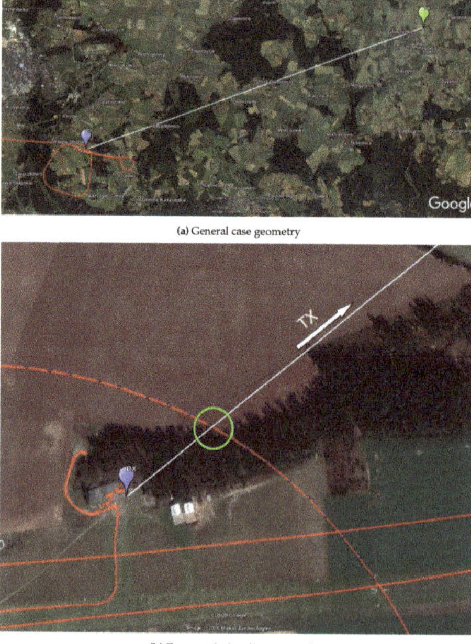

(a) General case geometry

(b) Zoom on the baseline cross point

Figure 2. Scenario geometry: transmitter–receiver baseline (white line) cross point marked by a green circle, aircraft trajectory in red.

During the passive FSR measurements, the radar receiving station was placed in an open space, on an airfield (see Figure 3). The location was chosen to also test FSR geometry in foliage penetration conditions. The receiver was placed close to the forest line, where the trees were in the direction of the transmitter of opportunity—see Figure 2b. The radar demonstrator was equipped with 6 antennas, but only two of them were used in the presented FSR experiment. During the trials the signal from both V- and H-polarized receiving antennas was gathered, however, due to the lack of significant differences between the results only the selected pair of the receiving channels was analyzed. Hence, in the further part of this paper the signal from V-polarized antennas mounted on a tripod mast at the height of ca. 3 m above the ground is presented. The parameters of the employed transmitter of opportunity are listed in Table 1.

Figure 3. Radar demonstrator during measurements.

Table 1. Main parameters of the transmitter of opportunity.

Name	"Lębork Skórowo Nowe"
Distance to receiver	27.8 km
Location height	96 m a.m.s.l
Mast height	93 m
EIRP	10.4 kW
Frequency	184.5 MHz
Signal bandwidth	7 MHz
Polarization	Vertical

During the measurements a cooperative target was used—a light Cessna aircraft (see Figure 4). At the moment of crossing the transmitter–receiver baseline, the aircraft flight parameters were as follows: the target altitude was 196 m above terrain level and velocity was 44 m/s. The distance between receiver and target was 280 m, the distance between transmitter and target was 27.5 km, and the distance between receiver and transmitter was 27.706 km.

Figure 4. The cooperative aerial target.

The composition of the receiving station was as follows. Antennas: commercial-off-the-shelf (COTS) 4-element Uda-Yagi, with directivity from 6 dBi to 8 dBi, operating in an upper VHF frequency band (170 MHz up to 230 MHz). Analog front-end: COTS channel amplifier, operating in 87–230 MHz frequency band, with 25 dB gain. Digital signal recorder (see Figure 5): Vector Signal Analyzer (VSA) based on National Instruments PXIe components, with six independent input channels synchronized coherently with GPS signal, operating in the frequency range 10 MHz–6.6 GHz with the maximum bandwidth of 50 MHz. More detailed description, of the radar demonstrator hardware, has been presented by Plotka et al. [19].

Figure 5. Digital multichannel signal recorder.

4.2. Results

Signals recorded by the passive radar demonstrator were processed with the passive FSR signal processing chain as presented in Figure 6.

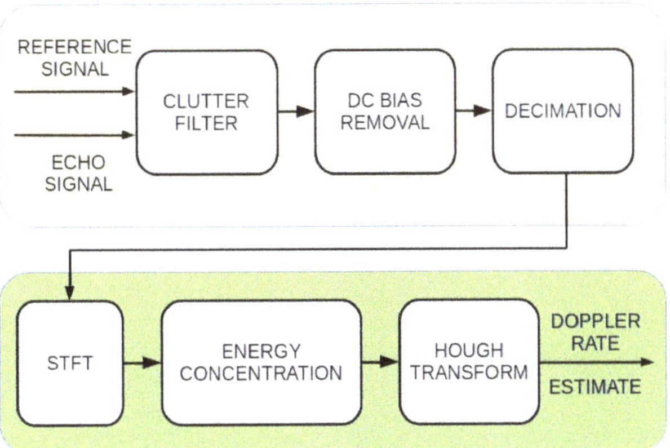

Figure 6. Passive FSR signal processing scheme.

At first, signals from two channels (reference and surveillance) were selected for further calculations. All further computations were performed on the signals' blocks with a 200 ms integration period (which resulted in a velocity resolution equal to 8.13 m/s). Next, a clutter filter was used for removing the reference signal from the surveillance signal [9]. This operation normally significantly reduces target echo power when reaching zero Doppler velocity. In order to limit the scale of this phenomenon, clutter removal filter coefficients were fixed for the time when the target was crossing the baseline [47]. An additional step of signal processing was the removal of direct current (DC) offset, which was achieved by subtracting the mean value from the signal after clutter filtering. The last step of the processing was a multi-stage decimation. According to the simulations carried out, the expected Doppler frequency was not greater than 100 Hz. The input sampling rate of the recorded signals was equal to 8 MHz, so the filtered signal might have been down-sampled a few thousand times without losing valuable information. This step considerably reduced the number of unnecessary subsequent calculations. It should be mentioned that the bistatic range resolution for the processed signal is equal to 37.5 m (this is the size of the first range cell).

Next, the signal was transformed into the TF domain using Equation (11) giving the classical signal distributions on the 2D TF plane, as shown in Figure 7. The Doppler history is clearly visible for the entire trajectory, however, some interference related to multipath propagation and clutter removal algorithm are apparent, especially at the point when the waveform changes the frequency sign. Additionally, the clutter cannot be suppressed at this point due to the presence of the useful signal that should not be filtered out.

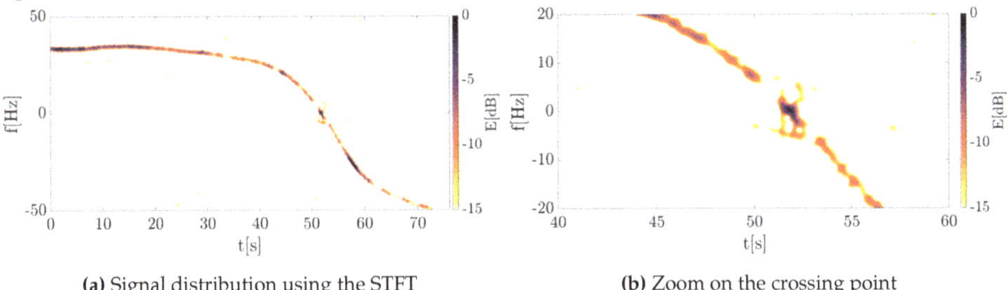

(a) Signal distribution using the STFT

(b) Zoom on the crossing point

Figure 7. The energy distributions of the measured signal obtained using the classical short-time Fourier transform (STFT)

The same signal was processed using the concept of energy concentration. The outcomes for this approach are depicted in Figure 8, where the significant concentration was obtained. In such a case, the component extraction and separation are obtainable even in the case of low SNR.

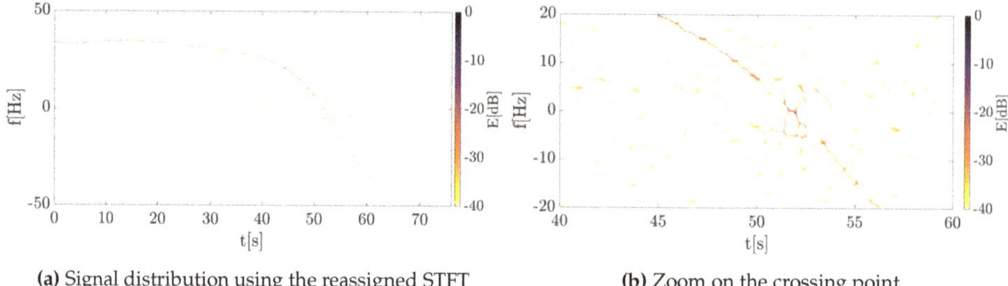

(a) Signal distribution using the reassigned STFT

(b) Zoom on the crossing point

Figure 8. The energy distributions of the measured signal obtained using the concentrated STFT.

Both classical and reassigned distributions were obtained using an 8192-point FFT and the Gaussian window of a standard deviation $\sigma = 0.2$. The shift of the window in consecutive steps the processing was equal to 1 sample in order to provide precise signal representation. Then, in accordance to the approach proposed by Ustalli et al. [20,26,28,29], the Hough transform was applied to estimate the signal Doppler rate as a straight line on the TF plane composed by the Doppler history of the signal near the zero-intersection point. The results are depicted in Figure 9.

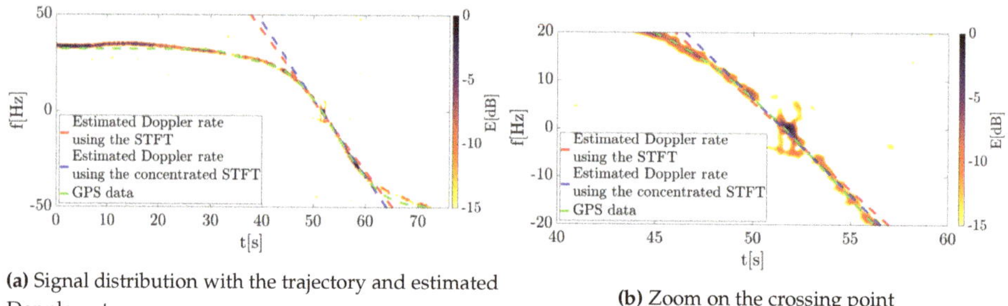

(a) Signal distribution with the trajectory and estimated Doppler rates

(b) Zoom on the crossing point

Figure 9. Results of the Hough transform obtained for the two methods investigated—the classical and the reassigned STFT.

As can be observed, both the classical and the concentrated distributions allowed the Doppler rate to be estimated. The selected sections of both distributions $f \in (-20, 20)$ Hz, $t \in (40, 60)$ s containing the most important parts of the signal were processed using the Hough transform, which gave results corresponding to the frequency slope at the interesting point. These values were estimated as follows: $f_S = -3.6392\frac{\text{Hz}}{\text{s}}$ for the classical STFT distribution and $f_R = -3.9873\frac{\text{Hz}}{\text{s}}$ for the reassigned spectrogram.

Both estimated lines coincide with the Doppler history, and the visual analysis of them does not give an answer as to which of them is more precise. Thus, in order to verify the correctness of the estimate, an additional line was created. The Hough transform was applied on the curve fragment derived from GPS data (see the green line in Figure 9), allowing the precise Doppler rate to be assessed. Next, the error between the reference (GPS data) and two estimated lines was computed. However, due to the limited precision of the DVB-T transmitter localization, as well as the smoothing of the GPS-based trajectory, the actual crossing point is mismatched. The additional sources of mismatch error may be connected with the GPS logger. This device's coordinates estimation accuracy is limited, and as well as its own location inside the aircraft also mattered. An another point which had an impact on the accuracy is the information about the transmitter (T_x) position. For the analysis, the authors took the T_x position from an open database and validated the T_x coordinates using Google Maps. However, the accuracy of the T_x position is also often given with a precision of several meters, which might have had an impact on the presented results. Therefore, an additional simulation was performed in which the geometry was appropriately modified by changing the transmitter position that aims to reduce this fault. After these modifications the error was eliminated, and the results for both the initial and modified trajectories are depicted in Figure 10. The plots show the absolute value of the estimation error δ.

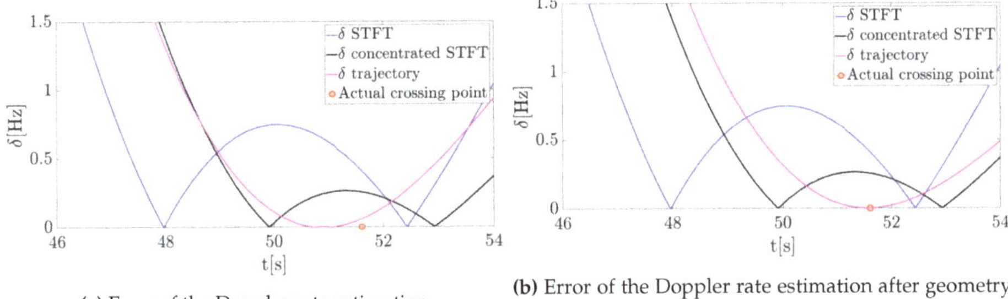

(a) Error of the Doppler rate estimation

(b) Error of the Doppler rate estimation after geometry correction

Figure 10. Absolute value of estimated errors of the Doppler rate estimation—initial error and after correction.

The proposed method allowing signal concentration in the TF to be obtained improved the precision of the Doppler rate estimation. As can be observed, the error was reduced for both parameters: the Doppler rate, and for the time when the target crossed the baseline. This result indicates the effectiveness of the proposed approach.

For the Hough transform, the computational complexity increases at a rate of $\mathcal{O}\left(A^{m-2}\right)$ where A denotes the size of the image space and m corresponds to the number of parameters applied in the processing pipeline. In the proposed method, the Hough transform is the same as in the classical approach with STFT. Therefore, the only difference results from the signal processing associated with the reassignment operation. For this reason, only the STFT and the reassigned STFT computational complexity are compared in Table 2 [40].

Table 2. Computational complexity for the STFT and the reassigned STFT in the $\mathcal{O}$ notation. N—amount of points in the FFT analysis, $K = \lceil M/H \rceil$—amount of time instants for which the FFT has to be applied for M—signal length in samples, and H—window shift in samples.

	STFT	Reassigned STFT
Computational complexity	$\mathcal{O}\left(KN\log_2(N)\right)$	$3 \cdot \mathcal{O}\left(KN\log_2(N)\right)$

The increase in computational complexity in the concentrated spectrogram technique results from the fact that 3 distributions are necessary to be implemented in accordance to Equation (20). Namely, the first distribution is the classical STFT with the original window according to Equation (11). The second one corresponds to Equation (16), and the last one to Equation (17). Hence, the precision of the Doppler rate can be improved at the expense of computational complexity. In fact, the processing time may be reduced through the manipulation of the processing parameters. For the purpose of this paper, the number of frequency bins and the window shift were assumed with some redundancy for high resolution of distributions. In practice, these parameters can be reduced to ensure fast processing. For the parameters defined above, the computation time for the spectrogram (t_S) and the reassigned spectrogram (t_R) was respectively $t_S = 0.42$ s and $t_R = 1641.09$ s, but for 1024 points of the FFT this time was reduced to $t_S = 0.17$ s and $t_R = 90.73$ s. For the purposes of calculations, a computer with an Intel i7-7700HQ 2.8 GHz processor, 16 GB DDR4 RAM, an SSD hard drive, and a 64-bit Windows 10 system was used. The calculations were performed in the Matlab environment. Consequently, by reducing the distribution quality and for the window shift $H = 2$, the processing time was additionally decreased to $t_S = 0.06$ s and $t_R = 29.62$ s with nearly the same readability of the distribution. This analysis shows how the processing time may be easily manipulated and reduced with almost conserved resolution, allowing enhanced estimation of the motion parameters to be performed.

5. Discussion and Conclusions

In this paper, the concept of applying the reassignment technique in passive FSR applications has been proposed. The main purpose of using this method was to enhance the readability of the energy distribution in the TF domain, which improved the result of the Hough transform and finally the precision of the Doppler rate estimation in the passive FSR system. In the considered case, the passive FSR system used a DVB-T VHF signal as a source of illumination, and the cooperating target (Cessna aircraft) crossed the baseline between the signal transmitter and the passive radar demonstrator. The investigation showed that the low-frequency illuminating signal allows detection in the specific geometry to be carried out and, additionally, by using the concept of energy concentration the signal processing pipeline may be improved. Such an improved energy representation can be utilized in further processing for different purposes, such as the estimation of the maneuvering target trajectory for the passive FSR configuration where the range information is ambiguous and the only data describing the target is its Doppler rate related to the velocity and the trajectory. The concept of using passive FSR systems may be implemented in various real-life systems known from the literature as well as in completely new applications. The authors consider realization of the technique in such applications:

- Border surveillance—the system may prevent to illegal transport and emigration outside the border crossing. In such a case, the illuminator of opportunity can be located both, in the same region as the receiver and in a foreign terrain which expands the possibility of the systems. In addition, passive FSR with other radars with rotating antenna as a source of illumination can be used in such scenario to detect and localize slow moving targets, which can be an extension of work described by Raja Abdullah et al. [48], where a cooperative transmitter was employed as an illuminator of opportunity.
- Debris detection—a passive FSR system can detect and estimate movement parameters of debris which may be of particular importance for space applications. In that case, the baseline can be established between a stationary receiver and the transmitter in the orbit working with one of the popular radiocommunication systems (Starlink, GPS, etc.). Such configuration may allow fast and precise debris detection in order to prevent other satellites destruction [49,50].
- Airport runway—the problem of airport runway security arises simultaneously with technology related to drones production and development. The passive FSR system may be deployed at the airport and detect non-cooperating objects appearing in the flight path of the aircraft. The system can be particularly useful for facilities not equipped with a transponder. In order not to employ another signal transmitter, an Airport Surveillance Radar with a rotating antenna may be used, which is routinely available at most airports [48].
- Ground aerial monitoring—where traffic is prohibited or where the velocity is strictly limited the passive FSR system and especially the algorithm proposed in this paper may be useful. The Doppler rate estimate may be an efficient solution for regions in which the velocity of cars has to be restricted and precisely monitored.
- Intruder detection—another application of the system in question is its usage in security systems for restricted areas, e.g., military or governmental areas. Implementation of the passive radar receiver in such a terrain when surrounding commercial transmitters (e.g., DVB-T) may improve detection capabilities and increase security.

In the future, the authors intend to work on the above applications and their implementation in passive FSR systems. An additional perspective is to investigate the possibility of applying real-time processing for the methods presented in this paper.

Author Contributions: Formal analysis, M.P. and K.A.; investigation, M.P. and K.A.; methodology, M.P., K.A., M.M., P.S., and K.K.; project administration, M.P.; software, M.P. and K.A.; supervision, M.M., P.S., and K.K.; validation, M.M., P.S., and K.K.; visualization, M.P. and K.A.; writing—original draft, M.P. and K.A.; writing—review and editing, M.M., P.S., and K.K. All authors have read and agreed to the published version of the manuscript.

Funding: This research received no external funding.

Conflicts of Interest: The authors declare no conflict of interest.

References

1. Kulpa, K.; Malanowski, M.; Samczyński, P. Passive radar: From target detection to imaging. In Proceedings of the 2019 IEEE Radar Conference (RadarConf), Boston, MA, USA, 22–26 April 2019; pp. 1–286.
2. Kulpa, K.; Malanowski, M. From Klein Heidelberg to Modern Multistatic Passive Radar. In Proceedings of the 2019 20th International Radar Symposium (IRS), Ulm, Germany, 26–28 June 2019; pp. 1–9.
3. Kuschel, H.; O'Hagan, D. Passive radar from history to future. In Proceedings of the 11-th INTERNATIONAL RADAR SYMPOSIUM, Vilnius, Lithuania , 16–18 June 2010; pp. 1–4.
4. Brisken, S.; Moscadelli, M.; Seidel, V.; Schwark, C. Passive radar imaging using DVB-S2. In Proceedings of the 2017 IEEE Radar Conference (RadarConf), Seattle, WA, USA, 8–12 May 2017; pp. 0552–0556.
5. Olsen, K.E.; Baker, C.J. FM-based Passive Bistatic Radar as a function of available bandwidth. In Proceedings of the 2008 IEEE Radar Conference, Rome, Italy, 26–30 May 2008; pp. 1–8.
6. Conti, M.; Moscardini, C.; Capria, A. Dual-polarization DVB-T passive radar: Experimental results. In Proceedings of the 2016 IEEE Radar Conference (RadarConf), Philadelphia, PA, USA, 2–6 May 2016; pp. 1–5.
7. Sun, H.; Tan, D.K.P.; Lu, Y. Aircraft target measurements using A GSM-based passive radar. In Proceedings of the 2008 IEEE Radar Conference, Rome, Italy, 26–30 May 2008; pp. 1–6.
8. Martelli, T.; Colone, F.; Lombardo, P. First experimental results for a WiFi-based passive forward scatter radar. In Proceedings of the 2016 IEEE Radar Conference (RadarConf), Philadelphia, PA, USA, 2–6 May 2016; pp. 1–6.
9. Malanowski M. *Signal Processing for Passive Bistatic Radar*; Artech-House: Norwood, MA, USA, 2019.
10. O'Hagan, D. W.; Setsubi, M.; Paine, S. Signal reconstruction of DVB-T2 signals in passive radar. In Proceedings of the 2018 IEEE Radar Conference (RadarConf18), Oklahoma City, OK, USA, 23–27 April 2018; pp. 1111–1116.
11. Wan, L.; Sun, L.; Wang, X.; Bi, G. A range-Doppler-angle estimation method for passive bistatic radar. In Proceedings of the 2018 International Conference on Signals and Systems (ICSigSys), Bali, Indonesia, 1–3 May 2018; pp. 213–218.
12. Jiabing, Z.; Liang, T.; Yi, H. Study on moving target detection to passive radar based on FM broadcast transmitter. *J. Syst. Eng. Electron.* **2007**, *18*, 462–468. [CrossRef]
13. Aubry, A.; Carotenuto, V.; De Maio, A.; Pallotta, L. Localization in 2D PBR With Multiple Transmitters of Opportunity: A Constrained Least Squares Approach. *IEEE Trans. Signal Process.* **2020**, *68*, 634–646. [CrossRef]
14. Aubry, A.; Carotenuto, V.; De Maio, A.; Pallotta, L. Joint Exploitation of TDOA and PCL Techniques for Two-Dimensional Target Localization. *IEEE Trans. Aeros. Electron. Syst.* **2020**, *56*, 597–609. [CrossRef]
15. Anastasio, V.; Farina, A.; Colone, F.; Lombardo, P. Cramer-Rao lower bound with Pd < 1 for target localisation accuracy in multistatic passive radar. *IET Radar Sonar Navig.* **2014**, *8*, 767–775 .
16. Cherniakov, M. *Basic Principles of Forward-Scattering Radars in Bistatic Radar: Principles and Practice, Part III*; Wiley: NewYork, NY, USA, 2007.
17. Willis, N. *Forward-Scatter Fences in Bistatic Radar*; SciTech Publishing: Raleigh, NC, USA, 2005.
18. TV Channel Frequencies. Available online: Available online: http://igorfuna.com/dvb-t/digital/tv-channel-frequencies (accessed on 26 April 2020).

19. Płotka, M.; Malanowski, M.; Samczyński, P.; Kulpa, K.; Abratkiewicz, K.; Passive Bistatic Radar Based on VHF DVB-T Signal. In Proceedings of the IEEE International Radar Conference, Washington, DC, USA, 27–30 April 2020; pp. 596–560.
20. Ustalli, N.; Di Lello, F.; Pastina, D.; Bongioanni, C.; Rainaldi, S.; Lombardo, P. Two-dimensional filter bank design for velocity estimation in Forward Scatter Radar configuration. In Proceedings of the 2017 18th International Radar Symposium (IRS), Prague, Czech Republic, 28–30 June 2017; pp. 1–10.
21. Pastina, D.; Contu, M.; Lombardo, P.; Gashinova, M.; De Luca, A.; Daniel, L.; Cherniakov, M. Target motion estimation via multi-node forward scatter radar system. *IET Radar Sonar Navig.* **2016**, *10*, 3–14. [CrossRef]
22. De Luca, A.; Contu, M.; Hristov, S.; Daniel, S.; Gashinova, M.; Cherniakov, M. FSR velocity estimation using spectrogram. In Proceedings of the 2016 17th International Radar Symposium (IRS), Krakow, Poland, 10–12 May 2016; pp. 1–5.
23. Abratkiewicz, K.; Krysik, P.; Gajo, Z.; Samczyński, P. Target Doppler Rate Estimation Based on the Complex Phase of STFT in Passive Forward Scattering Radar. *Sensors* **2019**, *19*, 3627. [CrossRef] [PubMed]
24. Toft, P.A. Using the generalized Radon transform for detection of curves in noisy images. In Proceedings of the 1996 IEEE International Conference on Acoustics, Speech, and Signal Processing Conference Proceedings, Atlanta, GA, USA, 9 May 1996; Volume 4, pp. 2219–2222.
25. van Ginkel M.; Hendriks C. L.; van Vliet L. *A Short Introduction to the Radon and Hough Transforms and How They Relate to Each Other, Number QI-2004-01 in the Quantitative Imaging Group Technical Report Series;* Delft University of Technology: Delft, The Netherlands, 2004.
26. Ustalli, N.; Pastina, D.; Lombardo, P. Kinematic parameters extraction from a single node Forward Scatter Radar configuration. In Proceedings of the 2018 19th International Radar Symposium (IRS), Bonn, Germany, 20–22 June 2018; pp. 1–10.
27. Ustalli, N.; Pastina, D.; Bongioanni, C.; Lombardo, P. Motion parameters estimation in dual-baseline Forward Scatter Radar configuration. In Proceedings of the International Conference on Radar Systems (Radar 2017), Belfast, UK, 23–26 October 2017; pp. 1–6.
28. Ustalli, N.; Pastina, D.; Lombardo, P. Target kinematic parameters estimation in single input multi output Forward Scatter Radar configuration. In Proceedings of the 2019 IEEE Radar Conference (RadarConf), Boston, MA, USA, 22–26 April 2019; pp. 1–6.
29. Ustalli, N.; Pastina, D.; Lombardo, P. Target Motion Parameters Estimation in Forward Scatter Radar. *IEEE Trans. Aeros. Electron. Syst.* **2020**, *56*, 226–248. [CrossRef]
30. Kodera, K.; de Villedary, C.; Gendrin, R. A new method for the numerical analysis of non-stationary signals. *Phys. Earth Planet. Inter.* **1976**, *12*, 142–150. [CrossRef]
31. Auger, F.; Flandrin, P. Improving the readability of time-frequency and time-scale representations by the reassignment method. *IEEE Trans. Signal Process.* **1995**, *43*, 1068–1089. [CrossRef]
32. Auger, F.; Flandrin, P.; Lin, Y.-T.; McLaughlin, S.; Meignen, S.; Oberlin, T.; Wu, H.-T. Time-Frequency Reassignment and Synchrosqueezing: An Overview. *IEEE Signal Process. Mag.* **2013**, *30*, 32–41. [CrossRef]
33. Xiao, J.; Flandrin, P. Multitaper Time-Frequency Reassignment for Nonstationary Spectrum Estimation and Chirp Enhancement. *IEEE Trans. Signal Process.* **2007**, *55*, 2851–2860. [CrossRef]
34. Fourer, D.; Auger, F.; Flandrin, P. Recursive versions of the Levenberg-Marquardt reassigned spectrogram and of the synchrosqueezed STFT. In Proceedings of the 2016 IEEE International Conference on Acoustics, Speech and Signal Processing (ICASSP), Shanghai, China, 20–25 March 2016; pp. 4880–4884.
35. Hahn, S.L. On the uniqueness of the definition of the amplitude and phase of the analytic signal. *Signal Process.* **2003**, *83*, 1815–1820. [CrossRef]
36. Czarnecki, K.; Fouan, D.; Achaoui, Y.; Mensah, S. Fast bubble dynamics and sizing. *J. Sound Vib.* **2015**, *356*, 48–60. [CrossRef]
37. Plante, F.; Meyer, G.; Ainsworth, W. A. Improvement of speech spectrogram accuracy by the method of reassignment. *IEEE Trans. Speech Audio Process.* **1998**, *6*, 282–287. [CrossRef]
38. Abratkiewicz, K.; Czarnecki, K.; Fourer, D.; Auger, F. Estimation of time-frequency complex phase-based speech attributes using narrow band filter banks. In Proceedings of the 2017 Signal Processing Symposium (SPSympo), Jachranka, Poland, 12–14 September 2017; pp. 1–6.
39. Czarnecki, K.; Leśniak, W. Bearing estimation using double frequency reassignment for a linear passive array. *Pol. Marit. Res.* **2017**, *24*, 26–35. [CrossRef]

40. Abratkiewicz, K.; Samczyński, P.; Fourer, D. A Comparison of the Recursive and FFT-based Reassignment Methods in micro-Doppler Analysis. In Proceedings of the IEEE Radar Conference, Florence, Italy, 21–25 September 2020. (accepted).
41. Manfredi, G.; Ovarlez, J.; Thirion-Lefevre, K. Features Extraction of the Doppler Frequency Signature of a Human Walking at 1 GHz. In Proceedings of the IGARSS 2019—2019 IEEE International Geoscience and Remote Sensing Symposium, Yokohama, Japan, 28 July–2 August 2019; pp. 2260–2263
42. Abratkiewicz, K.; Gromek, D.; Stasiak, K.; Samczyński, P. Time-Frequency Reassigned Micro-Doppler Signature Analysis Using the XY-DemoRad System. In Proceedings of the 2019 Signal Processing Symposium (SPSympo), Krakow, Poland, 17–19 September 2019; pp. 331–334.
43. Stevens, D.; Schuckers, S. A Novel Approach for the Characterization of FSK Low Probability of Intercept Radar Signals via Application of the Reassignment Method. In Proceedings of the 2014 IEEE Military Communications Conference, Baltimore, MD, USA, 6–8 October 2014; pp. 783–787.
44. Vehmas, R.; Jylhä, J.; Väilä, M.; Visa, A. ISAR imaging of non-cooperative objects with non-uniform rotational motion. In Proceedings of the 2016 IEEE Radar Conference (RadarConf), Philadelphia, PA, USA, 2–6 May 2016; pp. 1–6.
45. Miller, S.R.; Spanias, A.S.; Papandreou-Suppappola, A.; Santucci, R. Enhanced direction of arrival estimation via reassigned space-time-frequency methods. In Proceedings of the 2010 IEEE International Symposium on Circuits and Systems, Paris, France, 30 May–2 June 2010; pp. 2538–2541.
46. Swiercz, E. Application of the reassignment of time-frequency distributions to Doppler radar tomography imaging of a rotating multi-point object. In Proceedings of the 2016 17th International Radar Symposium (IRS), Krakow, Poland, 10–12 May 2016; pp. 1–5.
47. Krysik, P.; Kulpa, K.; Samczyński, P. GSM based passive receiver using forward scatter radar geometry. In Proceedings of the 2013 14th International Radar Symposium (IRS), Dresden, Germany, 19–21 June 2013; pp. 637–642.
48. Raja Abdullah, R.S.A.; Mohd Ali, A.; Rasid, M.F.A.; Abdul Rashid, N.E.; Ahmad Salah, A.; Munawar, A. Joint Time-Frequency Signal Processing Scheme in Forward Scattering Radar with a Rotational Transmitter. *Remote Sens.* **2016**, *8*, 1028. [CrossRef]
49. Radmard, M.; Bayat, S.; Farina, A.; Hajsadeghian, S.; Nayebi, M.M. Satellite-based forward scatter passive radar. In Proceedings of the 2016 17th International Radar Symposium (IRS), Krakow, Poland, 10–12 May 2016; pp. 1–4.
50. Gronowski, K.; Samczyński, P.; Stasiak, K.; Kulpa, K. First results of air target detection using single channel passive radar utilizing GPS illumination. In Proceedings of the 2019 IEEE Radar Conference (RadarConf), Boston, MA, USA, 22–26 April 2019; pp. 1–6.

Article

Noise Suppression for GPR Data Based on SVD of Window-Length-Optimized Hankel Matrix

Wei Xue [1,2,*], Yan Luo [1,2], Yue Yang [1,2] and Yujin Huang [1,2]

1 School of Automation, China University of Geosciences, Wuhan 430074, China

2 Hubei Key Laboratory of Advanced Control and Intelligent Automation for Complex Systems, Wuhan 430074, China

* Correspondence: xuew@cug.edu.cn; Tel.: +86-27-8717-5128

Received: 16 July 2019; Accepted: 31 August 2019; Published: 3 September 2019

Abstract: Ground-penetrating radar (GPR) is an effective tool for subsurface detection. Due to the influence of the environment and equipment, the echoes of GPR contain significant noise. In order to suppress noise for GPR data, a method based on singular value decomposition (SVD) of a window-length-optimized Hankel matrix is proposed in this paper. First, SVD is applied to decompose the Hankel matrix of the original data, and the fourth root of the fourth central moment of singular values is used to optimize the window length of the Hankel matrix. Then, the difference spectrum of singular values is used to construct a threshold, which is used to distinguish between components of effective signals and components of noise. Finally, the Hankel matrix is reconstructed with singular values corresponding to effective signals to suppress noise, and the denoised data are recovered from the reconstructed Hankel matrix. The effectiveness of the proposed method is verified with both synthetic and field measurements. The experimental results show that the proposed method can effectively improve noise removal performance under different detection scenarios.

Keywords: ground-penetrating radar; noise suppression; singular value decomposition; Hankel matrix; window length optimization

1. Introduction

Ground-penetrating radar (GPR) is a geophysical detecting instrument that transmits high-frequency electromagnetic wave and receives the reflections [1]. GPR has been widely used in several fields such as civil engineering, archaeology, geology, and military exploration [2–6] for its nondestructive, continuous, rapid, and efficient properties. Due to the effect of complex underground environment [7] and ultra-wide bandwidth receiver [8], the echoes of GPR contain significant noise. The noise collected by the system can easily mask the effective signals. Therefore, noise suppression is very important for improving the signal quality and interpretation accuracy.

Different approaches for GPR noise suppression have been reported to the literature [9–21]. The wavelet transform is a popular method for GPR data denoising [9,10], and it is simple and effective. However, the selection of the mother wavelet function, the decomposition level, and the threshold function still rely on subjective experiences. Frequency-wavenumber (F-K) filtering originating from seismic data denoising has also been applied to remove noise in GPR data [11,12] and can remove cross rebar reflections and ringing noise effectively. However, the filter design in the F-K domain is relatively complex and the method is only suitable for point targets. The ensemble empirical mode decomposition (EEMD) method is an improved empirical mode decomposition (EMD) method carrying out the EMD over an ensemble of the signal plus Gaussian white noise. The EEMD method can extract the effective signals components from noisy GPR data [13,14]. However, the EEMD method is time-consuming and incapable of processing the raw data with a low signal-to-noise ratio (SNR). The robust principle component analysis (RPCA) method can recover a low-rank matrix from noisy measurements and it has been employed to suppress the clutter and noise of GPR data [15–17]. However, the RPCA method is sensitive to the choice of thresholds. Singular value decomposition (SVD) is a convenient method to decompose a matrix, which can decompose GPR data into different subspaces that correspond to different components [18–21]. The noise can be suppressed by selecting components that contain effective signals to reconstruct GPR signals. Since each component corresponds to one singular value, the key problem of denoising is the selection of appropriate singular values corresponding to effective signals. A criterion based on the SNR of recovered data has been applied for GPR signal denoising [22], which shows better performance than the wavelet threshold denoising method. The local energy ratio rule has been used to remove background noise of GPR signals [23], which exhibits good robustness under different detection conditions. The fuzzy c-means (FCM) clustering rule has been used to extract multiple targets in heavily cluttered GPR images [24], which can accurately separate the overlapping boundaries of clutter, noise, and target signals and improve the performance of conventional SVD.

Although the denoising methods based on SVD are effective and easy to implement, they are designed to decompose a matrix (two-dimensional data) and cannot fully separate effective signals from the noise in one-dimensional data. To resolve this problem, the one-dimensional data can be transformed into many kinds of matrices, such as the Toeplitz matrix, cycle matrix, and Hankel matrix. The difference lies in the method of creating the matrix, which will affect signal processing of SVD. Among the matrices, SVD of the Hankel matrix can achieve a similar signal processing effect to the wavelet transform [25]. Therefore, SVD of the Hankel matrix is more suitable for noise suppression. A scheme based on SVD of the Hankel matrix has been used to reduce noise for radar cross-section (RCS) data [26], which can improve the accuracy of target recognition greatly. The Hankel matrix-based SVD can eliminate the false peak in processing an impulse signal with strong trend and enhance the SNR in the reconstructed signal [27], which helps to improve the fault diagnosis performance for rolling bearings. The SVD and Hankel matrix-based denoising process has also been applied to the ball bearing vibration signals in both time and frequency domain for the elimination of the background noise [28]. It was found that denoising in the frequency domain yields better fault identification results than the denoising in the time domain. The SVD method based on the Hankel matrix in the local frequency domain has been applied to eliminate random noise in GPR data [29], which can improve suppression of random noise around non-horizontal phase reflection events.

Although the aforementioned papers have proven the effectiveness of SVD of the Hankel matrix in noise suppression, little research has been conducted with respect to the influence of the Hankel matrix size on denoising performance. The size of the Hankel matrix depends on the length of the sliding window which affects the information quantity that can be extracted from this matrix [30]. Based on this previous research, this paper proposes SVD of a window-length-optimized Hankel matrix to suppress noise for GPR data. First, the Hankel matrix formed by one-dimensional GPR data is decomposed with SVD, and the fourth root of the fourth central moment (FRFCM) of singular values is used to select the optimal window length of the Hankel matrix. Then, one threshold is generated by the difference spectrum of singular values, which is used to select effective signal components. Finally, the Hankel matrix is reconstructed with singular values corresponding to effective signals to suppress noise, and the denoised data are recovered from the reconstructed Hankel matrix. The performance of the proposed method is verified with series of synthetic and field measurements. The experimental results of the proposed method are also compared with those of the conventional SVD method based on the local energy ratio rule and wavelet transform method. The results show that the proposed method can effectively improve the denoising performance for GPR data.

2. Methodology

2.1. Denoising Method Based on SVD of the Hankel Matrix

The two-dimensional GPR data can be denoted by $B \in R^{N \times L}$, where L is the number of traces and N is the number of sampling points in each trace. For the data of one trace (one-dimensional data)

X=[x(1),x(2), ... ,x(N)], a Hankel matrix can be formed by sliding a window over the corresponding vector [25], which can be written as

$$A = \begin{bmatrix} x(1) & x(2) & \cdots & x(n) \\ x(2) & x(3) & \cdots & x(n+1) \\ \vdots & \vdots & \vdots & \vdots \\ x(m) & x(m+1) & \cdots & x(N) \end{bmatrix} \quad (1)$$

where $m = N - n + 1, 1 < n \leq m < N, A \in R^{m \times n}$, and n is the window length.

The SVD of Hankel matrix A can be expressed as

$$A = USV^T \quad (2)$$

where $U \in R^{m \times m}$ and $V \in R^{n \times n}$ are the left singular and right singular orthogonal matrices, respectively. $S = diag(\sigma_1, \sigma_2, \ldots, \sigma_r)$ is a singular value matrix with $\sigma_1 \geq \sigma_2 \geq \ldots \geq \sigma_r \geq 0$, and $r = \min(m, n)$. According to the definition of Equation (1), the number of singular values r is equal to the window length n.

Then, Equation (1) can be written as

$$A = \sum_{i=1}^{r} \sigma_i u_i v_i^T = \sum_{i=1}^{n} \sigma_i u_i v_i^T \quad (3)$$

where $u_i \in R^{m \times 1}$ and $v_i \in R^{n \times 1}$. $u_i v_i^T \in R^{m \times n}$ is the single rank matrix, which is the ith eigen image of A. It is obvious that σ_i is actually the projection of matrix A on the basis $u_i v_i^T$.

As singular values are arranged in descending order, the first few larger singular values generally correspond to effective signals with strong correlations, while the smaller singular values correspond to the noise with weak correlation. Therefore, matrix A can be written as

$$A = \sum_{i=1}^{k} \sigma_i u_i v_i^T + \sum_{i=k+1}^{n} \sigma_i u_i v_i^T \quad (4)$$

where k is the demarcation point of singular values, and the first k singular values correspond to effective signals? Then the Hankel matrix with noise suppression can be reconstructed as

$$A_s = \sum_{i=1}^{k} \sigma_i u_i v_i^T \quad (5)$$

According to the construction rule of the Hankel matrix, the denoised one-dimensional data can be given by

$$X_s = [A_S(1,:),\ A_S(2:m,n)] \quad (6)$$

where $A_S(1,:)$ is the first row of matrix A_S and $A_S(2:m,n)$ is the last column without the first element.

2.2. Optimization Method of Window Length

The window length n is the only parameter of the Hankel matrix which not only affects the information quantity extracted from the matrix but also the performance of SVD. As an example, synthetic one-dimensional GPR data are used to analyze the effect of the window length n on the performance of SVD. The synthetic data are generated by the "gprMax" simulator [31].

Figure 1 shows the geometry of the simulation model for the scenario. The background medium is concrete. The relative permittivity and conductivity are 6 and 0.01, respectively. The target is a perfect metal cylinder, with 0.4-m diameter, which is buried at a depth of 0.6 m. The Ricker wavelet with a center frequency of 900 MHz is adopted. There are 80 traces in total and the trace interval is 0.035 m. The time window for each trace is 12 ns and each trace contains 2036 sampling points.

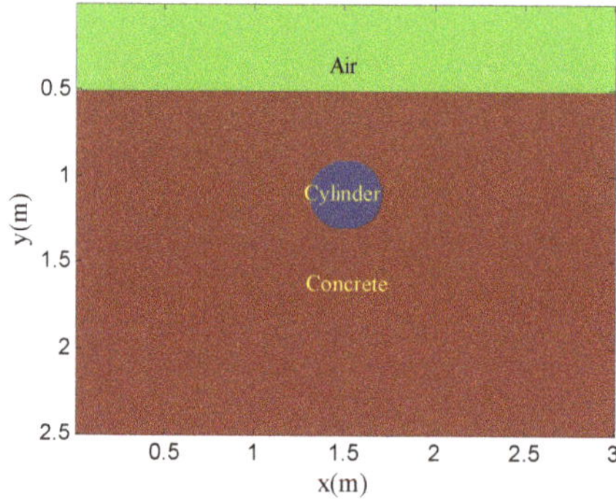

Figure 1. Geometry of the simulation model for point target detection.

The Gaussian white noise is added to the original GPR image and the SNR is −5.00 dB. Figure 2 shows the original GPR image and the noisy GPR image. Figure 3 shows the original data and noisy data of trace 38. The direct wave and target echoes are near the 250th and the 1300th sampling points, respectively. The noisy data are used to form the Hankel matrices with different window lengths, and SVD is applied to the Hankel matrices.

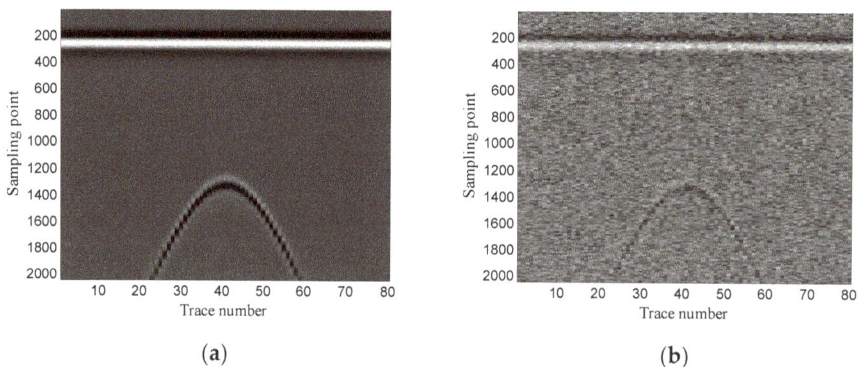

Figure 2. Synthetic ground-penetrating radar (GPR) image: (**a**) original image; (**b**) noisy image.

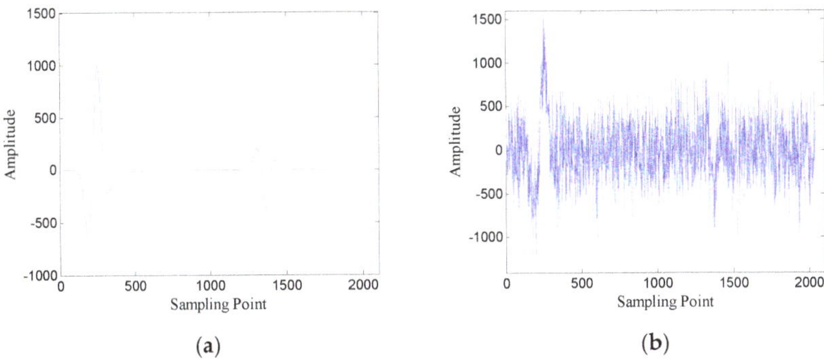

Figure 3. Data of trace 38: (**a**) original data; (**b**) noisy data.

Figure 4 shows the probability distribution of singular values for Hankel matrices with different window lengths. The few larger singular values corresponding to effective signals are distributed in a relatively wide range, and the distribution is sparse. However, the smaller singular values corresponding to the noise are distributed in a narrow range, and the distribution is approximately normal. Moreover, the window length has an obvious effect on the probability distribution of singular values.

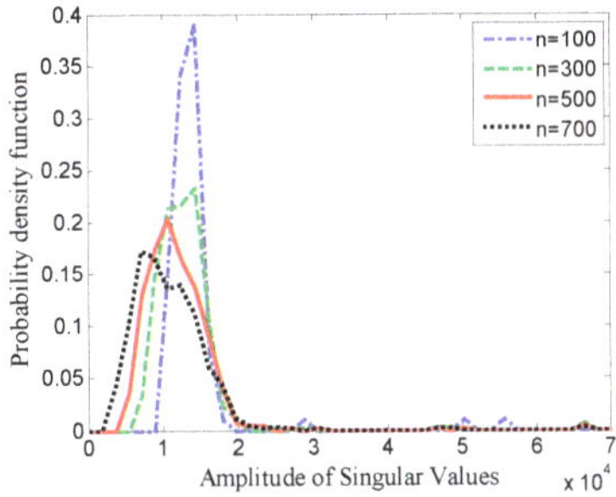

Figure 4. Probability distribution of singular values for Hankel matrices with different window lengths.

For noise suppression, when the distance between the distribution of singular values corresponding to effective signals and the distribution of singular values corresponding to the noise increases, it is easier to distinguish between effective signal components and noise components, which helps to improve noise removal performance. Based on the analysis of distribution characteristics of singular values in Figure 4, the fourth root of the fourth central moment (FRFCM) of singular values is proposed to measure the distance between the two distributions, which is defined by

$$P(n) = \left(\frac{1}{n}\sum_{i=1}^{n}(\sigma_i - \overline{\sigma})^4\right)^{\frac{1}{4}} \tag{7}$$

where n is the number of singular values, σ_i is the ith singular value, and $\overline{\sigma}$ is the mean of singular values.

In order to obtain optimal noise suppression performance, $P(n)$ should be maximized. Therefore, the optimal window length can be given by

$$n_{opt} = \underset{n}{\operatorname{argmax}}[P(n)] \tag{8}$$

2.3. Selection Method of Singular Values

The number of singular values selected results in a trade-off between noise suppression and recovery of the signal of interest. The selection methods based on SNR of recovered data [22] and local energy ratio [23] merely consider the energy of singular values, and their performance degrades when the SNR is relatively low. The selection method based on FCM clustering [24] uses a membership function to find suitable singular values corresponding to effective signals, which is relatively complex.

In order to obtain an efficient and accurate selection of singular values, the synthetic data in Section 2.2 are used to analyze the variation of singular values. Figure 3 shows the variation of singular values for Hankel matrices (the window length is 300) under different SNRs. For simplicity, only the first 80 singular values are shown in Figure 5.

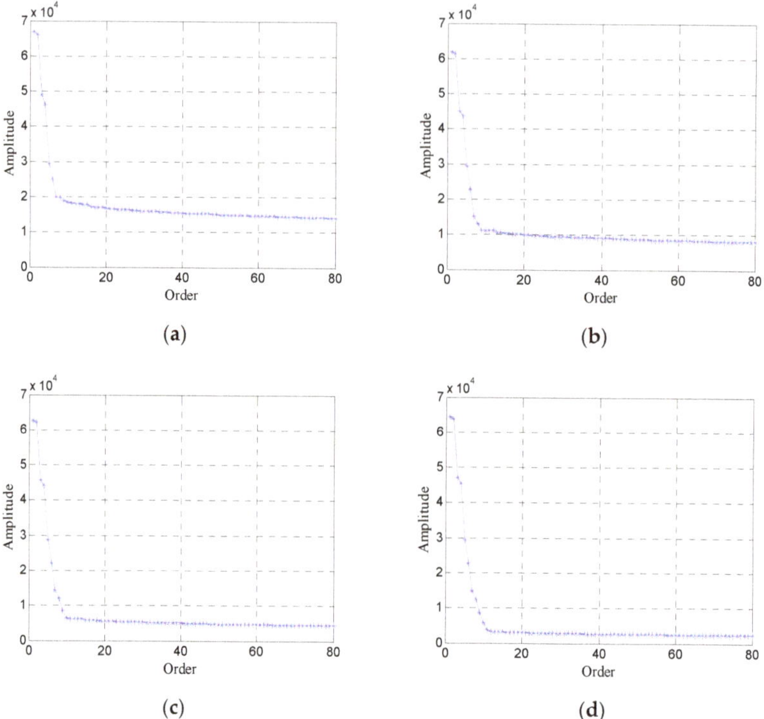

Figure 5. Variation of singular values for Hankel matrices under different signal-to-noise ratios (SNRs): (**a**) SNR = −5 dB; (**b**) SNR = 0 dB; (**c**) SNR = 5 dB; (**d**) SNR = 10 dB.

As shown in Figure 5, the first few singular values correspond to effective signals, and they are larger and decrease quickly with the increase of order; the remaining singular values correspond to the noise, and they are smaller and decrease slowly with the increase of order. For noise, when the SNR increases, the amplitude of singular values decreases obviously and the number of singular values also decreases slightly. For effective signals, when the SNR increases, the amplitude of singular values changes little and the number of singular values increases slightly.

Based on the analysis of variation characteristics of singular values, the difference spectrum of singular values is used to find the demarcation point between singular values corresponding to effective signals and singular values corresponding to the noise. The difference spectrum of singular values [32] can be defined as

$$b_i = \sigma_i - \sigma_{i+1} \quad i = 1, 2, \cdots r - 1 \tag{9}$$

where σ_i is the ith singular value and r is the number of singular values.

The mean of the difference spectrum of singular values is calculated, and a threshold is given by

$$T = \frac{\rho}{r-1}\sum_{i=1}^{r-1} b_i \tag{10}$$

where ρ is a weight coefficient that adjusts the threshold.

Then, the threshold T is used to select singular values corresponding to effective signals. To improve the accuracy of the selection, three adjacent difference spectra are compared with the threshold T to obtain the demarcation point

$$k_1 = i \big| \, b_i < T \; and \; b_{i+1} < T \; and \; b_{i+2} < T \quad i = 1, 2, \cdots, r-3 \tag{11}$$

where the first k_1 singular values correspond to effective signals.

For two-dimensional GPR data $B \in R^{N \times L}$, the noise suppression method based on SVD of a window-length-optimized Hankel matrix can be summarized by the following steps:

1. Select the data of one trace (one-dimensional data) from two-dimensional GPR data and use the one-dimensional data to form a Hankel matrix with a certain window length by Equation (1).

2. Decompose the Hankel matrix by Equation (3) and compute the FRFCM of singular values by Equation (7).

3. Repeat steps 1 and 2 for different window lengths and obtain the optimal window length by Equation (8).

4. For the Hankel matrix with optimal window length, calculate the difference spectrum of singular values and obtain a threshold by Equations (9) and (10).

5. Select the demarcation point between singular values corresponding to effective signals and singular values corresponding to the noise by Equation (11).

6. Reconstruct the denoised Hankel matrix with singular values corresponding to effective signals by Equation (5) and obtain the denoised one-dimensional data by Equation (6).

7. Repeat steps 1–6 for all the traces and implement noise removal for two-dimensional GPR data.

3. Results and Discussion

A series of synthetic and real data is used to evaluate the proposed method. In addition, the performance of the proposed method is also compared with those of the conventional SVD method based on the local energy ratio rule and the wavelet transform method. The synthetic data are also generated by the "gprMax" simulator [31] based on the finite difference time domain (FDTD) method [33]. All the programs are executed on a 3.60 _GHz CPU and 32_GB memory computer.

3.1. Synthetic Example 1

The example shows the scenario of point target detection. Figure 6 shows the geometry of the simulation model. The targets are three perfect conductor metal cylinders with 0.4 m diameter and they are buried at the same depth of 0.6 m. The interval of the three targets is 0.6 m. The transmitting antenna is placed in the air layer and excited by a Ricker wavelet with a center frequency of 900 MHz. There are 80 traces in total and the trace interval is 0.035 m. The time window for each trace is 12 ns and each trace contains 2036 sampling points. Figure 7 shows the original GPR image and the noisy GPR image (SNR= −5.00 dB).

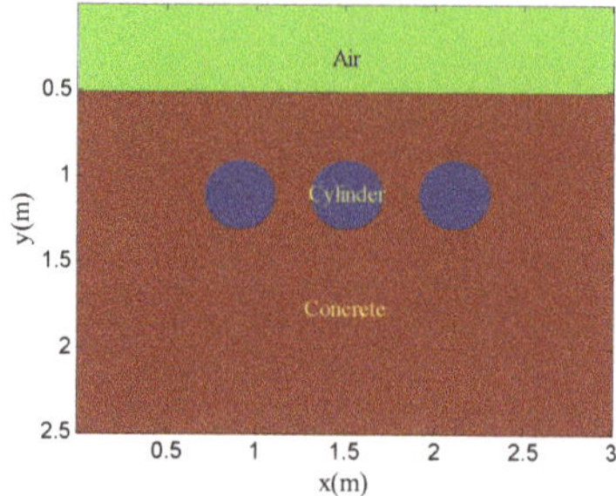

Figure 6. Geometry of the simulation model for point target detection.

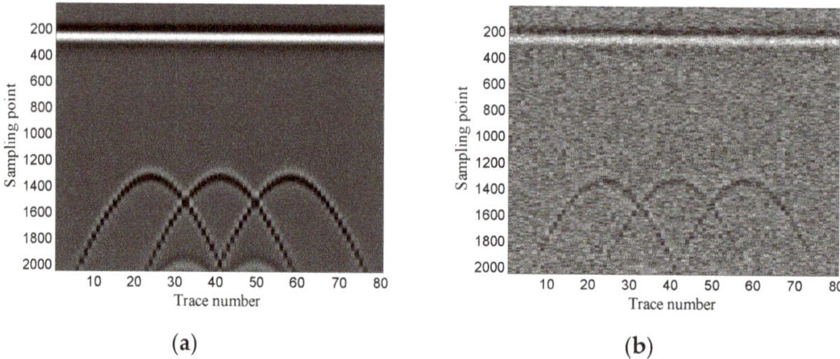

Figure 7. Synthetic GPR image: (**a**) original image; (**b**) noisy image.

First, the performance of the proposed method is analyzed using one-dimensional data. Figure 8 shows the data of trace 30.

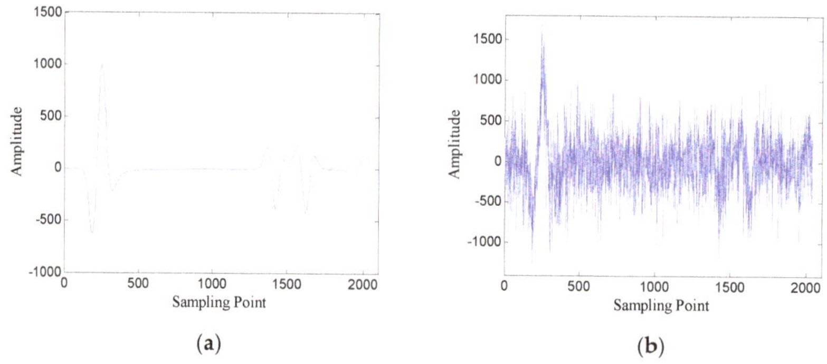

Figure 8. Data of trace 30: (**a**) original data; (**b**) noisy data (SNR = −4.62 dB).

Figure 9 shows the FRFCM of singular values for Hankel matrices with different window lengths. It can be seen that when the window length increases, the FRFCM of singular values first increases and then decreases and reaches the maximum when the window length is 250. Therefore, the optimal window length for the Hankel matrix is 250.

According to the selection method of singular values, the demarcation point k_1 is 6. Then the Hankel matrix is reconstructed with the first 6 singular values, and the denoised data are recovered from the reconstructed Hankel matrix.

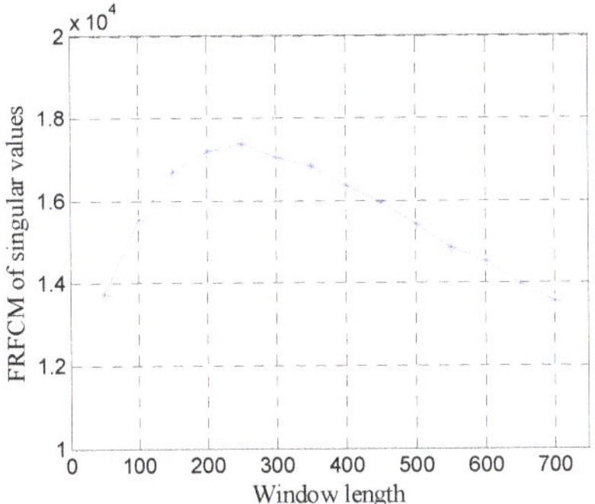

Figure 9. The fourth root of the fourth central moment (FRFCM) of singular values for Hankel matrices with different window lengths.

In order to verify the performance of the window length optimization method, the denoised results with several different window lengths are shown in Figure 10. When the window length is 100, the denoised data contain many burrs; when the window length is 250, the denoised data are relatively smooth; when the window length is 400 and 700, the denoised data also contain some noise. The results preliminarily verify the effectiveness of the window length optimization method.

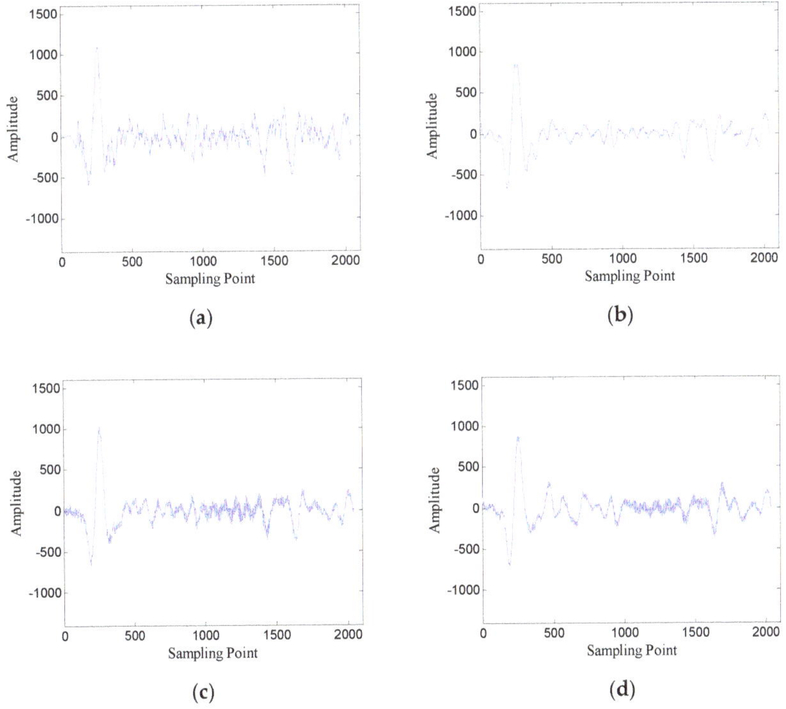

Figure 10. Denoised results with different window lengths for the data of trace 30: (**a**) $n = 100$ (SNR = 5.43 dB); (**b**) $n = 250$ (SNR = 7.45 dB); (**c**) $n = 400$ (SNR = 6.66 dB); (**d**) $n = 700$ (SNR = 5.19 dB).

In order to quantitatively analyze the performance of the window length optimization method, the SNR of denoised data with different window lengths is shown in Figure 11. The SNR exhibits a fluctuation similar to the FRFCM of singular values, and reaches the maximum 7.45 dB at the optimal window length (n=250), which shows that the window length optimization method can obtain the best noise removal performance for SVD of the Hankel matrix.

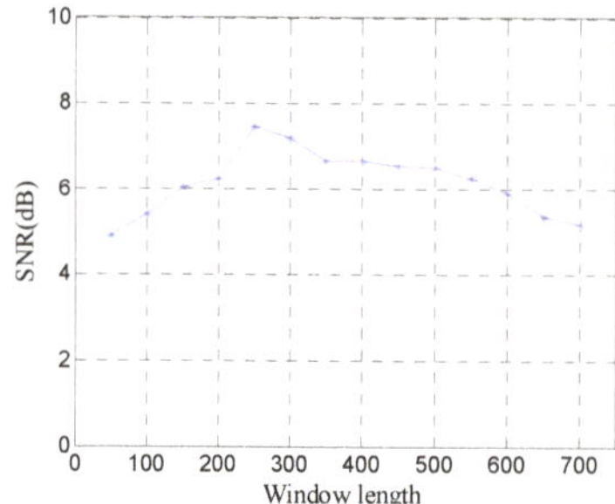

Figure 11. SNR of denoised data with different window lengths.

Then, the performance of the proposed method is verified using two-dimensional data. In addition, the experimental results of the proposed method are compared with those of the conventional SVD method based on the local energy ratio rule and the wavelet transform method. Figure 12 shows the denoised results of the three methods. As shown in Figure 12a the conventional SVD method can remove noise, but it also removes some of the target signals. As shown in Figure 12b, the wavelet transform method retains complete target signals, but it also retains a small amount of noise. As shown in Figure 12c the proposed method can retain complete target signals while removing more noise.

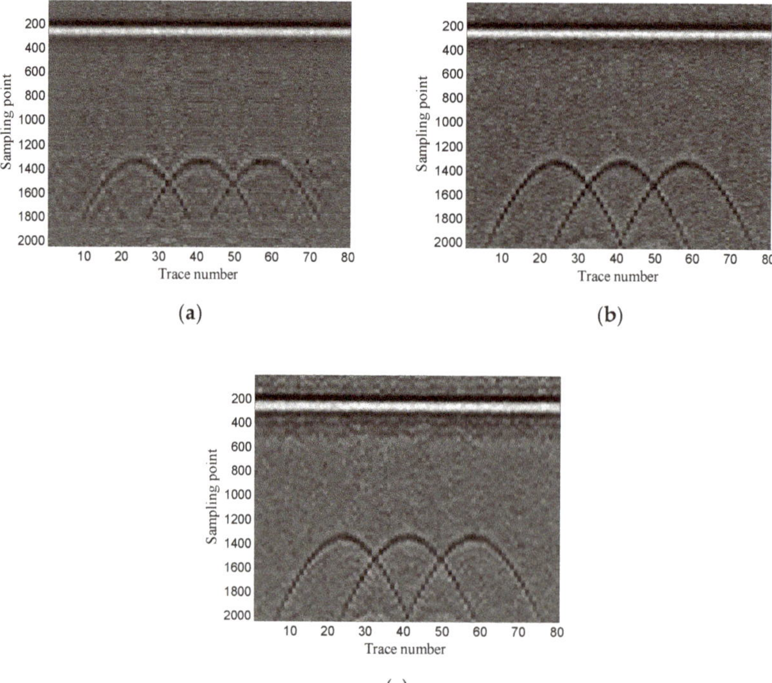

Figure 12. Denoised results of the three methods for a GPR image: (**a**) singular value decomposition (SVD) method based on the local energy ratio rule; (**b**) wavelet transform method; (**c**) proposed method.

Table 1 lists the SNR, processing time, and the amount of RAM memory required for the three methods. As shown in Table 1, the proposed method yields a higher SNR than the other two methods, and it also needs more processing time and larger RAM memory than the other two methods due to the calculation of SVD of the Hankel matrix for each one-dimensional data.

Table 1. Results of the three methods.

Method	SNR (dB)	Processing Time (s)	Amount of RAM Memory (MB)
SVD method based on local energy ratio rule	4.23	1.9	71
Wavelet transform method	7.08	2.31	39
Proposed method	7.55	4.17	99

3.2. Synthetic Example 2

The example shows the scenario of layer detection. Figure 13 shows the geometry of the simulation model. The model contains two layers: clay and sand. The transmitting antenna is placed in the air layer and excited by a Ricker wavelet with a center frequency of 900 MHz. There are 41 traces in total and the trace interval is 0.02 m. The time window for each trace is 10 ns and each trace contains 1696 sampling points. Figure 14 shows the original GPR image and the noisy GPR image (SNR= −5.00 dB).

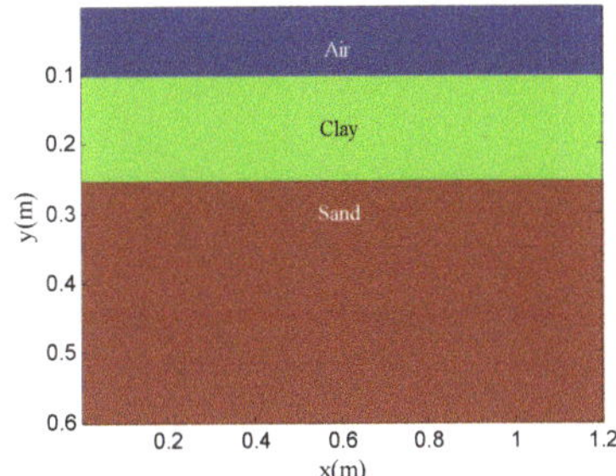

Figure 13. Geometry of the simulation model for layer detection.

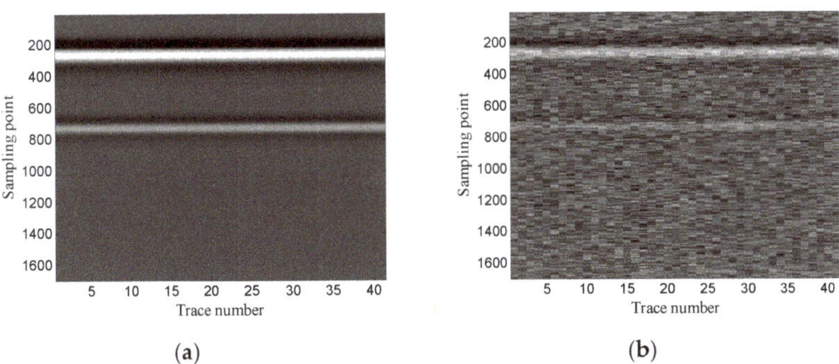

Figure 14. Synthetic GPR image: (**a**) original image; (**b**) noisy image.

First, the one-dimensional data are used to verify the performance of the proposed method. Figure 15 shows the data of trace 20. Figure 16 shows the FRFCM of singular values for Hankel matrices with different window lengths. Evidently, FRFCM reaches the maximum when the window length is 300. Therefore, the optimal window length for the Hankel matrix is 300. The demarcation point k_1 is set to 6 by the selection method of singular values. Then the Hankel matrix is reconstructed with the first 6 singular values, and the denoised data are recovered from the reconstructed Hankel matrix.

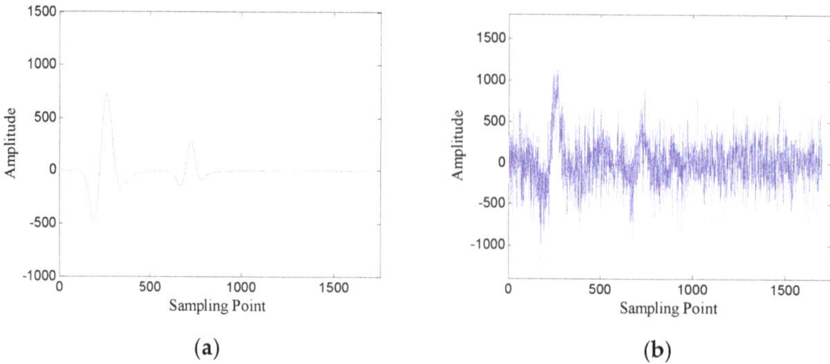

Figure 15. Data of trace 20: (**a**) original data; (**b**) noisy data (SNR = −5.12 dB).

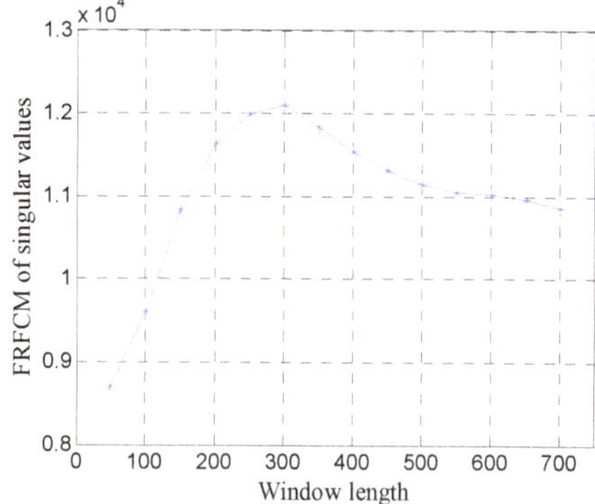

Figure 16. FRFCM of singular values for Hankel matrices with different window lengths.

The denoised results with several different window lengths are shown in Figure 17. As the figure shows, the optimal window length can obtain the best compromise between noise suppression and retaining effective signals.

In order to quantitatively analyze the performance of the window length optimization method, the SNR of denoised data with different window lengths is shown in Figure 18. The results further show the window length optimization method can achieve the best noise removal performance for SVD of the Hankel matrix.

Then, the two-dimensional data are used to verify the performance of the proposed method. The experimental results of the proposed method are also compared with those of the conventional SVD method based on the local energy ratio rule and wavelet transform method. Figure 19 shows the denoised results of the three methods. As shown in Figure 19a, the layer signals are relatively weak, and some horizontal noise is also introduced. Figure 19b shows that the layer signals are obvious, but a small amount of noise is also retained; and Figure 19c shows that the layer signals are relatively strong, and the noise is also removed more thoroughly.

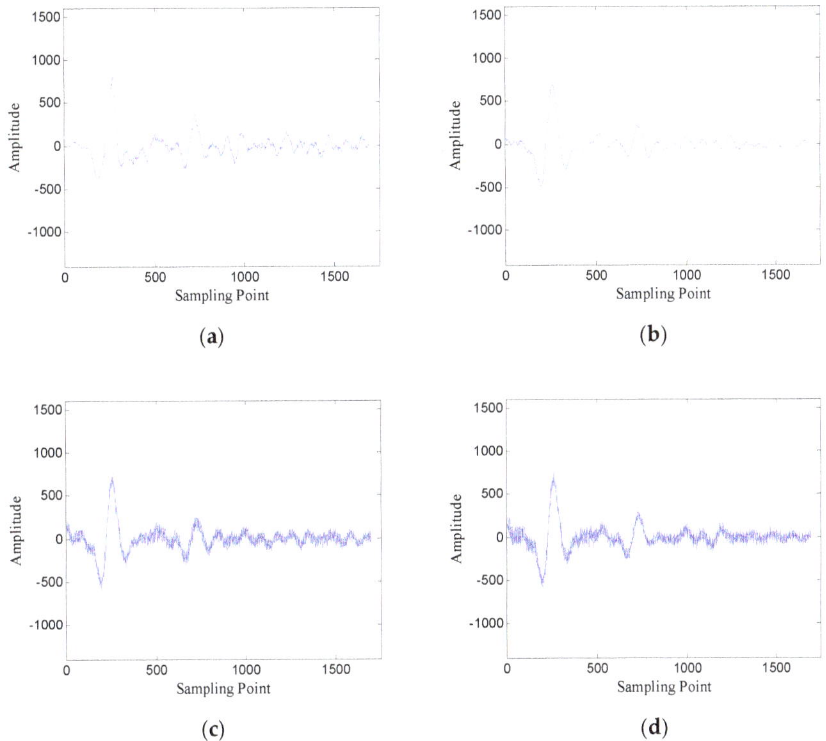

Figure 17. Denoised results with different window lengths for the data of trace 20: (**a**) n = 150 (SNR = 5.91 dB); (**b**) n = 300 (SNR = 7.86 dB); (**c**) n = 450 (SNR = 5.82 dB); (**d**) n = 600 (SNR = 5.40 dB).

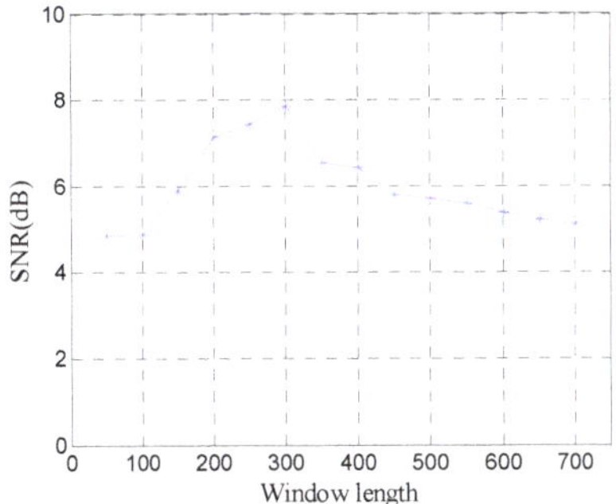

Figure 18. SNR of denoised data with different window lengths.

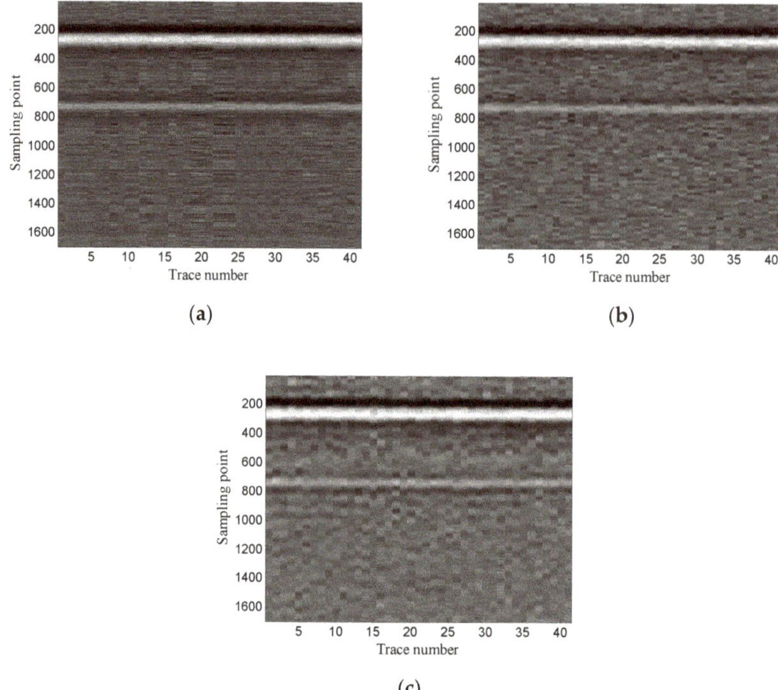

Figure 19. Denoised results of the three methods for a GPR image: (**a**) SVD method based on the local energy ratio rule; (**b**) wavelet transform method; (**c**) proposed method.

Table 2 lists the SNR, processing time, and the amount of RAM memory required for the three methods. Table 2 also shows that the proposed method yields a higher SNR and consumes more memory space compared with the other two methods.

Table 2. Results of the three methods.

Method	SNR (dB)	Processing Time (s)	Amount of RAM Memory (MB)
SVD method based on local energy ratio rule	5.6	0.78	48
Wavelet transform method	7.03	0.92	17
Proposed method	7.42	1.43	53

3.3. Synthetic Example 3

In this section, the performance of the proposed method is investigated in the presence of correlated noise. This example uses the same original GPR image as synthetic example 1. The autocorrelation function of the noise is an exponential function and the correlation length of the noise is 10. Figure 20 shows the original GPR image and the noisy GPR image (SNR = −5.00 dB).

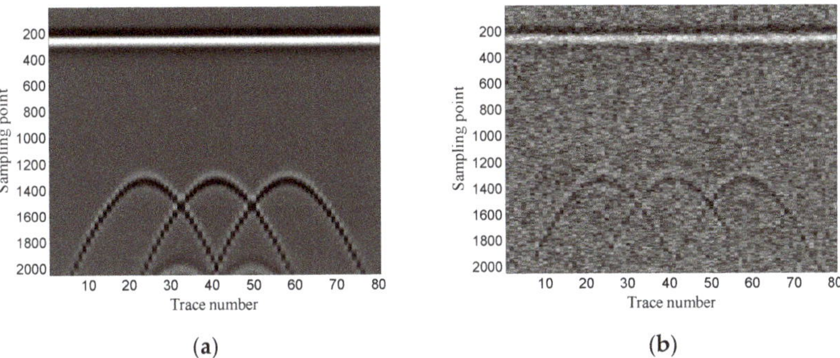

Figure 20. Synthetic GPR image: (**a**) original image; (**b**) noisy image.

First, the performance of the proposed method is analyzed using one-dimensional data. Figure 21 shows the data of trace 30. Figure 22 shows the FRFCM of singular values for Hankel matrices with different window lengths. In this case, it is evident that the value of FRFCM is greater than that in the case of white noise and the optimal window length for the Hankel matrix is 300. The Hankel matrix is reconstructed with the first eight singular values, and the denoised data are recovered from the reconstructed Hankel matrix.

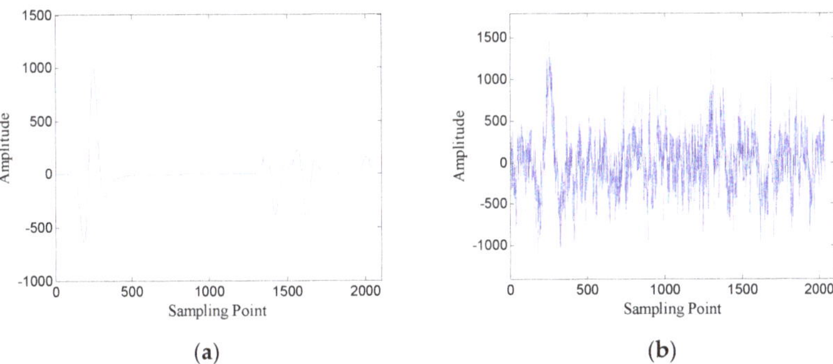

Figure 21. Data of trace 30: (**a**) original data; (**b**) noisy data (SNR = −4.55 dB).

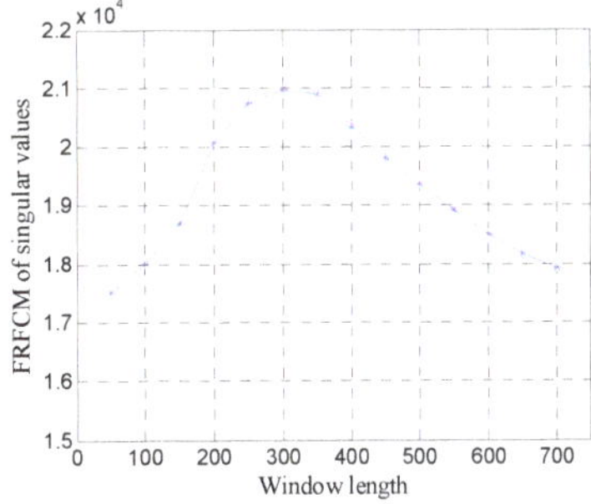

Figure 22. FRFCM of singular values for Hankel matrices with different window lengths.

Figure 23 shows the denoised results with several different window lengths. When the window length is 100, the denoised data contain some oscillating components; when the window length is 500 and 650, the denoised data also contain a lot of interference with large amplitude; when the window length is 300, the denoised data contain the least noise. The results confirm that the window length optimization method is also effective in the case of correlated noise.

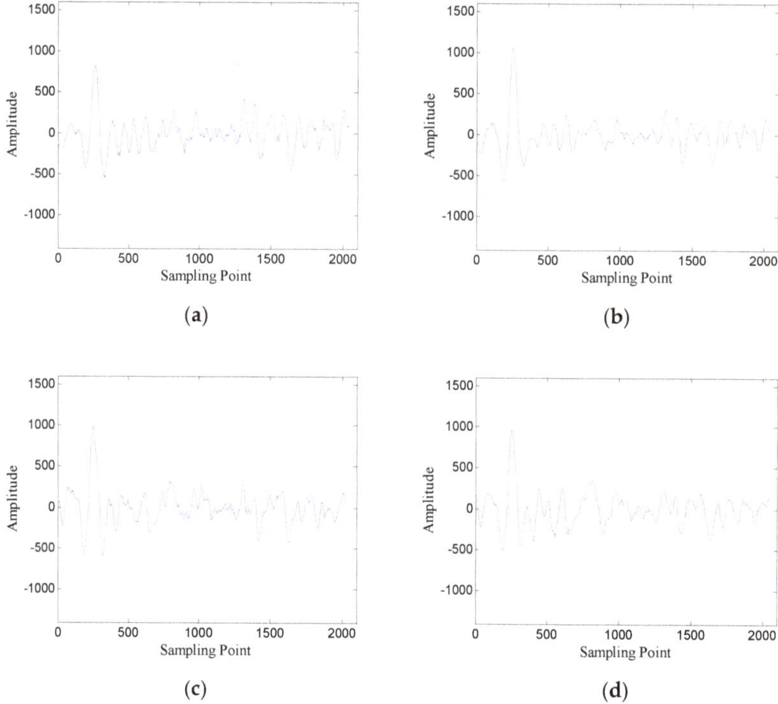

Figure 23. Denoised results with different window lengths for the data of trace 30: (**a**) $n = 100$ (SNR = 1.72 dB); (**b**) $n = 300$ (SNR = 4.31 dB); (**c**) $n = 500$ (SNR = 2.96 dB); (**d**) $n = 650$ (SNR = 2.39 dB).

Figure 24 shows the SNR of denoised data with different window lengths. The results show that SVD of the Hankel matrix obtains the best noise removal performance at the optimal window length ($n = 300$).

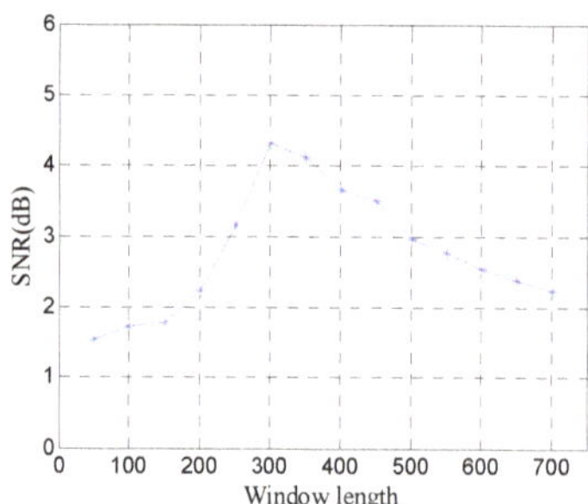

Figure 24. SNR of denoised data with different window lengths.

Then, the performance of the proposed method is verified using two-dimensional data. Moreover, the experimental results of the proposed method are compared with those of the conventional SVD method based on local energy ratio rule and the wavelet transform method. Figure 25 shows the denoised results of the three methods. As shown in Figure 25a the conventional SVD method removes some target signals while denoising. Figure 25b shows that the wavelet transform method also retains some noise while retaining target signals; and Figure 25c shows that the proposed method retains more target signals while removing more noise.

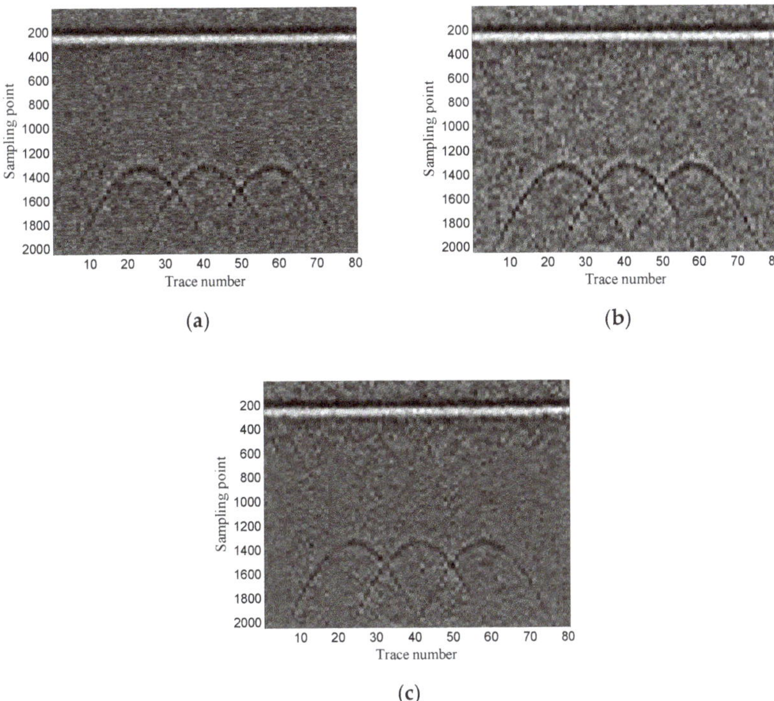

Figure 25. Denoised results of the three methods for a GPR image: (**a**) SVD method based on the local energy ratio rule; (**b**) wavelet transform method; (**c**) proposed method.

Table 3 lists the SNR, processing time, and the amount of RAM memory required for the three methods. Compared with the results of synthetic example 1, the SNR of the three methods all decreases due to the correlation of the noise. The proposed method yields an obviously higher SNR with an appropriate increase in processing time compared with the other two methods. In addition, the wavelet transform method requires larger RAM memory due to the correlation of the noise.

Table 3. Results of the three methods.

Method	SNR (dB)	Processing Time (s)	Amount of RAM Memory (MB)
SVD method based on local energy ratio rule	0.94	2.16	75
Wavelet transform method	2.1	2.41	51
Proposed method	4.21	4.19	101

3.4. Synthetic Example 4

In this section, the performance of the proposed method is also investigated in the presence of correlated noise. This example also uses the same original GPR image as synthetic example 1. The autocorrelation function of the noise is an exponential function and the correlation length of the noise is 20. Figure 26 shows the original GPR image and the noisy GPR image (SNR = −5.00 dB).

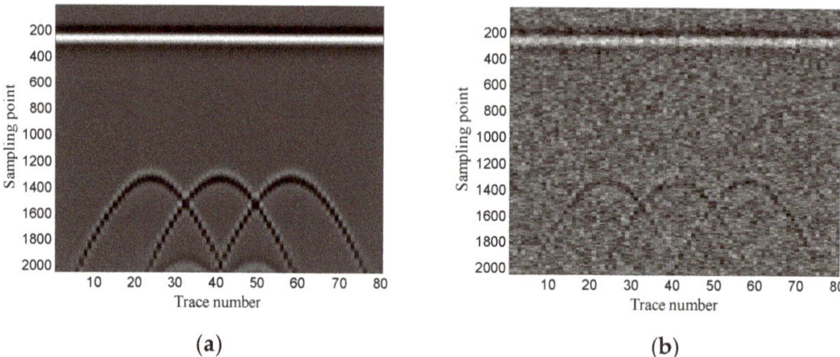

Figure 26. Synthetic GPR image: (**a**) original image; (**b**) noisy image.

First, the performance of the proposed method is analyzed using one-dimensional data. Figure 27 shows the data of trace 30. Figure 28 shows the FRFCM of singular values for Hankel matrices with different window lengths. Evidently, the correlation length of the noise increases, the value of FRFCM also increases, and the optimal window length for the Hankel matrix is 300. The Hankel matrix is reconstructed with the first nine singular values and the denoised data are recovered from the reconstructed Hankel matrix.

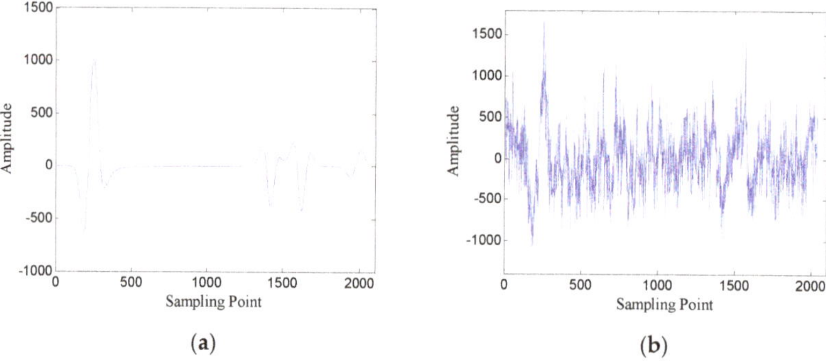

Figure 27. Data of trace 30: (**a**) original data; (**b**) noisy data (SNR = −4.49 dB).

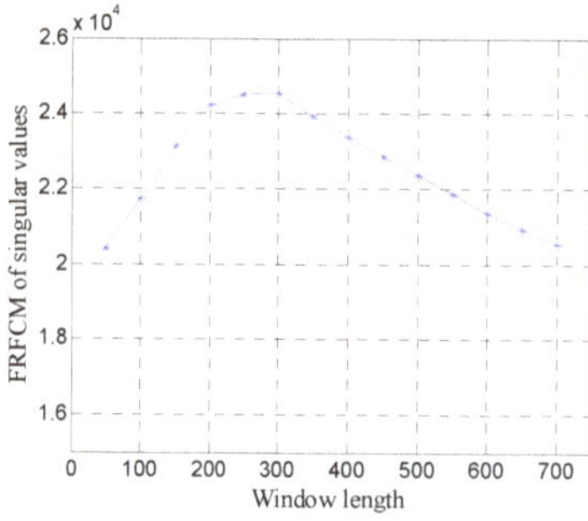

Figure 28. FRFCM of singular values for Hankel matrices with different window lengths.

Figure 29 shows the denoised results with several different window lengths. As the figure shows, denoised data of the optimal window length contain less noise than those of other window lengths.

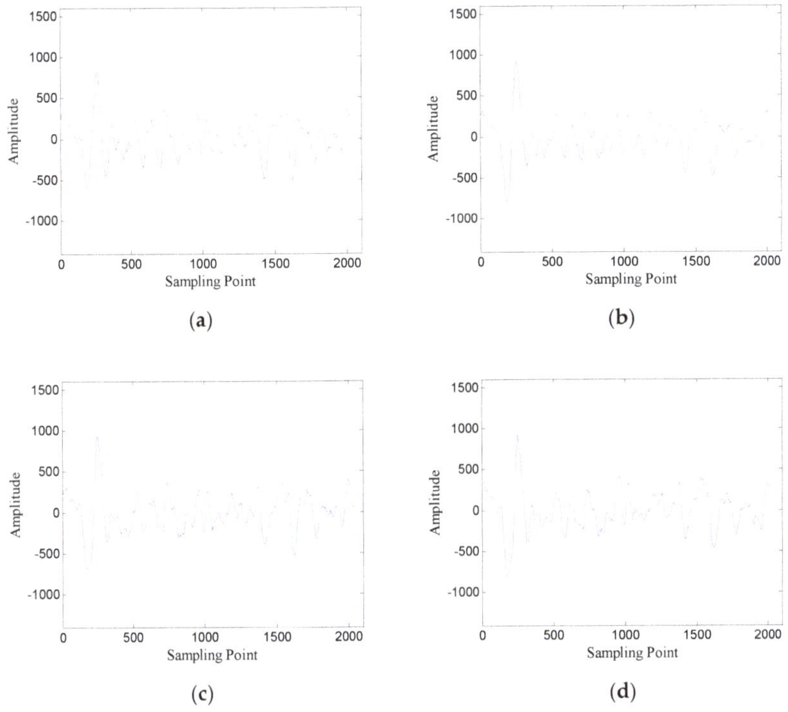

Figure 29. Denoised results with different window lengths for the data of trace 30: (**a**) n = 100 (SNR = 1.61 dB); (**b**) n = 300 (SNR = 3.02 dB); (**c**) n = 500 (SNR = 2.16 dB); (**d**) n = 650 (SNR = 1.66 dB).

Figure 30 shows the SNR of denoised data with different window lengths. The SNR exhibits more fluctuation, and it reaches maximum at the optimal window length (n = 300).

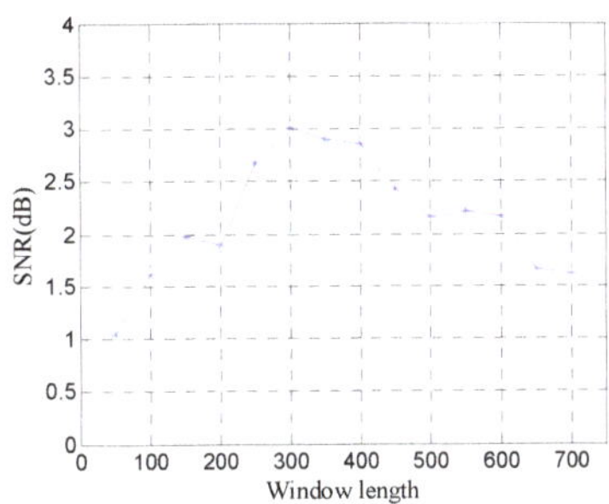

Figure 30. SNR of denoised data with different window lengths.

Then, the performance of the proposed method is verified using two-dimensional data. The experimental results of the proposed method are also compared with those of the conventional SVD method based on the local energy ratio rule and the wavelet transform method. Figure 31 shows the denoised results of the three methods. As shown in Figure 31, the conventional SVD method loses a lot of target signals; the wavelet transform method also retains a lot of noise while retaining target signals; the proposed method achieves a good compromise between retaining target signals and removing the noise.

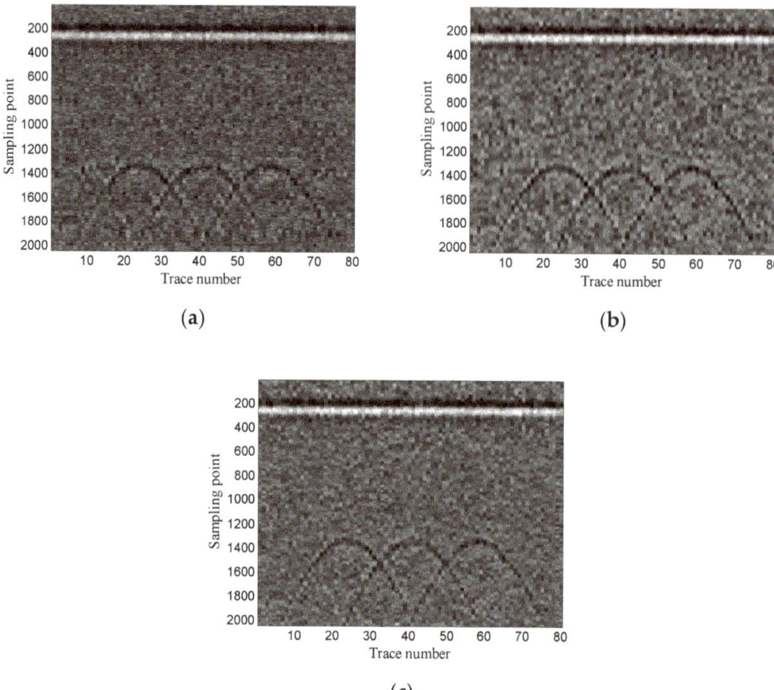

Figure 31. Denoised results of the three methods for a GPR image: (**a**) SVD method based on the local energy ratio rule; (**b**) wavelet transform method; (**c**) proposed method.

Table 4 lists the SNR, processing time, and the amount of RAM memory required for the three methods. Compared with the results of synthetic example 3, the increase of the correlation length of the noise obviously degrades the SNR of the three methods. The proposed method also achieves a higher SNR at the cost of the increasing processing time and more memory space compared with the other two methods.

Table 4. Results of the three methods.

Method	SNR (dB)	Processing Time (s)	Amount of RAM Memory (MB)
SVD method based on local energy ratio rule	0.02	1.94	74
Wavelet transform method	0.59	2.64	53
Proposed method	2.05	4.15	102

3.5. Field Measurements 1

The example shows the scenario of pipeline detection. The antenna center frequency is 400 MHz. There are 251 traces in total and each trace contains 301 sampling points. Figure 32 shows the original noisy GPR image. As the figure shows, there is a lot of noise around the target hyperbolic signals, which affects the target detection.

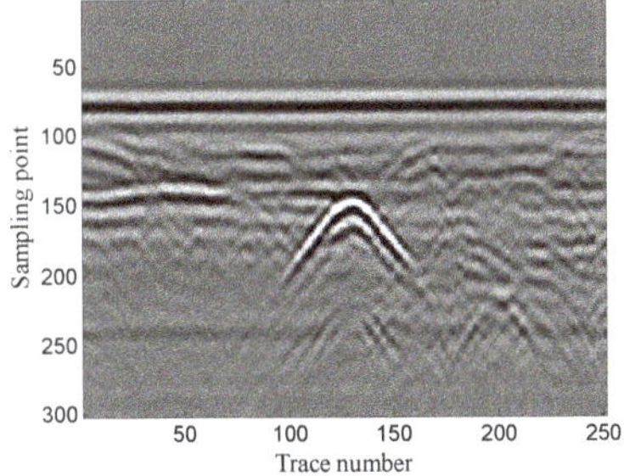

Figure 32. Real GPR image of pipeline detection.

The optimal window length for the Hankel matrix is 90. The denoised results of the three methods are shown in Figure 33. As shown in Figure 33a the conventional SVD method removes some of the noise, but it generates some false target hyperbolic signals. Figure 33b shows that the wavelet transform method removes most of the noise and retains complete target signals, but it introduces a small amount of vertical noise; Figure 33c shows that the proposed method removes most of the noise and preserves complete target signals without introducing any other signals. The results show that the proposed method achieves better noise removal performance than the other two methods and helps to detect the pipeline accurately.

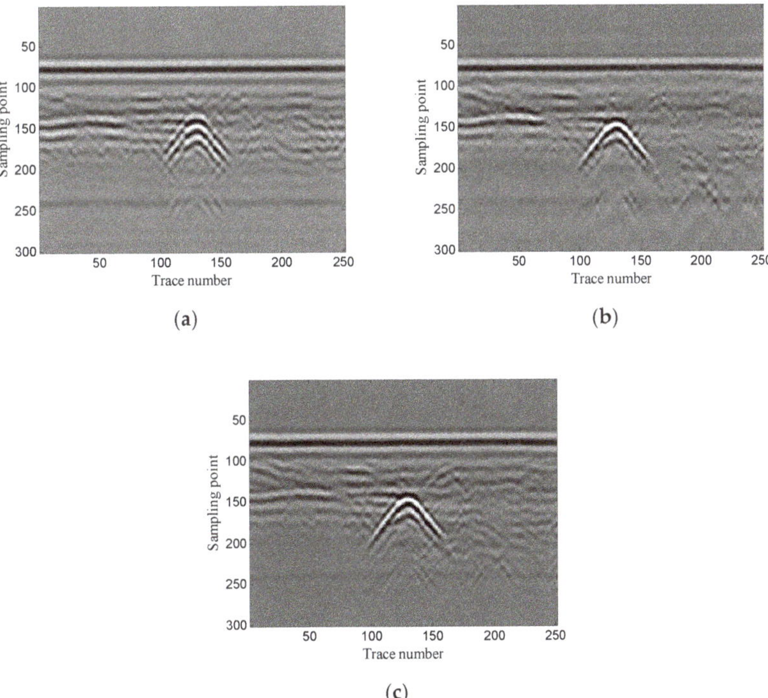

Figure 33. Denoised results of the three methods for a real GPR image: (**a**) SVD method based on the local energy ratio rule; (**b**) wavelet transform method; (**c**) proposed method.

The processing time of the conventional SVD method, the wavelet transform method, and the proposed method is 0.45 s, 1.32 s, and 1.77 s, respectively.

3.6. Field Measurements 2

The example shows the scenario of road layer detection. The antenna center frequency is 400 MHz. There are 46 traces in total and each trace contains 450 sampling points. Figure 34 shows the original noisy GPR image. As the figure shows, there is some horizontal noise around the layer signals, which interferences with the layer recognition.

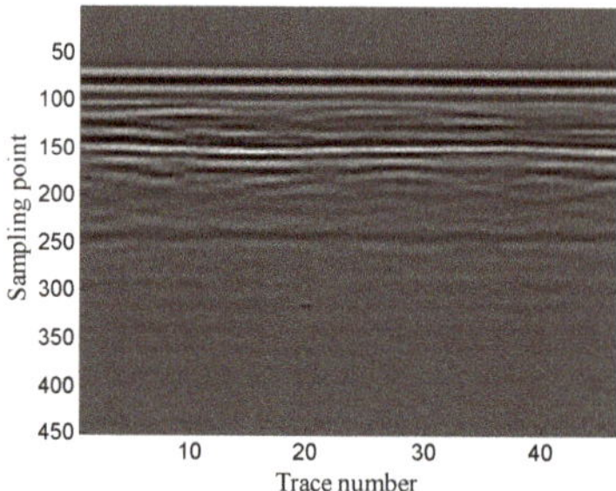

Figure 34. Real GPR image of road layer detection.

The optimal window length for the Hankel matrix is 110. The denoised results of the three methods are shown in Figure 35. As shown in Figure 35a, the conventional SVD method removes some of the noise, but it still retains some noise between the 80th and the 150th sampling points. Figure 35b shows that the wavelet transform method retains a small amount of noise between the 80th and the 150th sampling points, but it removes part of the layer signals near the 240th sampling point; Figure 35c shows that the proposed method removes most of the noise, and it retains the layer signals completely. The results show that the proposed method obtains the best noise removal performance and provides the best profile for layer detection.

The processing time of the conventional SVD method, the wavelet transform method, and the proposed method is 0.14 s, 0.47 s, and 0.67 s, respectively.

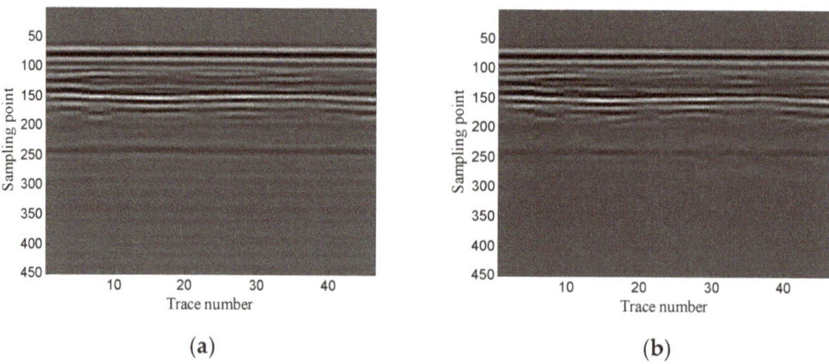

Figure 35. *Cont.*

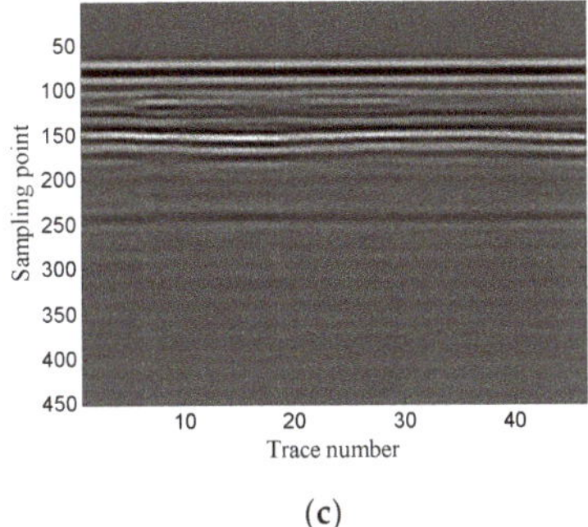

(c)

Figure 35. Denoised results of the three methods for a real GPR image: (**a**) SVD method based on the local energy ratio rule; (**b**) wavelet transform method; (**c**) proposed method.

4. Conclusions

In this paper, a method based on SVD of a window-length-optimized Hankel matrix is proposed to improve the noise suppression performance for GPR data. The fourth root of the fourth central moment of singular values is used to determine the window length of the Hankel matrix, which provides a solution to optimize the size of the Hankel matrix. Then, the difference spectrum of singular values is used to construct a threshold, which provides a solution to select singular values corresponding to effective signals.

The proposed method is verified by series of synthetic and practical data. The results show the proposed method can obtain the best noise removal performance for both white noise and correlated noise. The proposed method also achieves better denoising performance than the conventional SVD method based on the local energy ratio rule and wavelet transform method at the cost of the appropriate increases in processing time and memory space. Future work will investigate more efficient solutions to optimize SVD of the Hankel matrix to further improve noise removal performance.

Author Contributions: Investigation, Y.H.; methodology, W.X.; software, Y.L.; validation, Y.Y.; writing—original draft, W.X.; writing—review and editing, W.X.

Funding: This research was funded by the National Key Research and Development Program of China, grant number 2018YFC1503702; the Fundamental Research Funds for the Central Universities of China, grant number CUG2018JM15; and the State Scholarship Fund of China Scholarship Council, grant number 201806415018.

Conflicts of Interest: The authors declare no conflict of interest.

References

1. Jol, H. *Ground Penetrating Radar: Theory and Applications*; Elsevier Science: Amsterdam, The Netherlands, 2009; pp. 595–604.
2. Dong, Z.; Ye, S.; Gao, Y.; Fang, G.; Zhang, X.; Xue, Z.; Zhang, T. Rapid detection methods for asphalt pavement thicknesses and defects by a vehicle-mounted ground penetrating radar (GPR) system. *Sensors* **2016**, *16*, 2067. [CrossRef] [PubMed]
3. Zhao, W.; Forte, E.; Pipan, M.; Tian, G. Ground penetrating radar (GPR) attribute analysis for archaeological prospection. *J. Appl. Geophys.* **2013**, *97*, 107–117.
4. Chen, C.S.; Jeng, Y. GPR investigation of the near-surface geology in a geothermal river valley using contemporary data decomposition techniques with forward simulation modeling. *Geothermics* **2016**, *64*, 439–454. [CrossRef]
5. Manandhar, A.; Torrione, P.A.; Collins, L.M.; Morton, K.D. Multiple-instance hidden Markov model for GPR-based landmine detection. *IEEE Trans. Geosci. Remote Sens.* **2015**, *53*, 1737–1745. [CrossRef]

6. Giannakis, I.; Giannopoulos, A.; Warren, C. A Realistic FDTD Numerical Modeling Framework of Ground Penetrating Radar for Landmine Detection. *IEEE J. Sel. Top. Appl. Earth Obs. Remote Sens.* **2016**, *9*, 37–51. [CrossRef]
7. Gurbuz, A. Determination of Background Distribution for Ground-Penetrating Radar Data. *IEEE Geosci. Remote Sens. Lett.* **2012**, *9*, 544–548.
8. Vitebskiy, S.; Carin, L.; Ressler, M.A.; Le, F.H. Ultra-wideband, short-pulse ground-penetrating radar: Simulation and measurement. *IEEE Trans. Geosci. Remote Sens.* **1997**, *35*, 762–772. [CrossRef]
9. Baili, J.; Lahouar, S.; Hergli, M.; Al-Qadi, I.L.; Besbes, K. GPR signal de-noising by discrete wavelet transform. *Ndt E Int.* **2009**, *42*, 696–703. [CrossRef]
10. Javadi, M.; Ghasemzadeh, H. Wavelet analysis for ground penetrating radar applications: A case study. *J. Geophys. Eng.* **2017**, *14*, 1189–1202. [CrossRef]
11. Kim, J.H.; Cho, S.J.; Yi, M.J. Removal of ringing noise in GPR data by signal processing. *Geosci. J.* **2007**, *11*, 75–81. [CrossRef]
12. Wei, X.; Zhang, Y. Interference removal for autofocusing of GPR data from RC bridge decks. *IEEE J. Sel. Top. Appl. Earth Obs. Remote Sens.* **2015**, *8*, 1145–1151. [CrossRef]
13. Chen, C.S.; Jeng, Y. Nonlinear data processing method for the signal enhancement of GPR data. *J. Appl. Geophys.* **2011**, *75*, 113–123. [CrossRef]
14. Li, J.; Liu, C.; Zeng, Z.F.; Chen, L.N. GPR signal denoising and target extraction with the CEEMD method. *IEEE Geosci. Remote Sens. Lett.* **2015**, *12*, 1615–1619.
15. Song, X.; Xiang, D.; Zhou, K.; Su, Y. Fast Prescreening for GPR Antipersonnel Mine Detection via Go Decomposition. *IEEE Geosci. Remote Sens. Lett.* **2019**, *16*, 15–19. [CrossRef]
16. Tivive, F.H.C.; Bouzerdoum, A.; Abeynayake, C. GPR Target Detection by Joint Sparse and Low-Rank Matrix Decomposition. *IEEE Trans. Geosci. Remote Sens.* **2019**, *5*, 2583–2595. [CrossRef]
17. Song, X.; Xiang, D.; Zhou, K.; Su, Y. Improving RPCA-based clutter suppression in GPR detection of antipersonnel mines. *IEEE Geosci. Remote Sens. Lett.* **2017**, *8*, 1338–1342. [CrossRef]
18. Abujarad, F.; Nadim, G.; Omar, A. Clutter reduction and detection of landmine objects in ground penetrating radar data using singular value decomposition (SVD). In Proceedings of the 3rd International Workshop on Advanced Ground Penetrating Radar, Delft, The Netherlands, 2–3 May 2005; pp. 37–42.
19. Garcia-Fernandez, M.; Alvarez-Lopez, Y.; Arboleya-Arboleya, A.; Las-Heras, F.; Rodriguez-Vaqueiro, Y.; Gonzalez-Valdes, B.; Pino-Garcia, A. SVD-based clutter removal technique for GPR. In Proceedings of the 2017 IEEE International Symposium on Antennas and Propagation, San Diego, SA, USA, 9–14 July 2017; pp. 2369–2370.
20. Giannakis, I.; Xu, S.; Aubry, P.; Yarovoy, A.; Sala, J. Signal processing for landmine detection using ground penetrating radar. In Proceedings of the 2016 IEEE International Geoscience and Remote Sensing Symposium (IGARSS), Beijing, China, 10–15 July 2016; pp. 7442–7445.
21. Cagnoli, B.; Ulrych, T.J. Singular value decomposition and wavy reflections in ground-penetrating radar images of base surge deposits. *J. Appl. Geophys.* **2001**, *48*, 175–182. [CrossRef]
22. Liu, C.; Song, C.; Lu, Q. Random noise de-noising and direct wave eliminating based on SVD method for ground penetrating radar signals. *J. Appl. Geophys.* **2017**, *144*, 125–133. [CrossRef]
23. Shen, J.Q.; Yan, H.Z.; Hu, C.Z. Auto-selected rule on principal component analysis in ground penetrating radar signal denoising. *Chin. J. Radio* **2010**, *25*, 83–87.
24. Riaz, M.M.; Ghafoor, A. Ground penetrating radar image enhancement using singular value decomposition. In Proceedings of the IEEE International Symposium on Circuits and Systems, Beijing, China, 19–23 May 2013; pp. 2388–2391.
25. Zhao, X.; Ye, B. Similarity of signal processing effect between Hankel matrix-based SVD and wavelet transform and its mechanism analysis. *Mech. Syst. Signal. Process.* **2009**, *23*, 1062–1075. [CrossRef]
26. Lee, K.C.; Ou, J.S.; Fang, M.C. Application of SVD noise-reduction technique to PCA based radar target recognition. *Prog. Electromagn. Res.* **2008**, *81*, 447–459.
27. Qiao, Z.; Pan, Z. SVD principle analysis and fault diagnosis for bearings based on the correlation coefficient. *Meas. Sci. Technol.* **2015**, *26*, 085014. [CrossRef]
28. Golafshan, R.; Sanliturk, K.Y. SVD and Hankel matrix based de-noising approach for ball bearing fault detection and its assessment using artificial faults. *Mech. Syst. Signal. Process.* **2016**, *70*, 36–50. [CrossRef]

29. Bi, W.; Zhao, Y.; An, C.; Hu, S. Clutter elimination and random-noise denoising of GPR signals using an SVD method based on the Hankel matrix in the local frequency domain. *Sensors* **2018**, *18*, 3422. [CrossRef] [PubMed]
30. Kilundu, B.; Chiementin, X.; Dehombreux, P. Singular spectrum analysis for bearing defect detection. *J. Vib. Acoust.* **2011**, *133*, 051007. [CrossRef]
31. Warren, C.; Giannopoulos, A.; Giannakis, I. gprMax: Open source software to simulate electromagnetic wave propagation for ground penetrating radar. *Comp. Phys. Commun.* **2016**, *209*, 163–170. [CrossRef]
32. Wang, J.G.; Li, J.; Liu, Y.Y. An improved method for determining effective order rank of SVD denoising. *J. Vib. Shock* **2014**, *33*, 176–180.
33. Maloney, J.G.; Smith, G.S. A Study of Transient Radiation from the Wu-King Resistive Monopole—Fdtd Analysis and Experimental Measurements. *IEEE Trans. Antennas Propag.* **1993**, *41*, 668–676. [CrossRef]

Communication

New Concept of Combined Microwave Delay Lines for Noise Radar-Based Remote Sensors

Zenon Szczepaniak and Waldemar Susek *

Faculty of Electronics, Military University of Technology, gen. Sylwestra Kaliskiego St. No. 2, 00-908 Warsaw, Poland; zenon.szczepaniak@wat.edu.pl

* Correspondence: waldemar.susek@wat.edu.pl; Tel.: +48-261-839-831

Received: 31 August 2019; Accepted: 4 November 2019; Published: 6 November 2019

Abstract: Delay lines with a tunable length are used in a number of applications in the field of microwave techniques. The digitally-controlled analogue wideband delay line is particularly useful in noise radar applications as a precise detector of movement. In order to perform coherent reception in the noise radar, a delay line with a variable delay value is required. To address this issue, this paper comprises a new concept of a digitally-controlled delay line with a set of fine distance gates. In the paper, a solution for micro-movement detection is proposed, which is based on direct signal processing in the time domain with the use of a microwave analogue correlator. This concept assumes the use of a microwave analogue tapped delay line structure. It was found that the optimal solution for a noise radar with an analogue signal correlator is a combined delay line consisting of switched reference sections, a tapped delay line, and a precision phase shifter. The combined delay line presented in this paper is dedicated to serving as the adjustable reference delay for a noise radar intended for the detection of micro-movement. The paper contains the calculation results and delay line implementation for a given example. The new structure of the analogue tapped delay line with the calculation of optimal parameters is also presented. The precise detector of movement can be successfully used for the remote sensing of human vital signs (especially through-the-wall), e.g., breathing and heart beating, with the simultaneous determination of position.

Keywords: noise radar; radar signal processing techniques; analogue correlation; modern radar applications; delay line

1. Introduction

Delay lines with a tunable length are used in a number of applications in the field of microwave techniques [1–4]. The possible applications depend on the length of the line and generally they may be divided as follows: phase shifters, phase correctors, impedance tuning stubs, or time delay references.

The simplest form of a fixed microwave delay line is a section of a transmission line with a specified length, preferably without the effect of group velocity dispersion. In this case, the group delay T_g may be expressed by

$$T_g = \frac{L}{v_g}, \quad (1)$$

where L is the length of the transmission line and v_g is the group velocity, defined as

$$v_g = \frac{\partial \omega}{\partial \beta}. \quad (2)$$

In Equation (2), variable β is the propagation constant and ω is the angular frequency. For the frequency range, where the group velocity dispersion may be neglected, the group velocity may be approximated by the expression for phase velocity of $v_{\varphi} = \omega/\beta$ that gives

$$T_g = \frac{L\beta}{\omega}. \quad (3)$$

The time delay of a section of a transmission line is proportional to its physical length. In order to achieve large time delays, adequately long sections of the transmission line have to be used. Because the Relationship (3) also contains a phase constant and further electrical length βL, the time delay is a function of the parameters (especially permittivity) of the material filling of the transmission line. Therefore, there is a possibility of obtaining bigger time delays per unit length of the transmission line when it is filled with material with a high permittivity ε_r. Therefore, for a TEM line, one may write

$$T_g = \frac{\sqrt{\varepsilon_r}L}{c}. \quad (4)$$

An example of this relationship is as follows: one meter of free space propagation or a perfect TEM line with air filling corresponds to 3.33 ns of time delay. In order to have a properly working delay line set, the propagation of the delayed wave should not be affected by the group velocity dispersion. This requirement is fulfilled when TEM lines are used, for example, coaxial lines or a planar line microstrip or coplanar. The bandwidth of the transmitted signal should be limited in order to not excite an unwanted waveguide (not TEM) mode of propagation.

The digitally-tunable delay line allows a number of pre-defined values of the time delay or length corresponding to the smallest assumed delay step to be set [5]. In the case of analogue tuning of a line's electrical length, theoretically, it is possible to obtain any value of delay from a predefined delay range.

An electronically-controlled microwave delay line is part of an analogue correlation detector used in a noise radar. The most basic design of a noise radar consists of an analogue receiver and analogue delay line with the ability to adjust the time delay. The current development of noise radars mainly concerns the use of advanced techniques of digital signal processing in order to obtain fully-digital correlation receivers [6–8]. However, a noise radar with an analogue correlation receiver with a tunable reference delay line may still be a useful enough solution, especially for micro-movement detection with range determination. In comparison, a typical CW (Continuous Wave) Doppler radar only detects the micro-movement speed, e.g., breathing or heartbeat, without information about the distance to the measured object [9]. Ultra WideBand (UWB) radars including noise or pseudorandom noise-based radars are used in various applications, including remote vital sign detection for rescue, security, and medical care or diagnostics [10,11].

To perform coherent reception in noise radar, a delay line with a constant or variable delay value is required. By analyzing the possibility of micro-movement detection, which is also described in the literature by the term micro-Doppler detection [12], one may notice that current investigations concern the development of algorithms to perform digital signal processing in the baseband in the frequency domain (spectrum analysis). The main issue is finding the spectrum shift of the received signal with respect to the transmitted signal. This shift may be equal to an order of MHz for a spectrum width equal to dozens of MHz. It is very difficult to detect such a small shift and new specialized methods should be developed.

The research question is to find the optimized analogue microwave tunable delay line in order to adjust the operating point of the correlation detector. In this paper, a solution for micro-movement detection is proposed, which is based on direct signal processing in the time domain with the use of a microwave analogue correlator. Therefore, to address this issue, further sections of the paper comprise a new concept of a digitally-controlled delay line with a set of fine distance gates. This concept assumes the use of a combined set of three lines, including a new version of a tapped delay line. In order to verify the concept, the example of micro-movement detection is presented, with the following assumptions: target distance 12 m, micro-movement amplitude 1 mm, noise radar with bandwidth 1 GHz, and center frequency 6.5 GHz.

The combined delay line presented in this paper is dedicated to serving as the adjustable reference delay for a noise radar intended for the detection of micro-movement. This approach allows the measurement setup to be simplified and increases the possibility of micro-movement detection. The delay line's physical structure depends on the dedicated application of a given radar system. The source of micro-movement may be different, for example, it may be a vital activity of the human body or its organs, such as chest movement due to heartbeat and breathing.

2. Principle of Noise Radar Technology

Noise radars belong to radars which use random or pseudorandom signals for probing purposes and coherent detection techniques for receiving signals. Their fundamental parameters are the following: wide bandwidth, low power density, and high accuracy for distance and velocity measurements, which results from the properties of the ambiguity function of the wideband noise signal. A correlation receiver is a typical element of a noise radar. Coherent reception needs delay lines of constant or variable parameters to be applied in the receiving systems. The delay line allows one to memorize a sample of the transmitted noise signal for an amount of time delay resulting from the round-trip of the transmitted signal, from the radar transmitter to the target and back to the radar receiver. In general, the principle of the operation of a noise radar and its various structures has been widely described in the scientific literature, including the basic use of the radar, which is target range and radial velocity estimation [13–20].

The radar transmitter allows the noise signal from the frequency range, which is said to be from 1 to 18 GHz, and bandwidth of about 2 GHz, to be generated. The primary noise source in the transmitter may be realized by means of a semiconductor avalanche diode or a Zener's diode. The signal from the transmitter is fed to a transmitting antenna with the use of a directional coupler to the delay line. The signal collected by a receiving antenna is amplified and filtered in the front-end receiver and further correlated with a copy of the transmitted signal delayed by a delay line. When the time delay value T_{DL} is equal to the time T corresponding to the round-trip of the transmitted signal (from the radar to the target and back to the radar), a so-called "correlation peak" will appear at the output of the correlation detector. The existence of a distinguishable value of the signal $S_{OUT}(t)$ occurring for a given value of time delay T_{DL} provided by a delay line allows target detection and distance estimation according to the formula $R = cT_{DL}/2$.

For the following consideration, it is assumed that the transmitter generates a signal in the form of noise with a limited bandwidth and normal distribution, with an average value equal to zero and a variance equal to σ^2. The signal generated in the transmitter can be described by the Expression (5), whereas signals in the individual points of the system (Figure 1) are described by Relations (6) and (7):

$$S_T(t) = x(t)\cos(\omega_0 t) - y(t)\sin(\omega_0 t) \tag{5}$$

$$S_{DL}(t) = k_1 S_T(t - T_{DL}) \tag{6}$$

$$S_R(t) = k_2 S_T\left(t - T - \frac{2D(t)}{c}\right), \tag{7}$$

where T is the signal delay time on the radar-object-radar path, T_{DL} is the delay time of the delay line, $D(t)$ is the micro-movement of certain object parts, ω_0 is the median frequency of the band occupied by the noise signal, $x(t)$ and $y(t)$ are independent realizations of the stationary random process with a Gaussian distribution having an average value equal to zero, and k_1 and k_2 are the propagation coefficients.

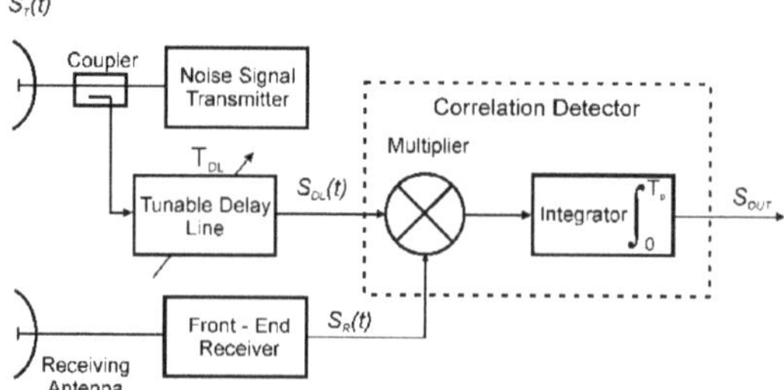

Figure 1. Outline of a noise radar.

As a result of the multiplication of signals described by (6) and (7), and afterwards, integration of the obtained products, the output signal $S_{OUT}(t)$ can be expressed in the following Form (8):

$$S_{OUT}(t) = A(\Delta T)\cos\left[\omega_0\left(\Delta T + \frac{2D(t)}{c}\right)\right], \tag{8}$$

where $\Delta T = T - T_{DL}$ and

$$A(\Delta T) = \frac{k_1 k_2}{T_p}\int_0^{T_p}(x(t-T)x(t-T_{DL}) + y(t-T)y(t-T_{DL}))\mathrm{d}t. \tag{9}$$

The result of Integration (9) is a value independent of time t. However, this value (i.e., integration result) depends on the time difference $T - T_{DL}$. When the value of T_{DL} is set by the delay line and the value of time T is constant (constant position of the target), the output signal from correlator S_{OUT} is also constant, and it reaches its maximum for $T = T_{DL}$. Another situation takes place when a target is making micro-movement, for example, that described by the harmonic expression $D(t) = D \times \cos(\omega_0 \times t)$, where D is the amplitude of this movement. Then, the round-trip time T is also harmonically dependent on t. In effect, the output signal from the integrating circuit also depends on t. This means that the integrator output signal S_{OUT} varies in time according to the varying distance from the radar to target and its frequency corresponds to the frequency of the target micro-movement.

There is the requirement that the micro-movement should be a slow-varying function of time compared to the integration time Tp of the integrating circuit. The value of micro-Doppler frequency must be lower than the cut-off frequency of the lowpass filter at the output of the correlator multiplying circuit.

Equation (8) describes the value of the correlation function for the noise signal transmitted and received by the noise radar [21–24].

3. New Concept of a Combined Microwave Delay Line with a Set of Fine Distance Gates

3.1. Combined Structure

In order to control the position of the detector operating point, an adjustable analogue delay system for micro-movement in a noise radar has to be used and it may be realized with the use of a combined delay line structure.

The two structures of microwave delay lines are as follows: the digitally-controlled cascaded line with switched delay sections and the analogue tapped delay line, which may be combined with one other. The resulting structure, formed by cascading, gains additional interesting features. The line with switched delay sections sets one time delay value from a predefined finite set, which may be called the coarse one. Furthermore, the delayed input signal enters the second line, which is the tapped delay line. The tapped line introduces several values of a smaller time delay for each tap output simultaneously.

These delays may be called the fine ones and they are added to the coarse time delay set by the first line. As a result, there is a comb of time delays corresponding to radar range gates (spread by unit time delay of the tapped line), which is switched up and down by the value of unit time delay of the digitally-controlled first line (coarse step). The fine time delays offset should be sorted to evenly cover the unit time delay of the coarse line.

The combined line mentioned above does not ensure that the optimal operating point on the detector characteristic (with the meaning of point P1 on the autocorrelation function) corresponds to one of the fine gate delays. In order to find the optimal operating point for detection, the third delay line has to be cascaded. This additional delay line is the adjustable one, preferably in the form of an analogue precision phase shifter.

The innovative concept of such a combined line is shown in Figure 2, with an example of specified time delay implementation.

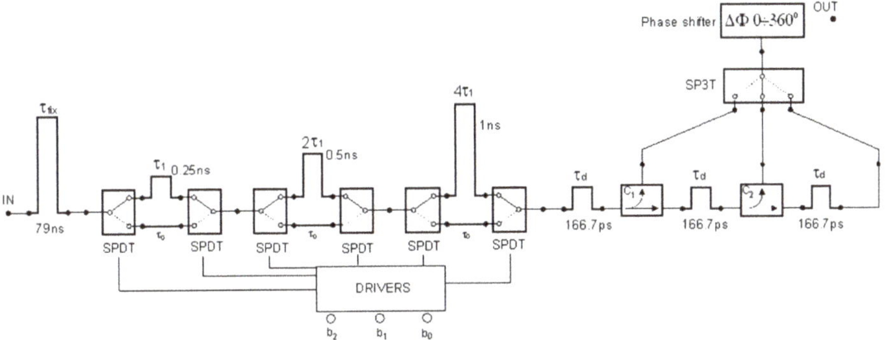

Figure 2. Example of a digitally-controlled delay line with a set of fine distance gates and an adjustable phase shifter for finding the optimal operating point of a correlation detector.

3.2. Analogue Tapped Cascaded Delay Line

Digitally-controlled microwave delay lines are considered here to realize a reference delay for a noise radar with analogue correlation. This type of radar performs analogue correlation of the received signal with a delayed version of the transmitted signal. However, compared to a radar with digital signal processing including signal cross-correlation, the analogue correlator only performs convolution for one selected value of time delay called the time gate.

In order to bring the functionality of a radar with an analogue correlator closer to the digital one, a special kind of delay line may be used, known as the analogue tapped delay line.

In general, the analogue tapped delay line, known from the literature, consists of a number of unit delay sections and microwave couplers, having the same coupling factor, cascaded as shown in Figure 3 [25]. One unit delay line with one coupler forms one delay line stage. Every tap, which is a coupler's coupled signal port, is the output of the delayed input signal, with a time delay equal to the unit time delay τ_1 multiplied by the stage number.

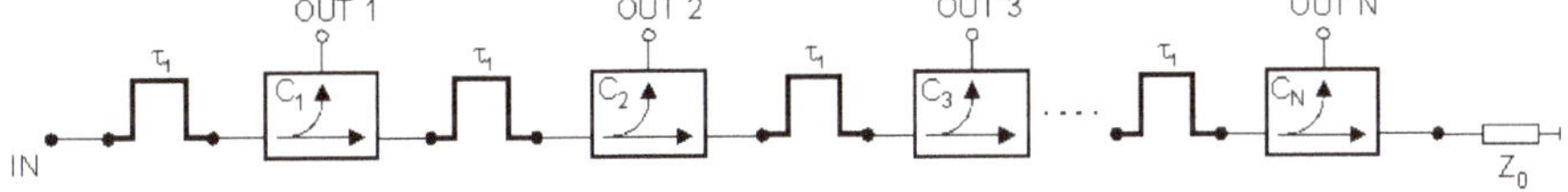

Figure 3. General scheme of an analogue tapped cascaded delay line.

According to this idea, every signal tap corresponds to one time gate and one signal correlator. The general scheme of an analogue tapped delay line is shown in Figure 3.

The application of this circuit causes a need to use several analogue correlator units equal to N, consisting of a signal mixer and a lowpass filter.

It is important to note that the analogue tapped delay line allows all signals corresponding to all time gates to be obtained quasi-simultaneously. Here, the term "quasi" means that the time gate output signals appear after subsequent unit time delay, but there is no need to use a signal switch (SPDT or SPST) to set the one desired value of time delay per one specified state of line. The analogue tapped cascaded delay line can be characterized by a number of features.

Advantages:

- Ability to obtain multiple values of the time delay quasi-simultaneously;
- Lack of unstable or transient states;
- Lack of microwave switches and their driving and biasing;
- Lower signal losses due to the absence of insertion losses of microwave switches.

Disadvantages:

- Need for high signal isolation between taps in order to minimize signal crosstalk;
- Maximal number of taps and therefore delay gates is a function of the signal coupling factor at each tap. In order to have a high number of taps, the coupling factor should be small, which results in the need for additional signal amplifying;
- Presence of insertion losses of microwave couplers;
- Different levels of output signal power at each tap appearing in descending order.

The design of a new concept of the analogue tapped delay line, which is proposed below, requires the design and optimization of dedicated microwave couplers with precisely chosen values of the coupling factor in order to obtain the same signal power level at each tap.

4. Numerical Validation

4.1. Optimization of the Correlation Detector Operating Point

A normalized Function (8) for T = 80 ns, 6–7 GHz band, and $D(t)$ = 0 is shown in Figure 4.

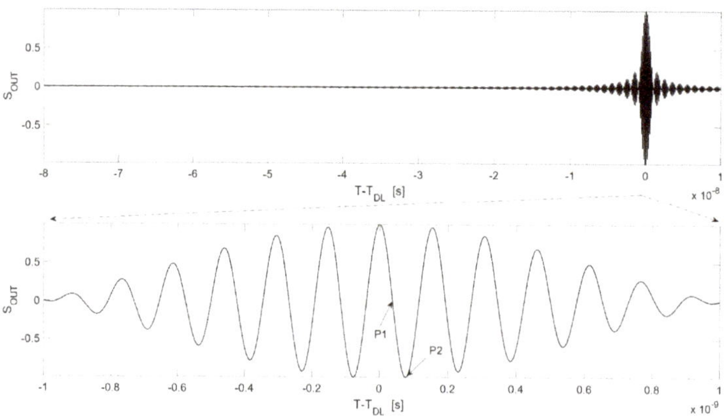

Figure 4. Plot of a normalized Function (8) for T = 80 ns, 6–7 GHz band, and $D(t)$ = 0.

As can be seen in Figure 4, the noise radar may be applied for micro-movement detection. The operating points P1 and P2 of the correlation detector can be selected from the operational range shown in Figure 4. An example of the output signal of the correlation detector $S_{OUT}(t)$ for P1 and P2 is shown in Figure 5. The assumptions for this calculation are as follows: distance to object R = 12 m, harmonic micro-movement $D(t)$ with amplitude 1 mm and frequency equal to 1 Hz, radar transmits noise signal with bandwidth B = 1 GHz, and center frequency f_0 = 6.5 GHz. In this case, the point P1 corresponds to delay T_{DL1} and P2 corresponds to delay T_{DL2}, which equal T_{DL1} = 80.0385 ns and T_{DL2} = 80.0770 ns. The use of a tunable microwave delay line is crucial in this case. The best operating point P1 for proper micro-movement detection is not placed for T = 80 ns, resulting from the distance between the radar and a target. Therefore, there is a need to correct the position of the operating point by introducing an additional value of time delay. It is possible to shift the operating point of the micro-movement detector due to the precise adjustment of time delay provided by the tunable delay line.

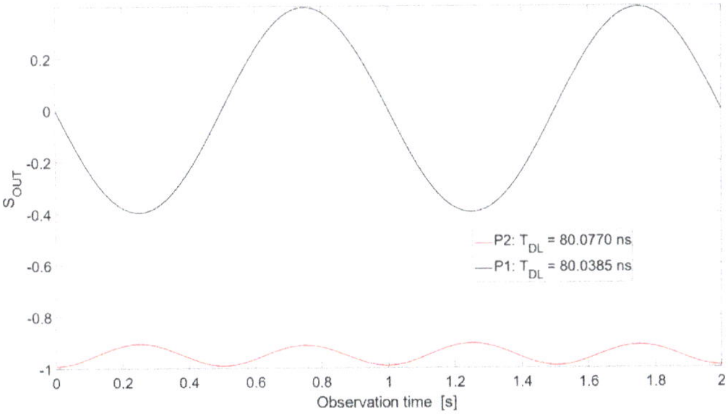

Figure 5. Illustration for the conversion of harmonic movement of an object to the output signal of the microwave correlation detector.

The operating point P1 is placed on the linear part of the detector characteristic and the detector output signal (black line in Figure 5) at this point properly corresponds to the shape and frequency of the micro-movement.

In contrast, the operating point P2 is placed on the nonlinear part of the detector characteristic and the detector output signal at this point (red line in Figure 5) incorrectly replicates the shape and frequency of the micro-movement. In particular, the micro-movement frequency is incorrect and equal to the doubled value of the proper frequency due to the even shape of the characteristic in the vicinity of the operating point.

4.2. Implementation of Time Delay Values

The example of the implementation of specified time delay values in a combined delay line structure, according to the proposed concept (Figure 2), assumes the case of micro-movement detection introduced above in Section 4.1.

The first line is the fixed delay line, which sets the time delay τ_{fix} corresponding to the distance 12 m lowered by the value of 1 ns, i.e., half of the autocorrelation function width (as shown in Figure 4). The digitally-controlled line with switched sections consists of three sections with the following time delay values: 0.25, 0.5, and 1 ns. It corresponds to a 3-bit control with unit time delay equal to 0.25 ns, and a maximal value of coarse time delay equal to 1.75 ns. Next, there is the tapped delay line with three taps and fine unit time delay equal to 166.7 ps. In effect, for every state of the digitally-controlled line, there are three values of fine gates shifted by 166.7 ps, which cover the range of 0.5 ns evenly.

The combined line presented in Figure 2 comprises one analogue phase shifter preceded by an SP3T switch. In this application, only one micro-Doppler detector is needed, and it is placed at the output of the phase shifter. This structure allows the optimal operating point for a correlation detector to be found in the case of a change of target position.

There is another variant of this solution that is possible when there are three correlation detectors connected to the subsequent output taps of the tapped line, and the phase shifter is placed between switched and tapped delay lines. In this case, there is no need for an SP3T switch (or in general, an N-way SPNT switch).

4.3. Calculations of Optimal Parameters of an Analogue Tapped Delay Line

According to the new concept, the main assumption for optimization of the tapped line design is that the same level of signal power is obtained at each tap, i.e., the coupled signal port.

Denoted in terms of power,

- Input power as x;
- Coupled signal power values as y_1 to y_N;
- Coupler's direct output power values as x_1 to x_N;
- Power coupling factors C_1 to C_N;

where N is the maximal number of taps, for lossless couplers, one may notice

$$x_1 = (1 - C_1)x \tag{10}$$

and

$$y_1 = C_1 x. \tag{11}$$

Substituting further for y_2 and x_2 results in

$$y_2 = C_2 x_1 = C_2(1 - C_1)x \tag{12}$$

and

$$x_1 = (1 - C_2)x_1 = (1 - C_1)(1 - C_2)x. \tag{13}$$

The considered situation is shown in Figure 6.

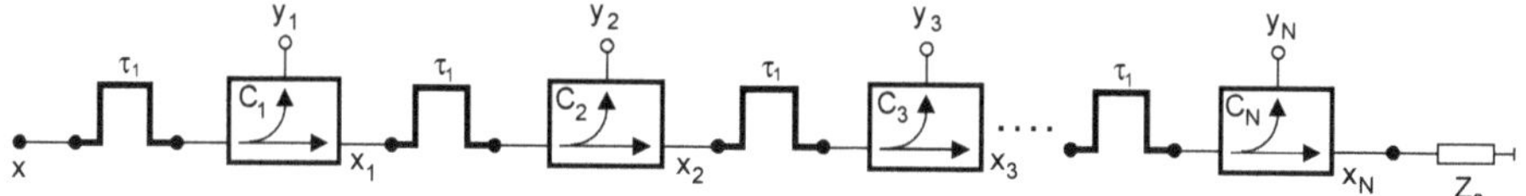

Figure 6. General scheme of a cascaded tapped delay line used in considerations, with variables denoted.

The condition for equal coupled signal powers is

$$y_1 = y_2 = \ldots y_N, \tag{14}$$

which gives the result

$$C_2 = \frac{C_1}{1 - C_1} \tag{15}$$

and

$$C_N = \frac{C_{N-1}}{1 - C_{N-1}}. \tag{16}$$

Finally, by substituting coupling factors for subsequent taps, one may obtain

$$C_N = \frac{C_1}{1 - (n-1)C_1}. \tag{17}$$

Because the last tap does need not a coupler, the value of C_N equals 1 and using Equation (17), one may find

$$C_N = \frac{1}{N+1-n}. \tag{18}$$

The equation above (18) expresses the value of the power coupling factor as a function of the tap number n for the assumed overall number N of taps. With the use of this equation, the example values of coupling factors were calculated and are presented in Table 1. The calculations were done for three values of the overall tap number, i.e., N = 4, 8, and 16.

Table 1. Calculated values of power coupling factors for values of the overall tap number: N = 4, 8, and 16.

$N = 4$	$N = 8$	$N = 16$	C_n	C_n (dB)
-	-	C_1	1/16	−12.04
-	-	C_2	1/15	−11.76
-	-	C_3	1/14	−11.46
-	-	C_4	1/13	−11.14
-	-	C_5	1/12	−10.79
-	-	C_6	1/11	−10.41
-	-	C_7	1/10	−10
-	-	C_8	1/9	−9.54
-	C_1	C_9	1/8	−9.03
-	C_2	C_{10}	1/7	−8.45
-	C_3	C_{11}	1/6	−7.78
-	C_4	C_{12}	1/5	−6.99
C_1	C_5	C_{13}	1/4	−6.02
C_2	C_6	C_{14}	1/3	−4.77
C_3	C_7	C_{15}	1/2	−3
C_4	C_8	C_{16}	1	0

When the design guidelines above are implemented, the power transmission factor for every tap output with respect to the main input of the whole delay line is the same and equals $1/N$. This is true for lossless unit delay lines and couplers; when these devices are lossy, the insertion losses accumulate and increase with the tap number.

Assuming that the constant tapped power rule is taken into account, the design methodology of the microwave tapped analogue delay line should follow

- Choice of unit time delay value;
- Choice of maximum number of taps (and unit delays) for the whole delay structure;
- Calculation of power coupling factors for subsequent taps C_1 to C_N;
- Choice of technology for physical realization of the unit delay line and couplers;
- Dedicated design or purchase of the unit delay line and set of couplers with specified, previously calculated, coupling factors.

Further optimization of the whole delay line structure, e.g., possible integration of coupling structures with circuits of unit delay lines, can be pursued in the case of one's own design (especially possible when planar technology is chosen).

5. Discussion

For very high frequencies in the microwave range, the simple analogue systems of the correlative detector with a controlled microwave delay line can be used, because digital realization of the micro-movement detector for microwave frequencies is very difficult and demands very high-speed analogue-to-digital conversion. As far as existing delay line technologies are concerned, the conclusion is that, in spite of constant development, there is no ideal solution, which would join the features like low losses, a high delay, a low cost, and small dimensions.

The choice of delay line solution depends on the expected application. One may use commercially available components or design one's own dedicated solutions. When the type and technology of unit delay lines is chosen, the whole structure of the delay line set should be considered in order to fulfill the requirements of the given project.

For the radar (especially noise one) with an analogue correlator, the optimal solution seems to be a combined delay line consisting of switched and tapped parts. The parameters like the number of control bits, number of delay states, and unit delay value are matched in order to optimize signal losses, bandwidth, structure complexity, maximal detection range, and range gate grid.

There are various possible applications of combined delay lines when noise radar is used:

1. Surveillance of an area defined by a set of subsequent range gates in order to detect the entrance of an object in the monitored area. Then, the combined delay line, which consists of a digitally-switched delay line and a cascaded tapped delay line without an SPNT switch and phase shifter(s), is a sufficient solution;
2. Solution as above assuming that the surveillance range is fixed. Then, there is no need for switching, the set of delay lines mentioned above can be simplified to one fixed delay line cascaded by a tapped delay line with the number of taps equal to the number of desired additional monitoring range gates. This solution may be especially suitable for systems designed to protect an object (vehicle, building) against an attack, for example, protection of a tank against Rocket-Propelled Grenades;
3. Monitoring the vital signs of a stationary person, for example, a patient in bed or a worker at the desk. In these cases, when the distance to the person in known, the fixed delay line may be used, followed by a tapped delay line with several delay stages. The tapped line delays cover the predicted delay value corresponding to the width of the autocorrelation function. The outputs of the tapped line may be connected to the SPNT switch and adjustable phase shifter or to a corresponding number of phase shifters directly.

The natural competition for analogue signal processing methods in a noise radar are digital techniques, which rely on calculation of the correlation function in the digital domain. This solution requires direct analog-to-digital (AD) conversion of the microwave signal in a given bandwidth. However, for very high frequencies in the microwave range, simple analogue systems of the correlative detector can still be used, because direct sampling of the microwave signal and digital realization of the autocorrelation function for these frequencies are very difficult and require extremely fast AD conversion.

The methods presented in this paper use an analogue correlation detection in the microwave band, in contrast to processing of the received signal in the primary band. This allows a movement using an internal structure of the correlation function of a noise signal to be precisely detected. The solution that uses the digitally-controlled switched delay line cascaded with the tapped delay line allows the features of synthesizing the big delay values (switched line) to be combined with simultaneously obtaining several values of the time delay (tapped line). An additional precision analogue phase shifter allows the optimal operating point of a micro-Doppler detector to be found.

Author Contributions: Conceptualization, Z.S. and W.S.; methodology, Z.S.; investigation, W.S.; writing—original draft preparation, Z.S.; writing—review and editing, W.S.

Funding: This research was funded by the Polish Ministry of Defense, grant number GBMON/13-996/2018.

Conflicts of Interest: The authors declare no conflicts of interest. The sponsors had no role in the design, execution, interpretation, or writing of the study.

References

1. Li, S.; Sun, H. Design of a 6–18GHz delay line. In Proceedings of the IEEE Electrical Design of Advanced Packaging and Systems Symposium (EDAPS), Hanzhou, China, 12–14 December 2011; pp. 1–3. [CrossRef]
2. Hu, F.; Mouthaan, K. A 1–21 GHz, 3-bit CMOS true time delay chain with 274 ps delay for ultra-broadband phased array antennas. In Proceedings of the European Microwave Conference (EuMC), Paris, France, 7–10 September 2015; pp. 1347–1350. [CrossRef]
3. Park, S.; Jeon, S. A 15–40 GHz CMOS True-Time Delay Circuit for UWB Multi-Antenna Systems. *IEEE Microw. Wirel. Compon. Lett.* **2013**, *23*, 149–151. [CrossRef]
4. Mabrouki, M.; Smida, A.; Ghayoula, R.; Gharsallah, A. A 4 bits Reflection type phase shifter based on Ga as FET. In Proceedings of the 2014 World Symposium on Computer Applications & Research (WSCAR), Sousse, Tunisia, 18–20 January 2014. [CrossRef]
5. Arzur, F.; Le Roy, M.; Pérennec, A.; Tanné, G.; Bordais, N. Hybrid architecture of a compact, low-cost and gain compensated delay line switchable from 1 m to 250 m for automotive radar target simulator. In Proceedings of the European Radar Conference (EURAD), Nuremberg, Germany, 11–13 October 2017; pp. 239–242. [CrossRef]
6. Meller, M. Fast Clutter Cancellation for Noise Radars via Waveform Design. *IEEE Trans. Aerosp. Electron. Syst.* **2014**, *50*, 2327–2334. [CrossRef]
7. Kulpa, J.S. Noise radar sidelobe suppression algorithm using mismatched filter approach. *Int. J. Microw. Wirel. Technol.* **2016**, *8*, 865–869. [CrossRef]
8. Kulpa, J.S.; Maslikowski, L.; Malanowski, M. Filter-Based Design of Noise Radar Waveform with Reduced Sidelobes. *IEEE Trans. Aerosp. Electron. Syst.* **2017**, *53*, 816–825. [CrossRef]
9. Łuszczyk, M.; Szczepaniak, Z. Non-contact breath sensor based on a Doppler detector. In *Computational Methods and Experimental Measurements*; Carlomagno, G.M., Brebbia, C.A., Eds.; WITT Press: Southampton, UK, 2011; ISBN 978-1-84564-540-3.
10. Sachs, J.; Helbig, M.; Herrmann, R.; Kmec, M.; Schilling, K.; Zaikov, E. Remote vital sign detection for rescue, security, and medical care by ultra-wideband pseudo-noise radar. *Ad Hoc Netw.* **2014**, *13*, 42–53. [CrossRef]
11. Bellizzi, G.; Bellizzi, G.G.; Ovidio, M.; Bucci, O.M.; Crocco, L.; Helbig, M.; Ley, S.; Sachs, J. Optimization of the Working Conditions for Magnetic Nanoparticle-Enhanced Microwave Diagnostics of Breast Cancer. *IEEE Trans. Biomed. Eng.* **2018**, *65*, 1607–1616. [CrossRef] [PubMed]
12. Bączyk, M.K.; Samczyński, P.; Kulpa, K.; Misiurewicz, J. Micro-Doppler signatures of helicopters in multistatic passive radars. *IET Radar Sonar Navig.* **2015**, *9*, 1276–1283. [CrossRef]
13. Narayanan, R.M.; Xu, X. Principles and applications of coherent random noise radar technology. *Proc. SPIE* **2003**, *5113*, 503–514.
14. Theron, I.P.; Walton, E.K.; Gunawan, S.; Cai, L. Ultrawide-band noise radar in the VHF/UHF band. *IEEE Trans. Antennas Propag.* **1999**, *47*, 1080–1084. [CrossRef]
15. Li, Z.; Narayanan, R. Doppler visibility of coherent ultrawideband random noise radar systems. *IEEE Trans. Aerosp. Electron. Syst.* **2006**, *42*, 904–916. [CrossRef]
16. Axelsson, S.R.J. Noise radar for range/Doppler processing and digital beamforming using low-bit ADC. *IEEE Trans. Geosci. Remote Sens.* **2003**, *41*, 2703–2720. [CrossRef]
17. Stephan, R.; Loele, H. Theoretical and practical characterization of a broadband random noise radar. In Proceedings of the Digest 2000 IEEE MTT-S International Microwave Symposium, Boston, MA, USA, 11–16 June 2000; pp. 1555–1558.
18. Dawood, M.; Narayanan, R.M. Receiver operating characteristics for the coherent UWB random noise radar. *IEEE Trans. Aerosp. Electron. Syst.* **2001**, *37*, 586–594. [CrossRef]
19. Lukin, K.A.; Vyplavin, P.L.; Zemlyaniy, O.V.; Palamarchuk, V.P.; Lukin, S.K. High Resolution Noise Radar without Fast ADC. *Int. J. Electron. Telecommun.* **2012**, *58*, 135–140. [CrossRef]
20. Savci, K.; Erdogan, A.Y.; Gulum, T.O. Software defined L-band noise radar demonstrator. In Proceedings of the 2016 17th International Radar Symposium (IRS), Krakow, Poland, 10–12 May 2016; pp. 1–4. [CrossRef]
21. Susek, W.; Stec, B.; Recko, C. Noise Radar with Microwave Correlation Receiver. *Acta Phys. Pol. A* **2010**, *119*, 483–487. [CrossRef]
22. Susek, W.; Stec, B. Broadband microwave corerelator of noise signals. *Metrol. Meas. Syst.* **2010**, *17*, 289–298. [CrossRef]

23. Susek, W.; Stec, B. Through-the-Wall Detection of Human Activities Using a Noise Radar with Microwave Quadrature Correlator. *IEEE Trans. Aerosp. Electron. Syst.* **2015**, *51*, 759–764. [CrossRef]
24. Stec, B.; Susek, W. Theory and Measurement of Signal-to-Noise Ratio in Continuous-Wave Noise Radar. *Sensors* **2018**, *18*, 1445. [CrossRef] [PubMed]
25. Mukti, P.H.; Schreiber, H.; Gruber, A.; Gadringer, M.E.; Bosch, W. A preliminary result on development of analog broadband tapped delay line for L-band applications. In Proceedings of the International Seminar on Intelligent Technology and Its Applications (ISITIA), Lombok, Indonesia, 28–30 July 2016; pp. 357–362. [CrossRef]

MDPI
St. Alban-Anlage 66
4052 Basel
Switzerland
Tel. +41 61 683 77 34
Fax +41 61 302 89 18
www.mdpi.com

Sensors Editorial Office
E-mail: sensors@mdpi.com
www.mdpi.com/journal/sensors

www.ingramcontent.com/pod-product-compliance
Lightning Source LLC
LaVergne TN
LVHW071609170726
843515LV00008B/2352
9783036509181